Texts and Monographs in Physics

Philippe A. Martin François Rothen

Many-Body Problems and Quantum Field Theory

An Introduction

Translated by Steven Goldfarb, Andrew Jordan
and Samuel Leach

Second Edition
With 102 Figures, 7 Tables and 23 Exercises

Springer

Professor Philippe A. Martin
Swiss Federal Institute of Technology
Institute of Theoretical Physics
1015 Lausanne
Switzerland

Professor François Rothen
University of Lausanne
Emeritus Professor
and
Swiss Federal Institute of Technology
Institute of Complex Matter Physics
1015 Lausanne
Switzerland

Translators:
Dr. Steven Goldfarb
CERN - PH
1211 Geneva 23
Switzerland

Dr. Andrew Noble Jordan
Dr. Samuel Leach
Department of Theoretical Physics
University of Geneva
Quai Ernest Ansermet, 24
1211 Geneva
Switzerland

Originally published in French under the title: *Problèmes à N-corps et champs quantiques*

ISSN 0172-5998

ISBN 3-540-21320-1 Second Edition Springer Berlin Heidelberg New York
ISBN 3-540-41153-4 First Edition Springer Berlin Heidelberg New York

Library of Congress Control Number: 2004103456

Springer is a part of Springer Science+Business Media

springeronline.com

Printed in Germany

Final processing: LE-TEX Jelonek, Schmidt & Vöckler GbR, Leipzig
Cover design: *design & production* GmbH, Heidelberg
Printed on acid-free paper 55/3141/di SPIN: 10965900 5 4 3 2 1 0

Foreword

An unusual aspect of this book is to bring together various subjects of physics rarely found in the same place. The first chapter not only recalls preliminary notions of quantum and classical physics but also serves as a preparation for the viewpoints developped throughout the whole book. Its presentation has been changed to better underline this perspective. This has led in turn to slight modifications at the beginning of Chap. 8.

Chapter 6 refers to nucleon pairing inside the nucleus. Its content differs greatly from that of the first edition. The description of nucleon pairing was borrowed from the concept of electron pairing in superconductivity. We have reduced the formal aspects of this description for the benefit of a more qualitative discussion of analogies and differences between the two systems. In addition, more emphasis is put on experimental results and theoretical arguments in favour of nucleon pairing. Fruitful discussions with Dr. Jan Jolie are acknowledged.

It is a pleasure to thank Drs. Andrew Jordan and Samuel Leach at the University of Geneva for their thorough revision of the English translation.

June 2004

Philippe A. Martin
François Rothen

Foreword to the First Edition

This text is a revised and augmented version of a course given to graduate and Ph.D. students in the context of the doctoral school for physics in the French-speaking part of Switzerland. This doctoral school provides a common teaching program for the universities of Bern, Fribourg, Geneva, Neuchâtel and Lausanne, as well as for the Swiss Federal Institute of Technology in Lausanne. The scope of the course should be sufficiently general to interest both experimentalists and theoreticians wishing to engage in research in condensed matter or nuclear and particle physics. The prerequisites are an introductory course to quantum mechanics and elements of classical electromagnetism and statistical mechanics.

Our main concern was how to maintain a reasonably broad level of knowledge for students with different orientations, in a world of research where the price of survival is extreme specialization and competitiveness. Is it still possible in the available time to provide a cultural education in physics by relatively elementary means and in an optimized form? We believe that this is an essential pedagogical duty. Attempting to meet this challenge has determined the conception of this book: each individual part of it is standard and without novelty but should belong, in our opinion, to the basic culture of every physicist; only their common organization in a single house of decent size might possibly be put to our credit.

We have tried to keep a balance between formal developments and the physical applications: in fact they cannot be separated insofar as mathematical methods develop naturally under the necessity of resolving physical questions. Concerning the applications, we have always given a short description of the phenomenological context so that the main information about physical facts is available from the start without recourse to other sources. In the formal developments, we adopt the usual notation of physicists, while aiming at mathematical precision. The reader is warmly encouraged to improve his practice of the formalism by checking and reproducing for himself the algebra given in the text. Some more extended exercises are proposed at the end of each chapter in order to illustrate additional aspects not introduced in the main text.

For each of the systems discussed in this book, we have tried to exhibit how the main physical ideas can be captured in a formalized description by the appropriate tools. In this spirit several important branches of physics are represented: solid state physics (cohesion and dielectric properties of the electron gas, phonons and electron–phonon interactions), low temperature physics (superconductivity and superfluidity), nuclear physics (pairing of nucleons), matter and radiation (interaction of atoms with the quantum-electromagnetic field), particle physics (interaction by exchanged intermediate particles, mass generation by the Higgs mechanism).

These choices could be considered rather conservative, compared to topical new developing areas. However we think that they still serve as indispensable paradigms for the understanding of any more advanced subject. Also, in keeping with our aim of offering a broad formative view to our readers, they enable us to illustrate similarities and differences between concepts stemming from various domains in physics. In this respect, the first chapter presents a parallel exposition of classical electromagnetism and classical elasticity, with the purpose of introducing and comparing the notions of photon and phonon. Moreover, quantum fields (Chap. 8) cannot be understood without a good knowledge of their classical analogues. Chapter 2 is devoted to a simple description of collective effects due to Bose and Fermi statistics. Bose condensation is described and the role of Fermi statistics for the stability of matter and in astrophysical objects is discussed. In the third chapter we develop the so-called second quantized formalism in full generality without reference to any particular system, so that it will be available in any situation where the number of particles varies. Chapters 4, 5 and 6 are devoted to the use of the variational method. It is hoped that the reader will appreciate the wide range of applications of the idea of fermionic pairing formulated in the BCS theory for superconductivity (Chap. 5) as well as for nuclear matter (Chap. 6). The relationship between superfluidity (Chap. 7) and superconductivity on the one hand, and collective excitations of the nuclei on the other, is put into perspective. The quantum-electromagnetic field serves as a model for other quantized matter fields in Chap. 8. The concept of gauge theory is introduced and the close analogy between the Higgs mechanism and the Meissner effect displayed. The method of Feynman graphs is explained in Chaps. 9 and 10, stressing again the existence of a common language for condensed matter and field theory. We essentially give the physical interpretation of diagrams without performing the corresponding more technical quantitative calculations. The analysis is restricted to ground-state properties: non-zero temperature Green functions and the alternative functional integration viewpoint are not considered.

The book is not aimed at the specialist in any of the addressed topics. In fact, no chapter is intended to provide the up-to-date knowledge necessary for an immediate fight on the battlefront of research. We refer in particular to the present state of relativistic quantum field theory since, without mentioning electroweak theory and chromodynamics, no presentation of the Lorentz group or of the Dirac equation can be found here. From the viewpoint of condensed matter, high-T_c is only briefly touched, mesoscopic physics and highly correlated fermions are not discussed. To the prospective particle physicist, the book can merely give a complementary education on the use of similar techniques in condensed-matter physics. Conversely, physicists belonging to the latter discipline, although they may be aware of the importance of field theory for particle physics, should learn about fundamental ideas underlying both domains. We therefore hope that our readers will discover a certain unity of thinking among different domains of physics. In this case, this book will not have been written in vain.

This book was written at the instigation of the Troisième Cycle de la Physique en Suisse Romande, and in particular of J.-J. Loeffel. We are grateful to the many colleagues who provided useful suggestions or enlightenment on various points, namely to the late P. Huguenin and to B. Jancovici, D. Pavuna, J.-P. Perroud and G. Wanders. V. Savona helped with the elaboration of the exercises at the end of each chapter. We thank R. Fernandez for encouraging us to translate the book. S. Goldfarb translated and typed the whole text, including the numerous equations; D. Watson helped us to formulate additional material; L. Klinger and L. Trento drew the figures; our thanks go to all these people for their contributions. Finally we are indebted to the Institute of Condensed-Matter Physics and the Physics Section of the University of Lausanne, as well as to the Physics Department of the Swiss Federal Institute of Technology, Lausanne, for financial support.

August 2001

Philippe A. Martin
François Rothen

Table of Contents

Reader's Guide

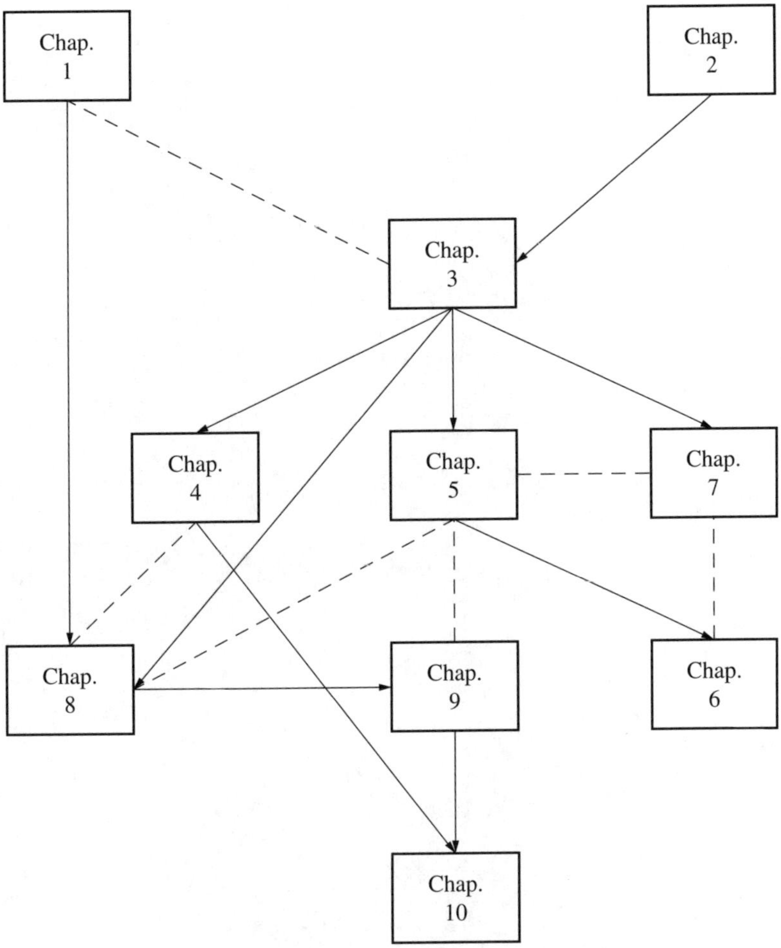

Logical organization of the material: the arrows indicate a recommended reading order

Chapter 1 offers an incomplete summary of quantum mechanics and of classical field theory. By consequence, it recalls notions which are useful, but which are generally part of an undergraduate repertoire. It is left to the reader to determine the necessity for a thorough study of its contents.

1. Classical Fields and Their Associated Particles

1.1 Introduction

Introductory quantum physics texts often begin with a historical review. A typical starting point is the analysis of the black-body radiation spectrum, conducted in 1900 by Max Planck and traditionally considered the birth of quantum physics. This is a natural choice, as it is on this occasion that Planck first introduced *his* constant h.

Such an historical approach, however, may lead to serious conceptual difficulties and does not necessarily develop a clear pedagogical logic. The black-body radiation problem immediately introduces a large number of particles, photons, and its analysis requires a thorough understanding of statistical physics, as well as quantum-field theory. Moreover, the photon, introduced by Einstein in 1905, obeys non-trivial Bose statistics. A more intuitive approach would be to first examine the problem of a single quantum particle before moving on to more complex systems. However, the model of the hydrogen atom, the first single-particle problem explicitly involving the Planck constant, was not created until 1913 by N. Bohr.

This fact reveals a profound reality: *there exists no quantum system that is strictly a single-particle problem.* Therefore, all problems involving a single particle (or a fixed number of particles) result from an approximation which is valid only if the energy of the system is weak.

Nevertheless, the formalism of quantum physics is equally well suited for single body problems as for analyses involving a large number of particles. For example its methods apply to the problem of a charged particle in a Coulomb potential, to the analysis of a collection of photons in a reflecting cavity, to the dynamics of nucleons in a nucleus or to electrons in a solid. However, although the passage from classical physics to quantum physics is well known for the case of a massive particle such as an electron, its generalization to electromagnetic radiation or to the vibrations of a crystal lattice is not so evident. Difficulties arise with the introduction of the wave–particle duality as it is expressed for the electron, the photon or the phonon.

The discovery of the electron is commonly attributed to J.-J. Thompson (1897). He recognized that, under the influence of magnetic fields, "cathode rays" behave like jets of particles, with their charge-to-mass ratios remaining constant. Millikan later determined the charge of these particles. So, classically, one could think of the electron as a particle characterized by its mass and kinematical properties. However, in 1923, an hypothesis of de Broglie attributed a wavelength to the electron and in 1926, Schrödinger gave his name to the famous equation which describes its wave-like character. This new quantum description, illustrated by the wave–particle duality, associationed a wavefunction with the electron obeying Schrödinger's equation of motion. Thus, the particle aspect of the electron was imposed before its wave aspect.

For the case of radiation, history followed the reverse path for reasons which were not accidental. Planck and Einstein associated the electromagnetic field with a quantum particle. The famous Planck relation $\Delta E = h\nu = \hbar\omega$ was introduced in 1900 to account for the black body spectrum. In Planck's mind, ΔE represented a minimal exchange of energy between radiation and matter. However in 1905, Einstein interpreted $h\nu$ as the energy of a constituent particle of radiation, the photon. Hence electromagnetic radiation's wave properties; recognized since the work of Young, Fresnel, Fizeau, Kirchhoff and Maxwell; simultaneously acquired the properties of a particle. Though Newton considered radiation as a flux of particles, he acted on an hypothesis (today's version of a model) that was not based on any experimental results. In this case, the wave aspect of radiation preceded its particle aspect.

Another fundamental difference between the electron and the photon is the latter has zero mass. For non-relativistic massive particles, the passage to quantum theory is achieved by applying the correspondence principle to Hamiltonian mechanics. Consider the case of an atom in a situation where the kinetic energy is small compared to the rest energy of the particles and small compared to the binding nuclear and ionization energies: the nucleus remains stable and the number of electrons does not change. Even if it is necessary to introduce spin and the Pauli exclusion principle, the particle composition of the atom (nucleons and electrons) remains unchanged. The passage to quantum mechanics via the correspondence principle affects the dynamics without changing either the number or the nature of the constituent particles of the atom.

For a relativistic particle of zero mass, such as the photon, the transition to quantum mechanics is less direct. For radiation inside a cavity, the electromagnetic field is continuously absorbed, emitted, or reflected by the walls of the vessel. The number of photons is not generally a constant of motion. The interaction of a photon with other particles cannot be represented by a potential energy. In relativistic physics, interactions are written in terms of exchanges of particles and collisions. The photon thus lacks a classical counterpart and its quantum description cannot follow the same path as an electron or an atom.

This chapter reviews two well-known subjects, the quantum harmonic oscillator and the classical electromagnetic field. We assume that the reader is familiar with them from his basic course in quantum mechannism and electromagnetism. Both systems are relevant for the formulation of the theory of many body problems and quantum fields. One cannot underrate their importance as subsequent developments heavily rely on relations presented in this introductory chapter. The opportunity is also taken to explain some aspects that will be useful in understanding more complex situations later on.

Classical physics offers another example of a system presenting a great analogy with electromagnetic waves: the elastic or acoustic waves which propagate through solid material. The concept of elastic waves will be needed in some of the following chapters, including those devoted to superconductivity and superfluidity of liquid helium. Consequently some properties of elastic waves are reviewed from the classical point of view as well as from the quantum aspect.

1.2 The Quantum Harmonic Oscillator

1.2.1 Review of Properties

In undergraduated courses, the quantum harmonic oscillator plays a very important role. It is physically relevant and very easy to solve, especially since its spectrum is completely discrete. However, it affords another interesting property, its energy levels are equidistant. This is a consequence of a special symmetry of the harmonic oscillator. If one disregards unimportant constants, its Hamiltonian is *invariant with respect to the interchange of momentum and position operators.*

The consequence of this obvious remark is important for the electromagnetic field as a quantum system and for more complex systems built upon the harmonic oscillator. It has far reaching consequences for quantum mechanics as a whole.

In one dimension, the Hamiltonian of the harmonic oscillator is written

$$H = \frac{p^2}{2m} + \frac{\gamma}{2}q^2 \quad , \tag{1.1}$$

where $p = -\mathrm{i}\hbar \mathrm{d}/\mathrm{d}x$ and q represents multiplication by x on the wavefunction $\varphi(x)$. The term $p^2/2m$ corresponds to the kinetic energy and $(\gamma/2)q^2$ corresponds to the potential energy, where γ is the spring constant. The operators of the momentum p and of the position q satisfy the commutation relation

$$[q, p] = \mathrm{i}\hbar \quad . \tag{1.2}$$

We can construct two new operators

$$P = (\gamma m)^{-1/4} p \quad \text{and} \quad Q = (\gamma m)^{1/4} q \tag{1.3}$$

which also preserve the commutation relation

$$[Q, P] = \mathrm{i}\hbar \quad . \tag{1.4}$$

Any such transformation which preserves the commutation relation is called *canonical*. The Hamiltonian can now be written in terms of the new operators

$$H = \frac{1}{2}\omega\left(P^2 + Q^2\right) \quad , \quad \text{where} \quad \omega = \left(\frac{\gamma}{m}\right)^{1/2} \quad . \tag{1.5}$$

From these operators we define the non-Hermitian operators

$$a^* = \frac{1}{\sqrt{2\hbar}}(Q - \mathrm{i}P) \tag{1.6}$$

and

$$a = \frac{1}{\sqrt{2\hbar}}(Q + \mathrm{i}P) \quad . \tag{1.7}$$

These are dimensionless operators with one being the Hermitian conjugate of the other. Transformations (1.6) and (1.7) are not canonical, as

$$[a, a^*] = I \quad , \tag{1.8}$$

where I is the identity operator. It is interesting, however, to express the Hamiltonian in terms of these operators

$$H = \hbar\omega\left(a^* a + \frac{1}{2}\right) \quad . \tag{1.9}$$

At this point we will review certain properties of the harmonic oscillator exploited later in the text.

Eigenstates

The Hamiltonian H has a non-degenerate ground state $|0\rangle$ defined by

$$a\,|0\rangle = 0 \quad . \tag{1.10}$$

The normalized eigenstates $|n\rangle$ of H are similarly non-degenerate and are obtained by successive operations of a^* on $|0\rangle$:

$$|n\rangle = \frac{(a^*)^n}{\sqrt{n!}}\,|0\rangle \quad , \quad n = 0, 1, 2, \ldots \quad . \tag{1.11}$$

From the following two relations

$$a^* |n\rangle = \sqrt{n+1}\,|n+1\rangle \tag{1.12}$$

and

$$a\,|n\rangle = \sqrt{n}\,|n-1\rangle \quad , \tag{1.13}$$

we see that

$$a^* a\,|n\rangle = a^* \sqrt{n}\,|n-1\rangle = n\,|n\rangle \quad , \tag{1.14}$$

so the Hamiltonian operator gives

$$H\,|n\rangle = \hbar\omega \left(n + \frac{1}{2}\right) |n\rangle \quad . \tag{1.15}$$

The interpretation of the operators a^* and a is clear: a^* *adds and* a *subtracts one quantum of energy* $\hbar\omega$*. The energy levels are equidistant, with two neighbouring levels separated by* $\hbar\omega$*.*

The vectors $|n\rangle$ form a complete orthonormal basis

$$\langle m|n\rangle = \delta_{m,n} = \begin{cases} 1 & m = n, \\ 0 & m \neq n, \end{cases} \qquad m, n = 0, 1, 2, \ldots \quad . \tag{1.16}$$

In this basis, a state $|\varphi\rangle$ has components

$$c_n = \langle n|\varphi\rangle \quad , \quad \text{where} \quad \sum_{n=0}^{\infty} |c_n|^2 < \infty \tag{1.17}$$

and may be expanded as

$$|\varphi\rangle = \sum_{n=0}^{\infty} |n\rangle\, c_n = \sum_{n=0}^{\infty} |n\rangle \langle n|\varphi\rangle \quad . \tag{1.18}$$

The completeness of the basis is expressed by the fact that the sum of the projection operators $|n\rangle\langle n|$ yields the identity operator I. That is,

$$\sum_{n=0}^{\infty} |n\rangle \langle n| = I \quad . \tag{1.19}$$

Similarly, the wavefunction $\varphi(x)$ associated with the state $|\varphi\rangle$ is given by the components of $|\varphi\rangle$ in the configuration representation

$$\varphi(x) = \langle x|\varphi\rangle \quad , \quad \text{where} \quad \int \mathrm{d}x\, |\varphi(x)|^2 < \infty \tag{1.20}$$

and

$$|\varphi\rangle = \int \mathrm{d}x\ |x\rangle\ \langle x|\varphi\rangle \quad . \tag{1.21}$$

Here, the vectors $|x\rangle$ are non-normalizable but formally satisfy relations analogous to (1.16) and (1.19):

$$\langle x|y\rangle = \delta(x-y) \tag{1.22}$$

where $\delta(x)$ is the Dirac "function". Furthermore, they satisfy the completeness relation

$$\int \mathrm{d}x\ |x\rangle\ \langle x| = I \quad . \tag{1.23}$$

The albeit uncountable set of vectors $\{|x\rangle\}$ is called the position basis.

From (1.3), (1.7) and (1.10), the ground state wavefunction $\varphi_0(x) = \langle x|0\rangle$ obeys the equation

$$\langle x|a|0\rangle = \frac{1}{\sqrt{2}}\left[\left(\frac{\hbar}{m\omega}\right)^{1/2}\frac{\mathrm{d}}{\mathrm{d}x} + \left(\frac{m\omega}{\hbar}\right)^{1/2} x\right]\varphi_0(x) = 0 \quad , \tag{1.24}$$

the solution is

$$\varphi_0(x) = \left(\frac{m\omega}{\hbar\pi}\right)^{1/4} \exp\left(-\frac{m\omega}{2\hbar}x^2\right) \quad . \tag{1.25}$$

In general, the wavefunctions of the eigenstates $|n\rangle$ can be expressed in terms of the *Hermite polynomials* $H_n(x)$:

$$\begin{aligned}\varphi_n(x) &= \left\langle x\left|\frac{(a^*)^n}{\sqrt{n!}}\right|0\right\rangle \\ &= \frac{1}{\sqrt{n!2^n}}\left(\frac{m\omega}{\pi\hbar}\right)^{1/4} H_n\left(\sqrt{\frac{m\omega}{\hbar}}x\right)\exp\left(-\frac{m\omega}{2\hbar}x^2\right) \quad .\end{aligned} \tag{1.26}$$

Uncertainty Relations

For the states $|n\rangle$, the oscillator has a well-determined quantum number $n \geq 0$. Moreover,

$$\Delta p \Delta q = \left(n + \frac{1}{2}\right)\hbar \tag{1.27}$$

so *if* $n \geq 1$ *the position–momentum uncertainty relation is not minimal.* This uncertainty relation can be derived from (1.7), (1.12) and (1.13) by first showing

$$(\Delta Q)^2 = \langle n|Q^2|n\rangle - (\langle n|Q|n\rangle)^2$$
$$= \frac{\hbar}{2}\left[\langle n|(a+a^*)^2|n\rangle - (\langle n|a+a^*|n\rangle)^2\right] = \frac{\hbar}{2}(2n+1) \tag{1.28}$$

and similarly

$$(\Delta P)^2 = \frac{\hbar}{2}(2n+1) \quad . \tag{1.29}$$

We arrive at (1.27) if we note

$$\Delta p \Delta q = \Delta P \Delta Q \quad . \tag{1.30}$$

Time Evolution

In the Heisenberg representation, the time evolution of the oscillator is given by

$$a^*(t) = \exp\left(\frac{\mathrm{i}tH}{\hbar}\right) a^* \exp\left(-\frac{\mathrm{i}tH}{\hbar}\right) = \mathrm{e}^{\mathrm{i}\omega t} a^* \quad . \tag{1.31}$$

One obtains this by solving the differential equation of motion for a^*

$$\frac{\mathrm{d}}{\mathrm{d}t} a^*(t) = \frac{\mathrm{i}}{\hbar}[H, a^*(t)] = \mathrm{i}\omega a^*(t) \quad . \tag{1.32}$$

It is useful to note $a^*(t)$ evolves with a "positive frequency" $\mathrm{e}^{\mathrm{i}\omega t}$ while $a(t)$ evolves with a "negative frequency" $\mathrm{e}^{-\mathrm{i}\omega t}$ so

$$a(t) = (a^*(t))^* = \mathrm{e}^{-\mathrm{i}\omega t} a \quad . \tag{1.33}$$

1.2.2 Coherent States

The quantum aspects of the harmonic oscillator emerge when we study the properties of the states $|n\rangle$, the eigenstates of the Hamiltonian. The uncertainty relation (1.27) exposes the non-classical concept that the values p and q cannot be measured simultaneously with arbitrary precision. There exists, however, another class of states for which the oscillator demonstrates a more classical aspect while remaining fully accountable to the laws of quantum physics. *Coherent states* play an important role in quantum optics. Additionally coherent states will allow us to view classical electrodynamics as a special case of the quantum theory of radiation in (Chap. 8).

A coherent state $|\alpha\rangle$ expanded in the basis $|n\rangle$ takes the form

$$|\alpha\rangle = \exp\left(-\frac{1}{2}|\alpha|^2\right) \sum_{n=0}^{\infty} \frac{\alpha^n}{\sqrt{n!}} |n\rangle \quad , \tag{1.34}$$

where α is any complex number. Note that $|\alpha\rangle$ is normalized to 1. The coherent states possess a number of remarkable properties.

The coherent states are eigenstates of the operator a with the eigenvalue α

$$\begin{aligned} a\,|\alpha\rangle &= \exp\left(-\frac{1}{2}|\alpha|^2\right)\sum_{n=0}^{\infty}\frac{\alpha^n}{\sqrt{n!}}a\,|n\rangle \\ &= \exp\left(-\frac{1}{2}|\alpha|^2\right)\sum_{n=1}^{\infty}\frac{\alpha^n}{\sqrt{n!}}\sqrt{n}\,|n-1\rangle = \alpha\,|\alpha\rangle \quad . \end{aligned} \tag{1.35}$$

In all coherent states $|\alpha\rangle$, the uncertainty is minimal

$$\Delta p \Delta q = \frac{\hbar}{2} \quad . \tag{1.36}$$

This is shown by taking into account (1.6) and (1.7), giving

$$\begin{aligned} (\Delta Q)^2 &= \langle\alpha|Q^2|\alpha\rangle - (\langle\alpha|Q|\alpha\rangle)^2 \\ &= \frac{\hbar}{2}\left[\langle\alpha|(a+a^*)^2|\alpha\rangle - (\langle\alpha|a+a^*|\alpha\rangle)^2\right] \\ &= \frac{\hbar}{2}\left[\alpha^2 + 2|\alpha|^2 + \alpha^{*2} + 1 - (\alpha+\alpha^*)^2\right] = \frac{\hbar}{2} \quad . \end{aligned} \tag{1.37}$$

The calculation of $(\Delta P)^2$ is analogous, giving

$$(\Delta P)^2 = \frac{\hbar}{2} \tag{1.38}$$

and thus (1.36). The ground state can be considered the coherent state which corresponding to $\alpha = 0$.

Correspondence with the Classical Oscillator

If one associates the phase space point (q_0, p_0) of the classical oscillator with α in the following manner

$$\alpha = \frac{1}{(2\hbar)^{1/2}}\left[(m\omega)^{1/2}q_0 + \mathrm{i}(m\omega)^{-1/2}p_0\right] \quad , \tag{1.39}$$

then the state $|\alpha\rangle$ is given in the position basis by

$$\varphi_\alpha(x) = \left(\frac{m\omega}{\hbar\pi}\right)^{1/4}\exp\left[-\frac{m\omega}{2\hbar}(x-q_0)^2 + \frac{\mathrm{i}p_0 x}{\hbar} - \frac{\mathrm{i}p_0 q_0}{2\hbar}\right] \quad . \tag{1.40}$$

We derive this by first expressing $\varphi_\alpha(x)$ in terms of $\varphi_n(x)$

$$\begin{aligned} \varphi_\alpha(x) = \langle x|\alpha\rangle &= \exp\left(-\frac{1}{2}|\alpha|^2\right)\sum_{n=0}^{\infty}\frac{\alpha^n}{\sqrt{n!}}\langle x|n\rangle \\ &= \exp\left(-\frac{1}{2}|\alpha|^2\right)\sum_{n=0}^{\infty}\frac{\alpha^n}{\sqrt{n!}}\varphi_n(x) \quad . \end{aligned} \tag{1.41}$$

From (1.26) we have

$$\varphi_\alpha(x) = \left(\frac{m\omega}{\pi\hbar}\right)^{1/4} \exp\left(-\frac{1}{2}|\alpha|^2 - \frac{m\omega x^2}{2\hbar}\right) \times \sum_{n=0}^{\infty} \frac{1}{n!}\left(\frac{\alpha}{\sqrt{2}}\right)^n H_n\left(\sqrt{\frac{m\omega}{\hbar}}x\right) \quad . \tag{1.42}$$

The Hermite polynomials are obtained from a generating function

$$\mathrm{e}^{-\lambda^2+2\lambda x} = \sum_{n=0}^{\infty} \frac{\lambda^n}{n!} H_n(x) \quad . \tag{1.43}$$

Substituting (1.43) into (1.42), we get

$$\varphi_\alpha(x) = \left(\frac{m\omega}{\pi\hbar}\right)^{1/4} \exp\left[-\frac{1}{2}\alpha(\alpha+\alpha^*) - \frac{m\omega}{2\hbar}x^2 + \left(\frac{2m\omega}{\hbar}\right)^{1/2} \alpha x\right] \quad , \tag{1.44}$$

which, after applying (1.39), yields (1.40).

The function $\varphi_\alpha(x)$ is a Gaussian distribution centered at q_0 with corresponding average momentum equal to p_0

$$\langle\alpha|q|\alpha\rangle = \int_{-\infty}^{\infty} \mathrm{d}x\, x|\varphi_\alpha(x)|^2 = q_0 \tag{1.45}$$

and

$$\langle\alpha|p|\alpha\rangle = -\mathrm{i}\hbar \int_{-\infty}^{\infty} \mathrm{d}x\, \varphi_\alpha^*(x) \frac{\mathrm{d}}{\mathrm{d}x} \varphi_\alpha(x) = p_0 \quad . \tag{1.46}$$

Evolution of the Coherent States

The coherent states remain coherent during the course of the evolution

$$\begin{aligned} \exp\left(-\frac{\mathrm{i}tH}{\hbar}\right)|\alpha\rangle &= \exp\left(-\frac{1}{2}|\alpha|^2\right) \sum_{n=0}^{\infty} \frac{\alpha^n}{\sqrt{n!}} \exp\left[-\mathrm{i}\left(n+\frac{1}{2}\right)\omega t\right] |n\rangle \\ &= \exp\left(-\frac{\mathrm{i}\omega t}{2}\right) |\mathrm{e}^{-\mathrm{i}\omega t}\alpha\rangle \quad . \end{aligned} \tag{1.47}$$

One can immediately verify that $|\mathrm{e}^{-\mathrm{i}\omega t}\alpha\rangle$ is an eigenstate of $a(t) = \mathrm{e}^{-\mathrm{i}\omega t}a$.

A coherent state, given in the position basis by function (1.40), evolves through time in such a way that the probability density $|\varphi_\alpha(x,t)|^2$ satisfies

$$|\varphi_\alpha(x,t)|^2 = \left(\frac{m\omega}{\pi\hbar}\right)^{1/2} \exp\left[-\frac{m\omega}{\hbar}\left(x - q(t)\right)^2\right] \quad , \tag{1.48}$$

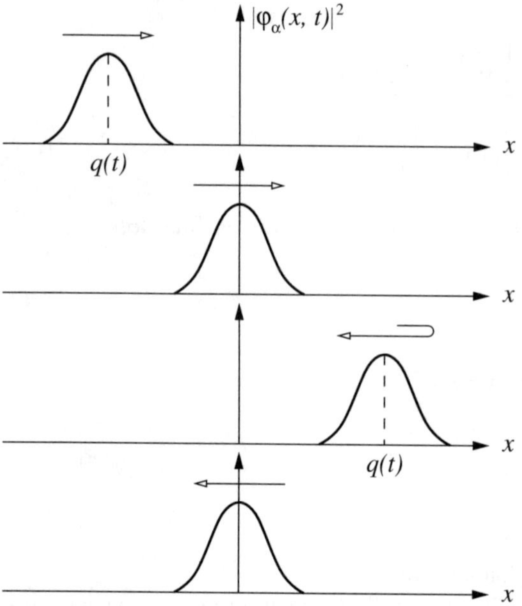

Fig. 1.1. The Gaussian wavepacket follows the classical motion without deforming

where

$$q(t) = q_0 \cos \omega t + \frac{p_0}{m\omega} \sin \omega t \tag{1.49}$$

describes the classical evolution of the harmonic oscillator. For verification let us compare it with (1.40) and express $\mathrm{e}^{-\mathrm{i}\omega t}\alpha$ with the help of (1.39). This gives

$$\begin{aligned} \mathrm{e}^{-\mathrm{i}\omega t}\alpha = \frac{1}{(2\hbar)^{1/2}} \Big[& (m\omega)^{1/2} \left(q_0 \cos \omega t + \frac{p_0}{m\omega} \sin \omega t \right) \\ & + \mathrm{i}(m\omega)^{-1/2} (p_0 \cos \omega t - m\omega q_0 \sin \omega t) \Big] \ . \end{aligned} \tag{1.50}$$

Thus, the Gaussian packet described in (1.40) follows the classical motion without deforming (Fig. 1.1).

Completeness of the Coherent States

According to (1.35), the operator a has an infinite number of eigenstates with complex eigenvalues. There is nothing contradictory in this as a is not an Hermitian operator, i.e. these states are not orthogonal to one another. From (1.34) it follows immediately that

$$\langle\alpha_1|\alpha_2\rangle = \exp\left(\alpha_1^*\alpha_2 - \frac{1}{2}|\alpha_1|^2 - \frac{1}{2}|\alpha_2|^2\right) \quad . \tag{1.51}$$

Nevertheless, the coherent states form a over complete set of states in the following sense

$$\frac{1}{\pi}\int \mathrm{d}^2\alpha\ |\alpha\rangle\langle\alpha| = I \tag{1.52}$$

where the integration is carried out over the entire complex plane α. This can be shown by noting that

$$\frac{1}{\pi}\int \mathrm{d}^2\alpha\ |\alpha\rangle\langle\alpha| = \sum_{n=0}^{\infty}\sum_{m=0}^{\infty}\frac{|n\rangle\langle m|}{\sqrt{n!m!}}\frac{1}{\pi}\int \mathrm{d}^2\alpha\ \exp(-|\alpha|^2)\alpha^{*m}\alpha^n \quad . \tag{1.53}$$

By substituting $\alpha = r\mathrm{e}^{\mathrm{i}\varphi}$ before integrating the right-hand side of (1.53), we can use (1.19) to arrive at (1.52).

Construction of the Coherent States from the Ground State

It is often convenient to construct the coherent states from the ground state $|0\rangle$. Combining (1.34) and (1.11), gives

$$\begin{aligned}|\alpha\rangle &= \exp\left(-\frac{1}{2}|\alpha|^2\right)\sum_{n=0}^{\infty}\frac{\alpha^n(a^*)^n}{n!}|0\rangle \\ &= \exp\left(-\frac{1}{2}|\alpha|^2\right)\exp(\alpha a^*)|0\rangle = D(\alpha)|0\rangle \quad ,\end{aligned} \tag{1.54}$$

where

$$\begin{aligned}D(\alpha) &= \exp(\alpha a^* - \alpha^* a) \\ &= \exp(\alpha a^*)\exp(-\alpha^* a)\exp\left(-\frac{1}{2}|\alpha|^2\right) \quad .\end{aligned} \tag{1.55}$$

When the two operators X and Y commute with their commutator

$$[X,[X,Y]] = [Y,[X,Y]] = 0 \quad , \tag{1.56}$$

we can write (Exercise 1 (iii))

$$\exp(X+Y) = \exp(X)\exp(Y)\exp\left(-\frac{1}{2}[X,Y]\right) \quad . \tag{1.57}$$

In this case, $[X,Y]$ is a multiple of the identity operator and we have

$$X = \alpha a^* \quad , \quad Y = -\alpha^* a \quad \text{and} \quad [X,Y] = |\alpha|^2 \quad , \tag{1.58}$$

giving (1.55) and (1.54) also because

$$e^{-\alpha^* a}|0\rangle = |0\rangle \quad . \tag{1.59}$$

Since the operator $\mathrm{i}(\alpha a^* - \alpha^* a)$ is Hermitian, $D(\alpha)$ is unitary.

An important property of the operator is it translates a by the complex quantity α,

$$D^*(\alpha)aD(\alpha) = a + \alpha \quad . \tag{1.60}$$

We can derive this by first considering the commutation

$$\begin{aligned}[a, D(\alpha)] &= \exp\left(-\frac{1}{2}|\alpha|^2\right)[a, e^{\alpha a^*}]e^{-\alpha^* a} = \exp\left(-\frac{1}{2}|\alpha|^2\right)\alpha e^{\alpha a^*}e^{-\alpha^* a} \\ &= \alpha D(\alpha) \quad .\end{aligned} \tag{1.61}$$

If we now multiply (1.61) on the left with $D^*(\alpha)$ and take into account the unitarity of $D(\alpha)$, we arrive at (1.60).

1.2.3 The Forced Oscillator

The classical equation of motion for a *forced oscillator* is

$$m\frac{d^2}{dt^2}q = -\gamma q + F(t) \quad , \tag{1.62}$$

where $F(t)$ is a given external force generally depending on time. The non-conservative Hamiltonian corresponding to (1.62) is given by

$$H(t) = \frac{p^2}{2m} + \frac{\gamma}{2}q^2 - qF(t) \quad . \tag{1.63}$$

In quantum physics, the Hamiltonian has the same form, with the interaction term $-qF(t)$ linear in a and a^* as shown in (1.3), (1.6) and (1.7). In place of (1.63), we can define a slightly more general Hamiltonian in terms of a and a^*, as follows

$$H(t) = \hbar\omega a^* a + f^*(t)a + f(t)a^* \quad , \tag{1.64}$$

where $f(t)$ is a given complex function of t. If $f(t)$ is real, (1.64) is then reduced to the form of (1.63), with $F(t) = -(2m\omega/\hbar)^{1/2}f(t)$.

The Evolution Operator

The evolution operator $U(t)$ satisfies the Schrödinger equation

$$\mathrm{i}\hbar\frac{d}{dt}U(t) = H(t)U(t) \quad \text{with} \quad U(0) = I \quad . \tag{1.65}$$

The evolution of the operator

$$a(t) = U^*(t) a U(t) \tag{1.66}$$

is thus driven by Heisenberg's equation of motion

$$\begin{aligned} \frac{\mathrm{d}}{\mathrm{d}t} a(t) &= \frac{\mathrm{i}}{\hbar} U^*(t)[H(t), a] U(t) = -\frac{\mathrm{i}}{\hbar} U^*(t) \left(\hbar\omega a + f(t) \right) U(t) \\ &= -\mathrm{i}\omega a(t) - \frac{\mathrm{i}}{\hbar} f(t) \quad . \end{aligned} \tag{1.67}$$

The solution of (1.67) which satisfies the initial condition $a(t=0) = a$ is

$$a(t) = \mathrm{e}^{-\mathrm{i}\omega t} a + \alpha(t) \quad , \tag{1.68}$$

where

$$\alpha(t) = -\frac{\mathrm{i}}{\hbar} \int_0^t \mathrm{e}^{-\mathrm{i}\omega(t-s)} f(s) \mathrm{d}s \quad . \tag{1.69}$$

Relations (1.33) and (1.60) show $a(t)$ can be expressed in the form

$$a(t) = \mathrm{e}^{\mathrm{i}\omega a^* a t} D^* \left(\alpha(t) \right) a D \left(\alpha(t) \right) \mathrm{e}^{-\mathrm{i}\omega a^* a t} \quad . \tag{1.70}$$

The time evolution of $a^*(t)$ is given by the same transformation. This is true of all polynomials constructed from $a(t)$ and $a^*(t)$, allowing us to conclude $U(t)$ and $D(\alpha(t))\,\mathrm{e}^{-\mathrm{i}\omega a^* a t}$ differ only by a phase factor

$$U(t) = \mathrm{e}^{\mathrm{i}\chi(t)} D \left(\alpha(t) \right) \mathrm{e}^{-\mathrm{i}\omega a^* a t} \quad . \tag{1.71}$$

The exact expression $\chi(t)$ is not important here.

Relation (1.71) has an important consequence: *if an oscillator is in its ground state at time $t = 0$, it will be excited by the force $f(t)$ to a coherent state at time t so*

$$U(t) \left| 0 \right\rangle = \mathrm{e}^{\mathrm{i}\chi(t)} D \left(\alpha(t) \right) \left| 0 \right\rangle = \mathrm{e}^{\mathrm{i}\chi(t)} \left| \alpha(t) \right\rangle \quad . \tag{1.72}$$

If we refer to (1.39), (1.45) and (1.46), we find the average values of q and p are given by

$$\langle 0 | U^*(t) q U(t) | 0 \rangle = \langle \alpha(t) | q | \alpha(t) \rangle = q_0(t) \quad , \tag{1.73}$$

$$\langle 0 | U^*(t) p U(t) | 0 \rangle = \langle \alpha(t) | p | \alpha(t) \rangle = p_0(t) \quad . \tag{1.74}$$

If $f(t)$ is real, we have

$$\begin{aligned} q_0(t) &= \left(\frac{2\hbar}{m\omega} \right)^{1/2} \mathrm{Re}\ \alpha(t) \\ &= \left(\frac{2}{\hbar m\omega} \right)^{1/2} \int_0^t \mathrm{d}s f(s) \sin\omega(s-t) \quad , \end{aligned} \tag{1.75}$$

$$p_0(t) = (2\hbar m\omega)^{1/2} \,\mathrm{Im}\, \alpha(t)$$
$$= -\left(\frac{2m\omega}{\hbar}\right)^{1/2} \int_0^t \mathrm{d}s f(s) \cos\omega(s-t) \quad . \tag{1.76}$$

Now we can verify $q_0(t)$ obeys

$$m\frac{\mathrm{d}^2}{\mathrm{d}t^2} q_0(t) + \gamma q_0(t) = -\left(\frac{2m\omega}{\hbar}\right)^{1/2} f(t) = F(t) \tag{1.77}$$

Thus, *a quantum oscillator submitted to an external force* $F(t)$ *and put into motion from its ground state will remain in a coherent state throughout the course of time.* According to (1.75), (1.76) and (1.77), its parameters $q_0(t)$ and $p_0(t)$ follow the corresponding classical laws of motion.

1.2.4 Normal Ordering

In a number of situations, it is necessary to evaluate the average value of monomials of the operators a and a^* of the form $aa^*a^*aaa^*\ldots$ This average is calculated in the ground state or in some coherent state. Such an evaluation is simple when all of the a^* operators are placed to the left of all of the a operators since, according to (1.35),

$$\langle\alpha|a^{*n}a^m|\alpha\rangle = \alpha^{*n}\alpha^m \quad . \tag{1.78}$$

Normal Product

If $aa^*a\ldots a^*$ is a given monomial comprising n a^* operators and m a operators, the *normal product* or the *Wick product* is defined by placing all of the a^* operators to the left of all of the a operators. The normal product associated to $aa^*a\ldots a^*$ is denoted

$$:a^*a\ldots a^*: = a^{*n}a^m \quad . \tag{1.79}$$

The problem consists of decomposing a given product of a and a^* operators into a sum of normal products using the commutation rule $[a, a^*] = 1$. For example

$$aa^* = a^*a + 1 = :aa^*: + 1 \quad . \tag{1.80}$$

In (1.80), we denote 1 as the identity operator.

We will now derive the general rule allowing one to make such a decomposition. Applying the appropriate generalization, it will play an important role in calculations of both many-body problems and quantum-field physics.

Contractions

We first define a *contraction* $\underbracket{aa^*}$ of a pair a and a^* by

$$\underbracket{aa^*} = \langle 0|aa^*|0\rangle \quad . \tag{1.81}$$

The contraction is therefore a number. It follows obviously that

$$\underbracket{aa^*} = \langle 0|aa^*|0\rangle = \langle 0|[a,a^*]|0\rangle = \langle 0|0\rangle = 1 \quad , \tag{1.82}$$

$$\underbracket{a^*a} = \langle 0|a^*a|0\rangle = 0 \quad \text{and} \quad \underbracket{aa} = \underbracket{a^*a^*} = 0 \quad . \tag{1.83}$$

The rule presented is: *any product of operators a^* and a is equal to the sum of the normal products obtained by applying contractions in all possible manners.*

$$\begin{aligned} &aa^*a\ldots a^* \\ &\qquad = :aa^*a\ldots a^*: && \text{(no contraction)} \\ &\qquad + :\underbracket{aa^*}a\ldots a^*: + \ldots + :\underbracket{aa^*a\ldots a^*}: + \ldots && \text{(one contraction)} \\ &\qquad + :\underbracket{aa^*}\underbracket{a\ldots a^*}: + \ldots && \text{(two contractions)} \\ &\qquad + \text{terms with three contractions} + \ldots \quad . \end{aligned} \tag{1.84}$$

In (1.84) the normal products with contractions are the normal products of the non-contracted operators multiplied by the contractions already carried out. That is,

$$:\underbracket{aa^*}a\ldots a^*: = \underbracket{aa^*}:a\ldots a^*: \quad . \tag{1.85}$$

Rather than to demonstrate this rule in the general case, it is more instructive to convince oneself of a few simple cases. Looking back at (1.80), for example, we see

$$aa^* = :aa^*: + \underbracket{aa^*} = :aa^*: + 1 \quad . \tag{1.86}$$

If we omit the vanishing contractions (1.83), we can easily calculate

$$\begin{aligned} aaa^*a^* &= :aaa^*a^*: + :a\underbracket{aa^*}a^*: + :\underbracket{aaa^*}a^*: + :a\underbracket{aa^*a^*}: \\ &\qquad + :\underbracket{aaa^*a^*}: + :\underbracket{a\underbracket{aa^*}a^*}: + :\underbracket{a\underbracket{a}a^*}\underbracket{a^*}: \\ &= a^{*2}a^2 + 4a^*a + 2 \quad . \end{aligned} \tag{1.87}$$

This rule provides a mechanism to carry out algebraic operations which are simple in principle but which would quickly become long calculations without its aid.

1.3 The Electromagnetic Field and the Photon

1.3.1 Maxwell's Equations

The electromagnetic field is one of the most important examples of a classical field. We shall first review the equations of the electromagnetic field, namely the Maxwell equations. We will be mainly interested in electromagnetic waves because their quantum counterpart automatically leads to the definition of the second particle ever discovered in physics, the photon. While the electron (the first discovered particle) can sometimes be handled in a purely classical frame, the photon as a massless particle, cannot.

As electromagnetic waves can be represented as a set of harmonic oscillators, the link between the two systems seems obvious. However, as previously discussed, there is also an important formal difference between them. The quantum harmonic oscillator discussed in Sect. 1.2 describes the oscillation of a single massive particle, while the harmonic oscillators occurring in the quantized electromagnetic field will represent the oscillations of an infinite set of eigenmodes.

Maxwell's equations are written in the following manner:

$$\nabla \cdot \boldsymbol{B} = 0 \quad , \qquad \nabla \times \boldsymbol{E} = -\frac{\partial \boldsymbol{B}}{\partial t} \quad , \tag{1.88}$$

$$\varepsilon_0 \nabla \cdot \boldsymbol{E} = \rho_e \quad , \qquad \nabla \times \boldsymbol{B} = \mu_0 \mathbf{j} + \frac{1}{c^2}\frac{\partial \boldsymbol{E}}{\partial t} \quad , \tag{1.89}$$

where $\boldsymbol{E}(\boldsymbol{x},t)$ and $\boldsymbol{B}(\boldsymbol{x},t)$ designate the *electric* and *magnetic fields*, respectively. We call the combination $(\boldsymbol{E},\boldsymbol{B})$ the electromagnetic field or simply "field". The quantities $\rho_e(\boldsymbol{x},t)$ and $\boldsymbol{j}(\boldsymbol{x},t)$ are the *charge density* and the *current density*. The speed of light in a vacuum is $c = (\varepsilon_0\mu_0)^{-\frac{1}{2}}$.

The *Lorentz force law*, describing the motion of a particle of charge e, mass m and velocity $\boldsymbol{v}$ in an electromagnetic field, completes relations (1.88) and (1.89). In the non-relativistic approximation, this law is written

$$m\frac{\mathrm{d}\boldsymbol{v}}{\mathrm{d}t} = e(\boldsymbol{E} + \boldsymbol{v} \times \boldsymbol{B}) \quad . \tag{1.90}$$

Equations for the conservation of charge and energy are derived from (1.88) and (1.89) as follows

$$\nabla \cdot \boldsymbol{j} + \frac{\partial \rho_e}{\partial t} = 0 \quad , \tag{1.91}$$

$$\nabla \cdot \boldsymbol{s} + \frac{\partial \mathrm{u}^{\mathrm{em}}}{\partial t} = -\boldsymbol{E} \cdot \boldsymbol{j} \quad . \tag{1.92}$$

The quantities u^{em} and $\boldsymbol{s}$ designate respectively the *density* and the *energy flux* of the field ($\boldsymbol{s}$ is called *Poynting's vector*). They are written in terms of the fields as

$$\mathrm{u}^{\mathrm{em}} = \frac{1}{2}(\varepsilon_0|\boldsymbol{E}|^2 + \mu_0^{-1}|\boldsymbol{B}|^2) \quad , \tag{1.93}$$

$$\boldsymbol{s} = \mu_0^{-1}(\boldsymbol{E} \times \boldsymbol{B}) \quad . \tag{1.94}$$

The interaction of the field with n charged non-relativistic point particles is defined by two rules: *each particle is a source of fields* $\boldsymbol{E}$ *and* $\boldsymbol{B}$*, according to (1.89), and each is acted upon by the Lorentz force (1.90).* It is thus necessary to specify the charge density and the current density in terms of the positions $\boldsymbol{q}_i(t)$ and the velocities $\boldsymbol{v}_i(t) = \mathrm{d}\boldsymbol{q}_i(t)/\mathrm{d}t$ of the particles of charge e_i. These relations are

$$\rho_e(\boldsymbol{x}, t) = \sum_{i=1}^{n} e_i \delta\left(\boldsymbol{x} - \boldsymbol{q}_i(t)\right) \tag{1.95}$$

$$\boldsymbol{j}(\boldsymbol{x}, t) = \sum_{i=1}^{n} e_i \boldsymbol{v}_i(t) \delta\left(\boldsymbol{x} - \boldsymbol{q}_i(t)\right) \quad . \tag{1.96}$$

The set of equations (1.88), (1.89), (1.90), (1.95) and (1.96) completely specifies the dynamics of the system comprised of the particles and the field. The classical non-relativistic interaction between matter and radiation is entirely described by these equations.

1.3.2 Gauge Transformations

Equation (1.90) is a Newtonian force law. In quantum physics, however, the description of the time evolution is based on the Hamiltonian formalism. So, to build a theory where the particles and the field are quantum objects, it is necessary to reformulate electrodynamics in a Hamiltonian form.

Toward this aim, it is useful to represent the fields $\boldsymbol{E}$ and $\boldsymbol{B}$ with the aid of a scalar potential V and a vector potential $\boldsymbol{A}$ such that

$$\boldsymbol{E} = -\frac{\partial \boldsymbol{A}}{\partial t} - \nabla V \quad \text{and} \quad \boldsymbol{B} = \nabla \times \boldsymbol{A} \quad . \tag{1.97}$$

In this form, the electromagnetic field automatically satisfies the first set of Maxwell's equations (1.88). Inserting (1.97) into the second set of Maxwell's equations (1.89), one obtains

$$\frac{\partial}{\partial t}(\nabla \cdot \boldsymbol{A}) + \nabla^2 V = -\varepsilon_0^{-1} \rho_e \tag{1.98}$$

and

$$\frac{1}{c^2}\frac{\partial^2}{\partial t^2}\boldsymbol{A} - \nabla^2 \boldsymbol{A} + \nabla\left(\nabla \cdot \boldsymbol{A} + \frac{1}{c^2}\frac{\partial V}{\partial t}\right) = \mu_0 \boldsymbol{j} \quad . \tag{1.99}$$

The set of equations (1.97), (1.98) and (1.99) are equivalent to (1.88) and (1.89). It is necessary, however, to note an important point.

Definition (1.97) does not fix the choice of V and $\boldsymbol{A}$ in a unique manner. That is, $\boldsymbol{E}$ and $\boldsymbol{B}$ do not change if we replace these potentials with new potentials V' and $\boldsymbol{A}'$ derived from V and $\boldsymbol{A}$ by the transformations

$$V' = V + \frac{\partial \chi}{\partial t} \quad \text{and} \quad \boldsymbol{A}' = \boldsymbol{A} - \nabla \chi \ , \tag{1.100}$$

where χ is a given function of space and time.

Transformations of the potentials (1.100) are called *gauge transformations*. Invariance of the electromagnetic field under these transformations is called gauge invariance. We have the freedom to chose χ to simplify (1.98) and (1.99). *When we chose a specific form of χ, we have chosen a gauge.* Two gauges are particularly important, the *Lorenz gauge* and the *Coulomb gauge* (also known as the radiation gauge or the transverse gauge).

The Lorenz gauge

In this gauge, χ is chosen to satisfy the *Lorenz condition*

$$\nabla \cdot \boldsymbol{A} + \frac{1}{c^2}\frac{\partial V}{\partial t} = 0 \ . \tag{1.101}$$

From (1.100) we have

$$\nabla \cdot \boldsymbol{A}' + \frac{1}{c^2}\frac{\partial V'}{\partial t} = \nabla \cdot \boldsymbol{A} + \frac{1}{c^2}\frac{\partial V}{\partial t} - \left(\nabla^2 \chi - \frac{1}{c^2}\frac{\partial^2 \chi}{\partial t^2}\right) \ . \tag{1.102}$$

While $\boldsymbol{E}$ and $\boldsymbol{B}$ are written in terms of potentials V and $\boldsymbol{A}$, which do not satisfy equation (1.101), equation (1.102) permits one to find a function χ such that the new potentials V' and $\boldsymbol{A}'$ satisfy (1.101). In this manner, it is always possible to satisfy the Lorenz condition. This is choosing the Lorenz gauge.

Equations (1.98) and (1.99) take the form of wave equations with sources

$$\frac{1}{c^2}\frac{\partial^2 V}{\partial t^2} - \nabla^2 V = \varepsilon_0^{-1} \rho_e \tag{1.103}$$

$$\frac{1}{c^2}\frac{\partial^2 \boldsymbol{A}}{\partial t^2} - \nabla^2 \boldsymbol{A} = \mu_0 \boldsymbol{j} \ . \tag{1.104}$$

Equations (1.101), (1.103) and (1.104) are equivalent to Maxwell's equations for $\boldsymbol{E}$ and $\boldsymbol{B}$ defined by (1.97). Although (1.103) and (1.104) may imply V and $\boldsymbol{A}$ are decoupled, they do not so since they have to satisfy the Lorenz condition (1.101).

The Lorenz condition does not yet determine the potentials in a unique manner. Any gauge transformation (1.100) requiring χ to satisfy

$$\frac{1}{c^2}\frac{\partial^2 \chi}{\partial t^2} - \nabla\chi^2 = 0 \tag{1.105}$$

leaves (1.101), (1.103) and (1.104) invariant. Transformations conforming to (1.105) are *restricted gauge transformations.*

The Lorenz gauge is often taken as a starting point for the formulation of relativistic quantum electrodynamics as (1.101), (1.103) and (1.104) are manifestly covariant under the Lorentz transformations.

The Coulomb Gauge

The Coulomb gauge is defined as

$$\nabla \cdot \boldsymbol{A} = 0 \quad . \tag{1.106}$$

This condition is always satisfiable. If $\boldsymbol{A}$ is any given vector potential, χ is chosen as

$$\nabla^2\chi = \nabla \cdot \boldsymbol{A} \quad . \tag{1.107}$$

Equation (1.100) shows that the transformed potential $\boldsymbol{A}'$ obeys (1.106). Equations (1.98) and (1.99) are written as

$$\nabla^2 V = -\varepsilon_0^{-1}\rho_e \quad , \tag{1.108}$$

$$\frac{1}{c^2}\frac{\partial^2 \boldsymbol{A}}{\partial t^2} - \nabla^2\boldsymbol{A} = \mu_0\boldsymbol{j} - \frac{1}{c^2}\frac{\partial \nabla V}{\partial t} \quad . \tag{1.109}$$

The Coulomb gauge takes its name from Poisson's equation (1.108) describing the electrostatic scalar potential. As in electrostatics, the solution to (1.108) is

$$V(\boldsymbol{x},t) = \frac{1}{4\pi\varepsilon_0}\int \mathrm{d}\boldsymbol{y}\frac{\rho_e(\boldsymbol{y},t)}{|\boldsymbol{x}-\boldsymbol{y}|} \quad . \tag{1.110}$$

V varies instantaneously when ρ_e varies anywhere in space. The theory of relativity is not contradicted as it only limits the speed of propagation of the electromagnetic field. The relevant equation (1.109) describes the radiation and corresponds to the dynamical degrees of freedom of the field.

1.3.3 Decomposition of the Field into Longitudinal and Transverse Components

A vector field $\boldsymbol{a}$ can be decomposed into *longitudinal* and *transverse* components

$$\boldsymbol{a} = \boldsymbol{a}_{\mathrm{L}} + \boldsymbol{a}_{\mathrm{T}} \tag{1.111}$$

such that

$$\nabla \times \boldsymbol{a}_{\mathrm{L}} = 0 \quad \text{and} \quad \nabla \cdot \boldsymbol{a}_{\mathrm{T}} = 0 \quad . \tag{1.112}$$

We call $\boldsymbol{a}_{\mathrm{L}}$ irrotational and $\boldsymbol{a}_{\mathrm{T}}$ divergence-free. It is unique, provided one specifies the boundary conditions for $\boldsymbol{a}$.

In the Coulomb gauge, relation (1.106) illustrates the vector potential is transverse or $\boldsymbol{A} = \boldsymbol{A}_{\mathrm{T}}$. If we consider the second term of (1.109), we can write

$$\mu_0 \boldsymbol{j}_{\mathrm{T}} = \mu_0 \boldsymbol{j} - \frac{1}{c^2}\frac{\partial \nabla V}{\partial t} \quad . \tag{1.113}$$

In this notation, $\boldsymbol{j}_{\mathrm{T}}$ is the transverse component of the current $\boldsymbol{j}$. One can demonstrate the divergence of $\boldsymbol{j}_{\mathrm{T}}$ is zero by taking the divergence of both sides of (1.113). Applying charge conservation (1.91) and Poisson's equation (1.108) respectively to the two terms on the right-hand side gives $\nabla \cdot \boldsymbol{j}_{\mathrm{T}} = 0$. Equation (1.109) can then be written

$$\frac{1}{c^2}\frac{\partial^2 \boldsymbol{A}}{\partial t^2} - \nabla^2 \boldsymbol{A} = \mu_0 \boldsymbol{j}_{\mathrm{T}} \quad . \tag{1.114}$$

Moreover, $\boldsymbol{B}$ is always transverse, whereas $\boldsymbol{E}$ can be formally decomposed into decoupled longitudinal and transverse components

$$\boldsymbol{E} = \boldsymbol{E}_{\mathrm{L}} + \boldsymbol{E}_{\mathrm{T}} \quad , \quad \boldsymbol{E}_{\mathrm{L}} = -\nabla V \quad \text{and} \quad \boldsymbol{E}_{\mathrm{T}} = -\frac{\partial \boldsymbol{A}}{\partial t} \quad . \tag{1.115}$$

The total energy of the field is derived from (1.93)

$$\begin{aligned} E^{\mathrm{em}} &= \int \mathrm{d}\boldsymbol{x} \mathrm{u}^{\mathrm{em}}(\boldsymbol{x}, t) = \frac{1}{2}\int \mathrm{d}\boldsymbol{x}\left(\varepsilon_0|\boldsymbol{E}|^2 + \mu_0^{-1}|\boldsymbol{B}|^2\right) \\ &= \frac{1}{2}\int \mathrm{d}\boldsymbol{x}\left(\varepsilon_0|\boldsymbol{E}_{\mathrm{T}}|^2 + \mu_0^{-1}|\boldsymbol{B}|^2\right) + \frac{1}{2}\int \mathrm{d}\boldsymbol{x}\varepsilon_0|\boldsymbol{E}_{\mathrm{L}}|^2 \quad . \end{aligned} \tag{1.116}$$

It should be noted that the energy, although quadratic in $\boldsymbol{E}$, is also the sum of transverse and longitudinal contributions. This is because the cross term $\varepsilon_0 \boldsymbol{E}_{\mathrm{L}} \cdot \boldsymbol{E}_{\mathrm{T}}$, once integrated over the entire volume, is equal to zero. Taking (1.106) into account,

$$\int \mathrm{d}\boldsymbol{x} \boldsymbol{E}_{\mathrm{L}} \cdot \boldsymbol{E}_{\mathrm{T}} = \int \mathrm{d}\boldsymbol{x} \nabla V \cdot \frac{\partial \boldsymbol{A}}{\partial t} = \int \mathrm{d}\boldsymbol{x} \left[\nabla \cdot \left(V \frac{\partial \boldsymbol{A}}{\partial t} \right) - V \frac{\partial}{\partial t} \nabla \cdot \boldsymbol{A} \right]$$
$$= \int \mathrm{d}\boldsymbol{\sigma} \cdot V \frac{\partial \boldsymbol{A}}{\partial t} = 0 \quad . \tag{1.117}$$

The surface integral is carried out over the boundary of the integration region. One usually chooses boundary conditions for A such that this integral vanishes. This will be the case for the choice of periodic boundary conditions to be introduced in Sect. 1.3.5. Using (1.110), the electrostatic energy may be expressed in the usual form

$$E^{\mathrm{es}} = \frac{1}{2} \int \mathrm{d}\boldsymbol{x} \varepsilon_0 |\boldsymbol{E}_{\mathrm{L}}|^2 = \frac{1}{2} \int \mathrm{d}\boldsymbol{x} \varepsilon_0 |\nabla V|^2 = -\frac{1}{2} \int \mathrm{d}\boldsymbol{x} \varepsilon_0 V \nabla^2 V$$
$$= \frac{1}{8\pi\varepsilon_0} \int \mathrm{d}\boldsymbol{x} \int \mathrm{d}\boldsymbol{y} \frac{\rho_e(\boldsymbol{x}, t) \rho_e(\boldsymbol{y}, t)}{|\boldsymbol{x} - \boldsymbol{y}|} \quad . \tag{1.118}$$

The advantage of the Coulomb gauge is that allows one to independently treat the transverse degrees of freedom (corresponding with the radiative part) and the longitudinal degrees of freedom (corresponding with the instantaneous Coulomb interaction). The photon, in this case, can be associated with the transverse field in a natural manner. Making the Coulomb gauge is preferred for problems involving the interaction of radiation with non-relativistic matter.

1.3.4 Hamiltonian of the Interaction of Radiation with Non-Relativistic Matter

To formulate a Hamiltonian version of the Lorentz force law (1.90), one begins with Maxwell's equations (1.88) and (1.89) chosen in the Coulomb gauge, and the charge and current densities described by (1.95) and (1.96) for n particles in motion. For simplicity, we assume each particle has the same charge e and mass m. The force law is then written

$$m \frac{\mathrm{d}\boldsymbol{v}_i}{\mathrm{d}t} = e \left[\boldsymbol{E}_{\mathrm{L}}(\boldsymbol{q}_i, t) + \boldsymbol{E}_{\mathrm{T}}(\boldsymbol{q}_i, t) + \boldsymbol{v}_i \times \boldsymbol{B}(\boldsymbol{q}_i, t) \right] \quad . \tag{1.119}$$

In (1.119) and following equations, $\boldsymbol{q}_i = \boldsymbol{q}_i(t)$ and $\boldsymbol{v}_i = \boldsymbol{v}_i(t)$, for $i = 1, 2, \ldots, n$, are the positions and velocities of the n particles at time t, respectively.

Accounting for (1.110), (1.115) and (1.95), the longitudinal component of the electric field evaluated at $\boldsymbol{x} = \boldsymbol{q}_i$ is given by

$$\boldsymbol{E}_{\mathrm{L}}(\boldsymbol{q}_i, t) = \frac{e}{4\pi\varepsilon_0} \sum_{\substack{j \\ (j \neq i)}} \frac{\boldsymbol{q}_i - \boldsymbol{q}_j}{|\boldsymbol{q}_i - \boldsymbol{q}_j|^3} \quad . \tag{1.120}$$

The infinite term $j = i$ corresponds to the self-energy of a moving charge and is omitted in this sum. The longitudinal field expressed by (1.120) does not explicitly depend on time. Its time dependence is implicit through the particle coordinates. In general, however, the transverse fields $\boldsymbol{E}_\mathrm{T}(\boldsymbol{q}_i, t)$ and $\boldsymbol{B}(\boldsymbol{q}_i, t)$, are time dependent both explicitly and via $\boldsymbol{q}_i = \boldsymbol{q}_i(t)$.

Hamilton's Equations

Considering the expression

$$H(\{\boldsymbol{q}_i, \boldsymbol{p}_i\}, t) = \sum_i \frac{1}{2m} |\boldsymbol{p}_i - e\boldsymbol{A}(\boldsymbol{q}_i, t)|^2 + \frac{1}{8\pi\varepsilon_0} \sum_{i \neq j} \frac{e^2}{|\boldsymbol{q}_i - \boldsymbol{q}_j|} \quad , \tag{1.121}$$

the first Hamilton equation gives the velocity

$$\frac{\mathrm{d}\boldsymbol{q}_i}{\mathrm{d}t} = \frac{\partial H}{\partial \boldsymbol{p}_i} = \frac{1}{m} [\boldsymbol{p}_i - e\boldsymbol{A}(\boldsymbol{q}_i, t)] = \boldsymbol{v}_i \quad . \tag{1.122}$$

The second Hamilton equation gives rise to the Lorentz force law. Denoting the components of the vectors with $\alpha, \beta = 1, 2, 3$, one derives

$$\begin{aligned} m \frac{\mathrm{d}^2 q_i^\alpha}{\mathrm{d}t^2} &= \frac{\mathrm{d}}{\mathrm{d}t} [p_i^\alpha - eA^\alpha(\boldsymbol{q}_i, t)] = -\frac{\partial H}{\partial q_i^\alpha} - e \frac{\mathrm{d}A^\alpha}{\mathrm{d}t}(\boldsymbol{q}_i, t) \\ &= e \sum_{\beta=1}^{3} v_i^\beta \frac{\partial A^\beta}{\partial q_i^\alpha} + \frac{e^2}{4\pi\varepsilon_0} \sum_{j \neq i}^{n} \frac{q_i^\alpha - q_j^\alpha}{|\boldsymbol{q}_i - \boldsymbol{q}_j|^3} - e \frac{\mathrm{d}A^\alpha}{\mathrm{d}t} \\ &= e \sum_{\beta=1}^{3} v_i^\beta \left(\frac{\partial A^\beta}{\partial q_i^\alpha} - \frac{\partial A^\alpha}{\partial q_i^\beta} \right) + eE_\mathrm{L}^\alpha(\boldsymbol{q}_i, t) - e \frac{\partial A^\alpha}{\partial t}(\boldsymbol{q}_i, t) \quad . \end{aligned} \tag{1.123}$$

Equation (1.119) is obtained by substituting $\boldsymbol{E}_\mathrm{T} = -\partial \boldsymbol{A}/\partial t$ and $\boldsymbol{B} = \nabla \times \boldsymbol{A}$.

In (1.121), the $(1/m)\,[\boldsymbol{p}_i - e\boldsymbol{A}(\boldsymbol{q}_i, t)]$ terms give rise to the transverse field contribution appearing in (1.123). These terms describe the interaction of the particles with the radiative part of the field. Moreover, (1.122) illustrates the momentum $m\boldsymbol{v}_i = m\mathrm{d}\boldsymbol{q}_i/\mathrm{d}t$ no longer coinciding with $\boldsymbol{p}_i$, the canonically conjugate variable to $\boldsymbol{q}_i$.

Conservation of Energy

As a result of the interaction, the energy of the particles or of the field taken in isolation is no longer constant. To obtain the total energy, it is necessary to add the energy of the radiation (the first term of (1.116)) to (1.121), as the Coulomb energy is already included. The total energy is thus

$$H^{\text{tot}} = H(\{\boldsymbol{p}_i, \boldsymbol{q}_i\}, t) + \frac{1}{2} \int \mathrm{d}\boldsymbol{x} \left[\varepsilon_0 |\boldsymbol{E}_\mathrm{T}(\boldsymbol{x}, t)|^2 + \mu_0^{-1} |\boldsymbol{B}(\boldsymbol{x}, t)|^2 \right] \quad . \tag{1.124}$$

It is now possible to verify H^{tot} does not depend on time. Including the self-energy of the charges, H^{tot} can be expressed from (1.122), (1.118) and (1.116) as

$$H^{\text{tot}} = \frac{m}{2} \sum_{i=1}^{n} |\boldsymbol{v}_i(t)|^2 + \int \mathrm{d}\boldsymbol{x} \mathrm{u}^{\text{em}}(\boldsymbol{x}, t) \quad . \tag{1.125}$$

It follows from the Lorentz law and from the charge current density (1.96)

$$\begin{aligned} \frac{\mathrm{d}}{\mathrm{d}t} H^{\text{tot}} &= e \sum_{i=1}^{n} \boldsymbol{v}_i \cdot \boldsymbol{E}_i + \frac{\mathrm{d}}{\mathrm{d}t} \int \mathrm{d}\boldsymbol{x} \mathrm{u}^{\text{em}}(\boldsymbol{x}, t) \\ &= \int \mathrm{d}\boldsymbol{x} \boldsymbol{E}(\boldsymbol{x}) \cdot \boldsymbol{j}(\boldsymbol{x}) + \int \mathrm{d}\boldsymbol{x} \frac{\partial}{\partial t} \mathrm{u}^{\text{em}}(\boldsymbol{x}, t) \\ &= -\int \mathrm{d}\boldsymbol{x} \nabla \cdot \boldsymbol{s} = -\int \mathrm{d}\boldsymbol{\sigma} \cdot \boldsymbol{s} \quad , \end{aligned} \tag{1.126}$$

where the last line results from the conservation law (1.92). The Poynting vector integrated over the surface of the system represents the energy flux through the same surface. When the flux is zero (1.126), the total energy is conserved.

1.3.5 Fourier Analysis of the Classical Free Field

In the Coulomb gauge, the free electromagnetic field satisfies the following equations:

$$\boldsymbol{E} = -\frac{\partial \boldsymbol{A}}{\partial t} \quad , \qquad \boldsymbol{B} = \nabla \times \boldsymbol{A} \quad , \qquad \nabla \cdot \boldsymbol{A} = 0 \quad , \tag{1.127}$$

$$\nabla^2 \boldsymbol{A} - \frac{1}{c^2} \frac{\partial^2 \boldsymbol{A}}{\partial t^2} = 0 \quad . \tag{1.128}$$

The solutions of the wave equation (1.128) are not defined in a unique manner unless boundary conditions are specified. Consider the field to be contained in a cube Λ of length L with the following periodic conditions:

$$\begin{aligned} \boldsymbol{A}(0, x_2, x_3) &= \boldsymbol{A}(L, x_2, x_3) \quad , \\ \boldsymbol{A}(x_1, 0, x_3) &= \boldsymbol{A}(x_1, L, x_3) \quad , \\ \boldsymbol{A}(x_1, x_2, 0) &= \boldsymbol{A}(x_1, x_2, L) \quad . \end{aligned} \tag{1.129}$$

It is more convenient to define these conditions than to assume the walls to be either conductive or reflective. Indeed, relations (1.129) allow for the existence

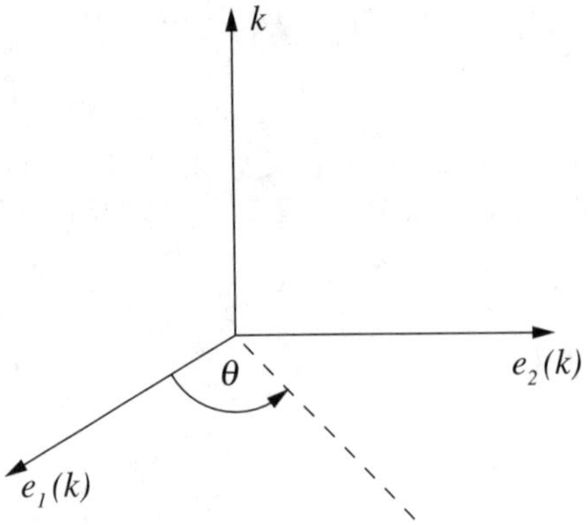

Fig. 1.2. The polarization vectors $\boldsymbol{e}_1(\boldsymbol{k})$ and $\boldsymbol{e}_2(\boldsymbol{k})$ are chosen real and orthogonal in the plane perpendicular to $\boldsymbol{k}$

of a non-zero energy flux (Poynting's vector) traversing the boundary of Λ. The region defined by Λ can thus be identified as a portion of space traversed by electromagnetic waves. In the limit $L \to \infty$, one finds the theory of a free field propagating in infinite space (see Sect. 2.1.3 for more on this subject).

Plane Waves Basis

Consider the periodic functions

$$\psi_{\boldsymbol{k}\lambda}(\boldsymbol{x}) = \boldsymbol{e}_\lambda(\boldsymbol{k})\frac{\mathrm{e}^{\mathrm{i}\boldsymbol{k}\cdot\boldsymbol{x}}}{L^{3/2}} \quad ,$$
$$\boldsymbol{k} = \left\{(k^1, k^2, k^3); k^\alpha = \frac{2\pi n^\alpha}{L}; \; n^\alpha \text{ integer}; \alpha = 1, 2, 3\right\} \tag{1.130}$$

where the two *polarization vectors* $\boldsymbol{e}_\lambda(\boldsymbol{k}), \lambda = 1, 2$ are chosen real and orthogonal in the plane perpendicular to $\boldsymbol{k}$ (Fig. 1.2). The vectors satisfy

$$\boldsymbol{k} \cdot \boldsymbol{e}_\lambda(\boldsymbol{k}) = 0 \tag{1.131}$$

and

$$\boldsymbol{e}_\lambda(\boldsymbol{k}) \cdot \boldsymbol{e}_{\lambda'}(\boldsymbol{k}) = \delta_{\lambda,\lambda'} = \begin{cases} 1 & \lambda = \lambda' \\ 0 & \lambda \neq \lambda' \end{cases} \quad . \tag{1.132}$$

The $\psi_{\boldsymbol{k}\lambda}(\boldsymbol{x})$ forms a complete set of vector functions orthogonal to each other

$$\begin{aligned} \langle \psi_{\boldsymbol{k}\lambda} | \psi_{\boldsymbol{k}'\lambda'} \rangle &= \frac{1}{L^3} \int_\Lambda \mathrm{d}\boldsymbol{x} \, \mathrm{e}^{-\mathrm{i}(\boldsymbol{k}-\boldsymbol{k}')\cdot\boldsymbol{x}} \boldsymbol{e}_\lambda(\boldsymbol{k}) \cdot \boldsymbol{e}_{\lambda'}(\boldsymbol{k}') \\ &= \delta_{\boldsymbol{k},\boldsymbol{k}'} \delta_{\lambda,\lambda'} \quad . \end{aligned} \tag{1.133}$$

It is always possible to expand the vector potential as a sum of functions with time-dependent coefficients $c_{\boldsymbol{k}\lambda}(t)$:

$$\boldsymbol{A}(\boldsymbol{x},t) = \sum_{\boldsymbol{k}\lambda} c_{\boldsymbol{k}\lambda}(t)\boldsymbol{\psi}_{\boldsymbol{k}\lambda}(\boldsymbol{x}) = \frac{1}{L^{3/2}}\sum_{\boldsymbol{k}} \boldsymbol{c}_{\boldsymbol{k}}(t)\mathrm{e}^{\mathrm{i}\boldsymbol{k}\cdot\boldsymbol{x}} \quad , \tag{1.134}$$

where

$$\boldsymbol{c}_{\boldsymbol{k}}(t) = \sum_{\lambda} c_{\boldsymbol{k}\lambda}(t)\boldsymbol{e}_{\lambda}(\boldsymbol{k}) \quad . \tag{1.135}$$

The transverse condition (1.131) is necessary and sufficient for the divergence of $\boldsymbol{A}$ to be zero.

The wave equation (1.128) is equivalent to the equation of the harmonic oscillator, as

$$\frac{\mathrm{d}^2}{\mathrm{d}t^2}\boldsymbol{c}_{\boldsymbol{k}}(t) + \omega_k^2\boldsymbol{c}_{\boldsymbol{k}}(t) = 0 \quad , \tag{1.136}$$

where $\omega_k = \omega_{\boldsymbol{k}} = c|\boldsymbol{k}|$. Since $\boldsymbol{A}$ is real the following relation between the $\boldsymbol{c}_{\boldsymbol{k}}$ is implied

$$\boldsymbol{c}^*_{\boldsymbol{k}}(t) = \boldsymbol{c}_{-\boldsymbol{k}}(t) \quad . \tag{1.137}$$

The general solution of (1.136) satisfying (1.137) is

$$\boldsymbol{c}_{\boldsymbol{k}}(t) = \boldsymbol{f}_{\boldsymbol{k}}\exp(-\mathrm{i}\omega_k t) + \boldsymbol{f}^*_{-\boldsymbol{k}}\exp(\mathrm{i}\omega_k t) = \boldsymbol{f}_{\boldsymbol{k}}(t) + \boldsymbol{f}^*_{-\boldsymbol{k}}(t) \quad , \tag{1.138}$$

where

$$\boldsymbol{f}_{\boldsymbol{k}}(t) = \boldsymbol{f}_{\boldsymbol{k}}\exp(-\mathrm{i}\omega_k t) \qquad \text{and} \qquad \boldsymbol{f}_{\boldsymbol{k}} = \sum_{\lambda=1}^{2} f_{\boldsymbol{k}\lambda}\boldsymbol{e}_{\lambda}(\boldsymbol{k}) \quad . \tag{1.139}$$

According to (1.134), (1.138) and (1.139), the vector potential expands

$$\boldsymbol{A}(\boldsymbol{x},t) = \sum_{\boldsymbol{k}\lambda}\left[f_{\boldsymbol{k}\lambda}\exp(-\mathrm{i}\omega_k t)\boldsymbol{\psi}_{\boldsymbol{k}\lambda}(\boldsymbol{x}) + f^*_{\boldsymbol{k}\lambda}\exp(\mathrm{i}\omega_k t)\boldsymbol{\psi}^*_{\boldsymbol{k}\lambda}(\boldsymbol{x})\right] \quad . \tag{1.140}$$

Initial conditions $\boldsymbol{A}(\boldsymbol{x},0)$ and $(\partial/\partial t)\boldsymbol{A}(\boldsymbol{x},0)$ entirely determine the coefficients $f_{\boldsymbol{k}\lambda}$.

Expression of the Energy and the Flux

It is possible to express the electromagnetic energy, given by

$$\begin{aligned} E^{\mathrm{em}} &= \int_{\Lambda}\mathrm{d}\boldsymbol{x}\,\frac{1}{2}(\varepsilon_0|\boldsymbol{E}|^2 + \mu_0^{-1}|\boldsymbol{B}|^2) \\ &= \int_{\Lambda}\mathrm{d}\boldsymbol{x}\,\frac{1}{2}\left(\varepsilon_0\left|\frac{\partial\boldsymbol{A}}{\partial t}\right|^2 + \mu_0^{-1}\,|\nabla\times\boldsymbol{A}|^2\right) \end{aligned} \tag{1.141}$$

in terms of the Fourier coefficients $f_{k\lambda}$. First, use of the identity

$$|\nabla \times \boldsymbol{A}|^2 = \sum_{\alpha=1}^{3} |\nabla A^\alpha|^2 - \nabla \cdot [(\boldsymbol{A} \cdot \nabla)\boldsymbol{A}] + (\boldsymbol{A} \cdot \nabla)(\nabla \cdot \boldsymbol{A}) \quad . \tag{1.142}$$

The last term is zero in the Coulomb gauge (1.106). Moreover, when (1.142) is substituted into (1.141), the second term of the right-hand side does not contribute as it is integrated and the contributions at the boundary of Λ cancel due to the periodic conditions. The electromagnetic energy is thus written as

$$E^{\mathrm{em}} = \frac{1}{2} \sum_{\alpha=1}^{3} \int_\Lambda \mathrm{d}\boldsymbol{x} \left[\varepsilon_0 \left(\frac{\partial A^\alpha}{\partial t} \right)^2 + \mu_0^{-1} |\nabla A^\alpha|^2 \right] \quad . \tag{1.143}$$

According to (1.134), the derivatives of the components of the vector potential are

$$\frac{\partial A^\alpha}{\partial t} = \frac{1}{L^{3/2}} \sum_k \frac{\mathrm{d}}{\mathrm{d}t} c_k^\alpha(t) \mathrm{e}^{\mathrm{i}\boldsymbol{k}\cdot\boldsymbol{x}} \quad , \tag{1.144}$$

$$\nabla A^\alpha = \frac{1}{L^{3/2}} \sum_k \mathrm{i}\boldsymbol{k} c_k^\alpha(t) \mathrm{e}^{\mathrm{i}\boldsymbol{k}\cdot\boldsymbol{x}} \quad . \tag{1.145}$$

Exploiting the orthogonality of the plane waves and taking into account (1.137) and (1.138), one finds

$$\begin{aligned}
\varepsilon_0 \int_\Lambda \mathrm{d}\boldsymbol{x} \sum_{\alpha=1}^{3} \left(\frac{\partial A^\alpha}{\partial t} \right)^2 &= \frac{\varepsilon_0}{2} \sum_{\boldsymbol{k}} \frac{\mathrm{d}\boldsymbol{c}_{\boldsymbol{k}}(t)}{\mathrm{d}t} \cdot \frac{\mathrm{d}\boldsymbol{c}_{-\boldsymbol{k}}(t)}{\mathrm{d}t} \\
&= \frac{\varepsilon_0}{2} \sum_{\boldsymbol{k}} \left[-\mathrm{i}\omega_k \boldsymbol{f}_{\boldsymbol{k}}(t) + \mathrm{i}\omega_k \boldsymbol{f}^*_{-\boldsymbol{k}}(t) \right] \cdot \left[-\mathrm{i}\omega_k \boldsymbol{f}_{-\boldsymbol{k}}(t) + \mathrm{i}\omega_k \boldsymbol{f}^*_{\boldsymbol{k}}(t) \right] \\
&= \frac{\varepsilon_0}{2} \sum_{\boldsymbol{k}} \omega_k^2 \left[\boldsymbol{f}^*_{-\boldsymbol{k}}(t) \cdot \boldsymbol{f}_{-\boldsymbol{k}}(t) + \boldsymbol{f}_{\boldsymbol{k}}(t) \cdot \boldsymbol{f}^*_{\boldsymbol{k}}(t) \right. \\
&\quad \left. - \boldsymbol{f}_{\boldsymbol{k}}(t) \cdot \boldsymbol{f}_{-\boldsymbol{k}}(t) - \boldsymbol{f}^*_{-\boldsymbol{k}}(t) \cdot \boldsymbol{f}^*_{\boldsymbol{k}}(t) \right] \quad .
\end{aligned} \tag{1.146}$$

Similarly, one derives

$$\begin{aligned}
\frac{1}{2\mu_0} \int_\Lambda \mathrm{d}\boldsymbol{x} \sum_{\alpha=1}^{3} |\nabla A^\alpha|^2 &= \frac{1}{2\mu_0} \sum_{\boldsymbol{k}} |\boldsymbol{k}|^2 \left[\boldsymbol{f}^*_{-\boldsymbol{k}}(t) \cdot \boldsymbol{f}_{-\boldsymbol{k}}(t) + \boldsymbol{f}_{\boldsymbol{k}}(t) \cdot \boldsymbol{f}^*_{\boldsymbol{k}}(t) \right. \\
&\quad \left. + \boldsymbol{f}_{\boldsymbol{k}}(t) \cdot \boldsymbol{f}_{-\boldsymbol{k}}(t) + \boldsymbol{f}^*_{-\boldsymbol{k}}(t) \cdot \boldsymbol{f}^*_{\boldsymbol{k}}(t) \right] \quad .
\end{aligned} \tag{1.147}$$

Combining (1.146) and (1.147) to form E^{em} and using $\omega_k = c|\boldsymbol{k}|, c = (\varepsilon_0\mu_0)^{-1/2}$, terms of the type $\boldsymbol{f}_{\boldsymbol{k}}(t) \cdot \boldsymbol{f}_{-\boldsymbol{k}}(t)$ and $\boldsymbol{f}^*_{-\boldsymbol{k}}(t) \cdot \boldsymbol{f}^*_{\boldsymbol{k}}(t)$ cancel each other out. The resulting expression is

$$\begin{aligned} E^{\text{em}} &= \varepsilon_0 \sum_k \omega_k^2 \left[\boldsymbol{f}_{-\boldsymbol{k}}^*(t) \cdot \boldsymbol{f}_{-\boldsymbol{k}}(t) + \boldsymbol{f}_{\boldsymbol{k}}(t) \cdot \boldsymbol{f}_{\boldsymbol{k}}^*(t) \right] \\ &= \varepsilon_0 \sum_k \omega_k^2 \left[\boldsymbol{f}_{\boldsymbol{k}}^*(t) \cdot \boldsymbol{f}_{\boldsymbol{k}}(t) + \boldsymbol{f}_{\boldsymbol{k}}(t) \cdot \boldsymbol{f}_{\boldsymbol{k}}^*(t) \right] \quad , \end{aligned} \tag{1.148}$$

where the second equality of (1.148) results from substituting $-\boldsymbol{k} \to \boldsymbol{k}$ in the first part of the sum. As a result of (1.139), it is evident that E^{em} does not depend on time. That is,

$$E^{\text{em}} = \varepsilon_0 \sum_{\boldsymbol{k}\lambda} \omega_k^2 (f_{\boldsymbol{k}\lambda}^* f_{\boldsymbol{k}\lambda} + f_{\boldsymbol{k}\lambda} f_{\boldsymbol{k}\lambda}^*) \quad . \tag{1.149}$$

The free electromagnetic energy is conserved.

The total energy flux $\boldsymbol{S} = \int_\Lambda \mathrm{d}\boldsymbol{x}\boldsymbol{s}$, where $\boldsymbol{s}$ is the Poynting vector (1.94), is calculated in an analogous manner, starting with (1.134)

$$\begin{aligned} \boldsymbol{S} &= \frac{1}{\mu_0} \int_\Lambda \mathrm{d}\boldsymbol{x} (\boldsymbol{E} \times \boldsymbol{B}) = \frac{1}{\mu_0} \int_\Lambda \mathrm{d}\boldsymbol{x} \left(-\frac{\partial \boldsymbol{A}}{\partial t} \right) \times (\nabla \times \boldsymbol{A}) \\ &= \varepsilon_0 c^2 \sum_{\boldsymbol{k}} \left(-\frac{\mathrm{d}\boldsymbol{c}_{\boldsymbol{k}}(t)}{\mathrm{d}t} \right) \times [-\mathrm{i}\boldsymbol{k} \times \boldsymbol{c}_{-\boldsymbol{k}}(t)] \\ &= \varepsilon_0 c^2 \sum_{\boldsymbol{k}} \mathrm{i}\boldsymbol{k} \left[\left(\frac{\mathrm{d}\boldsymbol{c}_{\boldsymbol{k}}(t)}{\mathrm{d}t} \right) \cdot \boldsymbol{c}_{-\boldsymbol{k}}(t) \right] \quad . \end{aligned} \tag{1.150}$$

One exploits the fact that the plane waves are orthogonal and $\boldsymbol{k} \cdot \boldsymbol{c}_{\boldsymbol{k}}(t) = 0$ as a result of (1.131) and (1.135). The derivation of (1.150) is similar to (1.149). Thus as with E^{em}

$$\boldsymbol{S} = \varepsilon_0 c^2 \sum_{\boldsymbol{k}\lambda} \boldsymbol{k} \omega_{\boldsymbol{k}} (f_{\boldsymbol{k}\lambda}^* f_{\boldsymbol{k}\lambda} + f_{\boldsymbol{k}\lambda} f_{\boldsymbol{k}\lambda}^*) \quad , \tag{1.151}$$

is a conserved quantity.

In (1.149) and (1.151), the order of factors has been preserved as it appeared in the calculation, with $\frac{1}{2}(f_{\boldsymbol{k}\lambda}^* f_{\boldsymbol{k}\lambda} + f_{\boldsymbol{k}\lambda} f_{\boldsymbol{k}\lambda}^*)$ taking the place of $|f_{\boldsymbol{k}\lambda}|^2$. This notation will be useful for a future study of the quantum field in Sect. 8.2.1.

1.3.6 Photons and Electromagnetic Waves

Relations (1.134)–(1.136) clearly illustrate the classical free field can be considered as a linear superposition of independent harmonic oscillators. Before quantizing the electromagnetic field (Chap. 8), another crucial point about the quantum harmonic oscillator must be made explicit.

Equidistant energy levels of the harmonic oscillator allows each energy level to be labeled by an integer n, as shown by (1.15). a and a^*, (1.12) and (1.13), transform a state vector n into another state vector m, integers n and m being arbitrary. The interpretation of the black body spectrum by Planck is usually identified with the dawn of quantum mechanics. However Planck did not interpret the energy quantum $h\nu$ as the footprint of a new particle. Planck interpreted $h\nu$ as a step between energy levels, although concepts of quantum mechanics were not existant. Einstein was more radical. He considered the black body radiation as a set of independent energy quanta with energy $h\nu$. This latter interpretation is also full accordance with the photoelectric effect.

This puzzle was solved by Dirac in 1927. There is a massless particle called photon. Its wave counterpart is the electromagnetic field. In fact, there is no ambiguity if one accepts the correspondence principle between classical and quantum mechanics allowing electromagnetic field to quantize in a very natural way (Chap. 8). Even though the resulting quantum field appears as a set of independent quantum oscillators, it also implies the existence of independent particles. They can be localized and are characterized by an energy, a momentum and helicity, a quantity analogous to the angular momentum.

The Photon

The *photon* γ is a particle of zero mass, energy E_γ and momentum $\boldsymbol{p}_\gamma$ given by the relations

$$E_\gamma = \hbar\omega_k \qquad \text{and} \qquad \boldsymbol{p}_\gamma = \hbar\boldsymbol{k} \quad . \tag{1.152}$$

whose equation of motion is given by the classical wave equation.

It is thus natural to express the energy and the flux of the electromagnetic field as that of a collection of photons. This is achieved by first introducing the dimensionless amplitudes $\alpha_{\boldsymbol{k}\lambda}$ defined by

$$f_{\boldsymbol{k}\lambda} = \left(\frac{\hbar}{2\varepsilon_0\omega_{\boldsymbol{k}}}\right)^{1/2} \alpha_{\boldsymbol{k}\lambda} \quad . \tag{1.153}$$

The expansion of the vector potential (1.140) is expressed in terms of the new amplitudes by

$$\begin{aligned} \boldsymbol{A}(\boldsymbol{x},t) = \left(\frac{\hbar}{\varepsilon_0 L^3}\right)^{1/2} \sum_{\boldsymbol{k}\lambda} \frac{\boldsymbol{e}_\lambda(\boldsymbol{k})}{\sqrt{2\omega_{\boldsymbol{k}}}} \{ & \alpha^*_{\boldsymbol{k}\lambda} \exp\left[-\mathrm{i}(\boldsymbol{k}\cdot\boldsymbol{x} - \omega_{\boldsymbol{k}} t)\right] \\ & + \alpha_{\boldsymbol{k}\lambda} \exp\left(\mathrm{i}(\boldsymbol{k}\cdot\boldsymbol{x} - \omega_{\boldsymbol{k}} t)\right)\} \end{aligned} \tag{1.154}$$

and it follows from (1.149) and (1.151) that

$$\begin{aligned} E^{\mathrm{em}} &= \sum_{\boldsymbol{k}\lambda} \hbar\omega_{\boldsymbol{k}} |\alpha_{\boldsymbol{k}\lambda}|^2 \quad , \\ \boldsymbol{S} &= c^2 \sum_{\boldsymbol{k}\lambda} \hbar\boldsymbol{k} |\alpha_{\boldsymbol{k}\lambda}|^2 \quad . \end{aligned} \tag{1.155}$$

These two relations motivate the interpretation of $|\alpha_{\boldsymbol{k}\lambda}|^2$ *as the mean number of photons with momentum* $\hbar\boldsymbol{k}$ *and polarization* λ. As a consequence,

$$\langle N\rangle = \sum_{\boldsymbol{k}\lambda} |\alpha_{\boldsymbol{k}\lambda}|^2 \tag{1.156}$$

represents the mean total number of photons associated with the classical field. From this, the energy and flux per photon are given by

$$\begin{aligned} \frac{E^{\mathrm{em}}}{\langle N\rangle} &= \sum_{\boldsymbol{k}\lambda} \hbar\omega_{\boldsymbol{k}} |\tilde{\alpha}_{\boldsymbol{k}\lambda}|^2 \quad , \\ \frac{1}{c^2}\frac{\boldsymbol{S}}{\langle N\rangle} &= \sum_{\boldsymbol{k}\lambda} \hbar\boldsymbol{k} |\tilde{\alpha}_{\boldsymbol{k}\lambda}|^2 \quad , \end{aligned} \tag{1.157}$$

where

$$\tilde{\alpha}_{\boldsymbol{k}\lambda} = \frac{\alpha_{\boldsymbol{k}\lambda}}{\sqrt{\langle N\rangle}} \qquad \text{and} \qquad \sum_{\boldsymbol{k}\lambda} |\tilde{\alpha}_{\boldsymbol{k}\lambda}|^2 = 1 \tag{1.158}$$

are amplitudes normalized to 1.

Photon States

To return to the usual formalism of quantum physics, it is necessary to introduce the state vectors $|\boldsymbol{k}\lambda\rangle$ in which the photon has a determined momentum and polarization. The states $|\boldsymbol{k}\lambda\rangle$ are vectors diagonalizing the momentum operator $\boldsymbol{p}_\gamma$ and the Hamiltonian H_γ of the photon. That is,

$$\boldsymbol{p}_\gamma\,|\boldsymbol{k}\lambda\rangle = \hbar\boldsymbol{k}\,|\boldsymbol{k}\lambda\rangle \quad , \tag{1.159}$$

$$H_\gamma\,|\boldsymbol{k}\lambda\rangle = \hbar\omega_{\boldsymbol{k}}\,|\boldsymbol{k}\lambda\rangle \quad , \tag{1.160}$$

$$\langle\boldsymbol{k}\lambda|\boldsymbol{k}'\lambda'\rangle = \delta_{\boldsymbol{k},\boldsymbol{k}'}\delta_{\lambda,\lambda'} \quad . \tag{1.161}$$

The Hilbert space of the photon $\mathcal{H}_\gamma$ is a collection of states $|\varphi\rangle$

$$\mathcal{H}_\gamma = \left\{ |\varphi\rangle\,;\, \sum_{\boldsymbol{k}\lambda} |\,\langle\boldsymbol{k}\lambda|\varphi\rangle\,|^2 < \infty \right\} \quad . \tag{1.162}$$

A review of the formalism of the quantum states of a particle can be found in Sects. 2.1.2 and 2.1.3.

The state of a single photon $|\varphi\rangle$, normalized to 1, has amplitudes

$$\langle\boldsymbol{k}\lambda|\varphi\rangle = \tilde{\alpha}_{\boldsymbol{k}\lambda} \qquad \text{and} \qquad \langle\varphi|\varphi\rangle = 1 \tag{1.163}$$

as defined by (1.153) and (1.158). From these expressions, the energy and the flux of the photon (1.157) are the mean value of the energy and the momentum of a photon in the state $|\varphi\rangle$ or

$$\frac{E^{\mathrm{em}}}{\langle N \rangle} = \sum_{\boldsymbol{k}\lambda} \hbar\omega_{\boldsymbol{k}} |\langle \boldsymbol{k}\lambda|\varphi\rangle|^2 = \langle\varphi|H_\gamma|\varphi\rangle \tag{1.164}$$

and

$$\frac{1}{c^2}\frac{\boldsymbol{S}}{\langle N \rangle} = \sum_{\boldsymbol{k}\lambda} \hbar\boldsymbol{k} |\langle \boldsymbol{k}\lambda|\varphi\rangle|^2 = \langle\varphi|\boldsymbol{p}_\gamma|\varphi\rangle \quad . \tag{1.165}$$

Relations (1.164) and (1.165) have a clear interpretation: *The photons of the classical field are all in the same individual state $|\varphi\rangle$ determined by the Fourier amplitudes of the field. The electromagnetic energy and the flux are respectively equal to the product of the energy and the momentum of one photon by the mean number of photons.* This interpretation will be confirmed and detailed in Sect. 8.2.5.

Helicity of the Photon

The probability amplitude $\alpha_{\boldsymbol{k}\lambda}$ was obtained classically as a Fourier transform – up to multiplicative factors – of the vector field $\boldsymbol{A}(\boldsymbol{x})$. From the transverse condition (1.106), only two components of the field $\boldsymbol{A}$ are independent. Two also happens to be the number of polarization states of the photon which are also transverse (relation (1.131) following directly from (1.106)).

It is tempting to associate the vectorial nature of the field and its three components in an intrinsic angular momentum (or spin) of the photon equal to one. This is complicated by the transversality of the field. As a consequence of (1.131), the probability amplitude $\langle \boldsymbol{k}\lambda|\varphi\rangle$ (1.163) is not the product of a function of $\boldsymbol{k}$ by a number only dependent on λ. This situation is tied to the masslessness of the photon. The spin of a massive particle is determined by examing how its wavefunction transforms in a reference frame in which the particle is at rest. Its orbital angular momentum is thus zero and all that remains is its spin. Travelling at the speed of light, however, the photon has no rest frame and this simple method is not applicable. It is possible, however, to proceed in another manner.

When a particle has a well-determined momentum $\boldsymbol{p}$, the projection of its orbital momentum $(\boldsymbol{x} \times \boldsymbol{p}) \cdot \hat{\boldsymbol{p}}$ along $\boldsymbol{p}$ is zero ($\hat{\boldsymbol{a}} = \boldsymbol{a}/|\boldsymbol{a}|$ designates the unit vector pointing in the direction of $\boldsymbol{a}$). Therefore, the projection $\boldsymbol{J} \cdot \hat{\boldsymbol{p}}/\hbar$ of the total angular momentum $\boldsymbol{J}$ along $\boldsymbol{p}$ involves only the intrinsic angular momentum of the particle. This quantity is called the *helicity*. It is a conserved quantity, since the total angular momentum is a constant of motion. Acting on a state $|\boldsymbol{k}\lambda\rangle$, $\boldsymbol{J} \cdot \hat{\boldsymbol{p}}/\hbar$ is the generator of rotations around the axis $\hat{\boldsymbol{k}}$. The operator U_θ which represents a rotation of angle θ around the direction of propagation $\hat{\boldsymbol{k}}$ acts as

$$U_\theta\,|\boldsymbol{k}\lambda\rangle = \sum_{\lambda'=1}^{2} R_{\lambda\lambda'}(\theta)\,|\boldsymbol{k}\lambda'\rangle = \mathrm{e}^{\mathrm{i}\theta\boldsymbol{J}\cdot\hat{\boldsymbol{k}}}\,|\boldsymbol{k}\lambda\rangle \quad . \tag{1.166}$$

The matrix of rotation $R_{\lambda\lambda'}(\theta)$ acts on the polarization vectors $\boldsymbol{e}_\lambda(\boldsymbol{k}), \lambda = 1, 2$ as a typical rotation in the plane normal to $\boldsymbol{k}$ (Fig. 1.2). It thus has the form

$$R(\theta) = \begin{pmatrix} \cos\theta & -\sin\theta \\ \sin\theta & \cos\theta \end{pmatrix} \quad , \tag{1.167}$$

when it is applied to a vector $(|\boldsymbol{k}1\rangle, |\boldsymbol{k}2\rangle)$.

To determine the helicity, it is convenient to use (1.166) and (1.167) for an infinitesimal rotation. Considering only the terms which are linear in θ, one obtains

$$I + \mathrm{i}\boldsymbol{J}\cdot\hat{\boldsymbol{k}}\theta = \begin{pmatrix} 1 & 0 \\ 0 & 1 \end{pmatrix} + \theta \begin{pmatrix} 0 & -1 \\ 1 & 0 \end{pmatrix} \quad . \tag{1.168}$$

Immediately evident, the matrix representation of the helicity operator $\boldsymbol{J}\cdot\hat{\boldsymbol{p}}/\hbar$ in the space of the polarization states is given by

$$\frac{1}{\hbar}\boldsymbol{J}\cdot\hat{\boldsymbol{p}} = \begin{pmatrix} 0 & \mathrm{i} \\ -\mathrm{i} & 0 \end{pmatrix} \quad . \tag{1.169}$$

This matrix has two eigenvalues, 1 and -1, and two orthonormal eigenvectors,

$$|\boldsymbol{k}\pm\rangle = \frac{1}{\sqrt{2}}(|\boldsymbol{k}1\rangle \pm \mathrm{i}\,|\boldsymbol{k}2\rangle) \quad . \tag{1.170}$$

These are the two possible helicity states of the photon. They correspond to the left and right circular polarization states of the classical field.

1.4 The Elastic Field and the Phonon

1.4.1 Elastic Waves and Elastic Energy

Elastic waves propagating through a solid medium show a striking analogy with electromagnetic waves. Recognition of the existence of these waves and the associated theory of elasticity are integral parts of classical physics and existed before the atomic hypothesis of matter was definitively imposed. From a formal point of view, these classical elastic waves arise in a manner very similar to electromagnetic waves and it is possible to associate them with an energy quantum, the phonon, which in this context plays an analogous role to that of the photon. Classical elastic energy (in linear elasticity) can then be expressed as the sum of the phonon energies (Sect. 1.4.4)

The analogy between electromagnetic and electric waves breaks down when we take into account the medium in which the waves propagate. While electromagnetic waves propagate in vacuum, elastic waves are themselves the vibrations of a crystal solid. To obtain the correct phonon spectrum, it is necessary to envision a crystal solid as a lattice of atoms or massive ions bound by forces which depend essentially on the electron distribution. As the electron is much lighter than the nucleus, the electronic core follows the latter in its motion such that, to first approximation, the solid may be considered as a lattice of massive particles with each of these particles characterised by its own canonical observables, its position and momentum.

If it is further accepted that the potential force between neighbouring ions is harmonic, the problem of the vibrations of the lattice in which the waves propagate is, at least in principle, completely solvable. In the limit of long wavelengths (longer than the distance between the ions) the solution of this problem becomes the same as that obtained by considering the solid as a continuum and the waves as modes of the continuum. In this sense, the phonon can be thought of as a quasi-particle: it has the attributes of a particle, but may be viewed as an excitation mode of an underlying material whose dynamics are known. The photon, on the other hand, is a particle in the sense that it can propagate in vacuum and cannot be interpreted today as the excitation mode of an ensemble of more elemantary particles. To underline this distinction, we present the treatment of classical and quantum linear harmonic chains in Sects. 1.4.5 and 1.4.6. Here, a comparative study of the photon and the phonon reveals similarities and differences of concepts inherent to both condensed matter physics (many-body problems) and quantum field theory.

From the macroscopic point of view, an elastic solid obeys Hooke's law,

$$\sigma^{\alpha\beta} = \sum_{\gamma,\delta=1}^{3} \lambda^{\alpha\beta\gamma\delta} u^{\gamma\delta} = \sigma^{\beta\alpha} \quad , \quad \alpha, \beta = 1, 2, 3 \quad , \tag{1.171}$$

which relates the *stress tensor* $\sigma^{\alpha\beta}$ to the *strain tensor* $u^{\alpha\beta}$ formed from the *displacement vector* $\boldsymbol{u}(\boldsymbol{x})$ measured relative to the equilibrium position $\boldsymbol{x}$ of an element of the material. The tensor $u^{\alpha\beta}$ is defined by

$$u^{\alpha\beta} = \frac{1}{2}\left(\frac{\partial u^{\beta}}{\partial x^{\alpha}} + \frac{\partial u^{\alpha}}{\partial x^{\beta}}\right) \quad . \tag{1.172}$$

The two tensors $\sigma^{\alpha\beta}$ and $u^{\alpha\beta}$ are symmetric. The Cartesian components are indexed here with Greek letters $\alpha, \beta, \gamma, \delta, \ldots$. In the following, the common convention of summation over repeated indices will be implied. In this manner, (1.171) can be rewritten more simply as

$$\sigma^{\alpha\beta} = \lambda^{\alpha\beta\gamma\delta} u^{\gamma\delta} \quad . \tag{1.173}$$

In the elastic limit considered here, $|u^{\alpha\beta}| \ll 1$ for all pairs α, β and the coefficients $\lambda^{\alpha\beta\gamma\delta}$ are constants. The *elastic modulus tensor* $\lambda^{\alpha\beta\gamma\delta}$ coincides with the collection of elastic moduli.

The equation of motion of a volume element of specific mass ρ_{M} is

$$\rho_{\mathrm{M}} \frac{\partial^2 u^{\alpha}}{\partial t^2} = \frac{\partial \sigma^{\alpha\beta}}{\partial x^{\beta}} \tag{1.174}$$

and implies

$$\rho_{\mathrm{M}}\frac{\partial^2 u^\alpha}{\partial t^2} = \lambda^{\alpha\beta\gamma\delta}\frac{\partial u^{\gamma\delta}}{\partial x^\beta} = \lambda^{\alpha\beta\gamma\delta}\frac{\partial^2 u^\delta}{\partial x^\beta \partial x^\gamma} \quad , \tag{1.175}$$

which visibly has the structure of a wave equation for the displacement vector. That is, there are vibrations and waves in the solid. In the elastic approximation, ρ_{M} is constant. The form of (1.175) is a consequence of the symmetries of $\lambda^{\alpha\beta\gamma\delta}$ (1.181) which result from the form of the elastic energy.

Elastic Energy Density

It is easy to determine the form of the elastic energy of a solid in the presence of vibrations or of waves satisfying (1.175). To do this, one takes inspiration from the form of (1.141) of the electromagnetic energy. The total elastic energy is then expressed as the integral of an elastic-energy density, such as

$$E^{\mathrm{el}} = \int \mathrm{d}\boldsymbol{x}\mathrm{u}^{\mathrm{el}}(\boldsymbol{x}) \quad . \tag{1.176}$$

One assumes that u^{el} depends on the first derivatives $\partial u^\alpha/\partial t$ and $u^{\alpha\beta}$ of the displacement field and then determines u^{el} such that E^{el} is conserved in time. This must be the case if the solid is unlimited and not deformed at infinity, so that

$$\frac{\mathrm{d}E^{\mathrm{el}}}{\mathrm{d}t} = \int \mathrm{d}\boldsymbol{x}\left[\frac{\partial \mathrm{u}^{\mathrm{el}}}{\partial\left(\frac{\partial u^\alpha}{\partial t}\right)}\frac{\partial^2 u^\alpha}{\partial t^2} + \frac{\partial \mathrm{u}^{\mathrm{el}}}{\partial u^{\alpha\beta}}\frac{\partial u^{\alpha\beta}}{\partial t}\right] = 0 \quad . \tag{1.177}$$

The symmetry of $u^{\alpha\beta}$ gives

$$\frac{\partial \mathrm{u}^{\mathrm{el}}}{\partial u^{\alpha\beta}}\frac{\partial u^{\alpha\beta}}{\partial t} = \frac{\partial \mathrm{u}^{\mathrm{el}}}{\partial u^{\alpha\beta}}\frac{\partial}{\partial x^\beta}\left(\frac{\partial u^\alpha}{\partial t}\right) \quad . \tag{1.178}$$

When the deformation disappears at infinity, it is possible to integrate (1.177) by parts, neglecting the contribution of the surface at infinity. From this, one gets

$$\int \mathrm{d}\boldsymbol{x}\left\{\frac{\partial \mathrm{u}^{\mathrm{el}}}{\partial\left(\frac{\partial u^\alpha}{\partial t}\right)}\frac{\partial^2 u^\alpha}{\partial t^2} - \left[\frac{\partial}{\partial x^\beta}\left(\frac{\partial \mathrm{u}^{\mathrm{el}}}{\partial u^{\alpha\beta}}\right)\right]\frac{\partial u^\alpha}{\partial t}\right\} = 0 \quad . \tag{1.179}$$

One should note that the periodic conditions at the boundary of a finite region ought to have the same effect on the integration by parts; such conditions, nevertheless, are not generally employed by classical elasticity.

One immediately sees (1.175) and (1.179) are equivalent provided

$$\mathrm{u}^{\mathrm{el}} = \mathrm{u}^{\mathrm{kin}} + \mathrm{u}^{\mathrm{pot}} = \frac{1}{2}\rho_{\mathrm{M}}\left|\frac{\partial \boldsymbol{u}}{\partial t}\right|^2 + \frac{1}{2}\lambda^{\alpha\beta\gamma\delta}u^{\alpha\beta}u^{\gamma\delta} \quad . \tag{1.180}$$

This form of the potential elastic energy density imposes the following minimal symmetries on the elastic tensor

$$\lambda^{\alpha\beta\gamma\delta} = \lambda^{\beta\alpha\gamma\delta} = \lambda^{\alpha\beta\delta\gamma} = \lambda^{\gamma\delta\alpha\beta} \quad . \tag{1.181}$$

In general, crystalline symmetries of the solid impose supplementary symmetries in $\lambda^{\alpha\beta\gamma\delta}$. The studies that follow are limited to an isotropic solid.

1.4.2 Elastic Waves and Energy of an Isotropic Solid

Isotropy strongly limits the form of $\mathrm{u}^{\mathrm{pot}}$ which can only depend on the two quadratic scalars $(u^{\alpha\alpha})^2$ and $u^{\alpha\beta}u^{\alpha\beta}$. Note that $u^{\alpha\alpha} = u^{11} + u^{22} + u^{33}$ is the same as the divergence $\nabla \cdot \boldsymbol{u}$ of the displacement. The elastic tensor of an isotropic solid can only be comprised of two independent components,

$$\mathrm{u}^{\mathrm{pot}} = \frac{\lambda}{2}(u^{\alpha\alpha})^2 + \mu u^{\alpha\beta}u^{\alpha\beta} \quad . \tag{1.182}$$

The *Lamé coefficients* λ and μ satisfy the inequalities

$$\lambda + \frac{2}{3}\mu > 0 \qquad \text{and} \qquad \mu > 0 \quad , \tag{1.183}$$

which assure stability in the event of an infinitesimal deformation. To verify this, one makes the decomposition

$$u^{\alpha\beta} = \tilde{u}^{\alpha\beta} + \frac{1}{3}\delta^{\alpha\beta}u^{\gamma\gamma} \quad , \tag{1.184}$$

where, by construction, the trace of the tensor $\tilde{u}^{\alpha\beta}$ vanishes. From this,

$$u^{\alpha\beta}u^{\alpha\beta} = \tilde{u}^{\alpha\beta}\tilde{u}^{\alpha\beta} + \frac{1}{3}(u^{\gamma\gamma})^2 \tag{1.185}$$

and (1.182) can be re-written as

$$\mathrm{u}^{\mathrm{pot}} = \frac{1}{2}\left(\lambda + \frac{2}{3}\mu\right)(u^{\alpha\alpha})^2 + \mu\tilde{u}^{\alpha\beta}\tilde{u}^{\alpha\beta} \quad . \tag{1.186}$$

Stability demands that $\mathrm{u}^{\mathrm{pot}}$ be a positive quadratic form of the the components of the tensor $u^{\alpha\beta}$, hence the inequalities of (1.183).

A derivation of the elastic moduli of an isotropic solid can be obtained by comparing (1.180) and (1.182). Taking into account (1.181), the following relations (without summation over repeated indices) hold:

$$\lambda^{\alpha\alpha\alpha\alpha} = \lambda + 2\mu \quad , \quad \alpha = 1,2,3 \quad ; \tag{1.187}$$

$$\lambda^{\alpha\alpha\beta\beta} = \lambda \quad , \quad \alpha \neq \beta \quad , \quad \alpha, \beta = 1,2,3 \quad ; \tag{1.188}$$

$$\lambda^{\alpha\beta\alpha\beta} = \mu \quad , \quad \alpha \neq \beta \quad , \quad \alpha, \beta = 1,2,3 \quad . \tag{1.189}$$

In vectorial notation, the elastic wave equation (1.175) takes the form

$$\rho_{\mathrm{M}}\frac{\partial^2 \boldsymbol{u}}{\partial t^2} = (\lambda + \mu)\nabla(\nabla \cdot \boldsymbol{u}) + \mu\nabla^2\boldsymbol{u} \quad . \tag{1.190}$$

Transverse and Longitudinal Waves

If one breaks down the displacement $\boldsymbol{u}$ into longitudinal $\boldsymbol{u}_\mathrm{L}$ and transverse $\boldsymbol{u}_\mathrm{T}$ components, the vectorial identity

$$\nabla^2 \boldsymbol{u} = \nabla(\nabla \cdot \boldsymbol{u}) - \nabla \times (\nabla \times \boldsymbol{u}) \tag{1.191}$$

for $\boldsymbol{u}_\mathrm{L}$ becomes

$$\nabla^2 \boldsymbol{u}_\mathrm{L} = \nabla(\nabla \cdot \boldsymbol{u}_\mathrm{L}) \quad . \tag{1.192}$$

Thus, (1.190) is equivalent to the longitudinal and transverse equations of propagation given by

$$\frac{\partial^2 \boldsymbol{u}_\mathrm{L}}{\partial t^2} = \frac{\lambda + 2\mu}{\rho_\mathrm{M}} \nabla^2 \boldsymbol{u}_\mathrm{L} = c_\mathrm{L}^2 \nabla^2 \boldsymbol{u}_\mathrm{L} \tag{1.193}$$

and

$$\frac{\partial^2 \boldsymbol{u}_\mathrm{T}}{\partial t^2} = \frac{\mu}{\rho_\mathrm{M}} \nabla^2 \boldsymbol{u}_\mathrm{T} = c_\mathrm{T}^2 \nabla^2 \boldsymbol{u}_\mathrm{T} \quad . \tag{1.194}$$

The longitudinal and transverse velocities are defined by

$$c_\mathrm{L} = \left(\frac{\lambda + 2\mu}{\rho_\mathrm{M}} \right)^{1/2} \tag{1.195}$$

and

$$c_\mathrm{T} = \left(\frac{\mu}{\rho_\mathrm{M}} \right)^{1/2} \quad . \tag{1.196}$$

The speed of propagation of the longitudinal waves is larger than that of the transverse waves as a result of (1.183). These inequalities, which assure that both c_L and c_T are real, give

$$c_\mathrm{L}^2 - c_\mathrm{T}^2 = \frac{1}{\rho_\mathrm{M}}(\lambda + \mu) > \frac{1}{\rho_\mathrm{M}} \left(\lambda + \frac{2}{3}\mu \right) > 0 \quad . \tag{1.197}$$

The longitudinal deformations have an important characteristic: They *are generally associated with a compression and an expansion, which is not the case for the transverse deformations.* One might consider, for example, an element of volume $\mathrm{d}V(\boldsymbol{x})$ and then calculate its extension $\mathrm{d}V'(\boldsymbol{x}') = \mathrm{d}V'(\boldsymbol{x} + \boldsymbol{u}(\boldsymbol{x}))$ after a deformation. If one supposes $\mathrm{d}V(\boldsymbol{x})$ to be a cube defined by the orthogonal vectors $\mathrm{d}\boldsymbol{x}_\lambda, \lambda = 1, 2, 3$, parallel to the three Cartesian axes, then

$$\mathrm{d}V(\boldsymbol{x}) = \mathrm{d}\boldsymbol{x}_1 \cdot (\mathrm{d}\boldsymbol{x}_2 \times \mathrm{d}\boldsymbol{x}_3) = \prod_{\alpha=1}^{3} \mathrm{d}x^\alpha \quad . \tag{1.198}$$

After the deformation, each of the vectors $\mathrm{d}\boldsymbol{x}_\lambda$ is transformed into a vector $\mathrm{d}\boldsymbol{x}'_\lambda$ given by

$$\begin{aligned}\mathrm{d}\boldsymbol{x}'_\lambda &= \boldsymbol{x} + \mathrm{d}\boldsymbol{x}_\lambda + \boldsymbol{u}\left(\boldsymbol{x} + \mathrm{d}\boldsymbol{x}_\lambda\right) - \left[\boldsymbol{x} + \boldsymbol{u}(\boldsymbol{x})\right] \\ &= \mathrm{d}\boldsymbol{x}_\lambda + \left(\mathrm{d}\boldsymbol{x}_\lambda \cdot \nabla\right)\boldsymbol{u}(\boldsymbol{x}) = \mathrm{d}\boldsymbol{x}_\lambda + \mathrm{d}x^\lambda \frac{\partial \boldsymbol{u}}{\partial x^\lambda}\end{aligned} \tag{1.199}$$

to first order in $\mathrm{d}\boldsymbol{x}_\lambda$. Note that there is no summation over λ in the last term. So, to first order in $\partial \boldsymbol{u}/\partial x^\lambda$, one obtains

$$\begin{aligned}\mathrm{d}V'(\boldsymbol{x}') &= \left(\mathrm{d}\boldsymbol{x}_1 + \mathrm{d}x^1 \frac{\partial \boldsymbol{u}}{\partial x^1}\right) \cdot \left[\left(\mathrm{d}\boldsymbol{x}_2 + \mathrm{d}x^2 \frac{\partial \boldsymbol{u}}{\partial x^2}\right) \times \left(\mathrm{d}\boldsymbol{x}_3 + \mathrm{d}x^3 \frac{\partial \boldsymbol{u}}{\partial x^3}\right)\right] \\ &\simeq \mathrm{d}V(\boldsymbol{x})\left(1 + \frac{\partial u^1}{\partial x^1} + \frac{\partial u^2}{\partial x^2} + \frac{\partial u^3}{\partial x^3}\right) = \mathrm{d}V(\boldsymbol{x})(1 + \nabla \cdot \boldsymbol{u}) \quad .\end{aligned} \tag{1.200}$$

This result is independent of the choice of the coordinate axes. Due to (1.112), only the longitudinal deformations are involved.

Elastic Potential Energy

The elastic potential energy is completely specified by (1.182). It is useful nevertheless to perform a transformation on the second term of the right-hand side of the equation

$$\mu u^{\alpha\beta} u^{\alpha\beta} = \frac{\mu}{2}\left(\frac{\partial u^\beta}{\partial x^\alpha} + \frac{\partial u^\alpha}{\partial x^\beta}\right)\left(\frac{\partial u^\beta}{\partial x^\alpha}\right) \tag{1.201}$$

by exploiting the symmetry of $u^{\alpha\beta}$ and then to integrate (1.201) over the entire volume. This is performed using integration by parts. As above, contributions to the integrand of the form $\partial f/\partial x^\alpha$, where f is any given tensor expression, do not contribute to the energy when transformed to integrals over the surface at infinity. They are thus omitted from the calculations that follow. To signify this, the $=$ sign is replaced by $\sim$. In this notation, a relation specified by $f \sim g$ signifies that f and g only differ by gradients and give equal contributions when integrated over all space. Hence

$$\begin{aligned}&\frac{1}{2}\left(\frac{\partial u^\beta}{\partial x^\alpha} + \frac{\partial u^\alpha}{\partial x^\beta}\right)\left(\frac{\partial u^\beta}{\partial x^\alpha}\right) \\ &\quad \sim -\frac{1}{2}(\boldsymbol{u} \cdot \nabla^2 \boldsymbol{u}) - \frac{1}{2}(\boldsymbol{u} \cdot \nabla)(\nabla \cdot \boldsymbol{u}) \\ &\quad = -(\boldsymbol{u} \cdot \nabla)(\nabla \cdot \boldsymbol{u}) + \frac{1}{2}\boldsymbol{u} \cdot \left[\nabla \times (\nabla \times \boldsymbol{u})\right] \\ &\quad \sim (\nabla \cdot \boldsymbol{u})^2 + \frac{1}{2}|\nabla \times \boldsymbol{u}|^2 \quad ,\end{aligned} \tag{1.202}$$

where the vectorial identities

$$\boldsymbol{a} \cdot \nabla b + b \nabla \cdot \boldsymbol{a} = \nabla \cdot (b\boldsymbol{a}) \tag{1.203}$$

and

$$\boldsymbol{a} \cdot (\nabla \times \boldsymbol{b}) - \boldsymbol{b} \cdot (\nabla \times \boldsymbol{a}) = \nabla \cdot (\boldsymbol{b} \times \boldsymbol{a}) \tag{1.204}$$

have been employed. Taking into account (1.180), (1.182) and (1.202) the total elastic energy can now be written as

$$\begin{aligned} E^{\mathrm{el}} &= \int \mathrm{d}\boldsymbol{x}\; \mathrm{u}^{\mathrm{el}}(\boldsymbol{x}) \\ &= \frac{1}{2} \int \mathrm{d}\boldsymbol{x} \left[\rho_{\mathrm{M}} \left| \frac{\partial \boldsymbol{u}}{\partial t} \right|^2 + (\lambda + 2\mu)(\nabla \cdot \boldsymbol{u})^2 + \mu |\nabla \times \boldsymbol{u}|^2 \right] \quad . \end{aligned} \tag{1.205}$$

If one breaks down $\boldsymbol{u}$ into longitudinal and transverse components, definitions (1.195) and (1.196) give

$$E^{\mathrm{el}} = \int \mathrm{d}\boldsymbol{x} \left(\mathrm{u}_{\mathrm{L}}^{\mathrm{el}} + \mathrm{u}_{\mathrm{T}}^{\mathrm{el}} \right) = E_{\mathrm{L}}^{\mathrm{el}} + E_{\mathrm{T}}^{\mathrm{el}} \quad , \tag{1.206}$$

where

$$E_{\mathrm{L}}^{\mathrm{el}} = \frac{1}{2} \rho_{\mathrm{M}} \int \mathrm{d}\boldsymbol{x} \left[\left| \frac{\partial \boldsymbol{u}_{\mathrm{L}}}{\partial t} \right|^2 + c_{\mathrm{L}}^2 (\nabla \cdot \boldsymbol{u}_{\mathrm{L}})^2 \right] \tag{1.207}$$

and

$$E_{\mathrm{T}}^{\mathrm{el}} = \frac{1}{2} \rho_{\mathrm{M}} \int \mathrm{d}\boldsymbol{x} \left[\left| \frac{\partial \boldsymbol{u}_{\mathrm{T}}}{\partial t} \right|^2 + c_{\mathrm{T}}^2 |\nabla \times \boldsymbol{u}_{\mathrm{T}}|^2 \right] \quad . \tag{1.208}$$

To derive (1.206), the relation

$$\frac{\partial \boldsymbol{u}_{\mathrm{T}}}{\partial t} \cdot \frac{\partial \boldsymbol{u}_{\mathrm{L}}}{\partial t} \sim 0 \tag{1.209}$$

was used. Note that the transverse field $\boldsymbol{a}_{\mathrm{T}}$ is the curl of a vector field $\boldsymbol{b}$; so, (1.209) is obtained as a consequence of (1.204).

1.4.3 Fourier Analysis of Elastic Waves

Transverse Waves

The energy of transverse elastic waves is formally analogous to that of the free electromagnetic field: the vector potential and the transverse field of the

Table 1.1. Comparison of electromagnetic wave properties with the analogous transverse elastic wave properties

Electromagnetic Waves	Transverse Elastic Waves
$\boldsymbol{A}$	$\boldsymbol{u}_\mathrm{T}$
$\nabla \cdot \boldsymbol{A} = 0$	$\nabla \cdot \boldsymbol{u}_\mathrm{T} = 0$
$\boldsymbol{B} = \nabla \times \boldsymbol{A}$	$\nabla \times \boldsymbol{u}_\mathrm{T}$
ε_0	ρ_M
$c = (\varepsilon_0\mu_0)^{-1/2}$	$c_\mathrm{T} = \sqrt{\mu/\rho_\mathrm{M}}$
$\mathrm{u}^\mathrm{em} = \frac{1}{2}\left(\varepsilon_0\lvert\boldsymbol{E}\rvert^2 + \mu_0^{-1}\lvert\boldsymbol{B}\rvert^2\right)$	$\mathrm{u}^\mathrm{el} = \frac{1}{2}\left(\rho_\mathrm{M}\lvert\partial\boldsymbol{u}_\mathrm{T}/\partial t\rvert^2 + \mu\lvert\nabla \times \boldsymbol{u}_\mathrm{T}\rvert^2\right)$

displacement $\boldsymbol{u}_\mathrm{T}$ play the same role. This is illustrated by comparing (1.128) to (1.194) and (1.141) to (1.208). See Table 1.1.

As in the case of free electromagnetic waves, it is useful to decompose the transverse displacement in a Fourier series. The displacement is defined in a cube of side L and using periodic boundary conditions. A similar derivation to Sect. 1.3.5 can be done with transverse plane waves (1.130) resulting

$$\boldsymbol{u}_\mathrm{T}(\boldsymbol{x},t) = \sum_{\boldsymbol{k}\lambda}\left[f_{\boldsymbol{k}\lambda}\exp(-\mathrm{i}\omega_{\mathrm{T}\boldsymbol{k}}t)\boldsymbol{\psi}_{\boldsymbol{k}\lambda}(\boldsymbol{x}) + f^*_{\boldsymbol{k}\lambda}\exp(\mathrm{i}\omega_{\mathrm{T}\boldsymbol{k}}t)\boldsymbol{\psi}^*_{\boldsymbol{k}\lambda}(\boldsymbol{x})\right] \quad , \tag{1.210}$$

$$E^\mathrm{el}_\mathrm{T} = \rho_\mathrm{M}\sum_{\boldsymbol{k}\lambda}(\omega_{\mathrm{T}\boldsymbol{k}})^2(f^*_{\boldsymbol{k}\lambda}f_{\boldsymbol{k}\lambda} + f_{\boldsymbol{k}\lambda}f^*_{\boldsymbol{k}\lambda}) \quad , \tag{1.211}$$

$$\omega_{\mathrm{T}\boldsymbol{k}} = \omega_{\boldsymbol{k}\lambda} = c_\mathrm{T}\lvert\boldsymbol{k}\rvert \quad , \quad \lambda = 1,2 \quad . \tag{1.212}$$

The order of the coefficients of $f_{\boldsymbol{k}\lambda}$ and $f^*_{\boldsymbol{k}\lambda}$ (1.211) is respected as in Sect. 1.3.5.

Longitudinal Waves

The total elastic energy (1.206) contains a longitudinal contribution E^el_L. Contrary to the case of the longitudinal Coulomb energy (1.118), which has no corresponding dynamical degree of freedom, the energy E^el_L is due to the longitudinal waves (1.193).

One can expand $\boldsymbol{u}_\mathrm{L}$ with the aid of the basis functions

$$\boldsymbol{\psi}_{\boldsymbol{k}3}(\boldsymbol{x}) = \frac{\boldsymbol{k}}{\lvert\boldsymbol{k}\rvert}\frac{1}{L^{3/2}}\mathrm{e}^{\mathrm{i}\boldsymbol{k}\cdot\boldsymbol{x}} = \boldsymbol{e}_3(\boldsymbol{k})\frac{1}{L^{3/2}}\mathrm{e}^{\mathrm{i}\boldsymbol{k}\cdot\boldsymbol{x}} \quad , \tag{1.213}$$

where the rotational component is zero. In a perfectly analogous manner to Sect. 1.3.5, one can write

$$u_{\mathrm{L}}(\boldsymbol{x},t) = \sum_{\boldsymbol{k}} [f_{\boldsymbol{k}3} \exp(-\mathrm{i}\omega_{\mathrm{L}\boldsymbol{k}}t)\boldsymbol{\psi}_{\boldsymbol{k}3}(\boldsymbol{x}) + f_{\boldsymbol{k}3}^* \exp(\mathrm{i}\omega_{\mathrm{L}\boldsymbol{k}}t)\boldsymbol{\psi}_{\boldsymbol{k}3}^*(\boldsymbol{x})] \quad . \tag{1.214}$$

It is sufficient to introduce (1.214) to (1.193) to establish that this last relation is satisfied, provided

$$\omega_{\mathrm{L}\boldsymbol{k}} = \omega_{\boldsymbol{k}3} = c_{\mathrm{L}}|\boldsymbol{k}| > \omega_{\mathrm{T}\boldsymbol{k}} \quad . \tag{1.215}$$

The longitudinal elastic energy can thus also take the form

$$E_{\mathrm{L}}^{\mathrm{el}} = \rho_{\mathrm{M}} \sum_{\boldsymbol{k}} (\omega_{\mathrm{L}\boldsymbol{k}})^2 (f_{\boldsymbol{k}3}^* f_{\boldsymbol{k}3} + f_{\boldsymbol{k}3} f_{\boldsymbol{k}3}^*) \quad . \tag{1.216}$$

To obtain (1.216), it is necessary to insert the expansion (1.214) into (1.207) and to make use of the orthogonality of the plane waves (1.213).

1.4.4 An Ensemble of Phonons and Classical Elastic Waves

In analogy to the photon, the *phonon* is defined as a quasi–particle of energy

$$E_{\mathrm{phonon}} = \hbar\omega_{\boldsymbol{k}\lambda} \quad , \quad \lambda = 1,2,3 \tag{1.217}$$

with quantum states (1.130), (1.213) and their linear combinations. The term *quasi-particle* refers to the fact that the phonon, a quantum of elastic energy, possesses certain attributes of a quantum particle, such as wavenumber and polarization states, which are well defined. From an experimental point of view, the inelastic diffusion of X-rays and neutrons in a solid provides evidence to support the particle aspect of the phonon. A neutron can excite or absorb an elastic wave of energy $\hbar\omega_{\boldsymbol{k}\lambda}$ and one may consider this phenomenon to be the creation or absorption of one or more phonons.

To continue the parallel comparison with electromagnetism, a dimensionless amplitude $\beta_{\boldsymbol{k}\lambda}$ can be introduced such that

$$f_{\boldsymbol{k}\lambda} = \left(\frac{\hbar}{2\rho_{\mathrm{M}}\omega_{\boldsymbol{k}\lambda}}\right)^{1/2} \beta_{\boldsymbol{k}\lambda} \quad , \quad \lambda = 1,2,3 \tag{1.218}$$

allowing the elastic energy to be written in the form

$$E^{\mathrm{el}} = \sum_{\boldsymbol{k}\lambda} \hbar\omega_{\boldsymbol{k}\lambda} |\beta_{\boldsymbol{k}\lambda}|^2 \quad . \tag{1.219}$$

It follows that $|\beta_{\boldsymbol{k}\lambda}|^2$ can be interpreted as the mean number of phonons of wavenumber $\boldsymbol{k}$ and polarization λ. An analysis similar to that of Sect. 1.3.6

can thus be followed to arrive at a result that is completely analogous: *the energy of the elastic field is equal to the product of the energy of one phonon and the mean number of phonons.*

While the phonon may play the same role for elastic waves as the photon does for electromagnetic waves, it is necessary to point out the fundamental difference between the two particles, their corresponding media of propagation. As electromagnetic waves propagate in the vacuum, the properties of a photon do not result from the dynamics of a material medium. That is, the photon is an elementary particle, with a well-determined momentum and helicity. On the other hand, elastic waves must propagate in a material which, in principle, has known dynamical properties. In a crystal, the behaviour of the waves and of the phonon is a measure of the interactions of the atoms or ions located in equilibrium at their lattice sites. In particular, *the quantum aspect of the phonon only reflects the quantum dynamics of the atoms or ions.* This discrete nature of the dynamics involves several differences with the continuum field theory description and should not be overlooked.

There is no physical motivation to introduce phonons of a wavelength shorter than the lattice constant a. The maximum wavenumber allowed k_{max}, in this case, is of the order $1/a$. Hence, contrary to the case of the photon, one cannot produce phonons with arbitrarily high energies. The linear dispersion relations (1.212) and (1.215) are correct only in the limit where the wavelength $2\pi/|\boldsymbol{k}|$ is much larger than a.

During the electron–phonon scattering process in a solid: $e^- +$phonon $\to$ $e^- +$ phonon (in analogy to Compton scattering), the total wavenumber is conserved up to a vector belonging to the reciprocal lattice. This is a consequence of the invariance of the crystal with respect to discrete translations of the lattice. In addition, the possible polarization states of the phonon are determined entirely by the crystallographic structure. There are thus no natural reasons to speak of an intrinsic momentum or spin of the phonon.

In terms of order-of-magnitude, one sees that $k_{\text{max}} \simeq 1/a \simeq 10^{10}\,\text{m}^{-1}$, so that the maximum energy of a phonon must be $\hbar\omega_{\text{max}} \simeq \hbar c_{\text{s}} k_{\text{max}} \simeq 10^{-3}\,\text{eV}$, where c_{s} is the average speed of sound. In comparison, the energy of an electron at the Fermi surface or the energy of a photon in the visible spectrum is on the order of $1\,\text{eV}$. Thus, when an electron or a photon absorbs or excites a phonon, its energy is only very slightly modified. The temperature necessary for a maximum-energy phonon to be thermally excited is $T = \hbar\omega_{\text{max}}/k_{\text{B}} \simeq 100\,\text{K}$. In this case, thermal neutrons are the best tool for studying phonons, as both objects have comparable energies and wavenumbers.

1.4.5 The Classical Linear Chain

The similarities and differences between the photon and the phonon expressed above become evident when one analyzes the simplest model of crystalline dynamics: the *linear harmonic chain.* This example allows one to see precisely how the phonon, introduced in Sect. 1.4.4, appears as a consequence of

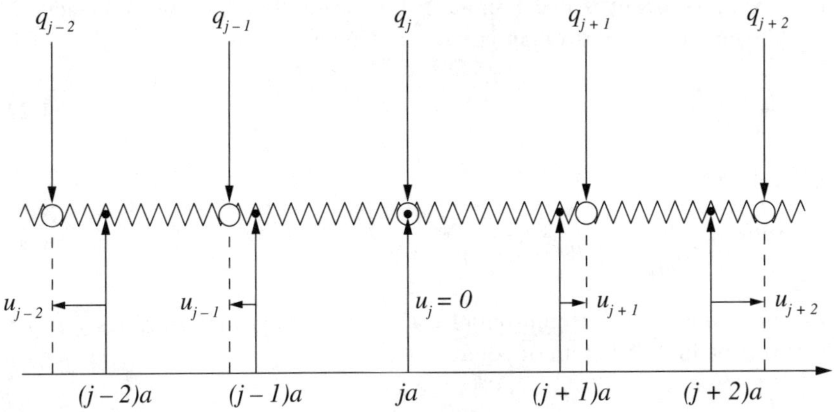

Fig. 1.3. Example of a chain of n particles coupled by a linear harmonic force

the quantum dynamics of the particles. In addition, this model provides an elementary prototype of a quantum field. This is exploited in Chap. 8, where the concept is further elaborated.

Consider n identical point particles situated in equilibrium on a linear periodic lattice of constant a. In this model, particle j occupies position $ja, j = 0, 1, \dots, n-1$ and, for each j, site $j+n$ is identical to site j (Fig. 1.3).

Position q_j and momentum p_j of particle j are classically written as

$$q_j(t) = ja + u_j(t) \tag{1.220}$$

and

$$p_j(t) = m\frac{\mathrm{d}u_j(t)}{\mathrm{d}t} \quad . \tag{1.221}$$

The quantity $u_j(t)$ is the displacement of the particle occupying site j at equilibrium. By hypothesis, $u_j(t)$ is purely longitudinal.

Each particle is allowed to harmonically interact with its closest neighbours; these interactions are represented by springs in Fig. 1.3. The Hamiltonian is written as

$$H = \sum_{j=0}^{n-1} \left(\frac{p_j^2}{2m} + \frac{m\omega^2}{2}(u_{j+1} - u_j)^2 \right) \tag{1.222}$$

with the periodicity condition for each j being

$$u_{j+n}(t) = u_j(t) \quad . \tag{1.223}$$

The spring constant is $m\omega^2$, and $\omega > 0$ has the dimension of frequency.

The classical Hamiltonian equations are written as

$$\frac{\mathrm{d}u_j}{\mathrm{d}t} = \frac{\partial H}{\partial p_j} = \frac{p_j}{m} \tag{1.224}$$

and

$$\frac{\mathrm{d}p_j}{\mathrm{d}t} = -\frac{\partial H}{\partial q_j} = -m\omega^2(2u_j - u_{j-1} - u_{j+1}) \quad . \tag{1.225}$$

The invariance of the Hamiltonian with respect to the translations $j \to j+1$ is apparent in the system of equations

$$\frac{\mathrm{d}^2 u_j}{\mathrm{d}t^2} = -\omega^2(2u_j - u_{j-1} - u_{j+1}) \quad , \quad j = 0, 1, \ldots, n-1 \tag{1.226}$$

obtained by taking the time derivative of (1.224) and substituting (1.225) on the right-hand side.

Normal Coordinates

Taking advantage of this symmetry, the *normal coordinates* $\{Q_k, P_k\}$ can be introduced as

$$u_j(t) = \sum_{k_r} \exp(\mathrm{i}k_r a j) Q_{k_r}(t) \tag{1.227}$$

and

$$p_j(t) = \sum_{k_r} \exp(\mathrm{i}k_r a j) P_{k_r}(t) \quad . \tag{1.228}$$

The periodic condition (1.223) is automatically satisfied if k_r is of the form

$$k_r = \frac{2\pi r}{an} \quad , \tag{1.229}$$

where r is an integer. If n is taken to be odd, sums (1.227) and (1.228) for the n values of k_r correspond to

$$r = 0, \pm 1, \pm 2, \ldots, \pm\frac{n-1}{2} \quad , \quad -\frac{\pi}{a} < k_r < \frac{\pi}{a} \quad . \tag{1.230}$$

The region defined by (1.230) is called the first Brillouin zone in solid state physics.

New variables Q_{k_r} and P_{k_r} are given by the inversion of (1.227) and (1.228). That is,

$$\frac{1}{n}\sum_{j=0}^{n-1}\exp(-\mathrm{i}k_s aj)u_j = \sum_{k_r}\frac{1}{n}\sum_{j=0}^{n-1}\exp\left[\mathrm{i}(k_r - k_s)aj\right]Q_{k_r} = Q_{k_s} \tag{1.231}$$

and similarly

$$\frac{1}{n}\sum_{j=0}^{n-1}\exp(-\mathrm{i}k_s aj)p_j = P_{k_s} \quad . \tag{1.232}$$

These are derived by noting that, when k_r and k_s are in the Brillouin zone,

$$\begin{aligned}\frac{1}{n}\sum_{j=0}^{n-1}\exp\left[\mathrm{i}(k_r - k_s)aj\right] &= \frac{1-\mathrm{e}^{2\mathrm{i}\pi(r-s)}}{1-\mathrm{e}^{2\mathrm{i}\pi(r-s)/n}} = \begin{cases}1 & r = s\\ 0 & r \neq s\end{cases}\\ &= \delta_{k_r,k_s} \quad , \quad -\frac{n-1}{2} < r \quad , \quad s < \frac{n-1}{2} \quad .\end{aligned} \tag{1.233}$$

One should note that Q_{k_r} and P_{k_r} are in general complex but satisfy

$$Q^*_{k_r} = Q_{-k_r} \quad \text{and} \quad P^*_{k_r} = P_{-k_r} \quad . \tag{1.234}$$

Taking into account these relations, the normal coordinates Q_{k_r} and P_{k_r} correspond to $2n$ new independent variables. The index r in k_r will be omitted from now on, as k will always be of the form (1.229) and kept in the Brillouin zone (1.230).

Solution of the Equations of Motion

The benefit of introducing the normal coordinates is that the equations of motion are decoupled. Multiplying (1.226) by $(1/n)\mathrm{e}^{-\mathrm{i}kaj}$ and then summing over j gives

$$\begin{aligned}\frac{\mathrm{d}^2 Q_k}{\mathrm{d}t^2} &= -\omega^2(2-\mathrm{e}^{\mathrm{i}ak}-\mathrm{e}^{-\mathrm{i}ak})Q_k\\ &= -4\omega^2\sin^2\left(\frac{ak}{2}\right)Q_k = -\omega_k^2 Q_k \quad ,\end{aligned} \tag{1.235}$$

where

$$\omega_k = 2\omega\left|\sin\left(\frac{ak}{2}\right)\right| \quad . \tag{1.236}$$

For $k \neq 0$, *the equation of motion (1.235) is that of an harmonic oscillator of frequency* ω_k. For $k = 0$, (1.235) is an equation of uniform motion. This

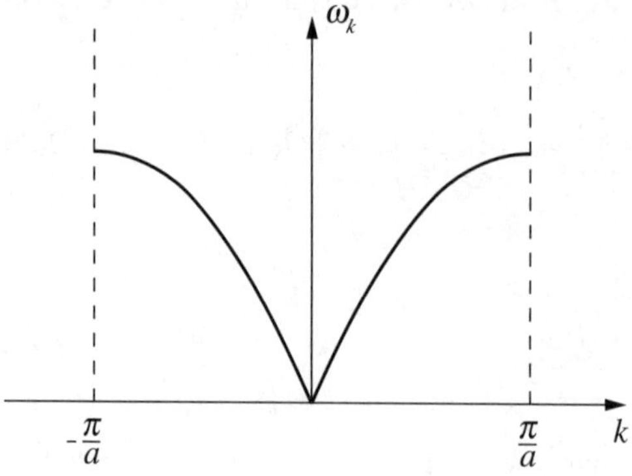

Fig. 1.4. Law of dispersion in the first Brillouin zone $|k| \le \pi/a$

is not surprising if one notes that in (1.231) Q_0 is just the position of the center of mass of the system.

The solution of (1.235) can be expressed with the help of (1.234) as

$$Q_k = f_k \exp(-\mathrm{i}\omega_{\boldsymbol{k}} t) + f^*_{-k} \exp(\mathrm{i}\omega_{\boldsymbol{k}} t) \quad , \tag{1.237}$$

where f_k is a complex constant. As a result, the general solution of the equations of motion (1.226) is a linear superposition of elementary solutions:

$$u_j(t) = \sum_{\boldsymbol{k}} \{f_k \exp[\mathrm{i}(kaj - \omega_{\boldsymbol{k}} t)] + f^*_k \exp[-\mathrm{i}(kaj - \omega_{\boldsymbol{k}} t)]\} \quad . \tag{1.238}$$

The displacement $u_j(t)$ is thus formed from a superposition of traveling waves of phase speed $c_k = \omega_{\boldsymbol{k}}/|k|$, where k plays the role of the wavenumber. This is the discrete analogy to the longitudinal elastic waves (1.214) introduced in Sect. 1.4.3. The dispersion law $\omega_{\boldsymbol{k}}$ is plotted in Fig. 1.4, where k is treated as a continuous variable in the first Brillouin zone $|k| \le \pi/a$.

For $|k| \to 0$, $\omega_{\boldsymbol{k}} \sim \omega a|k|$, corresponding to (1.215), the dispersion law of an elastic material with longitudinal velocity $c_{\mathrm{L}} = \omega a$. It is perfectly natural, in the limit of small $|k|$, to arrive at the law of elastic dispersion (1.215). Here, the corresponding waves have a wavelength $\lambda = 2\pi/|k| \gg a$ much greater than the lattice constant. These waves are thus insensitive to the discrete structure and the elastic theory for a continuous medium is valid.

The Form of the Hamiltonian

To express the Hamiltonian in terms of the normal coordinates, it is sufficient to substitute expansions (1.227) and (1.228) of u_j and p_j into (1.222). This gives

$$H = \frac{1}{2}\sum_{j=0}^{n-1}\sum_{kk'} \mathrm{e}^{\mathrm{i}(k-k')aj}\left[\frac{P_k P_{k'}^*}{m} + m\omega^2\left(\mathrm{e}^{\mathrm{i}ka}-1\right)\left(\mathrm{e}^{-\mathrm{i}k'a}-1\right)Q_k Q_{k'}^*\right]$$

$$= \sum_{\boldsymbol{k}} \frac{n}{2}\left(\frac{P_k P_k^*}{m} + m\omega_k^2 Q_k Q_k^*\right) = \frac{nP_0^2}{2m} + \sum_{k\neq 0} H_k \quad . \tag{1.239}$$

In the last calculation, the fact that u_j and p_j are real was used to expand alternatively in terms of Q_k and P_k or Q_k^* and P_k^*. Use was also made of (1.233). What is important here is that H is decomposed into a sum of terms containing of pairs P_k, P_k^* and Q_k, Q_k^* which are decoupled from each other. The first term of (1.239) is the kinetic energy of the center of mass.

1.4.6 The Quantum Linear Chain

The analysis of the preceding paragraph can be rewritten word for word for the quantum case, the only difference being that the atomic variables u_j and p_j become operators satisfying the following commutation rules:

$$[u_j, p_l] = \mathrm{i}\hbar\delta_{j,l} \tag{1.240}$$

and

$$[u_j, u_l] = [p_j, p_l] = 0 \quad . \tag{1.241}$$

The normal coordinates also become operators and the quantum Hamiltonian has the same form (1.239).

The contribution H_k is not exactly equivalent to that of the harmonic oscillator in that the operators Q_k and P_k do not refer to a single particle and are not Hermitian. It is nevertheless possible to define operators a_k and a_k^* with the same properties as those introduced in Sect. 1.2.1.

To begin with, it is useful to establish commutation relations satisfied by the normal coordinates. Starting with (1.231) and (1.232), then applying (1.240), (1.241) and (1.233), one finds

$$[Q_k, P_{k'}] = \frac{1}{n^2}\sum_{j,l=0}^{n-1} \mathrm{e}^{-\mathrm{i}kaj}\mathrm{e}^{-\mathrm{i}k'al}[u_j, p_l]$$

$$= \frac{\mathrm{i}\hbar}{n^2}\sum_{j=0}^{n-1}\mathrm{e}^{-\mathrm{i}(k+k')aj} = \frac{\mathrm{i}\hbar}{n}\delta_{k,-k'} \tag{1.242}$$

and

$$[Q_k, Q_{k'}] = [P_k, P_{k'}] = 0 \quad . \tag{1.243}$$

For all $k \neq 0$ the following operators are defined

$$a_k = \left(\frac{n}{2}\right)^{1/2} \left(\gamma_k Q_k + \frac{\mathrm{i}}{\hbar \gamma_k} P_k\right) \tag{1.244}$$

and

$$a_k^* = \left(\frac{n}{2}\right)^{1/2} \left(\gamma_k Q_{-k} - \frac{\mathrm{i}}{\hbar \gamma_k} P_{-k}\right) \quad , \tag{1.245}$$

where

$$\gamma_k = \left(\frac{m\omega_{\boldsymbol{k}}}{\hbar}\right)^{1/2} = \gamma_{-k} \quad .$$

Their corresponding commutation rules are derived from (1.242) and (1.243) to be

$$[a_k, a_{k'}^*] = \frac{n}{2}\left(-\frac{\mathrm{i}}{\hbar}[Q_k, P_{-k'}] + \frac{\mathrm{i}}{\hbar}[P_k, Q_{-k'}]\right) = \delta_{k,k'} \tag{1.246}$$

and

$$[a_k, a_{k'}] = [a_k^*, a_{k'}^*] = 0 \quad . \tag{1.247}$$

After a simple algebraic calculation, it follows from (1.244) and (1.245) that

$$a_k a_k^* + a_{-k}^* a_{-k} = \frac{n}{\hbar\omega_{\boldsymbol{k}}}\left(\frac{P_k P_k^*}{m} + m\omega_k^2 Q_k Q_k^*\right) \quad . \tag{1.248}$$

Summing (1.248) for all $k \neq 0$ and ignoring the center-of-mass energy, the Hamiltonian takes on the form

$$H = \frac{1}{2}\sum_{k\neq 0} \hbar\omega_{\boldsymbol{k}}\left(a_k^* a_k + a_k a_k^*\right) = \sum_{k\neq 0} \hbar\omega_{\boldsymbol{k}}\left(a_k^* a_k + \frac{1}{2}\right) \quad . \tag{1.249}$$

Note that the displacement (1.227) can be expressed in terms of operators a_k^* and a_k with the help of (1.244) and (1.245):

$$u_j(t) = \left(\frac{\hbar}{nm}\right)^{1/2} \sum_k \frac{1}{\sqrt{2\omega_k}} \{a_k^* \exp\left[-\mathrm{i}(kaj - \omega_k t)\right] + a_k \exp\left[\mathrm{i}(kaj - \omega_k t)\right]\} \quad . \tag{1.250}$$

The Hamiltonian H is the sum of independent H_k terms each of them formally identical to the Hamiltonian of the harmonic oscillator (1.9). According to (1.11) and (1.15), a complete basis of eigenvectors of H is provided by the states

$$|n_{k_1}, n_{k_2}, \ldots\rangle = |n_{k_1}\rangle\, |n_{k_2}\rangle \ldots \quad , \quad |n_k\rangle = \frac{(a_k^*)^{n_k}}{\sqrt{n_k!}}\, |0\rangle \quad , \tag{1.251}$$

where the n_k are non-negative integers. The corresponding eigenvalues of H are of the form

$$E = \sum_{k \neq 0} \hbar\omega_k \left(n_k + \frac{1}{2} \right) = \sum_{k \neq 0} \hbar\omega_k n_k + E_0 \quad , \tag{1.252}$$

where $E_0 = \frac{1}{2} \sum_{k \neq 0} \hbar\omega_k$ is the energy of the ground state.

Disregarding the E_0 term, the energy (1.252) of the quantum chain is similar to the classical energy (1.219). That is, *it may be written as a sum of the vibration energy quanta* $\hbar\omega_{\boldsymbol{k}}$. This interpretation is consistent with that of Sect. 1.4.4 if one calls $\hbar\omega_{\boldsymbol{k}}$ the energy of a phonon and n_k the number of phonons of wavenumber k. Generally, if $|\Phi\rangle$ is some quantum state of the chain, the mean energy is given by

$$\langle \Phi | H | \Phi \rangle = \sum_{k \neq 0} \hbar\omega_{\boldsymbol{k}} \, \langle \Phi | a_k^* a_k | \Phi \rangle + E_0 \tag{1.253}$$

and $\langle \Phi | a_k^* a_k | \Phi \rangle$ is thus the mean number of phonons in the state $|\Phi\rangle$. This formalism and its interpretation will be developed further in Chap. 3.

A similar study of the vibrational modes of a three-dimensional crystal lattice would be identical apart from the fact that each mode is characterized by a wavenumber vector $\boldsymbol{k}$ describing the first Brillouin zone in three-dimensional space and followed by a polarization index which takes on $3p$ values, where p is the number of sites of the primitive lattice cell.

Exercises

1. Normal Order

(i) Let λ be a real number. Show that

$$\mathrm{e}^{-\lambda a^* a} = :\mathrm{e}^{-(1-\mathrm{e}^{-\lambda}) a^* a}: \quad .$$

Hint: Use properties of the coherent states.

(ii) With $Q = \frac{1}{\sqrt{2}}(a + a^*)$, show that

$$:Q^n: = 2^{-n} H_n(Q) \quad ,$$

where H_n is the nth Hermite polynomial.

Hint: Use the generating function (1.43).

(iii) Establish formula (1.57).

Hint: Write differential equations for $\exp[\lambda(X+Y)]\exp(-\lambda Y)\exp(-\lambda X)$ and $\exp(\lambda Y)X\exp(-\lambda Y)$, λ a parameter.

2. Squeezed States

(i) Consider the unitary operator

$$S(\lambda) = \exp\left(\frac{\lambda}{2}(a^*)^2 - \frac{\lambda}{2}a^2\right) \quad ,$$

with λ a real number. Show that the operator a transforms as

$$S^*(\lambda)aS(\lambda) = a\cosh\lambda + a^*\sinh\lambda \quad .$$

Hint: Find a differential equation for $S^*(\lambda)aS(\lambda)$.

(ii) Define

$$|\lambda,\alpha\rangle = S(\lambda)D(\alpha)|0\rangle \quad ,$$

the state obtained by applying $S(\lambda)$ on a coherent state (squeezed states). Compute the mean square deviations of Q and P in this state

$$(\Delta Q)^2_\lambda = \frac{\hbar}{2}\mathrm{e}^{2\lambda} \quad , \quad (\Delta P)^2_\lambda = \frac{\hbar}{2}\mathrm{e}^{-2\lambda} \quad .$$

Comment: Squeezed states are characterized by the fact that the quantum fluctuations of one observable (say $(\Delta P)^2_\lambda$) can be strongly reduced for $\lambda > 0$ (at the expense of a growth of $(\Delta Q)^2_\lambda$ to maintain the validity of the uncertainty relation). Squeezed states of photons can be produced in non-linear quantum optics and used to reduce the noise due to quantum fluctuations in signal transmission.

3. Polarization

(i) Show that the photon state $\alpha_1|\boldsymbol{k},1\rangle + \alpha_2|\boldsymbol{k},2\rangle$ with α_1 and α_2 real corresponds to a linearly polarized plane wave having an angle $\theta = \arctan(\alpha_1/\alpha_2)$ with the polarization vector $\boldsymbol{e}_1$.

(ii) Show that the states $|\boldsymbol{k},\pm\rangle = (1/\sqrt{2})(|\boldsymbol{k},1\rangle \pm \mathrm{i}|\boldsymbol{k},2\rangle)$ correspond to left and right circularly polarized waves.

2. Fermions and Bosons

2.1 The Principle of Symmetrization

2.1.1 Identical Particles

Photons in a black-body cavity, electrons in a solid, nucleons in a nucleus, neutrons in a neutron star, and helium atoms in a superfluid are examples of systems of identical particles for which the methods of many-body problems may be applied. To define the concept of identical particles more precisely, it is useful to remark that a particle is characterized by two distinct types of properties. The first, called intrinsic, are those which do not depend on the dynamical state of the particle, such as the rest mass, the charge or the total spin. The extrinsic properties are those which depend on the state of the particle or on its preparation: the position and velocity, spin orientation and other internal parameters. The following definition will be adopted here: *two particles are considered identical if they possess the same intrinsic properties.*

While the application of this definition does not pose any problem for the elementary particles (those without internal structure), such as the photon or the electron, it is necessary to immediatly underline the *relative* nature of the concept of identity for all other cases. On one hand, observation may lead one to discover new properties, such as a more complex internal structure, that ought to be taken into account. On the other hand, attribution of the intrinsic properties of a particle often depends on the level of approximation chosen for the description.

As the proton and the neutron do not have the same electrical charge or exactly the same mass, these particles are different. However, if one neglects Coulomb interactions with respect to the nuclear forces, and puts aside the mass difference, one can consider the proton and the neutron to be two states of the same particle: the nucleon (isospin formalism). In light of these approximations, the distinction between the proton and the neutron loses its intrinsic nature. Equally, helium atoms can be treated as identical as long as they are all in the ground state, that is, in the temperature region which is low enough that the probability of excitation of the atoms is negligible. If, on the contrary, the thermal energy is raised high enough, some of the atoms may enter distinct excited states and the system can no longer be considered to be made up of identical particles. It is thus important when we speak of identical particles in many-body problems not to lose sight of the physical region of validity of the description.

From the mathematical point of view, to say that n particles are identical is to say that *all of the observables* $\mathcal{O}(\boldsymbol{x}_1, \boldsymbol{p}_1; \boldsymbol{x}_2, \boldsymbol{p}_2; \dots ; \boldsymbol{x}_n, \boldsymbol{p}_n)$ *of the particles are invariant under the exchange of their coordinates* $(\boldsymbol{x}_i, \boldsymbol{p}_i), i = 1, \dots , n$.[1] This is clearly the case for the kinetic energy

$$\mathtt{H}_0(n) = \sum_{i=1}^{n} \frac{|\boldsymbol{p}_i|^2}{2m} \tag{2.1}$$

as all of the masses are equal. The potential energy

$$\mathtt{V}(n) = \mathtt{V}(\boldsymbol{x}_1, \dots , \boldsymbol{x}_n) = \sum_{i<j}^{n} V(|\boldsymbol{x}_i - \boldsymbol{x}_j|) \tag{2.2}$$

(where $\sum_{i<j}$ signifies the sum over $n(n-1)/2$ pairs of particles) is also a symmetric function of its arguments as all of the pairs interact with the same potential $V_{ij} = V(|\boldsymbol{x}_i - \boldsymbol{x}_j|) = V_{ji}$. The same is true for the total momentum and the total angular momentum.

This symmetry plays a much more fundamental role in quantum physics than in classical physics. In classical dynamics, two identical particles can nevertheless be distinguished over the course of time if we follow their trajectories. These trajectories are continuous and uniquely specified by the initial conditions: the particles, although identical, are distinguishable. This is no longer the case in quantum mechanics, where identification of the particles by their initial conditions and trajectories is not possible, even in principle: *the particles become indistinguishable*.

In view of this situation, quantum particles may be divided into two classes, *bosons and fermions*, with the latter obeying the Pauli exclusion principle. Membership of a particle to one or the other of these two classes determines certain predominant aspects of its behaviour. These will be discussed in a qualitative manner in this chapter. On this topic, Ehrenfest said, "Take a piece of metal, or a rock. When we think of it, we are astonished to see that this piece of matter takes up such a large volume. In truth, the molecules are piled up one on top of the other, the same as the atoms in each molecule. But why are the atoms themselves so large? Answer: the Pauli principle, each electron is in a different state! That is why the atoms are so much larger than is necessary, and metal and rocks take up so much space."

2.1.2 One-Particle States

The states of a single particle occupy a very important place in the study of coupled systems of n particles both for physical reasons and for the construction of the formalism. In collision experiments, outgoing particles are detected

[1] For particles of non-zero spin, the spin variables are also included.

in their individual states: all of the information pertaining to the process of the collision is obtained from an analysis of single-particle asymptotic states. In condensed matter physics, it is normal to describe to first approximation the excitations of the system (such as phonons) as a collection of independent quasi-particles. Finally, in the mean-field method, the wavefunction for n particles is approximated by a product of appropriate single-particle states. The one-particle states are the "building blocks" of the construction of the second quantification formalism of Chap. 3.

The collection of states of a quantum particle form a Hilbert space $\mathcal{H}$ called the *single-particle space*. For a non-relativistic particle with no spin in infinite three-dimensional space, denoted by $\mathbb{R}^3$, $\mathcal{H}$ is the set of square integrable functions.

$$\mathcal{H} = \mathcal{L}^2(\mathbb{R}^3) = \{\varphi(\boldsymbol{x}); \int \mathrm{d}\boldsymbol{x} |\varphi(\boldsymbol{x})|^2 < \infty\} \tag{2.3}$$

with the scalar product

$$\langle \varphi | \psi \rangle = \int \mathrm{d}\boldsymbol{x} \varphi^*(\boldsymbol{x}) \psi(\boldsymbol{x}) \quad . \tag{2.4}$$

When the state is normalized to 1,

$$\langle \varphi | \varphi \rangle = \int \mathrm{d}\boldsymbol{x} |\varphi(\boldsymbol{x})|^2 = 1 \quad , \tag{2.5}$$

the expression $|\varphi(\boldsymbol{x})|^2 \mathrm{d}\boldsymbol{x}$ gives the probability to find the particle in an element of volume $\mathrm{d}\boldsymbol{x}$ centered at $\boldsymbol{x}$. If the particle possesses a spin s, its wavefunction will have $2s+1$ components $\varphi(\boldsymbol{x}, \sigma), \sigma = -s, -s+1, \ldots, s-1, s$, describing the possible orientations of the spin along a fixed axis. In the following, spin will only be introduced when necessary to describe physical effects.

Observables of the particle such as, for example, its position $\boldsymbol{q}$ and its momentum $\boldsymbol{p}$ are linear operators acting on $\mathcal{H}$:

$$(\boldsymbol{q}\varphi)(\boldsymbol{x}) = \boldsymbol{x}\varphi(\boldsymbol{x}) \tag{2.6}$$

$$(\boldsymbol{p}\varphi)(\boldsymbol{x}) = -\mathrm{i}\hbar \nabla \varphi(\boldsymbol{x}) \quad , \quad \nabla = \left(\frac{\partial}{\partial x^1}, \frac{\partial}{\partial x^2}, \frac{\partial}{\partial x^3} \right) \quad . \tag{2.7}$$

Operators $\boldsymbol{p}$ and $\boldsymbol{q}$ satisfy the canonical commutation relations

$$[q^\alpha, p^\beta] = \mathrm{i}\hbar \delta_{\alpha,\beta} \quad , \quad \alpha, \beta = 1, 2, 3 \quad . \tag{2.8}$$

In the Fourier (or momentum) representation,

$$\tilde{\varphi}(\boldsymbol{k}) = \frac{1}{(2\pi)^{3/2}} \int \mathrm{d}\boldsymbol{x} \mathrm{e}^{-\mathrm{i}\boldsymbol{k}\cdot\boldsymbol{x}} \varphi(\boldsymbol{x}) \quad , \tag{2.9}$$

the observable $\boldsymbol{p}$ acts as a multiplication operator

$$(\boldsymbol{p}\tilde{\varphi})(\boldsymbol{k}) = \hbar\boldsymbol{k}\tilde{\varphi}(\boldsymbol{k}) \tag{2.10}$$

and $|\tilde{\varphi}(\boldsymbol{k})|^2 \mathrm{d}\boldsymbol{k}$ is the probability to find the momentum of the particle in the volume $\hbar^3 \mathrm{d}\boldsymbol{k}$ centered at $\hbar\boldsymbol{k}$.

The evolution of the particle is governed by the Schrödinger equation

$$\mathrm{i}\hbar\frac{\partial}{\partial t}\varphi(\boldsymbol{x},t) = (H\varphi)(\boldsymbol{x},t) \quad . \tag{2.11}$$

The formal solution to (2.11) is given by the evolution operator $U(t)$:

$$\varphi(\boldsymbol{x},t) = \left(U(t)\varphi\right)(\boldsymbol{x}) \quad , \quad U(0) = I \quad . \tag{2.12}$$

When there are no external fields depending on time, $\{U(t), -\infty < t < +\infty\}$ forms a unitary group

$$U(t) = \exp\left(-\frac{\mathrm{i}tH}{\hbar}\right) \quad , \tag{2.13}$$

$$U^*(t) = U^{-1}(t) \quad , \quad U(t_1)U(t_2) = U(t_1 + t_2) \quad . \tag{2.14}$$

The Hamiltonian is obtained by analogy to the classical theory and the eigenvalue equation

$$(H\varphi)(\boldsymbol{x}) = \varepsilon\varphi(\boldsymbol{x}) \tag{2.15}$$

gives the energy spectrum of the particle. For a free particle, H reduces to the kinetic energy H_0 given by

$$H_0 = \frac{|\boldsymbol{p}|^2}{2m} = -\frac{\hbar^2}{2m}\nabla^2 \tag{2.16}$$

where ∇^2 is the Laplacian: $\sum_{\alpha=1}^{3}\left(\frac{\partial}{\partial x^\alpha}\right)^2$.

2.1.3 Periodic Boundary Conditions and the Thermodynamic Limit

In a scattering process, it is natural to consider the particles to be moving in the infinite space $\mathbb{R}^3$. The situation is different in condensed matter physics where one must deal with systems of non-zero density ρ. In this case, one must constrain the n particles to remain in a finite volume Λ of space, the volume of the sample, with $\rho = n/\Lambda$. The one-particle states in Λ are given by the set of all square integrable functions in Λ (which are zero outside of Λ):

$$\mathcal{H} = \mathcal{L}^2(\Lambda, \mathrm{d}\boldsymbol{x}) = \left\{\varphi(\boldsymbol{x}); \int_\Lambda \mathrm{d}\boldsymbol{x}|\varphi(\boldsymbol{x})|^2 < \infty\right\} \quad . \tag{2.17}$$

In order for the eigenvalue equation

$$(H_0\varphi)(\boldsymbol{x}) = -\frac{\hbar^2}{2m}\nabla^2\varphi(\boldsymbol{x}) = \varepsilon\varphi(\boldsymbol{x}) \tag{2.18}$$

to be well defined and the energy spectrum determined, *it is necessary to fix the conditions at the boundary of Λ for the differential operator ∇^2.*

Thermodynamic Limit

Here, we are concerned with situations where the volume Λ is much larger than that of the particles it contains and where the surface effects can be neglected. For electrons in a metallic sample of 1 cm^3, one has

$$\frac{\text{Bohr radius}}{L = 1\ \text{cm}} = 0.53 \times 10^{-8} \tag{2.19}$$

and for nucleons in a heavy nucleus

$$\begin{aligned}\frac{\text{nucleon radius}}{\text{nucleus radius}} &\simeq 1/3\ (\text{number of nucleons})^{-1/3} \\ &= \begin{cases} 0.09 & \text{Iron} \\ 0.05 & \text{Uranium} \end{cases} \quad .\end{aligned} \tag{2.20}$$

While the number of particles involved is very different for the two cases, the description of the system can be idealized by taking the *thermodynamic limit*, $n \to \infty, \Lambda \to \infty$ with $\rho = n/\Lambda$ fixed. *In this limit, one expects that those quantities which are described per unit volume or per particle, such as the density or the energy per particle, do not depend on the size or shape of Λ, or on the conditions of the wavefunctions at the boundary of Λ.* It is evident that this point necessitates a mathematical proof which can be provided in many cases. Its validity depends on the nature of the interactions and of the thermodynamic phase and will, for now, be accepted as fact for the systems to be studied below.

Periodic Conditions

The simplest and most convenient choice is to consider Λ to be a cube with sides of length L with the following *periodic boundary conditions*:

$$\begin{aligned}\varphi(0, x^2, x^3) &= \varphi(L, x^2, x^3) \\ \varphi(x^1, 0, x^3) &= \varphi(x^1, L, x^3) \\ \varphi(x^1, x^2, 0) &= \varphi(x^1, x^2, L) \quad .\end{aligned} \tag{2.21}$$

Equation (2.18) is with these boundary conditions easily solved. The energy eigenvalues are

$$\varepsilon_{\boldsymbol{k}} = \frac{\hbar^2}{2m} |\boldsymbol{k}|^2 \quad , \tag{2.22}$$

where $\boldsymbol{k}$ are the wavenumbers

$$\boldsymbol{k} = (k^1, k^2, k^3) = \left(\frac{2\pi r^1}{L}, \frac{2\pi r^2}{L}, \frac{2\pi r^3}{L} \right) \quad , \quad r^1, r^2, r^3 \text{ integers} \quad . \tag{2.23}$$

The eigenvectors, normalized to 1, are the plane waves

$$\psi_{\boldsymbol{k}}(\boldsymbol{x}) = \frac{1}{L^{3/2}} \mathrm{e}^{\mathrm{i}\boldsymbol{k}\cdot\boldsymbol{x}} \quad , \quad \boldsymbol{k}\cdot\boldsymbol{x} = \sum_{\alpha=1}^{3} k^\alpha x^\alpha \quad , \tag{2.24}$$

which satisfy orthogonality,

$$\langle \psi_{\boldsymbol{k}} | \psi_{\boldsymbol{k}'} \rangle = \frac{1}{L^3} \int_\Lambda \mathrm{d}\boldsymbol{x} \mathrm{e}^{\mathrm{i}(\boldsymbol{k}'-\boldsymbol{k})\cdot\boldsymbol{x}} = \delta_{\boldsymbol{k},\boldsymbol{k}'} \quad . \tag{2.25}$$

In Dirac's notation, $|\boldsymbol{k}\rangle$ is the state vector given in the position basis by the plane wave (2.24). The non-normalizable vectors $|\boldsymbol{x}\rangle$ can be introduced as in (1.22) and (1.23) so that the components in the plane wave basis are

$$\langle \boldsymbol{x} | \boldsymbol{k} \rangle = \psi_{\boldsymbol{k}}(\boldsymbol{x}) \quad , \tag{2.26}$$

where

$$\langle \boldsymbol{x} | \boldsymbol{x}' \rangle = \frac{1}{L^3} \sum_{\boldsymbol{k}} \mathrm{e}^{\mathrm{i}\boldsymbol{k}\cdot(\boldsymbol{x}-\boldsymbol{x}')} = \delta(\boldsymbol{x}-\boldsymbol{x}') \quad , \quad \boldsymbol{x}, \boldsymbol{x}' \text{ in } \Lambda \quad . \tag{2.27}$$

Each vector $|\varphi\rangle$ can be expanded in either of the two bases

$$|\varphi\rangle = \sum_{\boldsymbol{k}} |\boldsymbol{k}\rangle \langle \boldsymbol{k} | \varphi \rangle \quad ,$$
$$|\varphi\rangle = \int_\Lambda \mathrm{d}\boldsymbol{x} \, |\boldsymbol{x}\rangle \langle \boldsymbol{x} | \varphi \rangle \quad . \tag{2.28}$$

Periodic boundary conditions have the following advantage: the plane wave (2.24) is an eigenvector of the momentum $\boldsymbol{p} = -\mathrm{i}\hbar\nabla$

$$(\boldsymbol{p}\psi_{\boldsymbol{k}})(\boldsymbol{x}) = \hbar\boldsymbol{k}\psi_{\boldsymbol{k}}(\boldsymbol{x}) \quad , \tag{2.29}$$

from which one sees that *the particle has a well-defined momentum even in a finite volume* Λ. One should note that this is only true for periodic boundary conditions: points 0 and L are identified and one can imagine that the particle moves with momentum $\hbar\boldsymbol{k}$ on a one-dimensional line segment or circle (Fig. 2.1). In the large volume limit $L \to \infty$, the momentum recovers its usual definition, being the generator of the group of translations in space.

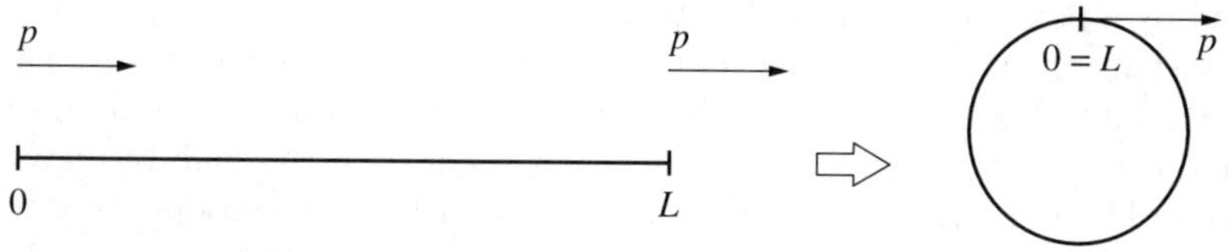

Fig. 2.1. Under periodic boundary conditions, with 0 and L identified, one can imagine that the particle moves with momentum $\hbar\boldsymbol{k}$ on a line segment or on a circle

Spacing of the Energy Levels

The energy levels are very close under normal conditions. The difference between two energy levels of an electron in a metal relative to the thermal energy can be estimated to be

$$\frac{\hbar^2/2m\big(2\pi(r+1)/L\big)^2 - \hbar^2/2m\big(2\pi r/L\big)^2}{k_{\mathrm{B}}T} = \frac{h^2}{2mk_{k_{\mathrm{B}}T}}\frac{2r+1}{L^2} \quad . \tag{2.30}$$

The wavenumber $|\boldsymbol{k}|$ of a conduction electron is of order $1/a$, where $a =$ interatomic distance $\simeq 1$ Å. Thus

$$r = \frac{kL}{2\pi} \simeq \frac{L}{2\pi a} \tag{2.31}$$

and for a typical value of $L = 1$ cm, the quantity becomes

$$\frac{h^2 r}{mk_{\mathrm{B}}TL^2} \simeq \frac{10^{-4}\mathrm{K}}{T} \quad , \tag{2.32}$$

which is small even at low temperature. As far as the ground state is concerned, the energy difference between two electronic levels is small relative to other typical energies of the system (photon energy, Zeeman energy).

For a system of finite volume with periodic boundary conditions, the evaluation of an extensive physical quantity A_{L} is frequently expressed in the form of a sum $A_{\mathrm{L}} = \sum_{\boldsymbol{k}} F(\boldsymbol{k})$ where F is a regular function. In light of the preceding arguments, *it is legitimate to approximate this sum by a corresponding integral.* That is,

$$\sum_{\boldsymbol{k}} F(\boldsymbol{k}) \simeq \frac{L^3}{(2\pi)^3}\int \mathrm{d}\boldsymbol{k}F(\boldsymbol{k}) \quad . \tag{2.33}$$

The density of A_{L} in the thermodynamic limit is thus given by

$$\lim_{L\to\infty}\frac{A_{\mathrm{L}}}{L^3} = \frac{1}{(2\pi)^3}\int \mathrm{d}\boldsymbol{k}F(\boldsymbol{k}) \quad . \tag{2.34}$$

As mentioned before, there is no need in principle to restrict the configuration space to a finite region when one is dealing with collision processes described by a field theory. It is nevertheless very convenient for the calculations to restrict the field to a cubic volume with periodic boundary conditions, as in Sect. 1.3.5. Here, use of periodic conditions is necessary to be able to assign momenta to the particles by describing them with the help of traveling waves. The limit $L \to \infty$ is taken at the end of the calculations. In this manner, it is possible to treat systems with non-zero densities (many-body problems) and collision processes with the same formalism.

Having made these remarks, it will be useful in future developments not to specify any one particular representation for the one-particle states, whether it be position, momentum or any other. It is better to consider the states as abstract vectors in single-particle space $\mathcal{H}$. This gives greater flexibility for application and allows one to choose in each case the representation best adapted for the physical discussion.

2.1.4 n-Particle States

The state of n non-relativistic spinless particles is given by a wave function $\Phi(\boldsymbol{x}_1, \ldots, \boldsymbol{x}_n)$ which is square integrable in infinite space (or in the hypervolume Λ^n if the particles are confined to Λ), and $|\Phi(\boldsymbol{x}_1, \ldots, \boldsymbol{x}_n)|^2 \mathrm{d}\boldsymbol{x}_1 \ldots \mathrm{d}\boldsymbol{x}_n$ is the joint probability to find particle 1 in $\mathrm{d}\boldsymbol{x}_1, \ldots,$ and particle n in $\mathrm{d}\boldsymbol{x}_n$. In addition, if $\tilde{\Phi}(\boldsymbol{k}_1, \ldots, \boldsymbol{k}_n)$ designates the Fourier transform of Φ, then $|\tilde{\Phi}(\boldsymbol{k}_1, \ldots, \boldsymbol{k}_n)|^2 \mathrm{d}\boldsymbol{k}_1 \ldots \mathrm{d}\boldsymbol{k}_n$ is the joint probability to find the momentum of particle 1 in $\hbar^3 \mathrm{d}\boldsymbol{k}_1, \ldots,$ and the momentum of particle n in $\hbar^3 \mathrm{d}\boldsymbol{k}_n$. The collection of all of these states forms the Hilbert space

$$\begin{aligned} \mathcal{H}^n &= \mathcal{L}^2(\mathbb{R}^{3n}) \\ &= \left\{ \Phi(\boldsymbol{x}_1, \ldots, \boldsymbol{x}_n);\ \int \mathrm{d}\boldsymbol{x}_1 \ldots \mathrm{d}\boldsymbol{x}_n |\Phi(\boldsymbol{x}_1, \ldots, \boldsymbol{x}_n)|^2 < \infty \right\} \quad . \end{aligned} \tag{2.35}$$

If the particles carry a spin s, the wavefunction also depends on variables $(\sigma_1, \ldots, \sigma_n), \sigma_i = -s, -s+1, \ldots, s-1, s$.

The Tensor Product

It is useful to know how to construct the n-particle states in $\mathcal{H}^n$ from the one-particle states, and independently of any particular representation. This is facilitated by introducing the more abstract notion of the *tensor product.*

To all families $|\varphi_i\rangle, i = 1, \ldots, n$, of the single-particle states in $\mathcal{H}$, one associates the state of n independent particles, denoted $|\varphi_1\rangle \otimes \ldots \otimes |\varphi_n\rangle = |\varphi_1 \otimes \ldots \otimes \varphi_n\rangle$. The first particle is in the state $|\varphi_1\rangle$, the second is in the state $|\varphi_2\rangle$, and the nth is in the state $|\varphi_n\rangle$. By definition, the corresponding wavefunction is given by the product of the individual wavefunctions

$$\langle \boldsymbol{x}_1, \ldots, \boldsymbol{x}_n | \varphi_1 \otimes \ldots \otimes \varphi_n \rangle = \varphi_1(\boldsymbol{x}_1) \ldots \varphi_n(\boldsymbol{x}_n) \quad . \tag{2.36}$$

The interpretation of (2.36) is that in the state $|\varphi_1 \otimes \ldots \otimes \varphi_n\rangle$ of the independent particles, the joint amplitude to find particle 1 in $\mathrm{d}\boldsymbol{x}_1, \ldots$, and particle n in $\mathrm{d}\boldsymbol{x}_n$ is the product of the individual amplitudes.

It follows from definition (2.36) that $|\varphi_1 \otimes \ldots \otimes \varphi_n\rangle$ is linear in each of its arguments

$$\begin{aligned} &|\varphi_1 \otimes \ldots \otimes (\varphi_j + \psi_j) \otimes \ldots \otimes \varphi_n\rangle \\ &\quad = |\varphi_1 \otimes \ldots \otimes \varphi_j \otimes \ldots \otimes \varphi_n\rangle + |\varphi_1 \otimes \ldots \otimes \psi_j \otimes \ldots \otimes \varphi_n\rangle \quad , \end{aligned} \tag{2.37}$$

$$|\varphi_1 \otimes \ldots \otimes \lambda\varphi_j \otimes \ldots \otimes \varphi_n\rangle = \lambda\, |\varphi_1 \otimes \ldots \otimes \varphi_j \otimes \ldots \otimes \varphi_n\rangle \quad . \tag{2.38}$$

The scalar product of $|\varphi_1 \otimes \ldots \otimes \varphi_n\rangle$ with $|\psi_1 \otimes \ldots \otimes \psi_n\rangle$ is the product of the scalar products in $\mathcal{H}$

$$\langle \varphi_1 \otimes \ldots \otimes \varphi_n | \psi_1 \otimes \ldots \otimes \psi_n\rangle = \langle \varphi_1|\psi_1\rangle \langle \varphi_2|\psi_2\rangle \ldots \langle \varphi_n|\psi_n\rangle \quad . \tag{2.39}$$

The vector space generated by all of the linear combinations of the vectors $|\varphi_1 \otimes \ldots \otimes \varphi_n\rangle$ and completed by the scalar product (2.39) forms a Hilbert space denoted $\mathcal{H} \otimes \ldots \otimes \mathcal{H} = \mathcal{H}^{\otimes n}$, which is by definition the tensor product of n copies of the single-particle space $\mathcal{H}$. If the vectors $\{|\psi_r\rangle, r = 1, 2, \ldots\}$ form a base of $\mathcal{H}$, one obtains a base of $\mathcal{H}^{\otimes n}$ by considering all of the vector products $|\psi_{r_1} \otimes \ldots \otimes \psi_{r_n}\rangle, r_1, r_2, \ldots = 1, 2, \ldots$. As all wavefunctions $\Phi(\boldsymbol{x}_1, \ldots, \boldsymbol{x}_n)$, even those which are not factorized, can be expanded in the basis formed from the product of one-particle states $|\psi_r\rangle$, one has

$$\Phi(\boldsymbol{x}_1, \ldots, \boldsymbol{x}_n) = \sum_{r_1 \ldots r_n} c_{r_1 \ldots r_n} \psi_{r_1}(\boldsymbol{x}_1) \ldots \psi_{r_n}(\boldsymbol{x}_n) \quad . \tag{2.40}$$

One sees that the space of wavefunctions for n particles (2.35) can be identified to the tensor product $\mathcal{H}^{\otimes n}$.

One-Body Observables

Observables of a system of n identical particles can be easily constructed from individual observables. If A is a single-particle observable in $\mathcal{H}$, one can define the action of A on the jth factor of the tensor product $\mathcal{H}^{\otimes n}$ as

$$\begin{aligned} &(I \otimes \ldots \otimes A \otimes \ldots \otimes I)\, |\varphi_1 \otimes \ldots \otimes \varphi_j \otimes \ldots \otimes \varphi_n\rangle \\ &\quad = |\varphi_1 \otimes \ldots \otimes A\varphi_j \otimes \ldots \otimes \varphi_n\rangle \end{aligned} \tag{2.41}$$

and its extension to all linear combinations. The notation $I \otimes \ldots \otimes A \otimes \ldots \otimes I$ indicates that the operator acts trivially on all of the factors of $\mathcal{H}^{\otimes n}$ except for the jth. In most cases, one can simply replace $I \otimes \ldots \otimes A \otimes \ldots \otimes I$ with A_j, with the index j indicating the factor on which A operates in $\mathcal{H}^{\otimes n}$. The corresponding extensive observable for the system of n particles is thus

$$\mathtt{A}(n) = \sum_{j=1}^{n} A_j \quad . \tag{2.42}$$

Such an observable is called a *one-body observable*. Simple examples include the total momentum $\mathtt{P}(n) = \sum_{j=1}^{n} \boldsymbol{p}_j$ and the kinetic energy $\mathtt{H}_0(n) = (1/2m)\sum_{j=1}^{n} |\boldsymbol{p}_j|^2$.

Two-Body Observables

If V is an observable in two-particle space $\mathcal{H} \otimes \mathcal{H}$, such as the interaction potential, one defines V_{ij} on $\mathcal{H}^{\otimes n}$ as the operator which acts as V on the ith and jth factors of $\mathcal{H}^{\otimes n}$ and trivially on the others. The corresponding observable for a system of n particles,

$$\mathtt{V}(n) = \sum_{i<j}^{n} V_{ij} \quad , \tag{2.43}$$

where the sum is made over the $n(n-1)/2$ pairs of particles, is called a *two-body observable*. As V is acting on two identical particles, V ought to be invariant under the exchange of the two particles. This implies that $V_{ij} = V_{ji}$, and one can write (2.43) as

$$\mathtt{V}(n) = \frac{1}{2}\sum_{i\neq j}^{n} V_{ij} \quad . \tag{2.44}$$

The total potential energy is an example of a two-body observable. In the same manner, one can define k-body observables for $k > 2$. These will not be used in the following sections.

2.1.5 Symmetrization

Permutation Operators

To express the invariance of the theory under the exchange of particles, it is useful to introduce the *permutation operators* P_π acting on $\mathcal{H}^{\otimes n}$

$$P_\pi \left|\varphi_1 \otimes \ldots \otimes \varphi_n\right\rangle = \left|\varphi_{\pi(1)} \otimes \ldots \otimes \varphi_{\pi(n)}\right\rangle \quad , \tag{2.45}$$

where $\pi = \begin{pmatrix} 1 & 2 & \ldots & n \\ \pi(1) & \pi(2) & \ldots & \pi(n) \end{pmatrix}$ is an element of the group of permutations $\mathcal{P}_n$ of n elements. The operator P_π permutes the order of the single-particle states, and the definition (2.45) extends over all of $\mathcal{H}^{\otimes n}$ by linearity. One can verify that P_π is a unitary representation of the group of permutations. It is immediately clear that

$$P_{\pi_1\pi_2} = P_{\pi_1}P_{\pi_2} \quad \text{and} \quad P_\pi^* = P_\pi^{-1} \quad . \tag{2.46}$$

To show unitarity, it is sufficient to demonstrate that P_π preserves the scalar product (2.39):

$$\begin{aligned}&\langle\varphi_1\otimes\ldots\otimes\varphi_n|P_\pi^*P_\pi|\psi_1\otimes\ldots\otimes\psi_n\rangle\\&\quad=\langle\varphi_{\pi(1)}|\psi_{\pi(1)}\rangle\ldots\langle\varphi_{\pi(n)}|\psi_{\pi(n)}\rangle\\&\quad=\langle\varphi_1|\psi_1\rangle\ldots\langle\varphi_n|\psi_n\rangle\\&\quad=\langle\varphi_1\otimes\ldots\otimes\varphi_n|\psi_1\otimes\ldots\otimes\psi_n\rangle\quad.\end{aligned}\tag{2.47}$$

Exchange Degeneracy and the Principle of Symmetrization

Systems of particles, as described in Sect. 2.1.1, are characterized by the fact that all of their observables $\mathtt{O}(n)$ remain unchanged under permutations of the individual variables of the particles. It is clear that this property is applicable to the one- and two-body observables (2.42) and (2.43). This is equivalent to stating that *these observables commute with all of the permutation operators* P_π

$$[\mathtt{O}(n),P_\pi]=0\quad\text{for all }\pi\text{ in }\mathcal{P}_n\quad.\tag{2.48}$$

Relation (2.48) has an important consequence. If $|\Phi\rangle$ is a state of a system of n particles, chosen for example as an eigenvector of the Hamiltonian and of a collection of observables which commute with $\mathtt{H}(n)$, then

$$\mathtt{H}(n)\,|\Phi\rangle=E\,|\Phi\rangle\quad,\tag{2.49}$$

so that it is also true that

$$\mathtt{H}(n)P_\pi\,|\Phi\rangle=EP_\pi\,|\Phi\rangle\quad\text{for all }\pi\text{ in }\mathcal{P}_n\quad.\tag{2.50}$$

Thus all of the states $P_\pi\,|\Phi\rangle$ and their linear combinations may be used to describe the same physical state (they are all eigenvectors for the same collection of observables with the same eigenvalues). This is called *exchange degeneracy*, and it is not difficult to see from examples that the predictions of the theory depend in general on the choice of a vector in the exchange degeneracy subspace. This ambiguity shows that the quantum mechanics of a system of n identical particles is not complete and predictive without a supplementary principle, the *principle of symmetrization: the states of n identical particles are either completely symmetric or completely antisymmetric with respect to permutations of the n particles.* In the symmetric case, the particles are called *bosons*, in the antisymmetric case, they are called *fermions.*

Symmetric and Antisymmetric States

From a given vector $|\Phi\rangle$ in $\mathcal{H}^{\otimes n}$ it is possible to construct (apart from a scalar factor) exactly one corresponding state which is totally symmetric or totally antisymmetric:

$$|\Phi\rangle_+ = S_+ |\Phi\rangle \quad , \tag{2.51}$$

$$|\Phi\rangle_- = S_- |\Phi\rangle \tag{2.52}$$

with

$$S_+ = \frac{1}{n!} \sum_{\pi \in \mathcal{P}_n} P_\pi \quad , \tag{2.53}$$

$$S_- = \frac{1}{n!} \sum_{\pi \in \mathcal{P}_n} (-1)^\pi P_\pi \quad , \tag{2.54}$$

where $(-1)^\pi$ is the signature of the permutation π. *Operators* S_+ *and* S_- *are two orthogonal projections in* $\mathcal{H}^{\otimes n}$

$$S_+^2 = S_+ = S_+^* \quad , \quad S_-^2 = S_- = S_-^* \quad , \quad S_+S_- = S_-S_+ = 0 \tag{2.55}$$

and they verify

$$P_\pi S_+ = S_+ P_\pi = S_+ \quad , \tag{2.56}$$

$$P_\pi S_- = S_- P_\pi = (-1)^\pi S_- \quad . \tag{2.57}$$

Equation (2.57) can be established from the group law (2.46) and the fact that the signature of a product of permutations $\pi\pi'$ is the product of the signatures,

$$\begin{aligned} P_\pi S_- &= \frac{1}{n!} \sum_{\pi' \in \mathcal{P}_n} (-1)^{\pi'} P_\pi P_{\pi'} = \frac{(-1)^\pi}{n!} \sum_{\pi' \in \mathcal{P}_n} (-1)^\pi (-1)^{\pi'} P_\pi P_{\pi'} \\ &= \frac{(-1)^\pi}{n!} \sum_{\pi' \in \mathcal{P}_n} (-1)^{\pi\pi'} P_{\pi\pi'} = \frac{(-1)^\pi}{n!} \sum_{\pi'' \in \mathcal{P}_n} (-1)^{\pi''} P_{\pi''} \\ &= (-1)^\pi S_- \quad . \end{aligned} \tag{2.58}$$

This equality results from the fact that $\mathcal{P}_n$ is a group: when π' runs in $\mathcal{P}_n$, $\pi\pi' = \pi''$ equally runs in all of the group $\mathcal{P}_n$. One can obtain (2.56) in the same manner (replace $(-1)^\pi$ with 1). Relation (2.55) follows from (2.53) and (2.54)

$$\begin{aligned} S_-^2 &= \frac{1}{n!} \sum_{\pi \in \mathcal{P}_n} (-1)^\pi P_\pi S_- = \frac{1}{n!} \sum_{\pi \in \mathcal{P}_n} (-1)^\pi (-1)^\pi S_- \\ &= \frac{1}{n!} \left(\sum_{\pi \in \mathcal{P}_n} 1 \right) S_- = S_- \quad , \end{aligned} \tag{2.59}$$

since $\mathcal{P}_n$ has $n!$ elements. Finally,

$$S_-S_+ = \frac{1}{n!}\sum_{\pi\in\mathcal{P}_n}(-1)^\pi P_\pi S_+ = \frac{1}{n!}\left(\sum_{\pi\in\mathcal{P}_n}(-1)^\pi\right)S_+ = 0 \quad , \tag{2.60}$$

since the number of even and odd permutations in $\mathcal{P}_n$ are equal. One can immediately conclude from (2.56) and (2.57) that $P_\pi\left|\Phi\right\rangle_+ = \left|\Phi\right\rangle_+$, $P_\pi\left|\Phi\right\rangle_- = (-1)^\pi\left|\Phi\right\rangle_-$; the states $\left|\Phi\right\rangle_+$ and $\left|\Phi\right\rangle_-$ are left invariant by all permutations (apart from the sign of $\left|\Phi\right\rangle_-$). *The principle of symmetrization thus completely removes the exchange degeneracy.*

According to this principle, the space of states of a system of n identical particles is not all of $\mathcal{H}^{\otimes n}$, but rather one of two subspaces:

$$\begin{aligned}\mathcal{H}_+^{\otimes n} &= S_+\mathcal{H}^{\otimes n} \quad , \\ \mathcal{H}_-^{\otimes n} &= S_-\mathcal{H}^{\otimes n} \quad .\end{aligned} \tag{2.61}$$

The quantum mechanics can be developed in each of the two subspaces in a coherent manner as long as (2.48) holds true, because they are left invariant by all physical observables as well as the evolution operator.

The following remark may help one to understand the meaning behind the principle of symmetrization. Given a simple permutation $P_{(12)}$ of particles 1 and 2,

$$P_{(12)}\left|\varphi_1\otimes\varphi_2\otimes\ldots\otimes\varphi_n\right\rangle = \left|\varphi_2\otimes\varphi_1\otimes\ldots\otimes\varphi_n\right\rangle \quad , \tag{2.62}$$

it is clear that $P_{(12)}^2 = I$ and $P_{(12)}^* = P_{(12)}$, so the eigenvalues of $P_{(12)}$ must necessarily be $+1$ or -1. As $P_{(12)}$ commutes with the Hamiltonian $\mathtt{H}$, $P_{(12)}$ can be diagonalized simultaneously with $\mathtt{H}$, hence the even or odd nature of the state under the permutation $P_{(12)}$ is conserved over the course of time. The principle of symmetrization carries two supplementary affirmations.

First of all, a state has the same parity under an exchange $P_{(ij)}$ of any pair (ij) of particles. In fact, this property is not required to break the exchange degeneracy. In general, $\mathcal{H}^{\otimes n}$ can be decomposed into a sum of invariant subspaces, corresponding to irreducible representations of the group of permutations. For $n \geq 3$, in addition to $\mathcal{H}_+^{\otimes n}$ and $\mathcal{H}_-^{\otimes n}$, there exist other irreducible subspaces with states of mixed parity exchange of particles. One can show that the ambiguity of the exchange degeneracy is removed by postulating that the states of the system belong to a fixed irreducible representation of the group of permutations. There is no obstacle in principle to developing the theory in any given irreducible subspace with mixed parity, called parastatistics in such cases. It is a fact of observation however, that in nature, systems appear to be either totally symmetric or totally antisymmetric. The second affirmation is that *the totally symmetric or totally antisymmetric character of the state does not depend on its preparation, but is an intrinsic attribute of the particle.*

Evidence shows furthermore that particles of spin 1 are bosons and that particles of spin 1/2 are fermions. Actually, it follows from Planck's law of black-body radiation that photons are bosons, and from analysis of atomic spectra that electrons must be fermions. It is nevertheless possible to establish a connection between spin and statistics in the frame of relativistic quantum-field theory. This will be outlined in Sect. 8.3.4.

2.1.6 Symmetry of Composite Particles

In certain circumstances, it is possible to attribute well-defined statistics to composite particles. Consider, for example, a state $|\Psi\rangle$ of a collection of mk nucleons which form m nuclei, each comprising k nucleons. If the thermal energy is much smaller than the energy of the binding nuclear force and the pressure is sufficiently low, the nuclei may be considered as localized individual objects. One can thus attribute a wave function $\Phi(\boldsymbol{x}_1, \ldots, \boldsymbol{x}_k) \equiv \Phi(1, \ldots, k)$ to a nucleus comprising k nucleons. This function only differs from zero when the distances $|\boldsymbol{x}_i - \boldsymbol{x}_j|, i, j = 1, \ldots, k$ between all of the pairs of nucleons do not exceed the nuclear diameter. All of the nuclei are in the same internal state, but can have different positions and momenta. By consequence, two states of the nuclei $|\Phi_i\rangle$ and $|\Phi_j\rangle$ differ only by the probability amplitude of their respective centers of masses $\boldsymbol{y}_i$ and $\boldsymbol{y}_j$.

If one neglects correlations between different nuclei, the wavefunction $\Psi(\boldsymbol{x}_1, \ldots, \boldsymbol{x}_{mk}) \equiv \Psi(1, \ldots, mk)$ of the mk nucleons constituting the m nuclei can be written as an antisymmetric product of m wavefunctions of the nuclei $\Phi\left((j-1)k+1, \ldots, jk\right), j = 1, \ldots, m$:

$$\begin{aligned} \Psi(1, \ldots, mk) = S_-\{&\Phi_1(1, \ldots, k)\Phi_2(k+1, \ldots, 2k) \\ &\ldots \Phi_m\left[(m-1)k+1, \ldots, mk\right]\} \quad . \end{aligned} \tag{2.63}$$

It is assumed that the wavefunctions of the nuclei are already antisymmetric. The mean value of observable $\mathtt{A}$ of the m nuclei in state $|\Psi\rangle$ is

$$\begin{aligned} \langle\Psi|A|\Psi\rangle &= \langle S_-[\Phi_1\Phi_2\ldots\Phi_m]|\,\mathtt{A}\,|S_-[\Phi_1\Phi_2\ldots\Phi_m]\rangle \\ &= \langle[\Phi_1\Phi_2\ldots\Phi_m]|\,\mathtt{A}\,|S_-[\Phi_1\Phi_2\ldots\Phi_m]\rangle \quad . \end{aligned} \tag{2.64}$$

Here, (2.55) is used, as well as the fact that S_- commutes with $\mathtt{A}$.

In the anti-symmetrization operation S_- to be carried out in (2.63), it is possible to distinguish between two types of permutations. The first type exchanges two nuclei, permuting all of the nucleons of which they are composed. The other type of permutation exchanges only certain nucleons belonging to different nuclei. Terms corresponding to the latter type of permutation do not contribute in a sensible manner to the calculation of the mean value (2.64). In effect, they all involve, in (2.64), a wavefunction Φ_i having arguments belonging to at least two different nuclei. One such term contributes to the mean value (2.64) one in the case where the two nucleons overlap, a highly improbable configuration for the pressure and temperature conditions presupposed. One can thus practically neglect in (2.63) permutations which partially exchange nucleons attached to different nuclei.

Under permutations which exchange entire nuclei, $\Psi(1, \ldots, mk)$ remains invariant or changes sign according to whether the nucleus has an even or an odd number of nucleons, since each fermion exchange causes a change of sign. Ignoring the internal structure of the nuclei, it is possible to simplify the description by writing $\Psi(1, \ldots, mk)$ in coordinates relative to the center of mass $\boldsymbol{y}_i$ of the nuclei. It is clear that the resulting wave function $\Psi(\boldsymbol{y}_1, \ldots, \boldsymbol{y}_m)$ has a well-defined boson or fermion symmetry. It is the same for weakly correlated nuclei described by a superposition of linear states (2.63). In this approximation, a ^{4}He nucleus (two protons and two neutrons) behaves as a boson and a ^{3}He nucleus (two protons and one neutron) behaves as a fermion.

Now suppose that electrons are placed around the ^{3}He or ^{4}He nuclei such that the resulting atoms are in their ground state. If the kinetic energy is weak enough that one can neglect the excitation probability of the atoms, the particles are still well localized and in the same internal state. The preceding argument thus still holds true for ^{3}He and ^{4}He atoms. As they are composed of 5 (6) fermions, the atoms behave as fermions (bosons). These arguments emphasize that *the symmetry of composite particles only makes sense relative to the order of magnitude of the energy available in the system.*

2.1.7 Occupation Number Representation

The symmetric or antisymmetric states of $\mathcal{H}_+^{\otimes n}$ or $\mathcal{H}_-^{\otimes n}$ corresponding to the products of one-particle states $|\varphi_i\rangle$ in $\mathcal{H}$ are written as

$$|\varphi_1, \varphi_2, \ldots, \varphi_n\rangle_\nu = S_\nu \, |\varphi_1 \otimes \varphi_2 \otimes \ldots \otimes \varphi_n\rangle \quad , \tag{2.65}$$

where the index ν takes the value 1 for bosons and -1 for fermions. In the symmetric case, the order of the states $|\varphi_i\rangle$ in $|\varphi_1, \ldots, \varphi_n\rangle_+$ is insignificant. In the antisymmetric case, $|\varphi_1, \ldots, \varphi_n\rangle_-$ changes sign under the permutation of two states $|\varphi_i\rangle$ and $|\varphi_j\rangle$:

$$|\varphi_1, \ldots, \varphi_i, \ldots, \varphi_j, \ldots, \varphi_n\rangle_- = - \, |\varphi_1, \ldots, \varphi_j, \ldots, \varphi_i, \ldots, \varphi_n\rangle_- \quad . \tag{2.66}$$

Note that such a vector vanishes if two states $|\varphi_i\rangle$ and $|\varphi_j\rangle$ are identical. This is Pauli's *exclusion principle: two fermions can never be found in the same individual quantum state.*

Even if the vectors $|\varphi_j\rangle$ are normalized, $\langle\varphi_j|\varphi_j\rangle = 1$, the states (2.65) resulting from a projection, are no longer normalized to 1. The scalar product of two such states must be, according to relations (2.55) and (2.39),

$$
\begin{aligned}
&{}_{\nu}\langle \varphi_1, \ldots, \varphi_n | \psi_1, \ldots, \psi_n \rangle_{\nu} \\
&\quad = \langle \varphi_1 \otimes \ldots \otimes \varphi_n | S_{\nu}^{*} S_{\nu} | \psi_1 \otimes \ldots \otimes \psi_n \rangle \\
&\quad = \langle \varphi_1 \otimes \ldots \otimes \varphi_n | S_{\nu} | \psi_1 \otimes \ldots \otimes \psi_n \rangle \\
&\quad = \frac{1}{n!} \sum_{\pi \in \mathcal{P}_n} (\nu)^{\pi} \, \langle \varphi_1 \otimes \ldots \otimes \varphi_n | \psi_{\pi(1)} \otimes \ldots \otimes \psi_{\pi(n)} \rangle \\
&\quad = \frac{1}{n!} \sum_{\pi \in \mathcal{P}_n} (\nu)^{\pi} \, \langle \varphi_1 | \psi_{\pi(1)} \rangle \ldots \langle \varphi_n | \psi_{\pi(n)} \rangle \quad . \qquad (2.67)
\end{aligned}
$$

In the fermion case, (2.67) is just the expansion of the determinant

$$
{}_{-}\langle \varphi_1, \ldots, \varphi_n | \psi_1, \ldots, \psi_n \rangle_{-} = \frac{1}{n!} \det\{\langle \varphi_i | \psi_j \rangle\} \quad . \qquad (2.68)
$$

In particular, the wavefunction of n independent fermions is given by the *Slater determinant*

$$
{}_{-}\langle \boldsymbol{x}_1, \ldots, \boldsymbol{x}_n | \psi_1, \ldots, \psi_n \rangle_{-} = \frac{1}{n!} \det\{\psi_j(\boldsymbol{x}_i)\} \quad . \qquad (2.69)
$$

Occupation Numbers

A particularly simple and useful description can be obtained when the single-particle states in (2.65) form a set $|\psi_{r_1}\rangle, \ldots, |\psi_{r_n}\rangle$ of states belonging to a fixed basis $\{|\psi_r\rangle\}$ of the one-particle space $\mathcal{H}$. The *occupation number states*, normalized to 1, are defined by

$$
|n_1, n_2, \ldots, n_r, \ldots\rangle_{+} = \frac{\sqrt{n!}}{\sqrt{n_1! \ldots n_r! \ldots}} S_{+} \, |\psi_{r_1} \otimes \ldots \otimes \psi_{r_n}\rangle \quad , \qquad (2.70)
$$

$$
|n_1, n_2, \ldots, n_r, \ldots\rangle_{-} = \sqrt{n!} \, S_{-} \, |\psi_{r_1} \otimes \ldots \otimes \psi_{r_n}\rangle \quad , \qquad (2.71)
$$

where n_r is the number of indices $r_1, \ldots, r_j, \ldots, r_n$ which are equal to r, that is, the number of particles which are in the individual state $|\psi_r\rangle$. In the fermion case (2.71), the indices are written in increasing order $r_1 < r_2 < \ldots < r_n$. *The integers n_r are called occupation numbers relative to the states $|\psi_r\rangle$.* For bosons, n_r can take on the values $0, 1, 2, \ldots, n$, whereas for the fermions, the values of n_r are limited to either 0 or 1 because of the Pauli principle. As the system has exactly n particles, one must in any case have $\sum_r n_r = n$. It is important not to confuse the notation of the states (2.65), (2.70) and (2.71). In (2.65), the one-particle states are arbitrary and thus must be explicitly noted. *In (2.70) and (2.71) the occupation numbers n_r correspond to a given fixed basis of the one-particle space.* A basis frequently used for calculations is the one in which the individual states are plane waves (2.24). The states of n particles are thus specified by the occupation numbers $n_{\boldsymbol{k}_1}, n_{\boldsymbol{k}_2}, \ldots, n_{\boldsymbol{k}_j}, \ldots$ corresponding to the wavenumbers $\boldsymbol{k}_1, \boldsymbol{k}_2, \ldots, \boldsymbol{k}_j, \ldots$.

If $|\psi_r\rangle$ are the eigenvectors of a single-particle observable A, such that

$$A\,|\psi_r\rangle = \alpha_r\,|\psi_r\rangle \quad , \tag{2.72}$$

then the states $|n_1, n_2, \dots, n_r, \dots\rangle_\nu$ are also eigenvectors of the one-body observable $\mathtt{A}(n)$ (2.42) for eigenvalues $\sum_r n_r\alpha_r$. It follows from definition (2.41) that

$$\begin{aligned}\mathtt{A}(n)\,|\psi_{r_1}\otimes\ldots\otimes\psi_{r_n}\rangle &= \sum_{j=1}^{n} A_j\,|\psi_{r_1}\otimes\ldots\otimes\psi_{r_n}\rangle \\ &= \left(\sum_r n_r\alpha_r\right)|\psi_{r_1}\otimes\ldots\otimes\psi_{r_n}\rangle \quad .\end{aligned} \tag{2.73}$$

As $\mathtt{A}(n)$ commutes with S_ν, the same relation necessarily applies for states $|n_1, n_2, \dots, n_r, \dots\rangle_\nu$

$$\begin{aligned}&\mathtt{A}(n)\,|n_1, n_2, \dots, n_r, \dots, \dots\rangle_\nu \\ &\quad = \left(\sum_r n_r\alpha_r\right)|n_1, n_2, \dots, n_r, \dots, \dots\rangle_\nu \quad .\end{aligned} \tag{2.74}$$

If the observable in question is the kinetic energy and if the $|\psi_r\rangle$ are plane waves (2.24), the possible values of the kinetic energy of the system of n particles are

$$E = \sum_{\boldsymbol{k}} n_{\boldsymbol{k}}\frac{\hbar^2|\boldsymbol{k}|^2}{2m} \quad , \quad \text{where} \quad \sum_{\boldsymbol{k}} n_{\boldsymbol{k}} = n \quad .$$

It is now left to show that states (2.70) and (2.71) are orthonormal:

$$_\nu\langle m_1, m_2, \dots, m_r, \dots|n_1, n_2, \dots, n_r, \dots\rangle_\nu = \delta_{m_1,n_1}\delta_{m_2,n_2}\cdots\delta_{m_r,n_r}\cdots \tag{2.75}$$

and form respectively bases of the $\mathcal{H}_\nu^{\otimes n}$ spaces, $\nu = \pm 1$. To see this, one calculates the scalar product (2.75) with the help of (2.67)

$$\begin{aligned}&_+\langle m_1, m_2, \dots, m_r, \dots|n_1, n_2, \dots, n_r, \dots\rangle_+ \\ &\quad = \frac{1}{\sqrt{m_1!\,m_2!\dots n_1!\,n_2!\dots}}\sum_{\pi\in\mathcal{P}_n}\langle\psi_{r_1}|\psi_{\pi(s_1)}\rangle\cdots\langle\psi_{r_n}|\psi_{\pi(s_n)}\rangle \quad .\end{aligned} \tag{2.76}$$

In $r_1, \dots, r_n$ there are n_r indices equal to r, and in $s_1, \dots, s_n$, there are m_r indices equal to r. In $\pi(s_1), \dots, \pi(s_n)$, there are also exactly m_r indices equal to r. As $\langle\psi_r|\psi_{\pi(s)}\rangle = 0$ if $r \neq \pi(s)$, the terms of the sum are equal to 1 if $\pi(s_1) = r_1, \dots, \pi(s_n) = r_n$ and 0 otherwise, and there are $n_1!\,n_2!\dots$ non-zero terms. One thus obtains orthonormality for the case of bosons. In the case of fermions, as n_r, m_r are equal to 0 or 1,

$$
\begin{aligned}
& {}_{-}\langle m_1, m_2, \ldots, m_r, \ldots | n_1, n_2, \ldots, n_r, \ldots \rangle_{-} \\
& \quad = \sum_{\pi \in \mathcal{P}_n} (-1)^{\pi} \langle \psi_{r_1} | \psi_{\pi(s_1)} \rangle \ldots \langle \psi_{r_n} | \psi_{\pi(s_n)} \rangle \quad . \qquad (2.77)
\end{aligned}
$$

This scalar product is zero if $(r_1, \ldots, r_n) \neq (s_1, \ldots, s_n)$. If $m_r = n_r$ and $r_1 = s_1, \ldots, r_n = s_n$, only the term corresponding to the identical permutation contributes and this term is equal to 1. This verifies the orthonormality.

To see that the states $|n_1, n_2, \ldots\rangle_\nu$ form a basis of $\mathcal{H}_\nu^{\otimes n}$, one notes that any vector $|\Phi\rangle$ in $\mathcal{H}_\nu^{\otimes n}$ is of the form $|\Phi\rangle = S_\nu |\Psi\rangle$ where $|\Psi\rangle$ belongs to $\mathcal{H}^{\otimes n}$. As the products $|\psi_{r_1} \otimes \ldots \otimes \psi_{r_n}\rangle$ form a basis of $\mathcal{H}^{\otimes n}$, $|\Psi\rangle$ may be expanded according to the rule

$$
|\Psi\rangle = \sum_{r_1 \ldots r_n} c_{r_1 \ldots r_n} |\psi_{r_1} \otimes \ldots \otimes \psi_{r_n}\rangle \quad , \qquad (2.78)
$$

so that

$$
|\Phi\rangle_\nu = \sum_{r_1 \ldots r_n} c_{r_1 \ldots r_n} S_\nu |\psi_{r_1} \otimes \ldots \otimes \psi_{r_n}\rangle \quad . \qquad (2.79)
$$

Taking into account definitions (2.70) and (2.71), expansion (2.79) shows that any state $|\Phi\rangle$ in $\mathcal{H}_\nu^{\otimes n}$ is a linear combination of the vectors $|n_1, n_2, \ldots\rangle_\nu$.

2.2 Degenerate Gases

2.2.1 The Ground State of n Bosons

First of all, we remark that bosons differ on an important point depending on whether or not they are *massive*. It is clear that massless bosons, such as photons, cannot form a thermodynamic phase in the usual sense with a non-zero mass density (one cannot associate a chemical potential to them). In addition, they are constantly undergoing the process of emission or absorption by matter, so they never constitute a system of a fixed number of particles. Their treatment thus requires an extension of the formalism to systems of variable numbers of particles. This will be described in Chap. 3. The same remarks apply to phonons. In this section, we will examine only massive bosons without spin and of a fixed number n, such as is the case for a sample of ^{4}He atoms in a container.

Suppose that the interactions between these n bosons are negligible. The energy of the system is uniquely defined by the one-particle Hamiltonian H with eigenvectors $|\psi_r\rangle$ and eigenvalues ε_r

$$
H |\psi_r\rangle = \varepsilon_r |\psi_r\rangle \quad , \quad r = 0, 1, 2, \ldots \quad . \qquad (2.80)
$$

By convention, the individual eigenvalues ε_r are classified in increasing order repeated as often as required by their multiplicity

$$\varepsilon_0 \leq \varepsilon_1 \leq \ldots \leq \varepsilon_r \leq \ldots \quad . \tag{2.81}$$

For a spinless particle, the single-particle ground state $|\psi_0\rangle$ is in general non-degenerate; this will be assumed here.

The ground state of n bosons is obtained by placing each of the bosons in its own individual minimal energy state. It is equally non-degenerate and according to (2.70) is given by

$$|n, 0, \ldots, 0, \ldots\rangle_+ = |\psi_0 \otimes \ldots \otimes \psi_0\rangle \quad , \tag{2.82}$$

where $|n, 0, \ldots, 0, \ldots\rangle_+$ is state (2.70) indexed by the occupation numbers $n_0 = n$ and $n_r = 0, r \neq 0$. The corresponding eigenvalue is $E_0 = n\varepsilon_0$ and the energy per unit volume

$$\mathrm{u}_0 = \frac{E_0}{\Lambda} = \frac{n}{\Lambda}\varepsilon_0 = \rho\varepsilon_0 \tag{2.83}$$

is proportional to the density ρ.

The important phenomenon which appears here is the *macroscopic occupation of a single one-particle state* $|\psi_0\rangle$, since $n_0 = \rho\Lambda$ is extensive. This is called *Bose condensation* into the state $|\psi_0\rangle$. One should note that this is not a condensation in the usual sense of the term. That is, it is not a density condensation in configuration space, as is the case for a gas–liquid transition, but rather as an accumulation of particles into one single-particle state.

If, for example, $H = H_0$ is the kinetic energy (2.16) with periodic boundary conditions, then $\varepsilon_0 = 0$. Furthermore each boson in the volume L^3 has the wavefunction $\psi_{\boldsymbol{k}=0}(\boldsymbol{x}) = 1/L^{3/2}$ corresponding to a uniform probability distribution in the volume. Generally, the minimal kinetic energy of a particle in a volume of linear dimension L corresponds to the smallest momentum compatible with the uncertainty principle, $|\boldsymbol{p}| \simeq \hbar/L$. The magnitude of u_0 is thus

$$\mathrm{u}_0 = C\rho\frac{\hbar^2}{m}\frac{1}{L^2} \quad , \tag{2.84}$$

where C is a numerical constant.

Temperature of Degeneracy

When the temperature is not strictly zero, thermal fluctuations lead to the population of energy states above the ground state. The closest neighbouring energy levels to E_0 are $(n-1)\varepsilon_0 + \varepsilon_1$, $(n-1)\varepsilon_0 + \varepsilon_2$, $(n-2)\varepsilon_0 + 2\varepsilon_1$, $(n-2)\varepsilon_0 + \varepsilon_1 + \varepsilon_2$, etc. The order evidently depends on relations between the one-particle energy eigenvalues. They can either be degenerate or not.

Population of the higher energy states $|\psi_r\rangle$ causes a decrease in the occupation rate of the ground state. The thermodynamics of a free (three-dimensional) Bose gas shows that *the occupation rate remains macroscopic*

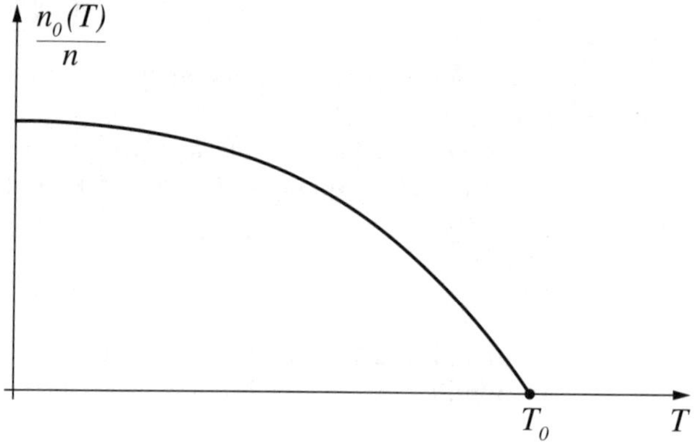

Fig. 2.2. Occupation rate of the ground state as a function of temperature for independent bosons of spin s

for temperatures below a certain characteristic temperature T_0 (Fig. 2.2). This phenomenon is called *Bose–Einstein condensation*. It can be shown that for independent bosons of spin s (in the limit $n \to \infty$)

$$\frac{n_0(T)}{n} = \begin{cases} 1 - (T/T_0)^{3/2} & , \quad T \le T_0 \\ 0 & , \quad T > T_0 \end{cases} \tag{2.85}$$

with the transition temperature given by

$$T_0 = 3.31 \frac{\hbar^2}{k_B m} \left(\frac{\rho}{g}\right)^{2/3} \quad , \quad \text{where} \quad g = 2s + 1 \quad . \tag{2.86}$$

In fact, the intensive quantity $k_B T_0$ can only depend on $\hbar$, m, and the density ρ. Under these conditions, dimensional analysis imposes the relation $k_B T_0 \simeq \hbar^2 \rho^{2/3}/m$ as the only possible combination of these parameters. The resulting temperature T_0 is called the degeneracy temperature. This is the temperature below which Bose statistics must be taken into account. For helium atoms, the degeneracy temperature can be calculated from (2.86) to be $T_0 = 3.2$ K.

Cold Atoms and Bose–Einstein Condensation

A direct observation of Bose–Einstein condensation in an almost perfect gas is an experimental challenge. Indeed at densities corresponding to atmospheric pressure, atoms normally freeze into a solid at temperatures well above the degeneracy temperature (2.86) (an exception is Helium which, at atmospheric pressure, remains liquid at all temperatures, showing the phenomenon of superfluidity discussed in Chap. 7). It is therefore necessary to be able to form a dilute atomic gas at very low temperature, less than 10^{-6} K. For comparison, the coldest intergalactic gases are millions of times too hot for Bose–Einstein condensation. It is therefore remarkable that Bose–Einstein condensation in a gaseous phase was observed for the first time in a vapor of ^{87}Rb atoms in 1995: the density was 2.5×10^{12} atoms per centimeter cube and the temperature 170×10^{-9} K. [2]

The choice of alkali atoms (such as sodium or rubidium) is well suited for the application of the cooling mechanisms that will be briefly described below. Moreover the atom–atom interactions are weak: the range of the interaction is about 10^{-6} cm, while at the required densities the interatomic spacing is about 10^{-4} cm so that the gas is almost perfect except for hard-core-type elastic collisions. The success in obtaining the condensate relies on the recent developments of efficient methods to cool atoms.

The first stage, laser cooling, is based on the Doppler effect. In sketchy terms the principle is as follows. One considers an atomic resonance of frequency ω_A and places the atom in a stationary wave created by two opposing lasers of frequency $\omega_\mathrm{L} < \omega_\mathrm{A}$. Because of the Doppler effect, an atom moving with velocity v perceives the higher frequency $\omega_\mathrm{L}(1 + v/c)$ from the wave propagating in the direction opposite to its motion and the lower frequency $\omega_\mathrm{L}(1 - v/c)$ from the wave propagating in the same direction. Since absorption of electromagnetic energy (namely photons) increases when the frequency gets closer to the resonance ω_A, absorption of photons from these two beams leads to an average effective viscous force $F = -\gamma v$ where γ is a friction coefficient. As a result the atom is slowed down and remains confined in-between the two lasers. Three dimensional confinement is achieved by the use of three orthogonal pairs of lasers. Atoms trapped in the intersection of the beams, forming a so-called optical molasses, arc cooled to about 40×10^{-6} K, but this is still 100 times too hot to lead to Bose–Einstein condensation. The temperature cannot be further lowered by this method because of the intrinsic quantum noise of photons. To get around this limitation, one replaces in a second stage the laser trap by a magnetic trap. Magnetic trapping works because atoms carry a magnetic moment $\boldsymbol{m}$: in an inhomogeneous magnetic field $\boldsymbol{B}(\boldsymbol{x})$ they are thus subject to a force $\boldsymbol{F}(\boldsymbol{x}) = \nabla(\boldsymbol{m} \cdot \boldsymbol{B}(\boldsymbol{x}))$. By suitable arrangement of the magnetic field, one can confine atoms by a potential which acts almost as an axially symmetric harmonic well. The last stage, evaporative cooling, consists in letting the most energetic atoms escape from the trap. This is done by an additional radio frequency magnetic field that drives higher-energy atoms outside of the trap. During this process it is important that the dilute remaining atoms still undergo enough elastic collisions to share energy and achieve thermal equilibrium.

The state of the gas is observed by turning off the confining magnetic field allowing for a ballistic expansion of the atomic cloud. The expanding cloud is

---

[2] M. H. Anderson, J. R. Ensher, M. R Matthews, C. E. Wieman and E. A. Cornell, Science, *269*, 198 (1995); E. A. Cornell and C. E. Wieman, Scientific American, 26 (March 1998).

then illuminated by a laser flash and the analysis of the scattered light determines the velocity distribution of the atoms in the cloud originating from the trap. The whole experiment is repeated at several temperatures. For temperatures above 170×10^{-9} K, the velocity distribution is essentially of Maxwellian form as predicted for a non-degenerate gas. At 170×10^{-9} K a sharp peak appears centered at zero velocity. As the temperature is further lowered, the peak becomes more and more pronounced on a diffuse background. This is a clear departure from the classical distribution and the signature of the occurrence of the Bose–Einstein condensate. There are two confirmations that the phenomenon is indeed quantum mechanical. As the temperature attains 50×10^{-9} K (corresponding to full condensation) the peak does not become an infinitely narrow $\delta(\boldsymbol{v})$-function, the zero temperature limit of the Maxwellian distribution. This is a manifestation of the uncertainty principle: the velocity distribution is given here by the square of the Fourier transform of the ground state wavefunction of an atom in the confining well. This function has a finite extension and so does its Fourier transform (an infinitely extended condensate would have a δ-peaked distribution, see however the discussion above (2.84)). Furthermore an examination of the shape of the peak reveals an anisotropy, reflecting an anisotropy of the confining potential, whereas the Maxwellian distribution is isotropic irrespective of the shape of the potential.

A Bose–Einstein condensate is a macroscopic quantum-mechanical fluid where all atoms are coherently described by the same wavefunction. Such systems should exhibit interference fringe patterns, like light or the electron in the two slit experiment. Interference fringes have been observed when two clouds of Bose condensate overlap.[3] Two separate condensates are prepared in a double well potential so that tunnelling through the barrier between the left and right wells is negligible. If $\psi_{\mathrm{L}}(\boldsymbol{x},t)$ and $\psi_{\mathrm{R}}(\boldsymbol{x},t)$ are the wavefunctions of the clouds originating from the left and right wells, the density of the overlapping clouds is given by $\rho(\boldsymbol{x},t) = |\psi_{\mathrm{L}}(\boldsymbol{x},t)+\psi_{\mathrm{R}}(\boldsymbol{x},t)|^2$, hence the phase difference between $\psi_{\mathrm{L}}(\boldsymbol{x},t)$ and $\psi_{\mathrm{R}}(\boldsymbol{x},t)$ will produce density interferences in space. The fringe period can be simply estimated. If two atoms of mass m originating at time $t = 0$ from the left and right well meet at time t inbetween the two sources (considered as two point sources at distance d), their relative velocity must be d/t and relative momentum md/t. By the de Broglie relation the wavelength of matter waves is $\lambda = h/p = ht/md$. All these predictions are in satisfactory agreement with the experimental observations.

---

[3] M. R. Andrews, C. G. Townsend, H. J. Miesner, D. S. Durfee, D. M. Kurn and W. Ketterle, Science, *275*, 637 (1997); Experiments in dilute atomic Bose–Einstein condensation, E. A. Cornell, J. R. Ensher and C. E. Wieman, Proceedings of the International School of Physics "Enrico Fermi" Course 140, Amsterdam IOS Press (1999); Theory of Bose–Einstein condensation in trapped gases, F. Dalfovo, S. Giorgini, L. P. Pitaevskii and S. Stringari, Rev. Mod. Phys. *71*, 463 (1999).

A Property of the Ground State of Interacting Bosons

When the particles interact, the ground state is given by the wavefunction $\varPhi_0(\boldsymbol{x}_1, \dots, \boldsymbol{x}_n)$ which no longer has the simple factorizable form of (2.82). However, if the range of interaction of the forces is small with respect to the mean distance between the particles, it is possible to give a description of the gas which is analogous to that of the perfect gas. This problem will be returned to in Chap. 7 with a study of superfluidity.

A general property which still holds for the wavefunction $\varPhi_0(\boldsymbol{x}_1, \dots, \boldsymbol{x}_n)$ of the ground state of interacting bosons is that it *does not have any zeros and thus has one constant sign*, arbitrarily chosen to be positive. This property is important for understanding the role of the statistics in relation with superfluidity. To show this property, it is useful to make the following argument, which can be derived rigorously from the point of view of the mathematics for a class of potentials $V(\boldsymbol{x}_1 - \boldsymbol{x}_2)$ which are sufficiently regular. Suppose that $\varPhi_0(\boldsymbol{x}_1, \dots, \boldsymbol{x}_n)$ is a normalized eigenstate with minimum energy E_0, then

$$\begin{aligned}\big(\mathtt{H}(n)\varPhi_0\big)(\boldsymbol{x}_1, \dots, \boldsymbol{x}_n) &= \left[-\frac{\hbar^2}{2m}\sum_{j=1}^{n}\nabla_j^2 + \sum_{i<j}^{n} V(\boldsymbol{x}_i - \boldsymbol{x}_j)\right]\varPhi_0(\boldsymbol{x}_1, \dots, \boldsymbol{x}_n) \\ &= E_0\varPhi_0(\boldsymbol{x}_1, \dots, \boldsymbol{x}_n) \quad , \end{aligned} \tag{2.87}$$

where

$$\begin{aligned}E_0 &= \langle\varPhi_0|\mathtt{H}(n)|\varPhi_0\rangle \\ &= \frac{\hbar^2}{2m}\int \mathrm{d}\boldsymbol{x}_1 \dots \int \mathrm{d}\boldsymbol{x}_n \sum_j |\nabla_j\varPhi_0(\boldsymbol{x}_1, \dots, \boldsymbol{x}_n)|^2 \\ &\quad + \int \mathrm{d}\boldsymbol{x}_1 \dots \int \mathrm{d}\boldsymbol{x}_n \left(\sum_{i<j}^{n} V(\boldsymbol{x}_i - \boldsymbol{x}_j)\right)|\varPhi_0(\boldsymbol{x}_1, \dots, \boldsymbol{x}_n)|^2 \quad . \end{aligned} \tag{2.88}$$

The form of the kinetic energy in (2.88) results from an integration by parts; contributions from the boundary compensate if periodic boundary conditions are chosen. Note that $\varPhi_0(\boldsymbol{x}_1, \dots, \boldsymbol{x}_n)$ can always be taken to be real: if this is not the case, a combination of $\varPhi_0(\boldsymbol{x}_1, \dots, \boldsymbol{x}_n) + \varPhi_0^*(\boldsymbol{x}_1, \dots, \boldsymbol{x}_n)$ can be formed which is real and which is a solution of (2.87) since $V(\boldsymbol{x}_1 - \boldsymbol{x}_2)$ is real.

Now further suppose that $\varPhi_0(\boldsymbol{x}_1, \dots, \boldsymbol{x}_n)$ changes sign in $\bar{\boldsymbol{x}}_i$ when one of its arguments is varied (assume that $\varPhi_0(\boldsymbol{x}_1, \dots, \boldsymbol{x}_n)$ and $\nabla_i\varPhi_0(\boldsymbol{x}_1, \dots, \boldsymbol{x}_n)$ do not simultaneously vanish, a property which is verifiable for potentials which do not contain strong singularities). The new function $\varPsi_0(\boldsymbol{x}_1, \dots, \boldsymbol{x}_n) = |\varPhi_0(\boldsymbol{x}_1, \dots, \boldsymbol{x}_n)|$ verifies $|\nabla_i\varPsi_0(\boldsymbol{x}_1, \dots, \boldsymbol{x}_n)|^2 = |\nabla_i\varPhi_0(\boldsymbol{x}_1, \dots, \boldsymbol{x}_n)|^2$ (see Fig. 2.3). According to (2.88), it thus corresponds to the same energy

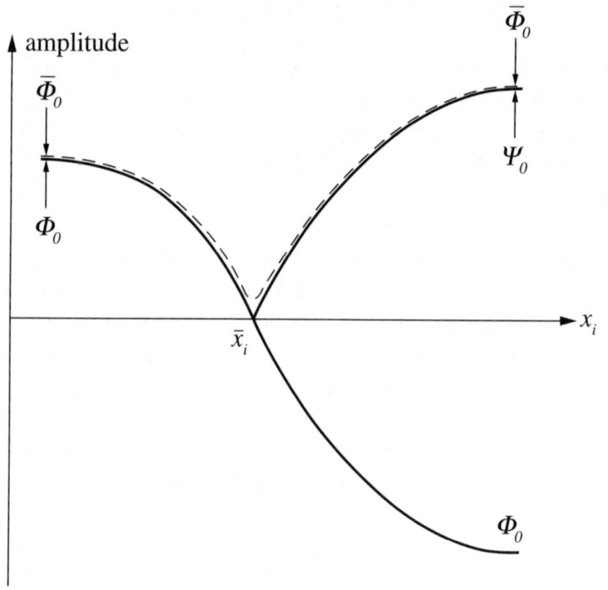

Fig. 2.3. The wavefunction of interacting bosons in the ground state has no zeros and can thus always be constructed to be positive

$$E_0 = \langle \Phi_0 | \mathtt{H}(n) | \Phi_0 \rangle = \langle \Psi_0 | \mathtt{H}(n) | \Psi_0 \rangle \quad . \tag{2.89}$$

Now consider the function $\bar{\Phi}_0(\boldsymbol{x}_1, \ldots, \boldsymbol{x}_n)$, identical to $\Psi_0(\boldsymbol{x}_1, \ldots, \boldsymbol{x}_n)$ but with the cusp in $\bar{\boldsymbol{x}}_i$ smoothed (Fig. 2.3). This operation has the effect of decreasing the derivative in the neighbouring region of $\bar{\boldsymbol{x}}_i$. If the potential remains finite, the kinetic energy is decreased without appreciably modifying the potential energy. From this, one obtains

$$\langle \overline{\Phi}_0 | \mathtt{H}(n) | \overline{\Phi}_0 \rangle < \langle \Psi_0 | \mathtt{H}(n) | \Psi_0 \rangle = E_0 \quad , \tag{2.90}$$

in contradiction with the fact that E_0 is the energy of the ground state. By consequence, $\Phi_0(\boldsymbol{x}_1, \ldots, \boldsymbol{x}_n)$ cannot change sign.

This argument shows moreover that the ground state is unique (up to a constant factor) and is totally symmetric. Two distinct ground states $|\Phi_0\rangle$ and $|\Phi_0'\rangle$ of energy E_0 could be chosen orthogonal, but as they each must have a constant sign, $\langle \Phi_0 | \Phi_0' \rangle = 0$ is impossible. Since the Hamiltonian commutes with all of the permutation operators and since $|\Phi_0\rangle$ is unique, $|\Phi_0\rangle$ is necessarily either completely symmetric or completely antisymmetric. The latter case is evidently excluded, since $|\Phi_0\rangle$ does not change sign.

2.2.2 The Ground State of n Fermions

The structure of the ground state of n massive independent fermions is determined by the application of the Pauli principle. As the individual energies ε_r

given by (2.80) are classified in increasing order, *it is necessary to place one fermion in each of the first n single particles states* $|\psi_r\rangle\,, r = 1, 2, \ldots, n$. The corresponding energy is thus

$$E_0 = \sum_{r=1}^{n} \varepsilon_r \quad , \tag{2.91}$$

where the ε_r are repeated as their multiplicity requires. The wave function of the ground state is the Slater determinant (2.69) formed from the first n states $|\psi_r\rangle$.

The Fermi Wavenumber and Energy

It is straightforward to evaluate E_0 for the case where the energy is purely kinetic (with no external field) and where the n fermions are in a cubic box with periodic boundary conditions. The individual energies are given by (2.22) and have a degeneracy equal to $g = 2s+1$ due to the $2s+1$ possible spin states. The occupation numbers $n_{\boldsymbol{k}\sigma}$ are indexed by the wavenumber $\boldsymbol{k}$ and the orientation of the spin σ. The individual plane wave states are occupied up to a maximum wave number, the *Fermi wavenumber* k_{F}, *defined by*

$$n_{\boldsymbol{k}\sigma} = \begin{cases} 1 & , \quad |\boldsymbol{k}| \le k_{\mathrm{F}} \\ 0 & , \quad |\boldsymbol{k}| > k_{\mathrm{F}} \end{cases} \quad . \tag{2.92}$$

The *Fermi sphere* is defined by the collection of $\boldsymbol{k}$ where $|\boldsymbol{k}| \le k_{\mathrm{F}}$. As there is exactly one particle per state $\boldsymbol{k}\sigma$, k_{F} is determined by the number of particles

$$n = g \sum_{|\boldsymbol{k}|}^{k_{\mathrm{F}}} 1 \tag{2.93}$$

and (2.91) can be written as

$$E_0 = g \sum_{|\boldsymbol{k}|}^{k_{\mathrm{F}}} \frac{\hbar^2}{2m} |\boldsymbol{k}|^2 \quad . \tag{2.94}$$

Taking the thermodynamic limit as in (2.34), one finds

$$\rho = \lim_{L\to\infty} \frac{n}{L^3} = \frac{g}{(2\pi)^3} \int_{|\boldsymbol{k}|\le k_{\mathrm{F}}} \mathrm{d}\boldsymbol{k} = \frac{g}{6\pi^2} k_{\mathrm{F}}^3 \tag{2.95}$$

$$\mathrm{u}_0 = \lim_{L\to\infty} \frac{E_0}{L^3} = \frac{g}{(2\pi)^3} \int_{|\boldsymbol{k}|\le k_{\mathrm{F}}} \mathrm{d}\boldsymbol{k} \frac{\hbar^2 |\boldsymbol{k}|^2}{2m} = g \frac{\hbar^2}{4\pi^2 m} \frac{k_{\mathrm{F}}^5}{5} \quad . \tag{2.96}$$

The relation between the energy density u_0 and the particle density ρ is obtained by eliminating k_{F} in (2.95) and (2.96)

$$\mathrm{u}_0 = \frac{3}{10}\left(\frac{6\pi^2}{g}\right)^{2/3}\frac{\hbar^2}{m}\rho^{5/3} \quad . \tag{2.97}$$

The *Fermi energy* ε_{F} can be defined here by

$$\varepsilon_{\mathrm{F}} = \frac{p_{\mathrm{F}}^2}{2m} = \frac{\hbar^2 k_{\mathrm{F}}^2}{2m} = \frac{\hbar^2}{2m}\left(\frac{6\pi^2\rho}{g}\right)^{2/3} \quad \text{where} \quad p_{\mathrm{F}} = \hbar k_{\mathrm{F}} \quad . \tag{2.98}$$

Comparing (2.97) and (2.84), one derives, up to a numerical factor,

$$\begin{aligned} \mathrm{u}_0^{\text{boson}} &\simeq \rho\frac{\hbar^2}{m}\frac{1}{L^2} \simeq \frac{n}{L^3}\frac{\hbar^2}{m}\left(\frac{1}{L}\right)^2 \\ \mathrm{u}_0^{\text{fermion}} &\simeq \rho^{5/3}\frac{\hbar^2}{m}\frac{1}{L^2} \simeq \frac{n}{L^3}\frac{\hbar^2}{m}\left(\frac{n^{1/3}}{L}\right)^2 \quad . \end{aligned} \tag{2.99}$$

This difference in behaviour between bosons and fermions is easy to interpret. When the exclusion principle does not apply (bosons), the whole volume L^3 of the sample is available for each particle. This implies a momentum on the order of $p_{\mathrm{B}} \simeq \hbar/L$ and an energy of $\varepsilon_{\mathrm{B}} \simeq \hbar^2/mL^2$. In the case of fermions, however, the available volume for a particle is constrained by the exclusion principle (volume per fermion $\simeq L^3/n$), notably increasing each particle's momentum $p_{\mathrm{F}} \simeq \hbar n^{1/3}/L$ and energy $\varepsilon_{\mathrm{F}} \simeq (\hbar^2/m)(n^{1/3}/L)^2$. *The essential difference with the bosons is that the kinetic energy density is no longer proportional to ρ but increases as $\rho^{5/3}$.* This is a consequence of the exclusion principle which requires that, as the number of particles increases, the fermions must occupy states of higher and higher kinetic energy.

Excited States

Excitations close to the ground state are obtained by removing a particle from inside the Fermi sphere with $|\boldsymbol{k}_1| < k_{\mathrm{F}}$ and placing it in a state $\boldsymbol{k}_2$ with $|\boldsymbol{k}_2| > k_{\mathrm{F}}$ (Fig. 2.4). The corresponding occupation numbers are

$$\begin{aligned} n_{\boldsymbol{k}\sigma} &= 1 \quad \text{if} \quad \boldsymbol{k}\sigma \neq \boldsymbol{k}_1\sigma_1 \quad , \quad |\boldsymbol{k}| < k_{\mathrm{F}} \quad , \\ n_{\boldsymbol{k}\sigma} &= 0 \quad \text{if} \quad \boldsymbol{k}\sigma \neq \boldsymbol{k}_2\sigma_2 \quad , \quad |\boldsymbol{k}| > k_{\mathrm{F}} \quad , \\ n_{\boldsymbol{k}_1\sigma_1} &= 0 \quad , \\ n_{\boldsymbol{k}_2\sigma_2} &= 1 \quad . \end{aligned} \tag{2.100}$$

The empty state left by an electron removed from the Fermi sphere is called an *electron hole* and the excitation (2.100) is said to create an *electron–hole pair*.

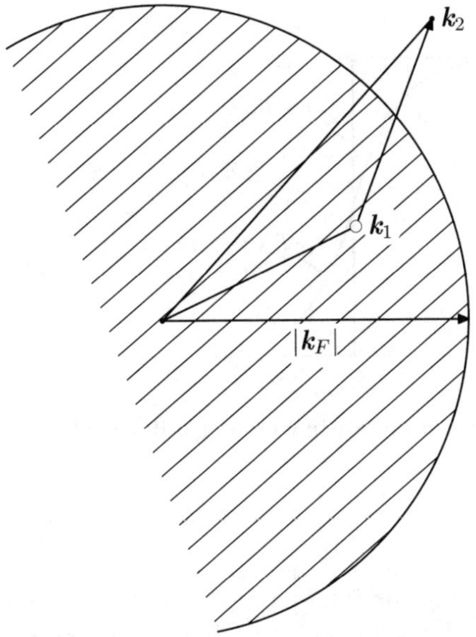

Fig. 2.4. The empty state left by an electron removed from the Fermi sphere is called an *electron hole*; excitation of an electron creates an *electron–hole pair*

In the ground state, the number of particles $n(\varepsilon)\mathrm{d}\varepsilon$ of energy between ε and $\varepsilon + \mathrm{d}\varepsilon$ is, according to (2.92), given by

$$n(\varepsilon) = \begin{cases} g & , \quad \varepsilon < \varepsilon_\mathrm{F} \\ 0 & , \quad \varepsilon > \varepsilon_\mathrm{F} \end{cases} \quad . \tag{2.101}$$

Excitations of the type (2.100) can be of thermal origin, or they can be the result of an interaction between electrons in the ground state, such as in the theory of superconductivity (Chap. 5). Their existence modifies the occupation rate of states with energy ε closely neighbouring ε_F as shown in Fig. 2.5.

It is useful to note that, *contrary to the case of a boson gas, a non-interacting Fermi gas does not undergo a phase transition.* It is possible, nevertheless, to define a temperature of degeneracy T_0 at the Fermi thermal energy $k_\mathrm{B}T_0 = \varepsilon_\mathrm{F}$. It follows from (2.98) that

$$T_0 = \frac{1}{2}(6\pi^2)^{2/3}\frac{\hbar^2}{k_\mathrm{B}m}\left(\frac{\rho}{g}\right)^{2/3} \tag{2.102}$$

with the same parameter dependence as (2.86), for the dimensional reasons previously discussed. Here, $T \ll T_0$ is the condition for Fermi statistics to play an important role.

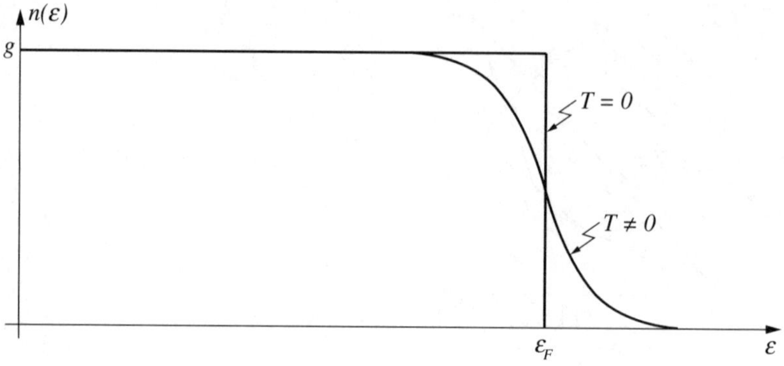

Fig. 2.5. Modification of ground state occupation rate in a Fermi gas due to thermal excitation of states with energy near ε_{F}

The Origin of Ferromagnetism

A quantitative study of the ground state for interacting fermions requires more elaborate techniques. These will be addressed in Chap. 3. It is possible, however, to discuss certain consequences of the Pauli principle from a qualitative point of view. To begin with, consider two electrons having an interaction which is independent of their spin. The wavefunction

$$\Phi_0(\boldsymbol{x}_1\sigma_1, \boldsymbol{x}_2\sigma_2) = \Phi_0(\boldsymbol{x}_1, \boldsymbol{x}_2)\chi(\sigma_1, \sigma_2) \tag{2.103}$$

can be factorized into an orbital part $\Phi_0(\boldsymbol{x}_1, \boldsymbol{x}_2)$ and a spin part $\chi(\sigma_1, \sigma_2)$ and must change sign under exchange of the two particles. This may be due to either an antisymmetric $\Phi_0(\boldsymbol{x}_1, \boldsymbol{x}_2)$ or an antisymmetric $\chi(\sigma_1, \sigma_2)$. There are thus two possible solutions which satisfy the Pauli principle:

$$\Phi_1(\boldsymbol{x}_1\sigma_1, \boldsymbol{x}_2\sigma_2) = \Phi_0^a(\boldsymbol{x}_1, \boldsymbol{x}_2)\chi^s(\sigma_1, \sigma_2) \quad , \tag{2.104}$$

$$\Phi_2(\boldsymbol{x}_1\sigma_1, \boldsymbol{x}_2\sigma_2) = \Phi_0^s(\boldsymbol{x}_1, \boldsymbol{x}_2)\chi^a(\sigma_1, \sigma_2) \quad , \tag{2.105}$$

where the index s (a) labels a symmetric (antisymmetric) wavefunction. As the interaction does not depend on spin, the spin functions are constructed with the help of the individual spinors $\chi_+(\sigma)$ and $\chi_-(\sigma)$ corresponding to eigenvalues $+1/2$ and $-1/2$ of the spin operator relative to a fixed direction. The antisymmetric combination is necessarily

$$\chi^a(\sigma_1, \sigma_2) = \left(1/\sqrt{2}\right)\left[\chi_+(\sigma_1)\chi_-(\sigma_2) - \chi_+(\sigma_2)\chi_-(\sigma_1)\right]$$

corresponding to total spin $s = 0$ (singlet state). The symmetric combination has three possibilities:

$$\begin{aligned} &\chi_+(\sigma_1)\chi_+(\sigma_2) \quad , \\ &\left(1/\sqrt{2}\right)\left[\chi_+(\sigma_1)\chi_-(\sigma_2) + \chi_+(\sigma_2)\chi_-(\sigma_1)\right] \quad \text{or} \\ &\chi_-(\sigma_1)\chi_-(\sigma_2) \end{aligned}$$

corresponding to eigenvalues of total spin $s = 1$ (triplet state).

It is thus only a matter of determining which of the two states, (2.104) or (2.105), has the lowest energy. It appears that the mean value of the electrostatic repulsion term between the the two electrons

$$\left\langle \Phi_0 \left| \frac{e^2}{4\pi\varepsilon_0|\boldsymbol{q}_1 - \boldsymbol{q}_2|} \right| \Phi_0 \right\rangle = \frac{e^2}{4\pi\varepsilon_0} \int \mathrm{d}\boldsymbol{x}_1 \int \mathrm{d}\boldsymbol{x}_2 \frac{|\Phi_0(\boldsymbol{x}_1, \boldsymbol{x}_2)|^2}{|\boldsymbol{x}_1 - \boldsymbol{x}_2|} \tag{2.106}$$

is least if $\Phi_0(\boldsymbol{x}_1, \boldsymbol{x}_2)$ is antisymmetric. In fact, since $\Phi_0^a(\boldsymbol{x}_1, \boldsymbol{x}_2)$ vanishes for $\boldsymbol{x}_1 = \boldsymbol{x}_2$, the probability is small to find the two electrons close to one another, where their potential energy is highest. Such an argument does not evidently hold for $\Phi_0^s(\boldsymbol{x}_1, \boldsymbol{x}_2)$. This effect is often dominant, but is not necessarily the rule, as it is also necessary to evaluate the kinetic energy and the interaction energy of the electrons with the nuclei or the ions. In any case, whenever $\Phi_0^a(\boldsymbol{x}_1, \boldsymbol{x}_2)$ yields the lowest possible energy, expression (2.104) corresponds to the ground state. This is the state with the maximum possible spin, $s = 1$.

These considerations can be extended to a wavefunction for n electrons $\Phi_0(\boldsymbol{x}_1\sigma_1, \dots, \boldsymbol{x}_n\sigma_n)$. In this case, however, the antisymmetric form of $\Phi_0(\boldsymbol{x}_1\sigma_1, \dots, \boldsymbol{x}_n\sigma_n)$ is not simply constructed as a product of entirely symmetric or entirely antisymmetric orbital and spin functions. This is because, for $n \geq 3$, $\chi(\sigma_1, \dots, \sigma_n)$ cannot be totally antisymmetric without vanishing since at least two of the indices $\sigma_1, \dots, \sigma_n$ must necessarily take on the same value $+1/2$ or $-1/2$. Nevertheless, it is possible to show that the wavefunction corresponding to the maximum total spin $s = n/2$ is "the most antisymmetric wavefunction possible" as far as the configurational part goes. It is in such a state that the repulsive Coulomb energy is lowest. Under favorable circumstances – that is, if the reduction of the potential energy is not compensated by an increase in the kinetic energy – the ground state has a non-zero spin per unit volume and thus a magnetic moment. It is this mechanism which gives rise to *ferromagnetism. Magnetic forces are thus not the origin of ferromagnetism, but rather Coulomb electrostatic forces acting in conjunction with the Pauli principle.*

2.2.3 Stability of Matter

Under normal conditions, matter is composed of electrons characterized by their charge and mass $(-e,m)$ and nuclei (Ze,M), obeying the laws of non-relativistic quantum mechanics and the Coulomb force. Here, only the case where a single chemical substance is present will be considered. Furthermore, the phrase *normal conditions* shall imply the assumption that the energy is low enough to neglect relativistic effects and nuclear reactions, and that the system will be studied on a scale where gravitational forces are negligible.

All properties of matter under normal conditions are in principle deductible from the Hamiltonian

$$\mathrm{H}(n,N) = \sum_{i=1}^{n} \frac{|\boldsymbol{p}_i|^2}{2m} + \sum_{j=1}^{N} \frac{|\boldsymbol{P}_j|^2}{2M} + \frac{1}{4\pi\varepsilon_0} \sum_{i<j}^{n} \frac{e^2}{|\boldsymbol{x}_i - \boldsymbol{x}_j|}$$
$$+ \frac{1}{4\pi\varepsilon_0} \sum_{i<j}^{N} \frac{Z^2 e^2}{|\boldsymbol{y}_i - \boldsymbol{y}_j|} - \frac{1}{4\pi\varepsilon_0} \sum_{i=1}^{n} \sum_{j=1}^{N} \frac{Z e^2}{|\boldsymbol{x}_i - \boldsymbol{y}_j|} \quad , \tag{2.107}$$

where n and N are respectively the number of electrons and nuclei of coordinates $(\boldsymbol{p}_i, \boldsymbol{x}_i), i = 1, \dots, n$, and $(\boldsymbol{P}_j, \boldsymbol{y}_j), j = 1, \dots, N$. Restricting the problem to that of globally neutral systems, the total charge of the electrons must equal that of the nuclei, so that

$$n = ZN \quad . \tag{2.108}$$

It is thus sufficient to index the Hamiltonian (2.107) by the number n only.

The stability of matter demands that the Hamiltonian have a lower bound which is proportional to the number of particles. If this were not the case, the energy per particle could become infinitely negative as the number of particles increased and the system would not have the normally observed thermodynamic behaviour. The stability of matter can be expressed by the fundamental inequality

$$\mathrm{H}(n) \geq -Bn \quad , \tag{2.109}$$

where B is a constant independent of n. As $\mathrm{H}(n)$ is an operator, the inequality (2.109) signifies that all of the mean values are greater than or equal to $-Bn$. It follows from neutrality (2.108) that (2.109) gives a lower bound which is proportional to the total number of particles $n + N = n(Z+1)/Z$. The constant B must be positive, since the quantum binding energy is negative.

A rigorous demonstration of (2.109) can be made, but it is not elementary.[4] Because of the singular character of the Coulomb potential, the energy of opposite sign charges (the last term of (2.107)) becomes infinitely negative when they approach each other. From an electrostatic point of view only, there is a collapse of the electrons onto the nuclei. However, by virtue of the uncertainty principle, a localized electron must acquire a very large kinetic energy, and the kinetic energy of n electrons (first term of (2.107)) is further increased as a consequence of the Fermi statistics. A simultaneous control of these effects makes it possible to obtain the desired bound. It is possible here to present a few qualitative arguments which lead to this result.

Suppressing the kinetic energy of the nuclei and the repulsive terms of the potential energy which are manifestly positive in (2.107), one obtains the following inequality,

$$\mathrm{H}(n) \geq \sum_{i=1}^{n} \frac{|\boldsymbol{p}_i|^2}{2m} - \frac{1}{4\pi\varepsilon_0} \sum_{i=1}^{n} \sum_{j=1}^{N} \frac{Z e^2}{|\boldsymbol{x}_i - \boldsymbol{y}_j|} = L^3(\mathrm{u}_0 + \mathrm{u}_{\mathrm{pot}}) \quad , \tag{2.110}$$

[4] E. Lieb, The stability of matter, Rev. Mod. Phys. *48*, 553 (1976).

where u_0 is the kinetic energy density of the electrons and

$$\mathrm{u}_{\mathrm{pot}} = -\frac{1}{4\pi\varepsilon_0 L^3}\sum_{i=1}^{n}\sum_{j=1}^{N}\frac{Ze^2}{|\boldsymbol{x}_i - \boldsymbol{y}_j|} \tag{2.111}$$

is the potential energy density of the charges of opposite sign.

To estimate $\mathrm{u}_{\mathrm{pot}}$, the system is assumed to be essentially homogeneous, with electron density ρ and resulting nuclear density ρ/Z. A fundamental property of charged systems is that of local neutrality: the energy is minimized if the electrons are arranged such that their charges compensate on average for the charges of the nuclei. The typical distance between them, λ is called the screening distance. This is evident when the system is composed of neutral entities, such as atoms or molecules, but is equally true for a plasma (an ionized gas). Beyond λ, the Coulomb potential is no longer effective, as the neutral entities interact with multipolar forces of shorter range. The screening distance λ is of the order of the distance between the nuclei

$$\lambda = \left(\frac{Z}{\rho}\right)^{1/3} \quad . \tag{2.112}$$

In such a state, the order of the magnitude of $\mathrm{u}_{\mathrm{pot}}$ can be obtained by replacing the sums in (2.111) by integrals over the densities $\rho = n/L^3 = ZN/L^3$ corresponding to a homogeneous system

$$\begin{aligned}\mathrm{u}_{\mathrm{pot}} &\simeq -\frac{1}{4\pi\varepsilon_0 L^3}\int_{L^3}\mathrm{d}\boldsymbol{x}\int_{\substack{L^3\\|\boldsymbol{x}-\boldsymbol{y}|\leq\lambda}}\mathrm{d}\boldsymbol{y}\frac{\rho^2 e^2}{|\boldsymbol{x}-\boldsymbol{y}|}\\ &\simeq -\frac{\rho^2 e^2}{\varepsilon_0}\int_0^{\lambda} x^2\mathrm{d}x\frac{1}{x} \simeq -\frac{\rho^2 e^2\lambda^2}{\varepsilon_0}\end{aligned} \tag{2.113}$$

hence

$$\mathrm{u}_{\mathrm{pot}} \simeq -\frac{e^2 Z^{2/3}\rho^{4/3}}{\varepsilon_0} \quad . \tag{2.114}$$

Concerning the order of magnitude calculations of (2.113) and the formulae which follow, $\simeq$ and $\gtrsim$ signify relations which are accurate up to a dimensionless numerical constant factor. This holds true for the following two paragraphs.

If the kinetic energy of the electrons is evaluated from formula (2.97) for a free gas, then

$$\mathtt{H}(n) \gtrsim \frac{\hbar^2}{m}\frac{n^{5/3}}{L^2} - \frac{Z^{2/3}e^2}{\varepsilon_0}\frac{n^{4/3}}{L} \quad . \tag{2.115}$$

A linear dimension L which minimizes the right-hand side of (2.115) can easily be determined for a fixed n to be

$$L \simeq \frac{\varepsilon_0 \hbar^2}{me^2} \frac{n^{1/3}}{Z^{2/3}} \simeq a_{\mathrm{B}} \frac{n^{1/3}}{Z^{2/3}} \tag{2.116}$$

where $a_{\mathrm{B}} = 4\pi\varepsilon_0\hbar^2/me^2$ is the Bohr radius. Inserting this value of L into (2.115) establishes that $\mathtt{H}(n)$ has an extensive lower bound proportional to $-(e^2/\varepsilon_0 a_{\mathrm{B}})Z^{4/3}n$. A rigorous estimation[5] determines that

$$\mathtt{H}(n) \geq -0.77 \frac{e^2}{\varepsilon_0 a_{\mathrm{B}}} Z^{4/3} n \quad . \tag{2.117}$$

One should note that the statistics of the nuclei do not come into play here, they could be bosons or fermions depending on whether their total spin is an integer or a half-integer. On the other hand, *the Fermi statistics of the electrons is essential.* If the electrons were replaced by some hypothetical bosons, the kinetic energy density, given by (2.84), must be replaced by $\hbar^2\rho/mL^2$ and inequality (2.115) would become (with $g = Z = 1$)

$$\mathtt{H}(n) \gtrsim \frac{\hbar^2}{m} \frac{n}{L^2} - \frac{e^2}{\varepsilon_0} \frac{n^{4/3}}{L} \quad . \tag{2.118}$$

Minimization of (2.118) gives $L = a_{\mathrm{B}} n^{-1/3}$, which leads to a non-extensive bound $\mathtt{H}(n) \gtrsim -(e^2/\varepsilon_0 a_{\mathrm{B}})n^{5/3}$. In fact, it is possible to establish that the energy of the ground state of this system of charged bosons decreases with n at least as quickly as $-n^{7/5}$.[6] *The increase of the electron's kinetic energy with the density to the 5/3 power plays an essential role for the stability of matter.*

2.2.4 Nucleo-Electronic Plasma at High Density

Fermionic matter is observed to behave quite differently at high density, depending on whether its interaction is short range or Coulomb. In the former case, the interaction energy $\mathrm{u}_{\mathrm{pot}} \simeq \rho^2$ dominates the kinetic energy $\mathrm{u}_0 \sim \rho^{5/3}$ and by consequence, the particles are always strongly coupled. For a *Coulomb nucleo-electronic plasma*, however, the opposite situation is produced: *the system becomes perfect at high density.* This remarkable property is due to the fact that the Coulomb energy $|\mathrm{u}_{\mathrm{pot}}| \simeq \rho^{4/3}$ grows slower than the fermion kinetic energy.

Here, conditions will be examined under which the Coulomb energy can be neglected. From (2.114) and (2.97),

$$|\mathrm{u}_{\mathrm{pot}}| \ll \mathrm{u}_0 \tag{2.119}$$

[5] W. Thirring, Commun. Math. Phys. *79*, 1 (1981).
[6] J.G. Conlon, E.H. Lieb and H.T. Yau, Commun. Math. Phys. *116*, 417 (1988).

implies that, for the electron density,

$$\rho \mathrel{\widetilde{\gg}} Z^2 \left(\frac{e^2 m_\mathrm{e}}{\hbar^2 \varepsilon_0}\right)^3 \simeq Z^2 a_\mathrm{B}^{-3} \quad . \tag{2.120}$$

Due to global neutrality, the number of electrons is equal to the number of protons. Note that in (2.120) and in the pages which follow, m_e will designate the mass of an electron in order to avoid any possible ambiguities. Since a nucleus typically contains about as many protons as neutrons, the average total mass $\overline{m}$ per electron is approximately

$$\overline{m} \simeq 2m_\mathrm{p} \quad , \tag{2.121}$$

where m_p is the proton mass. In (2.121), we neglect the electron mass compared to the proton mass, as well as the difference between the proton and neutron masses. Inequality (2.120) for mass density $\rho_\mathrm{M} = 2m_\mathrm{p}\rho$ thus becomes

$$\rho_\mathrm{M} \mathrel{\widetilde{\gg}} 2m_\mathrm{p} Z^2 a_\mathrm{B}^{-3} = 20\ Z^2\ \mathrm{g\ cm}^{-3} \quad . \tag{2.122}$$

Note that these formulae are only applicable to non-relativistic electrons, $\varepsilon_\mathrm{F} \ll m_\mathrm{e}c^2$, namely

$$p_\mathrm{F} = \hbar k_\mathrm{F} = \sqrt{2m_\mathrm{e}\varepsilon_\mathrm{F}} \ll \sqrt{2} m_\mathrm{e} c \quad . \tag{2.123}$$

With (2.95), this gives the condition

$$\rho_\mathrm{M} \simeq 2m_\mathrm{p}\rho \mathrel{\widetilde{\ll}} m_\mathrm{p} g \left(\frac{m_\mathrm{e} c}{\hbar}\right)^3 = 2 \times 10^6\ \mathrm{g\ cm}^{-3} \quad . \tag{2.124}$$

The density defined by (2.122) is not attained by ordinary matter under normal conditions. It is approached, however, for objects of astrophysical studies, such as white dwarfs and neutron stars. A *white dwarf* is the last stage of a star which has exhausted its nuclear fuel: it may be considered to consist of an extremely dense plasma of nucleons and electrons. For even larger densities, the nuclei can capture the electrons, transforming the protons into neutrons and emitting neutrinos. When this process of capture is sufficiently important, the system becomes a nuclear material almost exclusively composed of neutrons: this is a *neutron star* or pulsar. In both cases, one can neglect the Coulomb interaction, this applies for a white dwarf with density obeying (2.122) and for a neutron star composed essentially of neutral particles. It thus becomes necessary to take into account the gravitational energy.

2.2.5 Fermions and Gravitation

The Hamiltonian of n fermions of mass m acted upon only by gravity can be written analogously to (2.107) as

$$\mathtt{H}(n) = \sum_{i=1}^{n} \frac{|\boldsymbol{p}_i|^2}{2m} - G \sum_{i<j}^{n} \frac{m^2}{|\boldsymbol{x}_i - \boldsymbol{x}_j|} \quad , \tag{2.125}$$

where G is the gravitational constant.

The gravitational potential energy $\mathrm{u}_{\mathrm{pot}}$ can be estimated similarly to (2.113) with the only difference being that *the interaction is purely attractive so that there is no screening phenomenon.* Here, the only parameter with the dimension of length is the linear dimension R of the total system, giving

$$\mathrm{u}_{\mathrm{pot}} \simeq -G\rho^2 m^2 R^2 \tag{2.126}$$

and yielding an $\mathtt{H}(n)$ of order

$$\mathtt{H}(n) \simeq \frac{\hbar^2}{m} \frac{n^{5/3}}{R^2} - Gm^2 \frac{n^2}{R} \quad . \tag{2.127}$$

This quantity is minimized for

$$R \simeq \frac{\hbar^2}{Gm^3} n^{-1/3} \tag{2.128}$$

at which

$$\mathtt{H}(n) \simeq -\frac{G^2 m^5}{\hbar^2} n^{7/3} \quad . \tag{2.129}$$

This demonstrates that *the gravitational system contracts, decreasing its energy, when n grows infinitely.* That is, there is no stable state in the thermodynamic limit. The previous calculation makes sense, however, when applied to a system containing a *finite quantity of matter.* In this case, relation (2.128) determines the size of the system as a function of its total mass.

One possible application would be to estimate the order of magnitude of the radius of a neutron star and of a white dwarf. Assuming that all of the electrons have been captured, the mass of a neutron star is $M = 2nm_{\mathrm{p}}$ and (2.128) gives

$$R \simeq \frac{\hbar^2}{Gm_{\mathrm{p}}^{8/3}} M^{-1/3} \quad . \tag{2.130}$$

Taking $M = M_{\mathrm{sun}} = 2 \times 10^{30}$ kg, gives $R \simeq 10^3$ m. The mean distance between two neutrons is of the order $a = \rho^{-1/3}$ where $\rho \simeq M/m_{\mathrm{p}}R^3$ is the density of the nucleons. One thus obtains $a = 10$ Fermi; that is, the neutron star is one gigantic nucleus. It is possible to verify that in this nucleus gravitational interaction dominates nuclear interaction. With the result of Exercise 2(ii), $\mathrm{u}_{\mathrm{int,nucl}} \simeq \rho^2 v$ and $v = \int \mathrm{d}\boldsymbol{x} V(\boldsymbol{x})$ where $V(\boldsymbol{x})$ is the nucleon–nucleon potential. For nuclear forces, the range r_0 of this potential is of the order of magnitude of the distance between two nucleons, so $v \simeq r_0^3 v_0 \simeq \rho^{-1} v_0$. This gives $\mathrm{u}_{\mathrm{int,nucl}} \simeq \rho v_0$, where $v_0 \simeq 8$ MeV is a measure of the binding energy of a nucleon with its neighbours, hence

$$\frac{|\mathrm{u}_{\mathrm{pot,grav}}|}{\mathrm{u}_{\mathrm{int,nucl}}} \simeq \frac{G m_{\mathrm{p}}^2 n}{v_0 R} \simeq \frac{G^2 M^{4/3} m_{\mathrm{p}}^{11/3}}{\hbar^2 v_0} \simeq 10^3 \quad . \tag{2.131}$$

For a white dwarf, neglecting the electron mass with respect to the proton mass, the total energy is essentially that of the kinetic energy of the electrons and of the gravitational energy of the protons. The total mass is also approximated by $M = n(m_{\mathrm{e}} + m_{\mathrm{p}}) \simeq n m_p$. If $m = m_{\mathrm{e}}$ is substituted into the kinetic energy term of (2.127), one obtains a modified form of (2.130),

$$R \simeq \frac{\hbar^2}{G m_{\mathrm{e}} m_{\mathrm{p}}^{5/3}} M^{-1/3} \quad . \tag{2.132}$$

For $M = M_{\mathrm{sun}}$, one finds $R \simeq 10^7$ m.

It is still necessary to check whether this corresponds to the non-relativistic regime defined by (2.124). In fact, the mass density corresponding to (2.132) would be $\rho_{\mathrm{M}} = M/R^3 = 10^6$ g cm^{-3}, indicating that it would be useful here to take relativistic effects into account. It is instructive to examine the highly relativistic regime, where the kinetic energy of the electrons is given by $\mathrm{u}_0 \simeq \hbar c \rho^{4/3}$ (Exercise 2(ii)). In this case, the total energy has the form

$$\mathrm{H}_{\mathrm{M}} \simeq \left(\frac{\hbar c}{m_{\mathrm{p}}^{4/3}} M^{4/3} - G M^2 \right) \frac{1}{R} = \frac{f(M)}{R} \tag{2.133}$$

and $f(M)$ is zero for

$$M_{\mathrm{c}} = \frac{1}{m_{\mathrm{p}}^2} \left(\frac{\hbar c}{G} \right)^{3/2} \simeq 10^{30} \text{ kg} \quad . \tag{2.134}$$

If $M > M_{\mathrm{c}}$, $f(M)$ is negative and the star minimizes its energy by contracting ($R \to 0$) until other forms of interaction come into play. So, from a gravitational point of view only, there is no possible equilibrium. If $M < M_{\mathrm{c}}$, however, $f(M)$ is positive and the star can decrease its energy and its temperature by growing ($R \to \infty$). This would bring the system back to the non-relativistic regime, where there exists an equilibrium radius (2.132). *This reasoning shows that M_{c} is the upper limit of the mass of a white dwarf*, called the *Chandrasekhar limit*.

Exercises

1. The Role of Boundary Conditions

Consider a free particle in a finite interval $[0, L]$ with boundary conditions $\varphi(0) = \varphi(L) = 0$ (Dirichlet boundary conditions).

(i) Show that the eigenvalues and the eigenfunctions of (2.18) are

$$\varepsilon_n = \frac{\hbar^2}{2m}\left(\frac{n\pi}{L}\right)^2 \quad , \quad \psi_n(x) = \sqrt{\frac{2}{L}}\sin\frac{n\pi}{L}x \quad , \quad n = 1, 2, \dots \quad .$$

(ii) Show that the average momentum in these states vanishes.

(iii) Let

$$E_L = \frac{1}{Z_L}\sum_n \varepsilon_n \mathrm{e}^{-\varepsilon_n/(k_\mathrm{B}T)}$$

be the average kinetic energy of the particle at temperature T, with $Z_L = \sum_n e^{-\varepsilon_n/k_\mathrm{B}T}$ the partition function (k_B is the Boltzmann constant). Show that for both the periodic and the Dirichlet boundary conditions,

$$\lim_{L\to\infty} E_L = k_\mathrm{B}T/2$$

has the same infinite volume limit.

Hint: Show first in both cases that $Z_L \sim L/\lambda$ as $L \to \infty$, where $\lambda = \sqrt{2\pi\hbar^2/mk_\mathrm{B}T}$ is the thermal de Broglie wavelength.

Comment: In contrast with periodic boundary conditions, the eigenstates of a particle with Dirichlet boundary conditions are stationary waves that carry no momentum. However, average quantities (like the kinetic energy) are the same in the infinite volume limit.

2. Dimensional Analysis

Assume that the ground state energy density u_0 (kinetic) or u (potential) of a system of particles of mass m interacting with an integrable two-body potential $V(\boldsymbol{x})$ is of the form

$$u = A\hbar^\alpha \rho^\beta m^\gamma c^\delta v^\varepsilon \quad ,$$

with A a dimensionless constant, ρ the particle density, c the speed of light and $v = \int \mathrm{d}\boldsymbol{x} V(\boldsymbol{x})$.

(i) Show that dimensional analysis imposes

$$\alpha + \gamma + \varepsilon = 1 \quad , \quad 2\alpha - 3\beta + \delta + 5\varepsilon = -1 \quad , \quad \alpha + \delta + 2\varepsilon = 2 \quad .$$

(ii) Conclude that the kinetic energy density of a free non-relativistic Fermi gas must be of the form $u_0 = A(\hbar^2/m)\rho^{5/3}$. The kinetic energy of a free ultra-relativistic Fermi gas (rest mass of a particle negligible compared to its kinetic energy) is $u_0 = A\hbar c\rho^{4/3}$. The potential energy density (to first order in v) is $u = A\rho^2 v$. What happens for the kinetic energy density of a Bose gas?

Comment: Calculation of the ground state energy is in general a complex problem which requires perturbative techniques to be presented in subsequent chapters. It is nevertheless possible to obtain qualitative information on the dependence of the energy on particle number and volume by dimensional analysis.

3. Systems with Variable Particle Number

3.1 Introduction

In many situations, the number of particles of a given type is not exactly known or may vary during reactions taking place in the system. One example would be that of a photon gas. As previously mentioned, photons are continuously emitted and absorbed by matter and their number in general is not fixed. This is also the case for condensed-matter states, where interactions may modify the number of various excitation quanta, such as phonons. In high-energy physics, any number of particles may be created or annihilated during a collision process. Lastly, since the number of atoms or molecules constituting a macroscopic body can never be known exactly, it becomes useful to describe the body using the grand-canonical statistical ensemble, in which the number of particles is allowed to fluctuate.

It is thus useful to introduce a formalism which allows one to describe the states of an undetermined number of particles and interactions which do not conserve particle number. Such a formalism is known as second quantization. This terminology stems from historical reasons which will be presented in Sect. 8.4.1. It is necessary, however, to immediately qualify the slightly misleading terminology. Second quantization does not, in any manner, introduce a new theory nor any new physical principles which are different from those of the usual quantum mechanics. There is only one quantization: the one carried out when one considers particles as quantum objects. *The conceptual framework of the second quantization is thus that of the usual quantum mechanics.*

While the formalism of the second quantization is indispensable for the formulation of quantum fields (Chap. 8), its use also presents a number of advantages in condensed-matter physics. The general nature of the formalism provides immediate application to a large number of different physical situations. Once the particle type (photon, phonon, electron, ...) is defined, it becomes automatically possible to describe a collection of those particles of undetermined number. From the calculation point of view, the formalism naturally incorporates, with the aid of simple algebraic rules, the necessary but fastidious operations of symmetrization and anti-symmetrization. Even if the number of particles of the system is fixed, as will be seen in studies of superconductivity, of a nucleus, and of superfluidity, usage of this new formalism considerably facilitates the calculations.

3.2 Formalism of the Second Quantization

3.2.1 Fock Space

Given the space of single-particle states $\mathcal{H}$, the space $\mathcal{H}_\nu^{\otimes n}$ of n identical particles, properly symmetrized ($\nu = 1$ for bosons and $\nu = -1$ for fermions), was defined in Sect. 2.1.4. To construct the space of states of variable particle number, consider the collection of all $\mathcal{H}_\nu^{\otimes n}$, $n = 1, 2, \ldots$ and, in addition, the one-dimensional space $\mathcal{H}^0$ defined by

$$\mathcal{H}^0 = \{\lambda \, |\Phi(0)\rangle \, ; \lambda \in \mathbb{C}\} \tag{3.1}$$

comprised of one vector $|\Phi(0)\rangle$ assumed to be normalized to 1. The state $|\Phi(0)\rangle$ is called the *vacuum state*, and is denoted simply as $|\Phi(0)\rangle = |0\rangle$.

A state in which the number of particles is not fixed is given by a sequence

$$|\Phi\rangle = \{|\Phi(0)\rangle \, , |\Phi(1)\rangle \, , |\Phi(2)\rangle \, , \ldots \, , |\Phi(n)\rangle \, , \ldots\} = \{|\Phi(n)\rangle\}_n \quad , \tag{3.2}$$

where, for each n, $|\Phi(n)\rangle$ is a member of $\mathcal{H}_\nu^{\otimes n}$. The addition of two of these states $|\Phi\rangle = \{|\Phi(n)\rangle\}_n$ and $|\Psi\rangle = \{|\Psi(n)\rangle\}_n$ and their multiplication by a scalar λ are defined by $|\Phi\rangle + |\Psi\rangle = \{|\Phi(n)\rangle + |\Psi(n)\rangle\}_n$ and $\lambda \, |\Phi\rangle = \{\lambda \, |\Phi(n)\rangle\}_n$. The scalar product is given by

$$\langle \Phi | \Psi \rangle = \sum_{n=0}^{\infty} \langle \Phi(n) | \Psi(n) \rangle \quad , \tag{3.3}$$

where $\langle \Phi(n) | \Psi(n) \rangle$ is the scalar product in $\mathcal{H}_\nu^{\otimes n}$.

The collection of all vectors of the form (3.2) which are of finite norm

$$\langle \Phi | \Phi \rangle = \sum_{n=0}^{\infty} \langle \Phi(n) | \Phi(n) \rangle < \infty \tag{3.4}$$

forms a Hilbert space $\mathcal{F}_\nu(\mathcal{H})$ called a *Fock space*. Having defined a vector $|\Phi\rangle$ by the ensemble of its components in the $\mathcal{H}_\nu^{\otimes n}$ spaces, one can view the $\mathcal{H}_\nu^{\otimes n}$ as those subspaces in $\mathcal{F}_\nu(\mathcal{H})$ comprising the vectors $\{0, \ldots, 0, |\Phi(n)\rangle \, , 0, \ldots\}$ for which only the nth component is non-zero. These vectors, considered as elements of Fock space, will still be denoted simply as $|\Phi(n)\rangle$. Applying relation (3.3), their scalar product in Fock space vanishes for different particle numbers

$$\langle \Phi(n) | \Psi(m) \rangle = 0 \quad \text{for} \quad n \neq m \quad . \tag{3.5}$$

As a result, all pairs of these subspaces are orthogonal in $\mathcal{F}_\nu(\mathcal{H})$. One calls $\mathcal{F}_\nu(\mathcal{H})$ the direct sum of the $\mathcal{H}_\nu^{\otimes n}$ and $|\Phi\rangle$ is considered the sum $|\Phi\rangle = \sum_{n=0}^{\infty} |\Phi(n)\rangle$ of its orthogonal components $|\Phi(n)\rangle$. The notation $\mathcal{F}_\nu(\mathcal{H})$ emphasizes that when one chooses one particular $\mathcal{H}$ (determined by the physics of the problem), the associated Fock space is automatically given by the preceding construction.

Observables in Fock Space

If for each $n = 1, 2, \ldots$ there is a given observable $\mathtt{A}(n)$ of n particles, the corresponding *observable* $\mathtt{A}$ *acting on Fock space* is defined in a natural manner on each component of $|\Phi\rangle$

$$\mathtt{A}\,|\Phi\rangle = \sum_{n=0}^{\infty} \mathtt{A}(n)\,|\Phi(n)\rangle \quad . \tag{3.6}$$

The vacuum state is considered to be an eigenvector of all extensive observables with eigenvalue zero. That is, $\mathtt{A}(0) = 0$. In particular, the energy of the vacuum is zero,

$$\mathtt{H}\,|0\rangle = 0 \quad . \tag{3.7}$$

The same is true for the momentum, the angular momentum, and the total spin of the vacuum.

As the number of particles is undetermined, there exists a new observable in $\mathcal{F}_\nu(\mathcal{H})$, the *particle number operator*, defined by

$$\mathtt{N}\,|\Phi\rangle = \sum_{n=0}^{\infty} n\,|\Phi(n)\rangle \quad . \tag{3.8}$$

Clearly $|\Phi(n)\rangle$ is an eigenvector of $\mathtt{N}$ with eigenvalue n, and the $\mathcal{H}_\nu^{\otimes n}$ are the proper subspaces of $\mathtt{N}$. The mean number of particles in the state $|\Phi\rangle$ properly normalized ($\langle\Phi|\Phi\rangle = 1$), taking into account (3.3) and (3.8), is given by

$$\langle\Phi|\mathtt{N}|\Phi\rangle = \sum_{n=0}^{\infty} n\,\langle\Phi(n)|\Phi(n)\rangle \quad . \tag{3.9}$$

The square $\langle\Phi(n)|\Phi(n)\rangle$ of the norm of $|\Phi(n)\rangle$ is thus the probability to find n particles in the state $|\Phi\rangle$.

Interpretation of *the mean value of* $\mathtt{A}$ is clear,

$$\begin{aligned}\langle\Phi|\mathtt{A}|\Phi\rangle &= \sum_{n=0}^{\infty} \langle\Phi(n)|\mathtt{A}(n)|\Phi(n)\rangle \\ &= \sum_{n=0}^{\infty} \langle\Phi(n)|\Phi(n)\rangle\,\frac{\langle\Phi(n)|\mathtt{A}(n)|\Phi(n)\rangle}{\langle\Phi(n)|\Phi(n)\rangle}\end{aligned} \tag{3.10}$$

is the sum of the mean values of $\mathtt{A}(n)$ *for each system of* n *particles, multiplied by the probability* $\langle\Phi(n)|\Phi(n)\rangle$ *to have* n *particles.*

Basis of Fock Space

It is easy to construct a basis of $\mathcal{F}_\nu(\mathcal{H})$. Just as the product of one-particle states (2.36) generates $\mathcal{H}^{\otimes n}$, the collection of states $|\varphi_1, \dots, \varphi_n\rangle_\nu$ defined in (2.65), together with the vacuum state $|0\rangle$, generates $\mathcal{F}_\nu(\mathcal{H})$ when the $|\varphi_i\rangle$ are members of $\mathcal{H}$ and $n = 0, 1, 2, \dots$. As a result, to understand the action of an operator on $\mathcal{F}_\nu(\mathcal{H})$, it is sufficient to understand its action on $|\varphi_1, \dots, \varphi_n\rangle_\nu$ for all n.

In particular, one can construct a basis of $\mathcal{F}_\nu(\mathcal{H})$ starting from a given basis $\{|\psi_r\rangle\}$ of the one-particle space $\mathcal{H}$. To do this, one considers the set of vectors (2.70) or (2.71), but removes the restriction of fixed n in $\sum_r n_r = n$ by letting n be any non-negative integer, and imposing orthogonality if $\sum_r n_r \neq \sum_r n'_r$. The basis formed in this manner,

$$|n_1, n_2, \dots, n_r, \dots\rangle_\nu \quad , \quad \sum_r n_r < \infty \quad , \tag{3.11}$$

with $n_r = 0, 1, 2, \dots$ for bosons and $n_r = 0, 1$ for fermions, is the Fock space representation in occupation numbers of one-particle states $|\psi_r\rangle$.

3.2.2 Creation and Annihilation Operators

It is clear that the observables (3.6) leave the subspaces $\mathcal{H}_\nu^{\otimes n}$ invariant and commute with the particle-number operator. They cannot generate transitions between states of different particle number. To describe such transitions, it is necessary to introduce a new type of operator, the creation and annihilation operators. These operators play a fundamental role in the formalism of second quantization. They allow one to conveniently express interactions which do not conserve the number of particles, and their usage in many-body problems, as well as in field theory, allows one to considerably simplify many calculations.

The Creation Operator

To each one-particle state $|\varphi\rangle$ in $\mathcal{H}$, one associates the *creation operator* $a^*(\varphi)$ of one particle in the state $|\varphi\rangle$ defined in $\mathcal{F}_\nu(\mathcal{H})$ by

$$a^*(\varphi)\,|\varphi_1, \dots, \varphi_n\rangle_\nu = \sqrt{n+1}\,|\varphi, \varphi_1, \dots, \varphi_n\rangle_\nu \quad , \tag{3.12}$$

where $|\varphi_1, \dots, \varphi_n\rangle_\nu$ and $|\varphi, \varphi_1, \dots, \varphi_n\rangle_\nu$ are states (2.65) in $\mathcal{H}_\nu^{\otimes n}$ and $\mathcal{H}_\nu^{\otimes n+1}$ respectively. As the states (2.65) generate all of $\mathcal{F}_\nu(\mathcal{H})$, (3.12) defines $a^*(\varphi)$ by linearity on all $\mathcal{F}_\nu(\mathcal{H})$. *The interpretation of $a^*(\varphi)$ is clear: a particle in the individual state $|\varphi\rangle$ is added to the system of n particles without modifying their respective states.* One should also note that *after the action of $a^*(\varphi)$, the state of $n+1$ particles is again properly symmetrized.*

The Annihilation Operator

The *annihilation operator* $a(\varphi)$ of a particle in the state $|\varphi\rangle$ is defined as the Hermitian conjugate of the creation operator

$$a(\varphi) = (a^*(\varphi))^* \quad . \tag{3.13}$$

The notation $a^*(\varphi)$ is assigned conventionally to the creation operator rather than to the annihilation operator. This allows for some interesting analogies below. From definitions (3.12), (3.13), (2.37) and (2.38), it follows that $a^*(\varphi)$ *depends linearly on* $|\varphi\rangle$ *and that* $a(\varphi)$ *depends on* $|\varphi\rangle$ *in an anti-linear manner,*

$$\begin{aligned} a^*(\lambda_1\varphi_1 + \lambda_2\varphi_2) &= \lambda_1 a^*(\varphi_1) + \lambda_2 a^*(\varphi_2) \\ a(\lambda_1\varphi_1 + \lambda_2\varphi_2) &= \lambda_1^* a(\varphi_1) + \lambda_2^* a(\varphi_2) \quad . \end{aligned} \tag{3.14}$$

The action of $a(\varphi)$ on a state of n particles $|\varphi_1, \dots, \varphi_n\rangle_\nu$ is obtained by calculating the scalar product

$$\begin{aligned} &{}_\nu\langle \psi_1, \dots, \psi_m | a(\varphi) | \varphi_1, \dots, \varphi_n\rangle_\nu \\ &\quad = \sqrt{m+1}\, {}_\nu\langle \varphi, \psi_1, \dots, \psi_m | \varphi_1, \dots, \varphi_n\rangle_\nu \quad . \end{aligned} \tag{3.15}$$

Because of the orthogonality of the states of different particle numbers, this scalar product is zero for $m \neq n-1$. It is zero in particular for all m if $n = 0$. *The result is that the vacuum is an eigenvector of the* $a(\varphi)$ *with eigenvalue zero*, that is

$$a(\varphi)\,|0\rangle = 0 \tag{3.16}$$

for all $|\varphi\rangle$ *in* $\mathcal{H}$. If $m = n-1$ ($n \neq 0$), then one permutes φ with $\psi_1, \dots, \psi_{n-1}$ and φ is set equal to ψ_n. From (3.15) and (2.67) one finds

$$\begin{aligned} &{}_\nu\langle \psi_1, \dots, \psi_{n-1} | a(\varphi) | \varphi_1, \dots, \varphi_n\rangle_\nu \\ &\quad = \sqrt{n}(\nu)^{n-1}\, {}_\nu\langle \psi_1, \dots, \psi_n | \varphi_1, \dots, \varphi_n\rangle_\nu \\ &\quad = \frac{1}{\sqrt{n}} \frac{(\nu)^{n-1}}{(n-1)!} \sum_{\mathcal{P}_n} (\nu)^\pi \langle \psi_1 | \varphi_{\pi(1)}\rangle \dots \langle \psi_n | \varphi_{\pi(n)}\rangle \\ &\quad = \frac{1}{\sqrt{n}} \sum_{i=1}^{n} (\nu)^{i-1} \langle \psi_n | \varphi_i\rangle \\ &\qquad \times \frac{1}{(n-1)!} \sum_{\mathcal{P}_{n-1}} (\nu)^\pi \langle \psi_1 | \varphi_{\pi(1)}\rangle \dots \langle \psi_{n-1} | \varphi_{\pi(n-1)}\rangle \quad . \end{aligned} \tag{3.17}$$

To obtain (3.17), one notes that summing over permutations $\mathcal{P}_n$ is equivalent to summing over the $(n-1)!$ permutations which give $\pi(n) = i$ and then over $i = 1, \dots, n$. Therefore, in the last equality of (3.17), $\{\pi(1), \dots, \pi(n-1)\}$

represents a permutation of $\{1, \ldots, i-1, i+1, \ldots, n\}$. Taking into account (2.67), one thus obtains

$$\begin{aligned}
&{}_\nu\langle\psi_1, \ldots, \psi_{n-1}|a(\varphi)|\varphi_1, \ldots, \varphi_n\rangle_\nu \\
&\quad = \frac{1}{\sqrt{n}} \sum_{i=1}^{n} (\nu)^{i-1} \\
&\qquad \times {}_\nu\langle\psi_1, \ldots, \psi_{n-1}|\varphi_1, \ldots, \varphi_{i-1}, \varphi_{i+1}, \ldots, \varphi_n\rangle_\nu \langle\varphi|\varphi_i\rangle \quad . \qquad (3.18)
\end{aligned}$$

As this last relation holds for any state $|\psi_1, \ldots, \psi_{n-1}\rangle$, it must be true that

$$\begin{aligned}
&a(\varphi)\,|\varphi_1, \ldots, \varphi_n\rangle_\nu \\
&\quad = \frac{1}{\sqrt{n}} \sum_{i=1}^{n} (\nu)^{i-1}\, |\varphi_1, \ldots, \varphi_{i-1}, \varphi_{i+1}, \ldots, \varphi_n\rangle_\nu \langle\varphi|\varphi_i\rangle \quad . \qquad (3.19)
\end{aligned}$$

The annihilation operator $a(\varphi)$ decreases the number of particles by one unit, while preserving the symmetry of the state.

Commutation and Anticommutation Relations

Fundamental algebraic relations can now be determined between the creation and annihilation operators. Taking into account (2.66), one sees that

$$\begin{aligned}
a^*(\varphi)a^*(\psi)\,|\varphi_1, \ldots, \varphi_n\rangle_\nu &= \sqrt{(n+2)(n+1)}\,|\varphi, \psi, \varphi_1, \ldots, \varphi_n\rangle_\nu \\
&= \nu\sqrt{(n+2)(n+1)}\,|\psi, \varphi, \varphi_1, \ldots, \varphi_n\rangle_\nu \\
&= \nu a^*(\psi)a^*(\varphi)\,|\varphi_1, \ldots, \varphi_n\rangle_\nu \quad , \qquad (3.20)
\end{aligned}$$

from which it may be concluded that *the creation operators for bosons commute for all $|\varphi\rangle$ and $|\psi\rangle$ in $\mathcal{H}$*

$$a^*(\varphi)a^*(\psi) - a^*(\psi)a^*(\varphi) = 0 \qquad \text{(bosons)} \qquad (3.21)$$

and anti-commute for the fermions

$$a^*(\varphi)a^*(\psi) + a^*(\psi)a^*(\varphi) = 0 \qquad \text{(fermions)} \quad . \qquad (3.22)$$

Turning to the Hermitian conjugate relations, one establishes the same for *the annihilation operators*, which *commute for the bosons and anti-commute for the fermions,*

$$a(\varphi)a(\psi) - a(\psi)a(\varphi) = 0 \qquad \text{(bosons)} \qquad (3.23)$$

and

$$a(\varphi)a(\psi) + a(\psi)a(\varphi) = 0 \qquad \text{(fermions)} \quad . \qquad (3.24)$$

From (3.19) and (3.12), one obtains

$$\begin{aligned}
a^*(\varphi)a(\psi)\left|\varphi_1,\dots,\varphi_n\right\rangle_\nu
&= \sum_{i=1}^{n} (\nu)^{i-1} \left|\varphi,\varphi_1,\dots,\varphi_{i-1},\varphi_{i+1},\dots,\varphi_n\right\rangle_\nu \langle\psi|\varphi_i\rangle \\
&= \sum_{i=1}^{n} \left|\varphi_1,\dots,\varphi_{i-1},\varphi,\varphi_{i+1},\dots,\varphi_n\right\rangle_\nu \langle\psi|\varphi_i\rangle
\end{aligned} \tag{3.25}$$

and

$$\begin{aligned}
a(\psi)a^*(\varphi)\left|\varphi_1,\dots,\varphi_n\right\rangle_\nu
&= \sqrt{n+1}\,a(\psi)\left|\varphi,\varphi_1,\dots,\varphi_n\right\rangle_\nu \\
&= \left|\varphi_1,\dots,\varphi_n\right\rangle_\nu \langle\psi|\varphi\rangle \\
&\quad + \nu \sum_{i=1}^{n} \left|\varphi_1,\dots,\varphi_{i-1},\varphi,\varphi_{i+1},\dots,\varphi_n\right\rangle_\nu \langle\psi|\varphi_i\rangle \quad .
\end{aligned} \tag{3.26}$$

Comparing expressions (3.25) and (3.26), which must be true for all states $|\varphi_1,\dots,\varphi_n\rangle$, one finds that for bosons

$$a(\psi)a^*(\varphi) - a^*(\varphi)a(\psi) = \langle\psi|\varphi\rangle\,\mathtt{I} \qquad \text{(bosons)} \tag{3.27}$$

and for fermions

$$a(\psi)a^*(\varphi) + a^*(\varphi)a(\psi) = \langle\psi|\varphi\rangle\,\mathtt{I} \qquad \text{(fermions)} \quad . \tag{3.28}$$

In (3.27) and (3.28), $\mathtt{I}$ is the identity operator in $\mathcal{F}_\nu(\mathcal{H})$ which, from now on, will not be explicitly written. *Hence, the commutator or anti-commutator of a creation and an annihilation operator is a multiple of the identity operator in $\mathcal{F}_\nu(\mathcal{H})$, with the prefactor being the scalar product $\langle\psi|\varphi\rangle$ in the space of one-particle states.* These relations completely define the algebra of the creation and annihilation operators. They make it possible to uniquely express the states and the observables in $\mathcal{F}_\nu(\mathcal{H})$ with the help of these operators, automatically taking into account the statistics. In the following sections, most of the calculations can be reduced to repeated applications of rules (3.21), (3.23) and (3.27) for the bosons and (3.22), (3.24) and (3.28) for the fermions.

Creation and Annihilation Operators Associated to the States of a Single-Particle Basis

In the case when one chooses a fixed basis $\{|\psi_r\rangle\}$ in one-particle space, one simply writes $a_r^* = a^*(\psi_r)$ and $a_r = a(\psi_r)$. The preceding commutation or anticommutation relations become

$$a_r^* a_s^* \mp a_s^* a_r^* = 0 \quad , \tag{3.29}$$

$$a_r a_s \mp a_s a_r = 0 \tag{3.30}$$

and

$$a_r a_s^* \mp a_s^* a_r = \delta_{r,s} \tag{3.31}$$

with a −-sign for bosons or a +-sign for fermions.

In the position basis $|\boldsymbol{k}\sigma\rangle$, for example, one writes

$$a_{\boldsymbol{k}\sigma}^* = a^*(|\boldsymbol{k}\sigma\rangle) \tag{3.32}$$

for the creation operators of a particle of momentum $\hbar\boldsymbol{k}$ and spin $\hbar\sigma$, giving

$$a_{\boldsymbol{k}\sigma} a_{\boldsymbol{k}'\sigma'}^* \mp a_{\boldsymbol{k}'\sigma'}^* a_{\boldsymbol{k}\sigma} = \delta_{\boldsymbol{k},\boldsymbol{k}'} \delta_{\sigma,\sigma'} \quad . \tag{3.33}$$

In the position basis $|\boldsymbol{x}\sigma\rangle$, one writes

$$a^*(\boldsymbol{x},\sigma) = a^*(|\boldsymbol{x}\sigma\rangle) \tag{3.34}$$

for the creation operators of a particle of spin $\hbar\sigma$ at point $\boldsymbol{x}$, with the rules of commutation or anticommutation

$$a(\boldsymbol{x},\sigma)a^*(\boldsymbol{x}',\sigma') \mp a^*(\boldsymbol{x}',\sigma')a(\boldsymbol{x},\sigma) = \delta_{\sigma,\sigma'}\delta(\boldsymbol{x}-\boldsymbol{x}') \quad . \tag{3.35}$$

Expressing a one-particle state in either of the bases leads to the linear relations

$$\begin{aligned} a^*(\varphi) &= \sum_{\boldsymbol{k}\sigma} \langle \boldsymbol{k}\sigma|\varphi\rangle \, a_{\boldsymbol{k}\sigma}^* \quad , \\ a^*(\varphi) &= \int_{\Lambda} \mathrm{d}\boldsymbol{x} \sum_{\sigma} \langle \boldsymbol{x}\sigma|\varphi\rangle \, a^*(\boldsymbol{x},\sigma) \end{aligned} \tag{3.36}$$

and the corresponding Hermitian conjugate relations for $a(\varphi)$.

It is still necessary to remark that the simplicity of the algebraic relations comes from the judicious choice of the factor $\sqrt{n+1}$ in definition (3.12) of the creation operator. The choice is conventional, but completely adequate. In fact, it is possible to verify that it is the only choice which assures that the commutators (3.27) or anticommutators (3.28) be multiples of the identity operator. Comparing (3.27) with (1.8), one sees that for bosons $a(\varphi)$ and $a^*(\varphi)$ have the same commutation rules as those of the harmonic oscillator. This analogy is profound and will be discussed again in relation with quantum fields.

3.2.3 States of the Fock Space

Formation of States from the Vacuum

All states of n independent particles can be obtained by successive application of the creation operators on the vacuum. From definition (3.12), one can immediately see that

$$|\varphi_1, \dots, \varphi_n\rangle_\nu = \frac{1}{\sqrt{n}} a^*(\varphi_1) \dots a^*(\varphi_n) |0\rangle \quad . \tag{3.37}$$

This state is clearly symmetric or antisymmetric depending on whether the $a^*(\varphi_i)$ commute or anti-commute. All of the states in Fock space can thus be obtained by linear combinations of the states (3.37).

If $|n_1, \dots, n_r, \dots\rangle_\nu$ are the occupation-number states relative to the single-particle basis $\{|\psi_r\rangle\}$ and $a_r^* = a^*(|\psi_r\rangle)$, one sees from (2.70), (2.71) and (3.37) that

$$|n_1, \dots, n_r, \dots\rangle_\nu = \frac{(a_1^*)^{n_1}}{\sqrt{n_1!}} \dots \frac{(a_r^*)^{n_r}}{\sqrt{n_r!}} \dots |0\rangle \quad , \quad \begin{cases} n_r = 0, 1, 2, \dots & \text{(bosons)} \\ n_r = 0, 1 & \text{(fermions)} \end{cases} \quad . \tag{3.38}$$

One can deduce the actions of a_r^* and a_r. For the bosons,

$$a_r^* |n_1, \dots, n_r, \dots\rangle_+ = \sqrt{n_r + 1} |n_1, \dots, n_r + 1, \dots\rangle_+ \tag{3.39}$$

and

$$a_r |n_1, \dots, n_r, \dots\rangle_+ = \sqrt{n_r} |n_1, \dots, n_r - 1, \dots\rangle_+ \quad . \tag{3.40}$$

As $[a_r^*, a_s^*] = 0$, (3.39) follows immediately from (3.38). Repeated application of the identity $[A, BC] = [A, B]C + B[A, C]$ with the commutation rule $[a_r, a_r^*] = 1$ gives

$$a_r(a_r^*)^{n_r} = n_r(a_r^*)^{n_r - 1} + (a_r^*)^{n_r} a_r \quad . \tag{3.41}$$

It follows that, as $[a_r, a_s^*] = 0$ for $r \neq s$,

$$a_r |n_1, \dots, n_r, \dots\rangle_+ = n_r \frac{(a_1^*)^{n_1}}{\sqrt{n_1!}} \dots \frac{(a_r^*)^{n_r - 1}}{\sqrt{n_r!}} \dots |0\rangle + \frac{(a_1^*)^{n_1}}{\sqrt{n_1!}} \dots \frac{(a_r^*)^{n_r}}{\sqrt{n_r!}} \dots a_r |0\rangle \quad , \tag{3.42}$$

which gives (3.40) since $a_r |0\rangle = 0$. For the fermions, one obtains (with $a_r^2 = 0$ and $a_r a_s^* = -a_s^* a_r$ for $r \neq s$),

$$a_r^* |n_1, \dots, n_r, \dots\rangle_- = (1 - n_r)(-1)^{\gamma_r} |n_1, \dots, n_r + 1, \dots\rangle_- \tag{3.43}$$

and

$$a_r |n_1, \dots, n_r, \dots\rangle_- = n_r(-1)^{\gamma_r} |n_1, \dots, n_r - 1, \dots\rangle_- \quad , \tag{3.44}$$

where $\gamma_r = \sum_{s=1}^{r-1} n_s$. Finally, for both bosons and fermions, one sees that the occupation number states are eigenstates for all of the operators $a_r^* a_r$,

$$a_r^* a_r |n_1, \dots, n_r, \dots\rangle_\nu = n_r |n_1, \dots, n_r, \dots\rangle_\nu \quad , \tag{3.45}$$

with eigenvalues n_r.

States of Several Types of Particles

Representing states by application of the creation operators on the vacuum is easily generalized to the case where there are several types of particles present. If there are two types of particles, a and b, of statistics ν_a and ν_b, with single-particle spaces $\mathcal{H}^a$ and $\mathcal{H}^b$, the space of states of variable numbers of these particles is given by the tensor product $\mathcal{F}_{\nu_a}(\mathcal{H}^a) \otimes \mathcal{F}_{\nu_b}(\mathcal{H}^b)$. In this space, there is a unique vacuum state $|\Phi^a(0)\rangle \otimes |\Phi^b(0)\rangle$ which will be denoted $|0\rangle$. It is clear that this common vacuum state satisfies

$$a(\varphi^a)\,|0\rangle = a(\varphi^b)\,|0\rangle = 0 \tag{3.46}$$

under the action of the annihilators of both types of particles. One obtains n type a and m type b particle states by applying the same formula as (3.37)

$$\begin{aligned}|\varphi_1^a, \dots, \varphi_n^a, \varphi_1^b, \dots, \varphi_m^b\rangle \\ = \frac{1}{\sqrt{n!m!}} a^*(\varphi_1^a) \dots a^*(\varphi_n^a) a^*(\varphi_1^b) \dots a^*(\varphi_m^b)\,|0\rangle \quad .\end{aligned} \tag{3.47}$$

These states automatically have the correct symmetry properties according to the statistics of the two types of particles (assuming the operators belonging to the different particle types commute). Thus *one can conveniently construct the states of several types of particles of variable number: it is sufficient to give the vacuum state and the creators and annihilators with the commutation (anticommutation) rules appropriate for the statistics of the particles.*

Coherent States of Bosons

States of type (3.37) have a determined number of particles. That is, they fit the description of situations in which there is a known number of particles specified by their individual states. This is typical for the case of a scattering process in which the incoming particles are specified and the particles produced by the reaction are observed. There is another class of states which plays an important role in the analysis of a boson field, such as photons or phonons: the coherent states. These states consist of a superposition of n-particle states with $n = 0, 1, 2, \dots$ and can be defined in a completely analogous manner to those of the harmonic oscillator (Sect. 1.2.2). The following considerations apply to bosons. Generalizing (1.54),

$$|\varphi\rangle_{\text{coh}} = \exp\left(-\frac{1}{2}\,\langle\varphi|\varphi\rangle\right) \sum_{n=0}^{\infty} \frac{[a^*(\varphi)]^n}{n!}\,|0\rangle = \mathtt{D}(\varphi)\,|0\rangle \quad , \tag{3.48}$$

where

$$\begin{aligned}\mathtt{D}(\varphi) &= \exp\left[a^*(\varphi) - a(\varphi)\right] \\ &= \exp\left(-\frac{1}{2}\langle\varphi|\varphi\rangle\right)\exp\left[a^*(\varphi)\right]\exp\left[-a(\varphi)\right] \quad .\end{aligned} \tag{3.49}$$

The state $|\varphi\rangle_{\mathrm{coh}}$*, which must not be confused with a single-particle state, represents a superposition of n-particle states* ($n = 0, 1, 2, \dots$)*, with all the particles occupying the same individual state* $|\varphi\rangle$. It will be shown in Sect. 8.2.5 that a coherent state represents a state of photons (or phonons) associated with a classical electromagnetic (or acoustic) wave.

The coherent states (3.48) have the same properties as those of the harmonic oscillator, the most important being that they are eigenvectors of all of the annihilation operators (see (1.35)). From the commutation rule (3.27) for bosons,

$$[a(\psi), a^*(\varphi)^n] = n\,\langle\psi|\varphi\rangle\,[a^*(\varphi)]^{n-1} \quad . \tag{3.50}$$

Since $a(\psi)\,|0\rangle = 0$, one has

$$\begin{aligned}a(\psi)\,|\varphi\rangle_{\mathrm{coh}} &= \exp\left(-\frac{1}{2}\langle\varphi|\varphi\rangle\right)\sum_{n=0}^{\infty}\langle\psi|\varphi\rangle\,n\frac{(a^*(\varphi))^{n-1}}{n!}\,|0\rangle \\ &= (\langle\psi|\varphi\rangle)\,|\varphi\rangle_{\mathrm{coh}} \quad .\end{aligned} \tag{3.51}$$

In particular, if a_k is a plane-wave annihilator, then

$$a_{\boldsymbol{k}}\,|\varphi\rangle_{\mathrm{coh}} = \tilde{\varphi}(\boldsymbol{k})\,|\varphi\rangle_{\mathrm{coh}} \quad , \tag{3.52}$$

where $\tilde{\varphi}(\boldsymbol{k}) = \langle\boldsymbol{k}|\varphi\rangle$ is the Fourier amplitude of the single-particle state $|\varphi\rangle$.

In a coherent state, the amplitudes $\tilde{\varphi}(\boldsymbol{k})$ are well determined. This is true since, for all $\boldsymbol{k}$, the coherent state is an eigenvector of $a_{\boldsymbol{k}}$ with eigenvalue $\tilde{\varphi}(\boldsymbol{k})$. On the other hand, the number of bosons is not fixed, a fact easily verified by looking at (3.48). The existence in Fock space of fixed particle-number states (3.37) and fixed amplitude and phase states (3.48) is a manifestation of the wave–particle duality apparent in the second quantization formalism. Chapter 5, which treats the equivalent concept for fermions – and, more particularly, superconductivity – will offer another aspect of this duality.

3.2.4 Normal Order

As shown above, a state $|\Phi\rangle$ of Fock space can be represented by the action of certain combinations of creation operators on the vacuum. In the following it will be shown that an observable $\mathtt{A}$ can also be expressed in terms of the creators and annihilators. Consequently, the calculation of a mean value $\langle\Phi|\mathtt{A}|\Phi\rangle$ can always be reduced to the vacuum expectation value of polynomials of

$a(\varphi)$ and $a^*(\varphi)$. This calculation is facilitated by putting such polynomials in a normal order with the help of a rule, called Wick's theorem. This theorem is a generalization of the rule given in Sect. 1.2.4 on two points: it is equally valid for fermions and bosons and it is applicable to operators $a(\varphi)$ and $a^*(\psi)$ relative to different states $|\varphi\rangle$ and $|\psi\rangle$.

Normal Product

Consider a family of operators $\mathtt{A}_j$ which are creators or annihilators, or linear combinations

$$\mathtt{A}_j = a^*(\varphi_j) + a(\psi_j) \tag{3.53}$$

of creators and annihilators relative to any given single-particle states $|\varphi\rangle$ and $|\psi\rangle$. The *normal product* $:\mathtt{A}_1\mathtt{A}_2 \dots \mathtt{A}_n:$ is the product or sum of products obtained by putting all of the creators to the left of the annihilators. In the case of fermions, the products obtained must still be multiplied by the signature $(-1)^\pi$ of the permutation required by the rearrangement. For example, when $\mathtt{A}_1$ and $\mathtt{A}_2$ are of the form (3.53),

$$\begin{aligned} :\mathtt{A}_1\mathtt{A}_2: &= :(a^*(\varphi_1) + a(\psi_1))\,(a^*(\varphi_2) + a(\psi_2)): \\ &= a^*(\varphi_1)a^*(\varphi_2) + a(\psi_1)a(\psi_2) + a^*(\varphi_1)a(\psi_2) + \nu a^*(\varphi_2)a(\psi_1) \quad , \end{aligned} \tag{3.54}$$

where $\nu = 1$ for bosons and $\nu = -1$ for fermions. The rearrangement of the fermion operators is thus made as if they all anti-commute. It follows from the definition of the normal product that, for any permutation π of fermion operators

$$:\mathtt{A}_1\mathtt{A}_2 \dots \mathtt{A}_n: = (-)^\pi :\mathtt{A}_{\pi(1)}\mathtt{A}_{\pi(2)} \dots \mathtt{A}_{\pi(n)}: \quad . \tag{3.55}$$

Contractions

A *contraction* $\underbracket{\mathtt{A}_1\mathtt{A}_2}$ is defined to be the mean value of $\mathtt{A}_1\mathtt{A}_2$ in the vacuum

$$\underbracket{\mathtt{A}_1\mathtt{A}_2} = \langle 0|\mathtt{A}_1\mathtt{A}_2|0\rangle \tag{3.56}$$

and the normal product with contractions is defined by the relation

$$\begin{aligned} &:\mathtt{A}_1 \dots \mathtt{A}_{i-1}\underbracket{\mathtt{A}_i\mathtt{A}_{i+1} \dots \mathtt{A}_{j-1}\mathtt{A}_j}\mathtt{A}_{j+1} \dots: \\ &\quad = (\nu)^\pi \underbracket{\mathtt{A}_i\mathtt{A}_j} :\mathtt{A}_1 \dots \mathtt{A}_{i-1}\mathtt{A}_{i+1} \dots \mathtt{A}_{j-1}\mathtt{A}_{j+1} \dots: \quad , \end{aligned} \tag{3.57}$$

where, for fermions, $(-1)^\pi$ is the signature of the permutation which takes $(1, \dots, i, \dots, j, \dots, n)$ to $(i, j, 1, \dots, n)$. The normal product with contractions is, up to the sign, the normal product of the non-contracted operators multiplied by the contractions that have been carried out. *Wick's theorem* states that a *product of operators of type (3.53) is equal to the sum of the normal products obtained by carrying out all possible contractions*

$$\mathtt{A}_1 \dots \mathtt{A}_n = {:}\mathtt{A}_1 \dots \mathtt{A}_n{:} + {:}\underbracket{\mathtt{A}_1 \mathtt{A}_2} \dots \mathtt{A}_n{:} + \dots + {:}\underbracket{\mathtt{A}_1 \mathtt{A}_2 \dots \mathtt{A}_n}{:} + \dots$$
$$+ {:}\underbracket{\mathtt{A}_1 \mathtt{A}_2}\underbracket{\mathtt{A}_3 \mathtt{A}_4} \dots \mathtt{A}_n{:} + \dots \quad . \tag{3.58}$$

This theorem can be verified by taking the simple example

$$\mathtt{A}_1 \mathtt{A}_2 = {:}\mathtt{A}_1 \mathtt{A}_2{:} + \underbracket{\mathtt{A}_1 \mathtt{A}_2} \quad . \tag{3.59}$$

By linearity, it is sufficient to separately study the case where $\mathtt{A}_1$ and $\mathtt{A}_2$ are creation or annihilation operators. If $\mathtt{A}_1$ is a creator or if $\mathtt{A}_1$ and $\mathtt{A}_2$ are both annihilators, it follows from the definition of normal order and from (3.56) that $\mathtt{A}_1 \mathtt{A}_2 = {:}\mathtt{A}_1 \mathtt{A}_2{:}$ and $\underbracket{\mathtt{A}_1 \mathtt{A}_2} = 0$, so that (3.59) is verified. If $\mathtt{A}_1 = a(\psi)$ and $\mathtt{A}_2 = a^*(\varphi)$, (3.27) or (3.28) give

$$\mathtt{A}_1 \mathtt{A}_2 - {:}\mathtt{A}_1 \mathtt{A}_2{:} = a(\psi) a^*(\varphi) - \nu a^*(\varphi) a(\psi) = \langle \psi | \varphi \rangle \tag{3.60}$$

and, since $a(\psi) |0\rangle = 0$, it follows that

$$\underbracket{\mathtt{A}_1 \mathtt{A}_2} = \langle 0 | a(\psi) a^*(\varphi) | 0 \rangle = \langle \psi | \varphi \rangle \quad , \tag{3.61}$$

which again leads to (3.59).

As $\langle 0 | {:}\mathtt{A}_1 \dots \mathtt{A}_n{:} | 0 \rangle = 0$ for all cases, (3.58) gives the following result: *the mean value of* $\mathtt{A}_1 \dots \mathtt{A}_n$ *in the vacuum is equal to the sum of the terms where all pairs* $\mathtt{A}_i \mathtt{A}_j$ *are contracted*. As an example, taking (3.61) and sign rule (3.57),

$$\begin{aligned} &\langle 0 | a(\psi_1) a(\psi_2) a^*(\varphi_2) a^*(\varphi_1) | 0 \rangle \\ &= \underbracket{a(\psi_1)\underbracket{a(\psi_2) a^*(\varphi_2)} a^*(\varphi_1)} + a(\psi_1) a(\psi_2) a^*(\varphi_2) a^*(\varphi_1) \\ &= \langle \psi_1 | \varphi_1 \rangle \langle \psi_2 | \varphi_2 \rangle + \nu \langle \psi_1 | \varphi_2 \rangle \langle \psi_2 | \varphi_1 \rangle \quad . \end{aligned} \tag{3.62}$$

In (3.62) contractions which are manifestly equal to zero are omitted. *For fermions, the sign of a scheme of contractions is determined by the number of crossings of lines joining the pairs. The sign is positive (negative) if the number is even (odd).*

3.2.5 One-Body Operators

Observables in Fock space can be expressed in terms of the creation and annihilation operators. Consider a single-particle observable A and a one-body operator $\mathtt{A}(n) = \sum_{j=1}^{n} A_j$ associated to a system of n particles, defined by (2.42). As $\mathtt{A}(n)$ commutes with the permutation operators, the action of the corresponding observable $\mathtt{A}$ on states (2.65) of Fock space is

$$\mathtt{A}\,|\varphi_1,\dots,\varphi_n\rangle_\nu = \sum_{j=1}^{n} |\varphi_1,\dots,\varphi_{j-1},A\varphi_j,\varphi_{j+1},\dots,\varphi_n\rangle_\nu \quad . \tag{3.63}$$

It is possible to show that A can be written in terms of the creation and annihilation operators as

$$\mathtt{A} = \sum_r a^*(A\psi_r)a(\psi_r) \quad , \tag{3.64}$$

where $\{|\psi_r\rangle\}$ is some basis of single-particle space. From (3.25), the fact that the basis $\{|\psi_r\rangle\}$ is complete, and linearity, one obtains

$$\begin{aligned}\mathtt{A}\,|\varphi_1,\dots,\varphi_n\rangle_\nu &= \sum_r a^*(A\psi_r)a(\psi_r)\,|\varphi_1,\dots,\varphi_n\rangle_\nu \\ &= \sum_r\sum_{j=1}^{n} |\varphi_1,\dots,\varphi_{j-1},A\psi_r,\varphi_{j+1},\dots,\varphi_n\rangle_\nu\,\langle\psi_r|\varphi_j\rangle \\ &= \sum_{j=1}^{n} |\varphi_1,\dots,\varphi_{j-1},A\varphi_j,\varphi_{j+1},\dots,\varphi_n\rangle_\nu \quad ,\end{aligned} \tag{3.65}$$

which is precisely (3.63). This calculation shows that A is independent of the basis of one-particle space used in definition (3.64).

Introducing the matrix elements of A in the basis $\{|\psi_r\rangle\}$, that is $\langle r|A|s\rangle = \langle\psi_r|A|\psi_s\rangle$, and taking advantage of the linearity of $a^*(\varphi)$, (3.64) can be written as

$$\mathtt{A} = \sum_{rs} \langle r|A|s\rangle\, a_r^* a_s \quad , \tag{3.66}$$

with $a_r^* = a^*(\psi_r)$. If the $|\psi_r\rangle$ are chosen as eigenvectors of A, (3.66) simplifies to

$$\mathtt{A} = \sum_r \alpha_r\, a_r^* a_r \quad , \tag{3.67}$$

where the α_r are the eigenvalues of A. It follows from (3.45) that the occupation number states in the basis $\{|\psi_r\rangle\}$ are eigenvectors of A with eigenvalues $\sum_r \alpha_r n_r$.

Particle-Number Operator

As a first example, take $A = I$, the identity operator in one-particle space. The corresponding second quantization observable is the particle-number operator given by

$$\mathtt{N} = \sum_r a^*(\psi_r)a(\psi_r) \quad , \tag{3.68}$$

since, from (3.63), $\mathtt{N}$ has exactly the eigenvalue n for all states of n particles.

As the choice of the basis in (3.68) is arbitrary, one can write $\mathtt{N}$ in terms of the position basis $|\boldsymbol{x}\sigma\rangle$

$$\mathtt{N} = \int_\Lambda \mathrm{d}\boldsymbol{x} \sum_\sigma a^*(\boldsymbol{x},\sigma)a(\boldsymbol{x},\sigma) = \int_\Lambda \mathrm{d}\boldsymbol{x} n(\boldsymbol{x}) \quad , \tag{3.69}$$

where

$$n(\boldsymbol{x}) = \sum_\sigma n(\boldsymbol{x},\sigma) \quad ,$$
$$n(\boldsymbol{x},\sigma) = a^*(\boldsymbol{x},\sigma)a(\boldsymbol{x},\sigma) \quad . \tag{3.70}$$

The operator $n(\boldsymbol{x})$ thus represents the *particle density* at the point $\boldsymbol{x}$ and $n(\boldsymbol{x},\sigma)$ is the density of particles with spin σ. According to (3.66), $n(\boldsymbol{x},\sigma)$ can be considered the one-body operator which corresponds to the (singular) single-particle operator $A = \delta_{\boldsymbol{x},\sigma}(\boldsymbol{x}_1,\sigma_1)$ of multiplication by the Dirac function

$$\delta_{\boldsymbol{x},\sigma}(\boldsymbol{x}_1,\sigma_1)\varphi(\boldsymbol{x}_1,\sigma_1) = \delta(\boldsymbol{x}-\boldsymbol{x}_1)\delta_{\sigma,\sigma_1}\varphi(\boldsymbol{x}_1,\sigma_1) \quad . \tag{3.71}$$

In general, $a^*(\varphi)a(\varphi)$ is the number of particles in the state $|\varphi\rangle$.

Kinetic Energy

Other important examples of one-body observables, include the Hamiltonian and the momentum of a collection of free particles. According to (3.67), these are written as

$$\mathtt{H}_0 = \sum_{\boldsymbol{k}\sigma} \varepsilon_{\boldsymbol{k}} a^*_{\boldsymbol{k}\sigma} a_{\boldsymbol{k}\sigma} \tag{3.72}$$

and

$$\mathtt{P} = \sum_{\boldsymbol{k}\sigma} \hbar\boldsymbol{k} a^*_{\boldsymbol{k}\sigma} a_{\boldsymbol{k}\sigma} \quad , \tag{3.73}$$

where $\varepsilon_{\boldsymbol{k}}$ is the kinetic energy of the particle.

External Potential

Consider a system of particles under an external potential $V(\boldsymbol{x})$ which is independent of spin. As the matrix elements of V in the basis $|\boldsymbol{x}\sigma\rangle$ are

$$\langle\boldsymbol{x}\sigma|V|\boldsymbol{x}'\sigma'\rangle = \delta(\boldsymbol{x}-\boldsymbol{x}')\delta_{\sigma,\sigma'}V(\boldsymbol{x}) \quad , \tag{3.74}$$

(3.66) yields a second quantized representation of the external potential V

$$\mathsf{V} = \int_\Lambda \mathrm{d}\boldsymbol{x} V(\boldsymbol{x}) n(\boldsymbol{x}) \quad , \tag{3.75}$$

where $n(\boldsymbol{x})$ is the particle density (3.70). *Formula (3.75) is similar to the classical expression, except that $n(\boldsymbol{x})$ is the quantum density operator.*

In the basis of plane waves,

$$\langle \boldsymbol{k}\sigma | V | \boldsymbol{k}'\sigma' \rangle = \delta_{\sigma,\sigma'} \frac{1}{L^3} \tilde{V}(\boldsymbol{k} - \boldsymbol{k}') \quad , \tag{3.76}$$

where

$$\tilde{V}(\boldsymbol{k}) = \int_\Lambda \mathrm{d}\boldsymbol{x} \mathrm{e}^{-\mathrm{i}\boldsymbol{k}\cdot\boldsymbol{x}} V(\boldsymbol{x}) \quad . \tag{3.77}$$

From this, one gets

$$\begin{aligned} \mathsf{V} &= \frac{1}{L^3} \sum_{\boldsymbol{k}\boldsymbol{k}'} \sum_\sigma \tilde{V}(\boldsymbol{k} - \boldsymbol{k}') a^*_{\boldsymbol{k}\sigma} a_{\boldsymbol{k}'\sigma} = \frac{1}{L^3} \sum_{\boldsymbol{k}\boldsymbol{k}'} \sum_\sigma \tilde{V}(\boldsymbol{k}) a^*_{(\boldsymbol{k}'+\boldsymbol{k})\sigma} a_{\boldsymbol{k}'\sigma} \\ &= \frac{1}{L^3} \sum_{\boldsymbol{k}} \tilde{n}^*(\boldsymbol{k}) \tilde{V}(\boldsymbol{k}) \quad . \end{aligned} \tag{3.78}$$

In (3.78), the Fourier transform of the density operator

$$\begin{aligned} \tilde{n}(\boldsymbol{k}) &= \int_\Lambda \mathrm{d}\boldsymbol{x} \mathrm{e}^{-\mathrm{i}\boldsymbol{k}\cdot\boldsymbol{x}} \sum_\sigma a^*(\boldsymbol{x},\sigma) a(\boldsymbol{x},\sigma) = \sum_{\boldsymbol{k}'\sigma} a^*_{\boldsymbol{k}'\sigma} a_{(\boldsymbol{k}'+\boldsymbol{k})\sigma} \\ &= \sum_{\boldsymbol{k}'\sigma} a^*_{(\boldsymbol{k}'-\boldsymbol{k})\sigma} a_{\boldsymbol{k}'\sigma} = \tilde{n}^*(-\boldsymbol{k}) \end{aligned} \tag{3.79}$$

is introduced. Equation (3.79) is obtained with the help of the orthogonality of the plane waves and the transformation formulae (3.36). Relation (3.78) permits a graphical representation suggesting the type of process associated with this interaction (Fig. 3.1). *A particle in momentum state $\hbar\boldsymbol{k}'$ is "annihilated", then scattered by the potential $\tilde{V}(\boldsymbol{k} - \boldsymbol{k}')$ with a momentum transfer of $\hbar(\boldsymbol{k} - \boldsymbol{k}')$ and then finally "recreated" in momentum state $\hbar\boldsymbol{k}$.* An inhomogeneous external field destroys invariance under translations in space and hence conservation of momentum.

3.2.6 Free Evolution and Symmetries

Symmetries are known to be represented in one-particle space $\mathcal{H}$ by unitary groups of transformations in this space. If H is the Hamiltonian of one particle (assuming that external fields are time-independent), time translations are given by the group of operators $U(t) = \exp(-\mathrm{i}Ht/\hbar)$ in which $H/\hbar$ is the infinitesimal generator. The situation is similar for translations and rotations in space, where the generators are the momentum and total angular momentum, respectively. These symmetry operations can be similarly constructed in the Fock space.

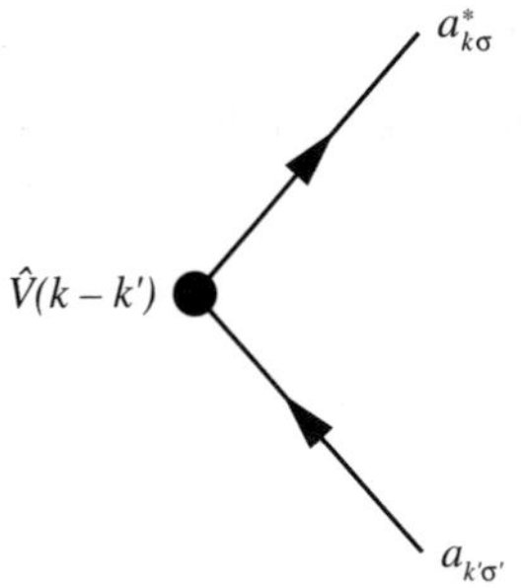

Fig. 3.1. A particle in momentum state $\hbar \boldsymbol{k}'$ is "annihilated", scattered by potential $\tilde{V}(\boldsymbol{k}-\boldsymbol{k}')$, then "recreated" in momentum state $\hbar \boldsymbol{k}$

Evolution of Non-Coupled Particles

The evolution of n particles without mutual interactions is written naturally with product states in $\mathcal{H}^{\otimes n}$ as

$$\mathtt{U}(t)(|\varphi_1\rangle \otimes \ldots \otimes |\varphi_n\rangle) = U(t)\,|\varphi_1\rangle \otimes \ldots \otimes U(t)\,|\varphi_n\rangle \quad . \tag{3.80}$$

Since $\mathtt{U}(t)$ commutes with the permutation operators, (3.80) transposes itself immediately for the properly symmetrized states (2.65)

$$\mathtt{U}(t)\,|\varphi_1, \ldots, \varphi_n\rangle_\nu = |U(t)\varphi_1, \ldots, U(t)\varphi_n\rangle_\nu \quad . \tag{3.81}$$

This formula defines obviously a unitary representation of time translations in Fock space $\mathcal{F}_\nu(\mathcal{H})$.

Writing $\mathtt{U}(t) = \exp(-\mathrm{i}\mathtt{H}t/\hbar)$ one obtains from (3.80) to first order in t

$$\begin{aligned}
&\left(\mathtt{I} - \frac{\mathrm{i}t}{\hbar}\mathtt{H}\right)|\varphi_1, \ldots, \varphi_n\rangle_\nu + \ldots \\
&= \left|\left(I - \frac{\mathrm{i}t}{\hbar}H\right)\varphi_1, \ldots, \left(I - \frac{\mathrm{i}t}{\hbar}H\right)\varphi_n\right\rangle_\nu + \ldots \\
&= |\varphi_1, \ldots, \varphi_n\rangle_\nu \\
&\quad - \frac{\mathrm{i}t}{\hbar}\left(|H\varphi_1, \ldots, \varphi_n\rangle_\nu + \ldots + |\varphi_1, \ldots, H\varphi_n\rangle_\nu\right) + \ldots \quad .
\end{aligned} \tag{3.82}$$

Therefore, the generator $\mathtt{H}$ of the evolution of independent particles in $\mathcal{F}_\nu(\mathcal{H})$ (the Hamiltonian of non-coupled particles)

$$\mathtt{H}\,|\varphi_1, \ldots, \varphi_n\rangle_\nu = \sum_{j=1}^{n} |\varphi_1, \ldots, \varphi_{j-1}, H\varphi_j, \varphi_{j+1}, \ldots, \varphi_n\rangle_\nu \tag{3.83}$$

is the one-body operator corresponding to the single-particle Hamiltonian H. This is a general result: *generators of symmetry transformations in $\mathcal{F}_\nu(\mathcal{H})$ are the one-body operators corresponding to generators of the same transformations in single-particle space.*

Evolution of the Creation and Annihilation Operators

From (3.12) and (3.81) it is possible to deduce that the evolution law for the creation and annihilation operators is

$$\begin{aligned} &\mathtt{U}^{-1}(t)a^*(\varphi)\,|\varphi_1,\dots,\varphi_n\rangle_\nu \\ &\quad = \sqrt{n+1}\,|U^{-1}(t)\varphi, U^{-1}(t)\varphi_1,\dots,U^{-1}(t)\varphi_n\rangle_\nu \\ &\quad = a^*\left(U^{-1}(t)\varphi\right)\mathtt{U}^{-1}(t)\,|\varphi_1,\dots,\varphi_n\rangle_\nu \quad , \end{aligned} \tag{3.84}$$

hence

$$\mathtt{U}^{-1}(t)a^*(\varphi)\mathtt{U}(t) = a^*\left(U^{-1}(t)\varphi\right) \quad . \tag{3.85}$$

All operations of symmetry can be expressed in this manner: *the transformed creation operator is the creation operator of the transformed state.* The infinitesimal transformation corresponding to (3.85) is obtained by writing this relation to first order in t

$$\left(\mathtt{I}+\frac{it}{\hbar}\mathtt{H}+\dots\right)a^*(\varphi)\left(\mathtt{I}-\frac{it}{\hbar}\mathtt{H}+\dots\right) = a^*\left[\left(\mathtt{I}+\frac{it}{\hbar}H+\dots\right)\varphi\right] \quad , \tag{3.86}$$

so that

$$[\mathtt{H}, a^*(\varphi)] = a^*(H\varphi) \quad . \tag{3.87}$$

One can verify, with the aid of the commutation (3.27) or anticommutation (3.28) relations that $\mathtt{H} = \sum_r a^*(H\psi_r)a(\psi_r)$ satisfies (3.87).

In the basis $\{|\psi_r\rangle\}$ which diagonalizes the one-particle Hamiltonian ($H\,|\psi_r\rangle = \varepsilon_r\,|\psi_r\rangle$), (3.85) becomes

$$\begin{aligned} \mathtt{U}^{-1}(t)a_r^*\mathtt{U}(t) &= a^*\left[U^{-1}(t)\psi_r\right] \\ &= a^*\left[\exp\left(\frac{i\varepsilon_r t}{\hbar}\right)\psi_r\right] = \exp\left(\frac{i\varepsilon_r t}{\hbar}\right)a_r^* \quad , \end{aligned} \tag{3.88}$$

where $a_r^* = a^*(\psi_r)$. In the same manner,

$$\mathtt{U}^{-1}(t)a_r\mathtt{U}(t) = \exp\left(-\frac{i\varepsilon_r t}{\hbar}\right)a_r \quad . \tag{3.89}$$

These evolution laws are identical to those of a harmonic oscillator with frequency $\varepsilon_r/\hbar$ (1.31). One should note that these laws are valid for fermions as well as for bosons.

Gauge Transformations of the First Kind

There exists a new symmetry in Fock space which corresponds to *conservation of particle number*. Consider a transformation which consists of multiplying all single-particle states by a constant phase factor $\mathrm{e}^{-\mathrm{i}\alpha}$. In comparison to (3.81), the operator $\mathtt{U}_\alpha$ in Fock space corresponding to this multiplication is written as

$$\mathtt{U}_\alpha \left|\varphi_1,\dots,\varphi_n\right\rangle_\nu = \left|U_\alpha\varphi_1,\dots,U_\alpha\varphi_n\right\rangle_\nu \quad , \tag{3.90}$$

where

$$U_\alpha \left|\varphi\right\rangle = \mathrm{e}^{-\mathrm{i}\alpha I}\left|\varphi\right\rangle \quad , \tag{3.91}$$

with I being the single-particle identity operator. To first order in α, $\mathtt{U}_\alpha = \mathtt{I} - \mathrm{i}\alpha\mathtt{B}$ determines the generator $\mathtt{B}$ of the transformation. Relations (3.82) and (3.83), with $\mathtt{H}t/\hbar$ replaced by $\mathtt{B}\alpha$, give

$$\begin{aligned}\mathtt{B}\left|\varphi_1,\dots,\varphi_n\right\rangle_\nu &= \sum_{j=1}^{n}\left|\varphi_1,\dots,\varphi_{j-1},I\varphi_j,\varphi_{j+1},\dots,\varphi_n\right\rangle_\nu \\ &= n\left|\varphi_1,\dots,\varphi_n\right\rangle_\nu \quad .\end{aligned} \tag{3.92}$$

Operator $\mathtt{B}$ is thus none other than the particle number operator $\mathtt{N}$, so that

$$\mathtt{U}_\alpha = \mathrm{e}^{-\mathrm{i}\alpha\mathtt{N}} \quad . \tag{3.93}$$

Continuing the analogy between $\mathtt{U}_\alpha$ and $\mathtt{U}(t)$, it is possible to write a relation which corresponds to (3.85),

$$\mathtt{U}_\alpha^{-1}a^*(\varphi)\mathtt{U}_\alpha = a^*(\mathrm{e}^{\mathrm{i}\alpha}\varphi) \quad . \tag{3.94}$$

Taking into account the linearity of $a^*(\varphi)$ and the anti-linearity of $a(\varphi)$ (shown in (3.14)), (3.94) can finally be written as

$$\mathrm{e}^{\mathrm{i}\alpha\mathtt{N}}a^*(\varphi)\mathrm{e}^{-\mathrm{i}\alpha\mathtt{N}} = \mathrm{e}^{\mathrm{i}\alpha}a^*(\varphi) \tag{3.95}$$

and

$$\mathrm{e}^{\mathrm{i}\alpha\mathtt{N}}a(\varphi)\mathrm{e}^{-\mathrm{i}\alpha\mathtt{N}} = \mathrm{e}^{-\mathrm{i}\alpha}a(\varphi) \quad . \tag{3.96}$$

Thus the transformation which consists of multiplying each single-particle state $\left|\varphi\right\rangle$ by the factor $\mathrm{e}^{-\mathrm{i}\alpha}$ is represented in Fock space $\mathcal{F}_\nu(\mathcal{H})$ by the operator $\mathtt{U}_\alpha = \mathrm{e}^{-\mathrm{i}\alpha\mathtt{N}}$. A transformation

$$\left|\varphi\right\rangle \to \mathrm{e}^{-\mathrm{i}\alpha}\left|\varphi\right\rangle \quad , \tag{3.97}$$

where α is a phase independent of space and time, is called a *gauge transformation of the first kind* (the second kind is reserved for gauge transformations of the electromagnetic field potential, discussed in Sect. 1.3.2).

Conservation of Particle Number

Invariance under gauge transformations of the first kind expresses conservation of particle number. The fact that the Hamiltonian does not induce any transitions between states of different particle number is equivalent to the commutation relation

$$[\mathtt{H}, \mathtt{N}] = 0 \quad , \tag{3.98}$$

which is itself equivalent to the invariance

$$e^{i\alpha\mathtt{N}}\mathtt{H}e^{-i\alpha\mathtt{N}} = \mathtt{H} \tag{3.99}$$

for all α.

A simple test can be used to recognize conservation of particle number. From (3.95) and (3.96), the creation and annihilation operators are multiplied by a phase factor and its complex conjugate under gauge transformations of the first kind. *As a result, a monomial comprising the same number of creators as annihilators is invariant.* So, any operator formed from such monomials conserves the number of particles. An important difference between many-body systems and fields is that, in a many-body problem, the number of massive (non-relativistic) particles is always conserved. In reactions between relativistic particles, the number can vary, but the total charge remains conserved. It will be shown in Sect. 8.3.2 that, in field theory, gauge invariance of the first kind expresses total conservation of charge, which remains valid for interactions of elementary particles.

3.2.7 Two-Body Operators

Consider an operator V acting in two-particle space $\mathcal{H} \otimes \mathcal{H}$. According to (2.43), a two-body observable acting on a state of n particles in Fock space is given by

$$\mathtt{V}\,|\varphi_1, \ldots, \varphi_n\rangle_\nu = \sum_{i<j}^{n} V_{ij}\,|\varphi_1, \ldots, \varphi_n\rangle_\nu \quad . \tag{3.100}$$

It is possible to write $\mathtt{V}$ in terms of the creation and annihilation operators with the formula

$$\mathtt{V} = \frac{1}{2} \sum_{r_1 r_2 s_1 s_2} \langle r_1 r_2 | V | s_1 s_2 \rangle \, a_{r_1}^* a_{r_2}^* a_{s_2} a_{s_1} \quad . \tag{3.101}$$

Elements of the matrix V

$$\langle r_1 r_2 | V | s_1 s_2 \rangle = \langle \psi_{r_1} \otimes \psi_{r_2} | V | \psi_{s_1} \otimes \psi_{s_2} \rangle \tag{3.102}$$

are relative to the tensor basis $\{|\psi_{r_1} \otimes \psi_{r_2}\rangle\}$ of $\mathcal{H} \otimes \mathcal{H}$ where $\{|\psi_r\rangle\}$ is a basis in one-particle space and $a^*_r = a^*(\psi_r)$. Note that invariance of V under the permutation $P_{(12)}$ of two particles implies $P^*_{(12)} V P_{(12)} = V$ and thus

$$\langle \psi_{r_1} \otimes \psi_{r_2} | V | \psi_{s_1} \otimes \psi_{s_2} \rangle = \langle \psi_{r_2} \otimes \psi_{r_1} | V | \psi_{s_2} \otimes \psi_{s_1} \rangle \quad . \tag{3.103}$$

To verify (3.101), (3.19) and (3.12) can be applied twice, carefully taking into account the order of the states in $|\varphi_1, \dots, \varphi_n\rangle_\nu$ and the signs, giving

$$\begin{aligned}
& a^*_{r_1} a^*_{r_2} a_{s_2} a_{s_1} |\varphi_1, \dots, \varphi_n\rangle_\nu \\
&\quad = \frac{a^*_{r_1} a^*_{r_2}}{\sqrt{n(n-1)}} \left\{ \sum_{i=1}^{n} \sum_{j=1}^{i-1} (\nu)^{i-1} (\nu)^{j-1} |\varphi_1, \dots, \varphi_{j-1}, \varphi_{j+1}, \dots \right. \\
&\qquad \dots \varphi_{i-1}, \varphi_{i+1}, \dots, \varphi_n\rangle_\nu \langle \psi_{s_1} | \varphi_i \rangle \langle \psi_{s_2} | \varphi_j \rangle \\
&\qquad + \sum_{i=1}^{n} \sum_{j=i+1}^{n} (\nu)^{i-1} (\nu)^{j} |\varphi_1, \dots, \varphi_{i-1}, \varphi_{i+1}, \dots \\
&\qquad \left. \dots \varphi_{j-1}, \varphi_{j+1}, \dots, \varphi_n\rangle_\nu \langle \psi_{s_1} | \varphi_i \rangle \langle \psi_{s_2} | \varphi_j \rangle \right\} \\
&\quad = \sum_{i=1}^{n} \sum_{j=1}^{i-1} |\varphi_1, \dots, \varphi_{j-1}, \psi_{r_2}, \varphi_{j+1}, \dots \\
&\qquad \dots \varphi_{i-1}, \psi_{r_1}, \varphi_{i+1}, \dots, \varphi_n\rangle_\nu \langle \psi_{s_1} \otimes \psi_{s_2} | \varphi_i \otimes \varphi_j \rangle \\
&\qquad + \sum_{i=1}^{n} \sum_{j=i+1}^{n} |\varphi_1, \dots, \varphi_{i-1}, \psi_{r_1}, \varphi_{i+1}, \dots \\
&\qquad \dots \varphi_{j-1}, \psi_{r_2}, \varphi_{j+1}, \dots, \varphi_n\rangle_\nu \langle \psi_{s_1} \otimes \psi_{s_2} | \varphi_i \otimes \varphi_j \rangle \quad .
\end{aligned} \tag{3.104}$$

This expression still must be summed over r_1, r_2, s_1, s_2 and multiplied by matrix elements $\langle r_1 r_2 | V | s_1 s_2 \rangle$. Carrying out this operation for terms of (3.104) with $j < i$ and exploiting the fact that the basis $\{|\psi_{s_1} \otimes \psi_{s_2}\rangle\}$ is complete, gives

$$\begin{aligned}
& \sum_{r_1 r_2 s_1 s_2} |\varphi_1, \dots, \varphi_{j-1}, \psi_{r_2}, \varphi_{j+1}, \dots, \varphi_{i-1}, \psi_{r_1}, \varphi_{i+1}, \dots, \varphi_n\rangle_\nu \\
&\qquad \times \langle \psi_{r_1} \otimes \psi_{r_2} | V | \psi_{s_1} \otimes \psi_{s_2} \rangle \langle \psi_{s_1} \otimes \psi_{s_2} | \varphi_i \otimes \varphi_j \rangle \\
&\quad = \sum_{r_1 r_2} |\varphi_1, \dots, \varphi_{j-1}, \psi_{r_2}, \varphi_{j+1}, \dots, \varphi_{i-1}, \psi_{r_1}, \varphi_{i+1}, \dots, \varphi_n\rangle_\nu \\
&\qquad \times \langle \psi_{r_1} \otimes \psi_{r_2} | V | \varphi_i \otimes \varphi_j \rangle \quad .
\end{aligned} \tag{3.105}$$

Applying the symmetry relation (3.103) and the definition (2.65), (3.105) becomes equivalent to

$$\sum_{r_1 r_2} |\varphi_1, \dots, \varphi_{j-1}, \psi_{r_1}, \varphi_{j+1}, \dots, \varphi_{i-1}, \psi_{r_2}, \varphi_{i+1}, \dots, \varphi_n\rangle_\nu$$
$$\times \langle \psi_{r_1} \otimes \psi_{r_2} | V | \varphi_j \otimes \varphi_i \rangle$$
$$= S_\nu \sum_{r_1 r_2} |\varphi_1 \otimes \dots \otimes \varphi_{j-1} \otimes \psi_{r_1} \otimes \varphi_{j+1} \otimes \dots$$
$$\dots \otimes \varphi_{i-1} \otimes \psi_{r_2} \otimes \varphi_{i+1} \otimes \dots \otimes \varphi_n\rangle \langle \psi_{r_1} \otimes \psi_{r_2} | V | \varphi_j \otimes \varphi_i \rangle$$
$$= S_\nu V_{ji} |\varphi_1 \otimes \dots \otimes \varphi_n\rangle \quad . \tag{3.106}$$

Note that the last line of (3.106) again follows from both the completeness of the basis and the action of V on particles i and j. A similar result is obtained for terms in (3.104) with $j > i$. Finally, combining (3.101), (3.105) and (3.106), and recalling that $V_{ij} = V_{ji}$, one finds

$$\mathsf{V} |\varphi_1, \dots, \varphi_n\rangle_\nu$$
$$= S_\nu \left(\frac{1}{2} \sum_{i=1}^{n} \sum_{j=1}^{i-1} V_{ji} |\varphi_1 \otimes \dots \otimes \varphi_n\rangle + \frac{1}{2} \sum_{i=1}^{n} \sum_{j=i+1}^{n} V_{ij} |\varphi_1 \otimes \dots \otimes \varphi_n\rangle \right)$$
$$= S_\nu \left(\frac{1}{2} \sum_{i \neq j}^{n} V_{ij} |\varphi_1 \otimes \dots \otimes \varphi_n\rangle \right) \quad , \tag{3.107}$$

which is identical to (3.100), since $\sum_{i \neq j}^{n} V_{ij}$ commutes with all of the permutation operators.

Position Representation

An important case to study is that of a two-body potential $V(\boldsymbol{x}_1 - \boldsymbol{x}_2)$ which is invariant under translations and independent of spin. In the basis $|\boldsymbol{x}\sigma\rangle$, V has matrix elements

$$\langle \boldsymbol{x}_1\sigma_1, \boldsymbol{x}_2\sigma_2 | V | \boldsymbol{x}_1'\sigma_1', \boldsymbol{x}_2', \sigma_2' \rangle$$
$$= V(\boldsymbol{x}_1 - \boldsymbol{x}_2) \delta(\boldsymbol{x}_1 - \boldsymbol{x}_1') \delta(\boldsymbol{x}_2 - \boldsymbol{x}_2') \delta_{\sigma_1, \sigma_1'} \delta_{\sigma_2, \sigma_2'} \quad . \tag{3.108}$$

Substituting $r = (\boldsymbol{x}, \sigma)$ and $a_r = a(\boldsymbol{x}, \sigma)$, (3.101) becomes

$$\mathsf{V} = \frac{1}{2} \int_\Lambda \mathrm{d}\boldsymbol{x}_1 \int_\Lambda \mathrm{d}\boldsymbol{x}_2 \sum_{\sigma_1 \sigma_2} V(\boldsymbol{x}_1 - \boldsymbol{x}_2)$$
$$\times a^*(\boldsymbol{x}_1, \sigma_1) a^*(\boldsymbol{x}_2, \sigma_2) a(\boldsymbol{x}_2, \sigma_2) a(\boldsymbol{x}_1, \sigma_1) \quad . \tag{3.109}$$

This takes on a particularly suggestive form if the particle density operator $n(\boldsymbol{x})$ (3.70) is introduced

$$\mathsf{V} = \frac{1}{2} \int_\Lambda \mathrm{d}\boldsymbol{x}_1 \int_\Lambda \mathrm{d}\boldsymbol{x}_2 V(\boldsymbol{x}_1 - \boldsymbol{x}_2) {:} n(\boldsymbol{x}_1) n(\boldsymbol{x}_2) {:} \quad . \tag{3.110}$$

Up to normal ordering, this expression has the same form as that of a classical potential energy, with the density operator (3.70) replacing the classical particle density.

Momentum Representation

It is often useful, for calculations, to express V in the plane-wave basis $|\boldsymbol{k}\sigma\rangle$ with periodic boundary conditions. The Fourier transform of $V(\boldsymbol{x})$ in a cubic box Λ is given by

$$\tilde{V}_L(\boldsymbol{k}) = \int_\Lambda \mathrm{d}\boldsymbol{x} \mathrm{e}^{-\mathrm{i}\boldsymbol{k}\cdot\boldsymbol{x}} V(\boldsymbol{x}) \tag{3.111}$$

with corresponding Fourier series

$$V_L(\boldsymbol{x}) = \frac{1}{L^3} \sum_{\boldsymbol{k}} \mathrm{e}^{\mathrm{i}\boldsymbol{k}\cdot\boldsymbol{x}} \tilde{V}_L(\boldsymbol{k}) \quad . \tag{3.112}$$

Note that $V_L(\boldsymbol{x})$ is a periodic function with period L and is only equal to $V(\boldsymbol{x})$ if $\boldsymbol{x}$ is in Λ. In the limit of infinite volume, however, $\tilde{V}_L(\boldsymbol{k})$ approaches the Fourier transform $\tilde{V}(\boldsymbol{k})$ of $V(\boldsymbol{x})$ in the whole space,

$$\lim_{L\to\infty} \tilde{V}_L(\boldsymbol{k}) = \int \mathrm{d}\boldsymbol{x} \mathrm{e}^{-\mathrm{i}\boldsymbol{k}\cdot\boldsymbol{x}} V(\boldsymbol{x}) = \tilde{V}(\boldsymbol{k}) \tag{3.113}$$

and

$$\lim_{L\to\infty} V_L(\boldsymbol{x}) = V(\boldsymbol{x}) \quad . \tag{3.114}$$

Using the transformation $\langle \boldsymbol{x}\sigma|\boldsymbol{k}\sigma'\rangle = L^{-3/2}\delta_{\sigma,\sigma'}\mathrm{e}^{\mathrm{i}\boldsymbol{k}\cdot\boldsymbol{x}}$, one can obtain the matrix elements of V_L

$$\begin{aligned}
&\langle \boldsymbol{k}_1\sigma_1, \boldsymbol{k}_2\sigma_2|V|\boldsymbol{k}_1'\sigma_1', \boldsymbol{k}_2'\sigma_2'\rangle \\
&= \frac{\delta_{\sigma_1,\sigma_1'}\delta_{\sigma_2,\sigma_2'}}{(L^3)^2} \int_\Lambda \mathrm{d}\boldsymbol{x}_1 \int_\Lambda \mathrm{d}\boldsymbol{x}_2 \mathrm{e}^{-\mathrm{i}(\boldsymbol{k}_1\cdot\boldsymbol{x}_1+\boldsymbol{k}_2\cdot\boldsymbol{x}_2)} \\
&\quad \times V_L(\boldsymbol{x}_1-\boldsymbol{x}_2)\mathrm{e}^{\mathrm{i}(\boldsymbol{k}_1'\cdot\boldsymbol{x}_1+\boldsymbol{k}_2'\cdot\boldsymbol{x}_2)} \\
&= \frac{\delta_{\sigma_1,\sigma_1'}\delta_{\sigma_2,\sigma_2'}}{L^3} \sum_{\boldsymbol{k}} \tilde{V}_L(\boldsymbol{k}) \left(\frac{1}{L^3}\int_\Lambda \mathrm{d}\boldsymbol{x}_1 \mathrm{e}^{\mathrm{i}(\boldsymbol{k}+\boldsymbol{k}_1'-\boldsymbol{k}_1)\cdot\boldsymbol{x}_1}\right) \\
&\quad \times \left(\frac{1}{L^3}\int_\Lambda \mathrm{d}\boldsymbol{x}_2 \mathrm{e}^{\mathrm{i}(\boldsymbol{k}_2'-\boldsymbol{k}_2-\boldsymbol{k})\cdot\boldsymbol{x}_2}\right) \\
&= \delta_{\sigma_1,\sigma_1'}\delta_{\sigma_2,\sigma_2'} \frac{1}{L^3} \sum_{\boldsymbol{k}} \tilde{V}_L(\boldsymbol{k})\delta_{\boldsymbol{k}_1,\boldsymbol{k}_1'+\boldsymbol{k}}\delta_{\boldsymbol{k}_2,\boldsymbol{k}_2'-\boldsymbol{k}} \quad .
\end{aligned} \tag{3.115}$$

When (3.115) is introduced into (3.101) with $r = (\boldsymbol{k}\sigma)$ and $a_r^* = a_{\boldsymbol{k}\sigma}^*$, the second-quantized operator corresponding to the potential $V_L(\boldsymbol{x})$ takes the form

$$\mathsf{V} = \frac{1}{2L^3} \sum_{\boldsymbol{k}_1\boldsymbol{k}_2\boldsymbol{k}} \sum_{\sigma_1\sigma_2} \tilde{V}_L(\boldsymbol{k}) a^*_{(\boldsymbol{k}_1+\boldsymbol{k})\sigma_1} a^*_{(\boldsymbol{k}_2-\boldsymbol{k})\sigma_2} a_{\boldsymbol{k}_2\sigma_2} a_{\boldsymbol{k}_1\sigma_1} \quad . \tag{3.116}$$

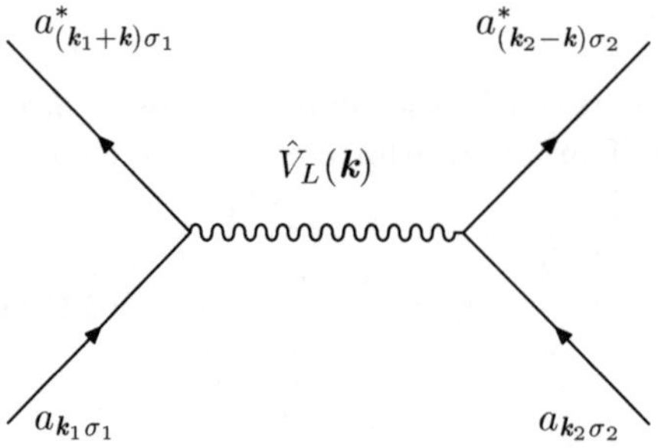

Fig. 3.2. Two particles of momentum $\hbar\boldsymbol{k}_1$ and $\hbar\boldsymbol{k}_2$ scattering with momentum transfer $\hbar\boldsymbol{k}$

One can equally write $\mathtt{V}$ with the aid of the Fourier transform $\tilde{n}(\boldsymbol{k})$ (3.79) of the density operator in a manner analogous to (3.110)

$$\mathtt{V} = \frac{1}{2L^3}\sum_{\boldsymbol{k}} \tilde{V}_L(\boldsymbol{k}){:}\tilde{n}^*(\boldsymbol{k})\tilde{n}(\boldsymbol{k}){:} \quad . \tag{3.117}$$

Formula (3.116) can be described graphically (Fig. 3.2) as *two particles of momentum $\hbar\boldsymbol{k}_1$ and $\hbar\boldsymbol{k}_2$ "scattering" with a momentum transfer of $\hbar\boldsymbol{k}$. As a consequence of the invariance of the two-body potential to translations of the center of mass, the total momentum $\hbar(\boldsymbol{k}_1+\boldsymbol{k}_2)$ is conserved throughout the process.*

One should note that the two-body interaction (3.109) does not coincide exactly with the form of $\mathtt{V}$ in (3.116). In the latter, two particles interact with the potential $V_L(\boldsymbol{x}_1-\boldsymbol{x}_2)$ which is not identical to the given potential $V(\boldsymbol{x}_1-\boldsymbol{x}_2)$ unless $\boldsymbol{x}_1-\boldsymbol{x}_2$ is in Λ (in general $\boldsymbol{x}_1-\boldsymbol{x}_2$ covers a cube of side $2L$ when $\boldsymbol{x}_1$ and $\boldsymbol{x}_2$ cover Λ). Nevertheless, as a result of (3.113) and (3.114), one expects to obtain the same predictions in the limit of infinite volume, starting with either form (3.109) or (3.116) of $\mathtt{V}$ (a point which requires a mathematical verification for each case). Anticipating the result, one could immediately replace $\tilde{V}_L(\boldsymbol{k})$ in (3.116) by its limit $\tilde{V}(\boldsymbol{k})$. The form (3.116) of $\mathtt{V}$ is particularly useful because it explicitly expresses the conservation of total momentum.

Matrix Elements of V in the Occupation Number Representation

As an example, one can calculate the diagonal matrix elements of $\mathtt{V}$ in the occupation number states for a system of fermions

$$
\begin{aligned}
&{}_-\langle n_1,\dots,n_r,\dots|\mathsf{V}|n_1,\dots,n_r,\dots\rangle_- \\
&= \frac{1}{2}\sum_{r_1r_2s_1s_2} {}_-\langle n_1,\dots,n_r,\dots|a^*_{r_1}a^*_{r_2}a_{s_2}a_{s_1}|n_1,\dots,n_r,\dots\rangle_- \\
&\quad\times\langle r_1r_2|V|s_1s_2\rangle \quad . \qquad (3.118)
\end{aligned}
$$

The matrix elements of $a^*_{r_1}a^*_{r_2}a_{s_2}a_{s_1}$ in sum (3.118) are the scalar products of $a_{s_2}a_{s_1}|n_1,\dots,n_r,\dots\rangle_-$ with $a_{r_2}a_{r_1}|n_1,\dots,n_r,\dots\rangle_-$. Up to an overall sign, $a_{r_2}a_{r_1}|n_1,\dots,n_r,\dots\rangle_-$ is equal to $n_{r_1}n_{r_2}|n_1,\dots,n_{r_1}-1,\dots,n_{r_2}-1,\dots\rangle_-$ for the case where $r_1\neq r_2$, see (3.44). Therefore, the matrix elements in (3.118) are only non-zero provided $s_1=r_1$, $s_2=r_2$, or $s_1=r_2$, $s_2=r_1$. To determine the sign of the result, note that if $s_1=r_1$, $s_2=r_2$ with $s_1\neq s_2$, then

$$
\begin{aligned}
&a^*_{r_1}a^*_{r_2}a_{r_2}a_{r_1}|n_1,\dots,n_r,\dots\rangle_- \\
&= a^*_{r_2}a_{r_2}a^*_{r_1}a_{r_1}|n_1,\dots,n_r,\dots\rangle_- \\
&= n_{r_1}n_{r_2}|n_1,\dots,n_r,\dots\rangle_- \qquad (3.119)
\end{aligned}
$$

while if $s_1=r_2$, $s_2=r_1$ with $s_1\neq s_2$, then

$$
\begin{aligned}
&a^*_{r_1}a^*_{r_2}a_{r_1}a_{r_2}|n_1,\dots,n_r,\dots\rangle_- \\
&= -a^*_{r_1}a_{r_1}a^*_{r_2}a_{r_2}|n_1,\dots,n_r,\dots\rangle_- \\
&= -n_{r_1}n_{r_2}|n_1,\dots,n_r,\dots\rangle_- \quad . \qquad (3.120)
\end{aligned}
$$

Therefore, in all cases, one has

$$
\begin{aligned}
&{}_-\langle n_1,\dots,n_r,\dots|a^*_{r_1}a^*_{r_2}a_{s_2}a_{s_1}|n_1,\dots,n_r,\dots\rangle_- \\
&= (\delta_{r_1,s_1}\delta_{r_2,s_2}-\delta_{r_1,s_2}\delta_{r_2,s_1})n_{r_1}n_{r_2} \quad . \qquad (3.121)
\end{aligned}
$$

Finally, inserting (3.121) into (3.118) gives

$$
\begin{aligned}
&{}_-\langle n_1,\dots,n_r,\dots|\mathsf{V}|n_1,\dots,n_r,\dots\rangle_- \\
&= \frac{1}{2}\sum_{r_1r_2}(\langle r_1r_2|V|r_1r_2\rangle-\langle r_1r_2|V|r_2r_1\rangle)n_{r_1}n_{r_2} \quad . \qquad (3.122)
\end{aligned}
$$

The second term of the right-hand side of (3.122) is called the *exchange contribution* to the energy.

3.2.8 Reduced Density Matrices and Correlations

Often the state vector of a quantum system is not completely known. One example is the state of a radiation field produced by a thermal source; such a field is the result of the incoherent and random emission of atoms from the source. Another example is the polarization state of a beam of particles prepared for a scattering experiment, which can be partially undetermined. Finally, a system in a state of thermal equilibrium at temperature T has a probability proportional to $\exp(-E/k_{\mathrm{B}}T)$ to be found in an eigenstate of the Hamiltonian of energy E. It is hence not completely determined. *For all of these cases, an adequate description of the quantum state is provided by the density operator* ρ. It is a positive, Hermitian operator with trace equal to 1,

$$\rho > 0 \quad , \quad \rho = \rho^* \quad \text{and} \quad \mathrm{Tr}\rho = 1 \quad , \tag{3.123}$$

and is defined such that the mean value of some observable $\mathtt{A}$ is given by

$$\langle \mathtt{A} \rangle = \mathrm{Tr}\rho\mathtt{A} \quad . \tag{3.124}$$

If $\rho = \rho^2$, ρ *is the projector on a vector state* $|\Phi\rangle$ (determined up to a phase factor) and $\mathrm{Tr}\rho\mathtt{A} = \langle\Phi|\mathtt{A}|\Phi\rangle$ reduces to the usual quantum mean value. On the other hand, if $\rho > \rho^2$, then the density operator represents a genuine *statistical mixture* of vector states.

These concepts apply equally to states of systems with variable particle numbers. For example, the grand canonical partition function of a system of massive particles at thermal equilibrium characterized by the inverse temperature $\beta = 1/k_{\mathrm{B}}T$ and the chemical potential μ

$$Z(\beta,\mu) = 1 + \sum_{n=1}^{\infty} \mathrm{e}^{\beta\mu n} \mathrm{Tr}_n \left\{\exp\left[-\beta \mathtt{H}(n)\right]\right\} \quad , \tag{3.125}$$

where $\mathtt{H}(n)$ is the Hamiltonian for n particles and $\mathrm{Tr}_n\left\{\exp\left[-\beta\mathtt{H}(n)\right]\right\}$ is the canonical partition function, with the trace taken in the space of n particles $\mathcal{H}_\nu^{\otimes n}$. It follows from the definition (3.6) of the Hamiltonian in Fock space $(\mathtt{H} = \sum_{n=0}^{\infty} \mathtt{H}(n), \mathtt{H}(0) = 0)$ and from the definition of the particle number operator (3.8) that

$$Z(\beta,\mu) = \mathrm{Tr}\left\{\exp\left[-\beta(\mathtt{H} - \mu\mathtt{N})\right]\right\} \quad , \tag{3.126}$$

where the trace is taken in Fock space $\mathcal{F}_\nu(\mathcal{H})$. The equivalence of (3.125) and (3.126) results from the fact that $\mathtt{H}$ and $\mathtt{N}$ commute and it is possible to calculate the trace in $\mathcal{F}_\nu(\mathcal{H})$ with the help of a diagonal basis of $\mathtt{N}$ (the trace is independent of the choice of basis). With (3.126), it is possible to define the grand canonical density operator in $\mathcal{F}_\nu(\mathcal{H})$ to be

$$\rho = \frac{1}{Z(\beta,\mu)} \exp\left[-\beta(\mathtt{H} - \mu\mathtt{N})\right] \quad . \tag{3.127}$$

Density operator (3.127) describes the equilibrium state of massive particles, in which the mass density is controlled by the chemical potential μ. As for particles of zero mass (phonons, photons), their thermodynamic state is uniquely characterized by the temperature. For such particles, expressions (3.126) and (3.127) remain valid if $\mu = 0$.

One-Body Reduced Density Matrix

In principle, knowledge of the density operator in Fock space gives all of the microscopic information available regarding a system, allowing one to calculate the mean values of all of the observables. In practice, however, only certain simple observables are accessible from an experiment. These are, for example, one-body observables, such as the kinetic energy, the spin, the number of particles, and two-body observables, such as the interaction energy.

If ρ is a density matrix in Fock space, the mean value of a one-body observable $\mathtt{A}$ corresponding to the single-particle observable A takes a form which follows from (3.66)

$$\begin{aligned}\langle \mathtt{A}\rangle &= \mathrm{Tr}\rho\mathtt{A} = \sum_{rs} \langle r|A|s\rangle \, \mathrm{Tr}\rho a_r^* a_s \\ &= \sum_{rs} \langle r|A|s\rangle \, \langle s|\rho^{(1)}|r\rangle = \mathrm{Tr}_1\rho^{(1)} A \quad , \end{aligned} \tag{3.128}$$

where the matrix $\rho^{(1)}$ is defined with elements

$$\langle s|\rho^{(1)}|r\rangle = \mathrm{Tr}\rho a_r^* a_s \quad . \tag{3.129}$$

It is clear that one can consider $\rho^{(1)}$ as an operator in one-particle space $\mathcal{H}$, represented by (3.129) in the basis $\{|\psi_r\rangle\}$. The operator $\rho^{(1)}$ is Hermitian (this can be shown using the property of cyclic permutation under the trace)

$$\begin{aligned}\langle s|\rho^{(1)}|r\rangle^* &= (\mathrm{Tr}\rho a_r^* a_s)^* = \mathrm{Tr}(\rho a_r^* a_s)^* = \mathrm{Tr} a_s^* a_r \rho \\ &= \mathrm{Tr}\rho a_s^* a_r = \langle r|\rho^{(1)}|s\rangle \end{aligned} \tag{3.130}$$

and positive since, from (3.128), the mean value of any positive observable is positive. The normalization of $\rho^{(1)}$ is given by

$$\mathrm{Tr}_1\rho^{(1)} = \sum_r \mathrm{Tr}\rho a_r^* a_r = \mathrm{Tr}\rho\mathtt{N} = \langle \mathtt{N}\rangle \quad . \tag{3.131}$$

Thus, $\rho^{(1)}$ has all of the properties of a density matrix in one-particle space, except for the fact that its normalization is fixed by the mean number of particles. A matrix $\rho^{(1)}$ is called a *one-body reduced density matrix. Relation (3.128) shows that knowledge of* $\rho^{(1)}$ *is sufficient to predict the mean values of all of the one-body observables.*

The representation of $\rho^{(1)}$ in position space is given by the expression

$$\langle \boldsymbol{x}\sigma|\rho^{(1)}|\boldsymbol{x}'\sigma'\rangle = \mathrm{Tr}\rho a^*(\boldsymbol{x}',\sigma') a(\boldsymbol{x},\sigma) \quad . \tag{3.132}$$

Its diagonal component

$$\langle \boldsymbol{x}\sigma|\rho^{(1)}|\boldsymbol{x}\sigma\rangle = \mathrm{Tr}\rho n(\boldsymbol{x},\sigma) = \langle n(\boldsymbol{x},\sigma)\rangle \tag{3.133}$$

is, according to (3.70), the mean density of particles of spin σ at point $\boldsymbol{x}$. This can be verified for the case where the state is a vector $|\Phi(n)\rangle$ in n-particle space. Since $n(\boldsymbol{x},\sigma)$ is the one-body operator which corresponds to multiplication by the Dirac function (3.71),

$$n(\boldsymbol{x},\sigma)\Phi(\boldsymbol{x}_1\sigma_1,\dots,\boldsymbol{x}_n\sigma_n) = \sum_{j=1}^{n}\delta(\boldsymbol{x}-\boldsymbol{x}_j)\delta_{\sigma,\sigma_j}\Phi(\boldsymbol{x}_1\sigma_1,\dots,\boldsymbol{x}_n\sigma_n) \quad . \tag{3.134}$$

As $|\Phi(\boldsymbol{x}_1\sigma_1,\dots,\boldsymbol{x}_n\sigma_n)|^2$ is a symmetric function of its arguments, it follows that

$$\begin{aligned}\langle\Phi(n)|n(\boldsymbol{x},\sigma)|\Phi(n)\rangle &= \int \mathrm{d}\boldsymbol{x}_1\dots\mathrm{d}\boldsymbol{x}_n \sum_{\sigma_1\dots\sigma_n}\left(\sum_{j=1}^{n}\delta(\boldsymbol{x}-\boldsymbol{x}_j)\delta_{\sigma,\sigma_j}\right)|\Phi(\boldsymbol{x}_1\sigma_1\dots\boldsymbol{x}_n\sigma_n)|^2 \\ &= n\int \mathrm{d}\boldsymbol{x}_2\dots\mathrm{d}\boldsymbol{x}_n\sum_{\sigma_2\dots\sigma_n}|\Phi(\boldsymbol{x}\sigma,\boldsymbol{x}_2\sigma_2,\dots,\boldsymbol{x}_n\sigma_n)|^2\end{aligned} \tag{3.135}$$

which, up to a multiplicative factor n, is *the probability to find any one of the particles of spin σ at $\boldsymbol{x}$.*

Two-Body Reduced Density Matrix

If $\mathtt{V}$ is the two-body operator corresponding to the two-particle observable V, its mean value, from (3.101), is given by

$$\begin{aligned}\mathrm{Tr}\rho\mathtt{V} &= \frac{1}{2}\sum_{r_1r_2s_1s_2}\langle r_1r_2|V|s_1s_2\rangle\,\mathrm{Tr}\rho a^*_{r_1}a^*_{r_2}a_{s_2}a_{s_1} \\ &= \frac{1}{2}\mathrm{Tr}_2\rho^{(2)}V \quad ,\end{aligned} \tag{3.136}$$

where the matrix $\rho^{(2)}$ has elements

$$\langle s_1s_2|\rho^{(2)}|r_1r_2\rangle = \mathrm{Tr}\rho a^*_{r_1}a^*_{r_2}a_{s_2}a_{s_1} \tag{3.137}$$

in the tensor product basis $(1/\sqrt{2})(\psi_{r_1}\otimes\psi_{r_2}+\nu\psi_{r_2}\otimes\psi_{r_1})$ of two-particle space, with the appropriate symmetry. One can consider $\rho^{(2)}$ as an operator in two-particle space $\mathcal{H}_\nu^{\otimes 2}$. It is Hermitian, positive, and its normalization is given by

$$\mathrm{Tr}_2\rho^{(2)} = \sum_{rs}\mathrm{Tr}\rho a^*_r a^*_s a_s a_r = \langle\mathtt{N}(\mathtt{N}-1)\rangle \quad . \tag{3.138}$$

Knowledge of the *two-body reduced density matrix* $\rho^{(2)}$ is sufficient to *calculate the mean values of all two-body observables.*

In position space, $\rho^{(2)}$ has elements

$$\begin{aligned}&\langle \boldsymbol{x}_1\sigma_1, \boldsymbol{x}_2\sigma_2|\rho^{(2)}|\boldsymbol{x}_1'\sigma_1', \boldsymbol{x}_2'\sigma_2'\rangle \\ &\quad = \mathrm{Tr}\rho a^*(\boldsymbol{x}_1', \sigma_1')a^*(\boldsymbol{x}_2', \sigma_2')a(\boldsymbol{x}_2, \sigma_2)a(\boldsymbol{x}_1, \sigma_1) \quad ,\end{aligned} \tag{3.139}$$

of which the diagonal component

$$\begin{aligned}\Gamma(\boldsymbol{x}_1\sigma_1, \boldsymbol{x}_2\sigma_2) &= \langle \boldsymbol{x}_1\sigma_1, \boldsymbol{x}_2\sigma_2|\rho^{(2)}|\boldsymbol{x}_1\sigma_1, \boldsymbol{x}_2\sigma_2\rangle \\ &= \langle :n(\boldsymbol{x}_1, \sigma_1)n(\boldsymbol{x}_2, \sigma_2):\rangle\end{aligned} \tag{3.140}$$

has an important physical interpretation: it represents *the density correlations.* This can be verified for a vector $|\Phi(n)\rangle$ in n particle space. It follows from the definition given in Sect. 3.2.7 that $\frac{1}{2}{:}n(\boldsymbol{x}_1, \sigma_1)n(\boldsymbol{x}_2, \sigma_2){:}$ is the two-body operator which corresponds to multiplication by a product of Dirac functions in two-particle space,

$$\begin{aligned}&\delta_{\boldsymbol{x}_1,\sigma_1}(\boldsymbol{y}_1, \tau_1)\delta_{\boldsymbol{x}_2,\sigma_2}(\boldsymbol{y}_2, \tau_2)\Phi(\boldsymbol{y}_1\tau_1, \boldsymbol{y}_2\tau_2) \\ &\quad = \delta(\boldsymbol{x}_1 - \boldsymbol{y}_1)\delta_{\sigma_1,\tau_1}\delta(\boldsymbol{x}_2 - \boldsymbol{y}_2)\delta_{\sigma_2,\tau_2}\Phi(\boldsymbol{y}_1\tau_1, \boldsymbol{y}_2\tau_2) \quad .\end{aligned} \tag{3.141}$$

It follows that

$$\begin{aligned}&\langle\Phi(n)|{:}n(\boldsymbol{x}_1, \sigma_1)n(\boldsymbol{x}_2, \sigma_2){:}|\Phi(n)\rangle \\ &\quad = \int \mathrm{d}\boldsymbol{y}_1 \ldots \mathrm{d}\boldsymbol{y}_n \sum_{\tau_1\ldots\tau_n} \left(\sum_{i\neq j}^{n} \delta(\boldsymbol{x}_1 - \boldsymbol{y}_i)\delta_{\sigma_1,\tau_i}\delta(\boldsymbol{x}_2 - \boldsymbol{y}_j)\delta_{\sigma_2,\tau_j}\right) \\ &\qquad \times |\Phi(\boldsymbol{y}_1\tau_1, \ldots, \boldsymbol{y}_n\tau_n)|^2 \\ &\quad = n(n-1)\int \mathrm{d}\boldsymbol{y}_3 \ldots \mathrm{d}\boldsymbol{y}_n \sum_{\tau_3\ldots\tau_n} |\Phi(\boldsymbol{x}_1\sigma_1, \boldsymbol{x}_2\sigma_2, \boldsymbol{y}_3\tau_3, \ldots, \boldsymbol{y}_n\tau_n)|^2 \quad .\end{aligned} \tag{3.142}$$

This last expression, which results from the symmetry of $|\Phi(\boldsymbol{y}_1\tau_1, \ldots, \boldsymbol{y}_n\tau_n)|^2$, is precisely proportional to the *joint probability to find any two particles of spin σ_1 and σ_2 at $\boldsymbol{x}_1$ and $\boldsymbol{x}_2$, respectively.* The density correlation function obtained by summing (3.140) over all of the spin states,

$$\Gamma(\boldsymbol{x}_1, \boldsymbol{x}_2) = \langle{:}n(\boldsymbol{x}_1)n(\boldsymbol{x}_2){:}\rangle = \sum_{\sigma_1\sigma_2} \langle{:}n(\boldsymbol{x}_1, \sigma_1)n(\boldsymbol{x}_2, \sigma_2){:}\rangle \quad , \tag{3.143}$$

allows one to calculate the mean value of the interaction energy (3.110)

$$\langle \mathsf{V}\rangle = \frac{1}{2}\int \mathrm{d}\boldsymbol{x}_1\mathrm{d}\boldsymbol{x}_2 V(\boldsymbol{x}_1 - \boldsymbol{x}_2)\Gamma(\boldsymbol{x}_1, \boldsymbol{x}_2) \quad . \tag{3.144}$$

3.2.9 Correlations in Free Fermi and Bose Gases

It is instructive to evaluate the reduced density matrices for the independent fermion or boson gas. Suppose that the particles occupy the basis states $\{|\psi_r\rangle\}$ of the one-particle space $\mathcal{H}$. The state of n fermions or n bosons is thus an occupation number state $|n_1, \ldots, n_r, \ldots\rangle_\nu$ with $\sum_r n_r = n$.

The Non-Diagonal Part of the One-Body Density Matrix

One can easily obtain the one-body reduced density matrix with the help of (3.45)

$$\langle s|\rho^{(1)}|r\rangle = {}_\nu\langle n_1, \ldots, n_r, \ldots |a_r^* a_s|n_1, \ldots, n_r, \ldots\rangle_\nu = \delta_{r,s} n_r \quad . \tag{3.145}$$

The calculation of $\langle \boldsymbol{x}\sigma|\rho^{(1)}|\boldsymbol{x}'\sigma'\rangle$ follows immediately if one uses the completeness of the basis $\{|\psi_r\rangle\}$ to expand the vector $|\boldsymbol{x}\sigma\rangle$

$$|\boldsymbol{x}\sigma\rangle = \sum_r |r\rangle \langle r|\boldsymbol{x}\sigma\rangle \quad , \quad \langle \boldsymbol{x}\sigma|r\rangle = \psi_r(\boldsymbol{x}, \sigma) \quad . \tag{3.146}$$

Taking into account (3.145), one finds

$$\begin{aligned}\langle \boldsymbol{x}\sigma|\rho^{(1)}|\boldsymbol{x}'\sigma'\rangle &= \sum_{rs} \langle \boldsymbol{x}\sigma|r\rangle \langle r|\rho^{(1)}|s\rangle \langle s|\boldsymbol{x}'\sigma'\rangle \\ &= \sum_r n_r \langle \boldsymbol{x}\sigma|r\rangle \langle r|\boldsymbol{x}'\sigma'\rangle \\ &= \sum_r n_r \psi_r(\boldsymbol{x}, \sigma)\psi_r^*(\boldsymbol{x}', \sigma') \quad .\end{aligned} \tag{3.147}$$

For the case where the single-particle states are plane waves $\psi_{\boldsymbol{k}\tau}(\boldsymbol{x}, \sigma) = L^{-3/2}\delta_{\sigma,\tau}\mathrm{e}^{\mathrm{i}\boldsymbol{k}\cdot\boldsymbol{x}}$, (3.147) becomes

$$\langle \boldsymbol{x}\sigma|\rho^{(1)}|\boldsymbol{x}'\sigma'\rangle = \frac{\delta_{\sigma,\sigma'}}{L^3} \sum_{\boldsymbol{k}} n_{\boldsymbol{k}\sigma} \mathrm{e}^{\mathrm{i}\boldsymbol{k}\cdot(\boldsymbol{x}-\boldsymbol{x}')} \quad . \tag{3.148}$$

Now, consider a gas of non-interacting fermions in the ground state, with

$$n_{\boldsymbol{k}\sigma} = \begin{cases} 1 & , \quad |\boldsymbol{k}| \le k_\mathrm{F} \\ 0 & , \quad |\boldsymbol{k}| > k_\mathrm{F} \end{cases} \quad .$$

Taking the limit of infinite volume, gives

$$\begin{aligned}\langle \boldsymbol{x}\sigma|\rho^{(1)}|\boldsymbol{x}'\sigma'\rangle &= \delta_{\sigma,\sigma'} \frac{1}{(2\pi)^3} \int_{|\boldsymbol{k}|\le k_\mathrm{F}} \mathrm{d}\boldsymbol{k} \mathrm{e}^{-\mathrm{i}\boldsymbol{k}\cdot(\boldsymbol{x}-\boldsymbol{x}')} \\ &= \delta_{\sigma,\sigma'} \frac{k_\mathrm{F}^3}{2\pi^2} \left(\frac{j_1(k_\mathrm{F} r)}{k_\mathrm{F} r} \right) \quad , \quad r = |\boldsymbol{x} - \boldsymbol{x}'| \quad ,\end{aligned} \tag{3.149}$$

where $j_1(r) = \sin r/r^2 - \cos r/r$ is the Bessel function of order 1. Note that $\lim_{r\to 0} j_1(r)/r = 1/3$. Summing the diagonal component of (3.149) over the spin states, one finds the density (2.95) (with $g = 2$)

$$\rho = \sum_{\sigma} \langle \boldsymbol{x}\sigma | \rho^{(1)} | \boldsymbol{x}\sigma \rangle = \frac{1}{3\pi^2} k_{\mathrm{F}}^3 \quad . \tag{3.150}$$

On the other hand, *the non-diagonal elements vanish when* $r = |\boldsymbol{x} - \boldsymbol{x}'| \to \infty$

$$\lim_{|\boldsymbol{x}-\boldsymbol{x}'|\to\infty} \langle \boldsymbol{x}\sigma | \rho^{(1)} | \boldsymbol{x}'\sigma' \rangle = 0 \quad . \tag{3.151}$$

For non-interacting bosons, (taken here to have zero spin) the situation changes, as the population of a microscopic state can be extensive. The state $\boldsymbol{k} = 0$ can have a macroscopic occupation rate $n_0 = \rho_0 L^3$. Isolating this term in (3.148) and taking the limit of infinite volume gives

$$\langle \boldsymbol{x} | \rho^{(1)} | \boldsymbol{x}' \rangle = \rho_0 + \frac{1}{(2\pi)^3} \int \mathrm{d}\boldsymbol{k} n_{\boldsymbol{k}} \mathrm{e}^{-\mathrm{i}\boldsymbol{k}\cdot(\boldsymbol{x}-\boldsymbol{x}')} \quad . \tag{3.152}$$

The diagonal component $\langle \boldsymbol{x} | \rho^{(1)} | \boldsymbol{x} \rangle$ is the total density ρ and the second term of (3.152) is the contribution to ρ of the excited states. For the non-diagonal elements,

$$\lim_{|\boldsymbol{x}-\boldsymbol{x}'|\to\infty} \langle \boldsymbol{x} | \rho^{(1)} | \boldsymbol{x}' \rangle = \rho_0 \quad , \tag{3.153}$$

since the integrand of the second term of (3.152) is a rapidly oscillating function of $|\boldsymbol{x} - \boldsymbol{x}'|$. The difference between the two gases lies in the behaviour (3.151) and (3.153) of the non-diagonal elements of $\rho^{(1)}$. *The fact that the latter approaches a non-zero value when* $|\boldsymbol{x} - \boldsymbol{x}'| \to \infty$ *in a boson gas is a characteristic manifestation of the Bose–Einstein condensation.*

Induced Correlations by Fermi Statistics

The diagonal component of the two-body reduced density matrix for independent fermions can be calculated using relation (3.146) in (3.139) and (3.140), then applying (3.121). This gives

$$\begin{aligned}
&\Gamma(\boldsymbol{x}_1\sigma_1, \boldsymbol{x}_2\sigma_2) \\
&\quad = \sum_{r_1 r_2 s_1 s_2} {}_{-}\langle n_1, \dots, n_r, \dots | a_{r_1}^* a_{r_2}^* a_{s_2} a_{s_1} | n_1, \dots, n_r, \dots \rangle_{-} \\
&\qquad \times \langle \boldsymbol{x}_1\sigma_1 | r_1 \rangle \langle \boldsymbol{x}_2\sigma_2 | r_2 \rangle \langle s_2 | \boldsymbol{x}_2\sigma_2 \rangle \langle s_1 | \boldsymbol{x}_1\sigma_1 \rangle \\
&\quad = \sum_{r_1 r_2} n_{r_1} n_{r_2} (| \langle \boldsymbol{x}_1\sigma_1 | r_1 \rangle |^2 | \langle \boldsymbol{x}_2\sigma_2 | r_2 \rangle |^2 \\
&\qquad - \langle \boldsymbol{x}_1\sigma_1 | r_1 \rangle \langle r_1 | \boldsymbol{x}_2\sigma_2 \rangle \langle \boldsymbol{x}_2\sigma_2 | r_2 \rangle \langle r_2 | \boldsymbol{x}_1\sigma_1 \rangle) \\
&\quad = \langle \boldsymbol{x}_1\sigma_1 | \rho^{(1)} | \boldsymbol{x}_1\sigma_1 \rangle \langle \boldsymbol{x}_2\sigma_2 | \rho^{(1)} | \boldsymbol{x}_2\sigma_2 \rangle - | \langle \boldsymbol{x}_1\sigma_1 | \rho^{(1)} | \boldsymbol{x}_2\sigma_2 \rangle |^2 \quad ,
\end{aligned} \tag{3.154}$$

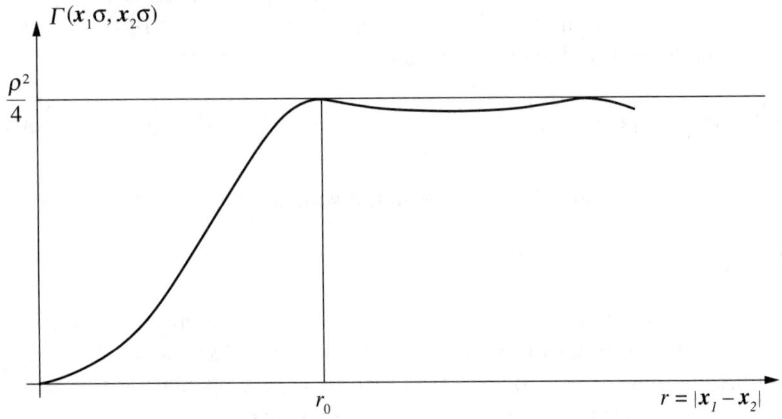

Fig. 3.3. The probability to find two fermions with the same spin at distance $r = |\boldsymbol{x}_1 - \boldsymbol{x}_2|$ becomes very small for $r < r_0$

where the last line follows from a comparison with (3.147). The density correlation is entirely determined by the one-body matrix density. This fact is characteristic of uncoupled particles only and is not true if they interact. In fact, for independent particles of classical statistics, the correlation function is factorizable and is equal to the product of the densities, corresponding to the first term of (3.154). The second term of (3.154) is a *quantum correlation effect induced by the Fermi statistics.* For the case of free fermions in their ground state, one explicitly obtains from (3.149)

$$\Gamma(\boldsymbol{x}_1\sigma_1, \boldsymbol{x}_2\sigma_2) = \left(\frac{\rho}{2}\right)^2 \left[1 - \delta_{\sigma_1,\sigma_2} 9 \left(\frac{j_1(k_\mathrm{F} r)}{k_\mathrm{F} r}\right)^2\right] \quad , \quad r = |\boldsymbol{x}_1 - \boldsymbol{x}_2| \quad . \tag{3.155}$$

For particles in different spin states, the exclusion principle, as expected, has no effect. On the other hand, if $\sigma_1 = \sigma_2 = \sigma$, $\Gamma(\boldsymbol{x}_1\sigma, \boldsymbol{x}_2\sigma)$ vanishes when $\boldsymbol{x}_1 = \boldsymbol{x}_2$ and remains small within a sphere of radius r_0 given by the first zero of function j_1 (Fig. 3.3)

$$k_\mathrm{F} r_0 = 4.49 \quad . \tag{3.156}$$

The probability to find two particles with the same spin orientation at a distance less than r_0 is very low: the exclusion principle acts as an effective repulsive force within r_0. For metallic electrons, k_F^{-1} is approximately 1 Å, so r_0 is of the order of 4 to 5 Å.

3.3 Quantum Physics and the Concept of a Perfect Gas

Experimental evidence of Bose–Einstein condensation was discussed in Sect. 2.2.1. The importance of this spectacular purely quantum phenomenon vindicates the attempts made in order to observe it. Now at temperatures for which quantum effects become important, there is a more direct reason why gases exhibit properties of perfect gases provided that their elements are fermions.

In classical thermodynamics, the recipe to create a perfect gas is well known. One takes a real gas and dilutes it at a fixed temperature. At some point, as the gas expands, the mean distance between the molecules passes beyond the limit of the intermolecular potential. In such a gas, quantum effects do not play an important role, at least not in the sense understood above. The discussion of Sect. 3.2.9 suggests, however, another means to achieve a (nearly) perfect gas in which the quantum properties are manifest: the exclusion principle can be used if the gas molecules are fermions. To understand this better, it is useful to have a good qualitative understanding of the nature of collisions between two fermions. A good example for such a study is a gas of atoms of the light helium isotope, ^{3}He, the properties of which will be discussed in Chap. 7.

In 1939, Alvarez and Cornog showed with the aid of a cyclotron that helium, in the state found naturally in air or other gases, contains small quantities of the light isotope ^{3}He. The nucleus of ^{3}He, containing two protons and one neutron, has spin 1/2. The atom also has spin 1/2, which contains two 1s electrons of opposite spin. The associated magnetic moment is very small, implying that the magnetic energy due to the alignment of the moments of two neighbouring atoms is completely negligible. In other words, the interaction potential of two helium atoms is practically indifferent to the nature of the isotopes of which they are composed. One can thus describe the interaction between two helium atoms without specifying the isotopes.

This interaction can be described with the help of a highly repulsive potential acting at short distance (hard central core), weakly attractive at intermediate distances (several Å) and vanishing everywhere else. It is isotropic to very good approximation, since direct interaction between the spins is negligible. In this case, collision theory applies to the effective cross-section as the sum of contributions with each corresponding to a fixed value l of the orbital angular momentum. At very low energy, the only significant contribution is due to the s waves ($l = 0$). In classical terms, this signifies that the atoms do not "see" each other unless the collision is direct. For all other impact parameters, the interaction is unnoticed.

However, as discussed in Sect. 2.2.2 regarding the origin of ferromagnetism, two atoms of ^{3}He (fermions because of their spin) with parallel or antiparallel spins have a very large importance as to the dynamics of their collision. This is not tied to the form of the potential, since it is relatively indifferent to the spin, but to the exclusion principle. If the spins of the two atoms are parallel and in the same direction (one says they are aligned), the orbital wavefunction, expressed in terms of relative variables, is antisymmetric. This is dictated by the exclusion principle.

To interpret this characteristic of low energy collisions, it is possible to use the de Broglie wavelength λ_B. For an atom, a wavelength λ_B is defined by the de Broglie relation

$$p = \frac{h}{\lambda_B} \quad , \tag{3.157}$$

where p is the momentum measured in the center of mass. If $\lambda_B \ll b_0$ (where b_0 is the range of the potential), a number of l values contribute to the effective cross-section. If $\lambda_B \gg b_0$ (low energy), the wave aspect dominates and one returns to the case where only the s waves play a role.

For a non-degenerate gas, the notion of the de Broglie wave is still generally useful and (3.157) is used, replacing p by its mean value $\bar{p}$ such that the equipartition of energy gives

$$\frac{\bar{p}^2}{2m} = \frac{3}{2} k_B T \quad . \tag{3.158}$$

The mean de Broglie wavelength is thus

$$\bar{\lambda}_B = \frac{h}{\sqrt{2\pi m k_B T}} \quad , \tag{3.159}$$

which characterizes the gas. The numerical coefficient used in (3.159) conforms to current usage and differs slight with what is obtained substituting (3.157) into (3.158).

When the gas is formed with atoms of the heavy isotope ^{4}He, the qualitative description of the collisions given above is valid. It is also true with two atoms of ^{3}He having opposite spins. However, when the spins of the light isotope are parallel and in the same direction, there can be a spectacular effect when $\bar{\lambda}_B \gg b_0$. From this, the contribution to the effective cross-section of values of $l \geq 1$ is very small. On the other hand, the exclusion principle forces the wavefunction to be antisymmetric. So, contributions to the effective cross-section corresponding to even l (notably $l = 0$) vanish. In other words, atoms of parallel spin have a negligible probability to be at a distance short enough to feel the effects of their interaction. That is, the two atoms are "transparent" to each other and the effective cross-section falls to values much smaller than in the other cases.

Polarized ^{3}He Gas

The low value of the effective cross-section notably reduces the number of collisions and hence the viscosity of the gas. For this to have macroscopic effects on ^{3}He gases, it is necessary to meet the following conditions:

– The gas must be diluted enough so that the number of collisions of more than two bodies is greatly reduced.
– All of the spins must be aligned: this is called polarized helium three (^{3}He $\uparrow$). This situation can be arranged with the use of a magnetic field. The problem is to avoid the flipping of spin, highly probable at the walls of the container.
– The temperature must be low enough so that the de Broglie wavelength effectively exceeds the range of the potential b_0. Taking into account (3.159) and the fact that the range is several Å, it is only at temperatures on the order of 1–2 K where the condition $\bar{\lambda}_{\mathrm{B}} > b_0$ can be realized.

One effectively observes that ^{3}He $\uparrow$ approaches a perfect gas below 2 K. Its viscosity is significantly reduced when compared to ^{3}He at the same temperature. It was C. Lhuillier and F. Laloë who first suggested polarizing ^{3}He in order to lower its viscosity.[1]

Exercises

1. Second Quantized Spin Operators

Let C^2 be the two-dimensional state space of spin 1/2. Construct the second quantized spin operator S on $\mathcal{F}(C^2)$ and verify that for fermions as well as for bosons its components $\mathsf{S}_\mu, \mu = 1, 2, 3$, obey the commutation rules of angular momenta.

2. Algebra of Creation and Annihilation Operators

Show that if an operator O on $\mathcal{F}(\mathcal{H})$ commutes with all the $a(\varphi)$ and all the $a^*(\varphi), \varphi \in \mathcal{H}$, then $\mathsf{O} = \lambda \mathsf{I}$ is necessarily a multiple of the identity operator.

Hint: Remember that the states (3.37) generate $\mathcal{F}(\mathcal{H})$ and that the vacuum state is unique.

Comment: The $a(\varphi)$ and $a^*(\varphi), \varphi \in \mathcal{H}$, generate a "complete" algebra: all operators in Fock space can, in principle, be expressed in terms of them.

3. The Hubbard Hamiltonian

The description of the many-particle system in terms of the full second quantized Hamiltonian given by the sum of the kinetic energy (3.72) and the potential energy (3.109) may be complicated. One has sometimes recourse to simplified model Hamiltonians that capture some essential physical mechanisms (examples are the reduced Hamiltonian (5.99) of superconductivity

[1] C. Lhuillier and F. Laloë, J. de Physique *40*, 239 (1979).

and the Bogoliubov Hamiltonian (7.74) of superfluidity). The physics of electrons in an atomic lattice Λ is often studied with the help of the Hubbard Hamiltonian. The following simplifications are introduced:

– Electrons can "hop" from one atom to a nearest neighbour atom with (real) hopping transition amplitude $\kappa > 0$. The kinetic energy reads accordingly

$$\mathsf{H}_0^{\text{Hubbard}} = -\kappa \sum_{\langle \boldsymbol{i},\boldsymbol{j}\rangle \in \Lambda} \sum_{\sigma} a^*_{\boldsymbol{i},\sigma} a_{\boldsymbol{j},\sigma} \quad ,$$

where $a^*_{\boldsymbol{j},\sigma}, a_{\boldsymbol{j},\sigma}$ are the creation and annihilation operators for electrons at the site indexed by $\boldsymbol{j}$ and the sum runs over all next neighbouring pairs of lattice sites.

– The only two-body interaction consists of a Coulomb repulsion when two electrons are found on the same atom

$$\mathsf{V}^{\text{Hubbard}} = U \sum_{\boldsymbol{j} \in \Lambda} n_{\boldsymbol{j}\uparrow} n_{\boldsymbol{j}\downarrow} \quad , \quad U > 0 \quad ,$$

with $n_{\boldsymbol{j}\uparrow}, n_{\boldsymbol{j}\downarrow}$ the electronic densities (3.70) at site $\mathbf{j}$ and spin $\sigma = \uparrow, \downarrow$.

(i) For the one-dimensional (infinite) lattice $\Lambda = \{jd, j = \ldots, -1, 0, 1 \ldots\}$, one has

$$\mathsf{H}_0^{\text{Hubbard}} = \sum_{\sigma} \int_{-\pi/d}^{\pi/d} \mathrm{d}k\, \varepsilon(k) a^*(k,\sigma) a(k,\sigma) \quad , \quad \varepsilon(k) = -2\kappa \cos kd \quad ,$$

where $a(k,\sigma) = \sqrt{d/2\pi} \sum_j \mathrm{e}^{-\mathrm{i}kjd} a_{j,\sigma}$ are Fourier transforms of the $a_{j,\sigma}$ and $a(k,\sigma)a^*(k',\sigma') + a^*(k,\sigma)a(k',\sigma') = \delta_{\sigma,\sigma'}\delta(k-k')$.

Compare this with the kinetic energy (3.72). Compare V^{Hubbard} with the potential energy (3.109).

(ii) Show that the particle number is conserved.

(iii) Let $\mathsf{S}_{\boldsymbol{j},\mu} = (\hbar/2)\sum_{\sigma,\sigma'} a^*_{\boldsymbol{j},\sigma} (\tau_\mu)_{\sigma,\sigma'} a_{\boldsymbol{j},\sigma'}$ be the μ-component ($\mu = 1,2,3$) of the spin operator at site $\boldsymbol{j}$, with τ_μ the Pauli matrices

$$\tau_1 = \begin{pmatrix} 0 & 1 \\ 1 & 0 \end{pmatrix} \quad , \quad \tau_2 = \begin{pmatrix} 0 & -i \\ i & 0 \end{pmatrix} \quad , \quad \tau_3 = \begin{pmatrix} 1 & 0 \\ 0 & -1 \end{pmatrix} \quad .$$

Compute

$$|\mathbf{S}_{\boldsymbol{j}}|^2 = \sum_{\mu=1}^{3} (\mathsf{S}_{\boldsymbol{j},\mu})^2 = \hbar^2 \left(\frac{1}{4}(n_{\boldsymbol{j}\uparrow} + n_{\boldsymbol{j}\downarrow}) - \frac{3}{2} n_{\boldsymbol{j}\uparrow} n_{\boldsymbol{j}\downarrow} \right) \quad .$$

Conclude that the Hubbard Hamiltonian can be expressed in terms of spin operators by

$$\mathtt{H}^{\text{Hubbard}} = -\kappa \sum_{\langle \boldsymbol{i},\boldsymbol{j}\rangle \in \Lambda} \sum_{\sigma} a^*_{\boldsymbol{i},\sigma} a_{\boldsymbol{j},\sigma} - \frac{2}{3\hbar^2} U \sum_{\boldsymbol{j}} |\mathbf{S}_{\boldsymbol{j}}|^2 + \frac{U}{6}\mathtt{N}$$

and that the total spin $\mathbf{S} = \sum_{\boldsymbol{j}\in\Lambda} \mathbf{S}_{\boldsymbol{j}}$ is a conserved quantity.

Hint: Use the relation $\sum_{\mu=1}^{3} (\tau_\mu)_{\sigma,\sigma'} (\tau_\mu)_{\rho,\rho'} = 2\delta_{\sigma,\rho'}\delta_{\sigma',\rho} - \delta_{\sigma,\sigma'}\delta_{\rho,\rho'}$.

4. Electron Gas

4.1 The Hartree–Fock Method

4.1.1 The Variational Principle

The principal properties of a system of n particles at zero temperature can be deduced from knowledge of the energy E_0 of the ground-state wavefunction $\Phi_0(\boldsymbol{x}_1\sigma_1, \ldots, \boldsymbol{x}_n\sigma_n)$

$$\mathtt{H}\,|\Phi_0\rangle = E_0\,|\Phi_0\rangle \quad . \tag{4.1}$$

It is evident that when the particles interact and n is large, solving the eigenvalue equation (4.1) can present considerable difficulty. For this reason, it is useful to examine a variety of approximation techniques. The two most important are *perturbation theory* and the *variational method.* The next three chapters are based on the variational method. The following is a brief outline of this method.

Let E_r and $|\Phi_r\rangle$ be the energies and eigenstates of the Hamiltonian $\mathtt{H}$

$$\mathtt{H}\,|\Phi_r\rangle = E_r\,|\Phi_r\rangle \quad , \quad r = 0, 1, 2, \ldots \quad , \tag{4.2}$$

with the energy eigenvalues written in increasing order and repeated according to their multiplicity. If $|\Phi\rangle$ is any state normalized to 1, the following fundamental inequality holds

$$\langle\Phi|\mathtt{H}|\Phi\rangle \geq E_0 \quad . \tag{4.3}$$

For all states, the mean value of the energy cannot be less than the energy of the ground state. To see this, expand $|\Phi\rangle$ in the basis of eigenstates $\{|\Phi_r\rangle\}$

$$|\Phi\rangle = \sum_{r\geq 0} c_r\,|\Phi_r\rangle \quad , \quad \sum_{r\geq 0} |c_r|^2 = \langle\Phi|\Phi\rangle = 1 \quad , \tag{4.4}$$

so that

$$\begin{aligned} \langle\Phi|\mathtt{H}|\Phi\rangle &= \sum_{r,s\geq 0} c_r^* c_s\,\langle\Phi_r|\mathtt{H}|\Phi_s\rangle \\ &= \sum_{r\geq 0} |c_r|^2 E_r \geq E_0 \sum_{r\geq 0} |c_r|^2 = E_0 \quad , \end{aligned} \tag{4.5}$$

since $E_0 \leq E_r, r = 1, 2, \dots$. Given a class of states $\mathcal{C}$, the best approximation of E_0 with $|\Phi\rangle$ belonging to $\mathcal{C}$ is obtained by minimizing $\langle \Phi|\mathtt{H}|\Phi\rangle$ in $\mathcal{C}$. Success depends on a *judicious choice* of the variational space $\mathcal{C}$. This choice is guided by the physics of the problem and should lead to feasible calculations.

The relative advantages of perturbation theory and the variational method are different. Perturbation theory is *systematic*. That is, it provides an algorithm for calculating the energy corrections as a power series of the interaction strength. The variational method, however, presents greater *flexibility*. It allows for the inclusion of physical insight in the choice of trial functions. It can be applied in cases where E_0 cannot be obtained perturbatively, (see Chap. 5, the BCS theory of superconductivity). However, the error associated with the E_0 calculation is very difficult to estimate.

In Sects. 4.1.2, 4.2.1 and 4.2.2, the *Hartree–Fock* method is presented, with its application to an electron gas. *The main idea of this method is to replace the many-body problem by an effective one-body problem as a first approximation.* As seen in Sect. 2.2.2, a state of n non-interacting fermions is an anti-symmetrical product of eigenfunctions of the single-particle Hamiltonian. One can note the success of numerous theories which treat the electrons as independent particles. For examples of the electron gas of a metal, one can cite Sommerfeld's theory and Bloch's theory. In the case of the nucleus, the shell model treats the nucleons as being independent and interacting with an average potential from the other nucleons (Chap. 6). These models justify a priori the idea of the Hartree–Fock method, which is applied equally to the atomic nucleus.

The variational problem of Hartree–Fock consists of assuming that the state of mutually interacting particles can still be expressed as the product of effective single-particle functions, determined in a manner that minimizes the total energy. This implies that, for fermions, the class of trial functions, $\mathcal{C}$ is constituted of all n-particle states of the form

$$|\Phi\rangle = a^*(\varphi_1)a^*(\varphi_2)\dots a^*(\varphi_n)\,|0\rangle \quad , \tag{4.6}$$

where the $|\varphi_j\rangle, j = 1, 2, \dots, n$, belong to single-particle space $\mathcal{H}$ and are orthonormal

$$\langle \varphi_i|\varphi_j\rangle = \delta_{i,j} \quad . \tag{4.7}$$

The two conditions (4.6) and (4.7) define the variational space $\mathcal{C}$ when the $|\varphi_j\rangle$ run over $\mathcal{H}$. Constraint (4.7) assures that the $|\varphi_j\rangle$ are linearly independent: this must be the case for any non-zero $|\Phi\rangle$. As the $|\varphi_j\rangle$ are members of a basis of $\mathcal{H}$, $|\Phi\rangle$ is in fact a vector in occupation numbers relative to this basis and, by consequence, is normalized to 1, according to (2.75).

Before establishing the equations which must satisfy the states $|\varphi_j\rangle$, it is useful to distinguish for which systems the Hartree–Fock method can be applied. Three such systems are particularly important: *the atom, the electron gas in a solid, and the nucleus.* The case of the atom was tackled at the very beginning of quantum physics where one studies the energy levels of n electrons submitted to a mutual Coulomb interaction force and the Coulomb force of the nucleus (assumed to be static because of its relatively large mass). The electron gas differs from the atom in that the positive charge, also considered to be static, is extended over all space, so as to be localized on the sites of an ionic lattice or approximated by a uniform density. In the latter model, because of the homogeneity of the density, the Hartree–Fock variational problem provides an exact solution which is given by the plane waves (Sect. 4.2.2). In the case of the nucleus, the method is hindered by the lack of precise knowledge of the interaction potential. As the nucleon is a composite particle, the information which is used to determine the effective potential is provided by calculations of quantum field theory or from experimental studies of scattering.

4.1.2 The Hartree–Fock Equations

Consider a system of n fermions governed by the n-particle Hamiltonian

$$\mathtt{H}(n) = \sum_{j=1}^{n} H_j + \frac{1}{2}\sum_{i\neq j}^{n} V_{ij} \quad . \tag{4.8}$$

The single-particle Hamiltonian

$$H = \frac{|\boldsymbol{p}|^2}{2m} + V^{\text{ext}}(\boldsymbol{q}) \tag{4.9}$$

comprises the kinetic energy and the effect of an external potential $V^{\text{ext}}(\boldsymbol{q})$. To simplify the notation, it is useful to introduce the collective variable $q = (\boldsymbol{x}, \sigma)$ and to symbolize the sum over position space and the spins by $\int \mathrm{d}q \ldots = \int \mathrm{d}\boldsymbol{x} \sum_\sigma \cdots$. The two-body interaction $V_{ij} = V(q_i, q_j)$ acts multiplicatively on the position and spin.

The Variational Problem

To formulate the variational problem with states of the form (4.6) and (4.7), it is necessary to express $\langle \Phi|\mathtt{H}(n)|\Phi\rangle$ in terms of the individual states $|\varphi_j\rangle$. This is straightforward considering that the $|\varphi_j\rangle$ are members of a basis of $\mathcal{H}$ and $|\Phi\rangle$ is a state in the occupation number representation. Each one-particle state $|\varphi_j\rangle$ is occupied once and only once. It follows from (3.122) that

$$\begin{aligned}\langle \Phi|\mathtt{H}(n)|\Phi\rangle = \sum_{j=1}^{n} \langle \varphi_j|H|\varphi_j\rangle + \frac{1}{2}\sum_{j,m=1}^{n} \big(&\langle \varphi_j \otimes \varphi_m|V|\varphi_j \otimes \varphi_m\rangle \\ &- \langle \varphi_j \otimes \varphi_m|V|\varphi_m \otimes \varphi_j\rangle\big) \quad .\end{aligned} \tag{4.10}$$

To account for the n^2 constraints (4.7),[1] one can introduce n^2 *Lagrange parameters* ε_{jm}, $j, m = 1, 2, \ldots, n$, as well as the functional

$$F(|\varphi_1\rangle, \ldots, |\varphi_n\rangle) = \langle\Phi|\mathtt{H}(n)|\Phi\rangle - \sum_{j,m=1}^{n} \varepsilon_{jm}(\langle\varphi_j|\varphi_m\rangle - \delta_{j,m}) \quad . \tag{4.11}$$

It is thus a matter of finding the minimum of F when the $|\varphi_j\rangle$ vary in $\mathcal{H}$: a necessary condition is that F remains stationary under all variations. A variation of $|\varphi_j\rangle$ is obtained by replacing $|\varphi_j\rangle$ by $|\varphi_j\rangle + \lambda\,|\psi_j\rangle$ where $|\psi_j\rangle$ is an arbitrary vector in $\mathcal{H}$ and λ a parameter. The stationarity of F demands

$$\frac{\mathrm{d}}{\mathrm{d}\lambda} F(|\varphi_1\rangle + \lambda\,|\psi_1\rangle, \ldots, |\varphi_n\rangle + \lambda\,|\psi_n\rangle) = 0 \quad , \tag{4.12}$$

when $\lambda = 0$ for all choices of $|\psi_j\rangle$. Performing this operation on (4.11) gives

$$\begin{aligned} 0 = &\sum_{j=1}^{n} \langle\psi_j|H|\varphi_j\rangle \\ &+ \frac{1}{2}\sum_{j,m=1}^{n} \big(\langle\psi_j \otimes \varphi_m|V|\varphi_j \otimes \varphi_m\rangle + \langle\varphi_j \otimes \psi_m|V|\varphi_j \otimes \varphi_m\rangle \\ &- \langle\psi_j \otimes \varphi_m|V|\varphi_m \otimes \varphi_j\rangle - \langle\varphi_j \otimes \psi_m|V|\varphi_m \otimes \varphi_j\rangle\big) \\ &- \sum_{j,m=1}^{n} \varepsilon_{mj}\,\langle\psi_j|\varphi_m\rangle + \text{complex conjugate} \quad . \end{aligned} \tag{4.13}$$

By employing the invariance of V under the exchange of two particles (3.103) and permuting indices j, m in the third and fifth terms of (4.13), this equation becomes

$$\begin{aligned} 0 = &\sum_{j=1}^{n} \langle\psi_j|H|\varphi_j\rangle \\ &+ \sum_{j,m=1}^{n} \big(\langle\psi_j \otimes \varphi_m|V|\varphi_j \otimes \varphi_m\rangle - \langle\psi_j \otimes \varphi_m|V|\varphi_m \otimes \varphi_j\rangle\big) \\ &- \sum_{j,m=1}^{n} \varepsilon_{jm}\,\langle\psi_j|\varphi_m\rangle + \text{complex conjugate} \quad . \end{aligned} \tag{4.14}$$

As the $|\psi_j\rangle$ are arbitrary, (4.14) is still valid if $|\psi_j\rangle$ is replaced with $\mathrm{i}\,|\psi_j\rangle$. Furthermore, when $|\psi_j\rangle$ are all zero, except for one value of the index j, (4.14) implies

---

[1] The fact that condition (4.7) is symmetric vis-a-vis the exchange of indices i and j implies that only the symmetric part of ε_{jm} plays a role.

$$\langle\psi_j|H|\varphi_j\rangle + \sum_{m=1}^{n}\left(\langle\psi_j\otimes\varphi_m|V|\varphi_j\otimes\varphi_m\rangle - \langle\psi_j\otimes\varphi_m|V|\varphi_m\otimes\varphi_j\rangle\right)$$
$$= \sum_{m=1}^{n}\varepsilon_{jm}\langle\psi_j|\varphi_m\rangle \quad , \quad j=1,2,\ldots,n \quad . \tag{4.15}$$

The scalar products are explicitly written as

$$\int \mathrm{d}q\psi_j^*(q)\left(-\frac{\hbar^2\nabla^2}{2m}+V^{\mathrm{ext}}(q)\right)\varphi_j(q)$$
$$+\int \mathrm{d}q\psi_j^*(q)\int \mathrm{d}q'V(q,q')\left(\sum_{i=1}^{n}|\varphi_i(q')|^2\varphi_j(q)\right)$$
$$-\int \mathrm{d}q\psi_j^*(q)\int \mathrm{d}q'\left(\sum_{i=1}^{n}\varphi_i(q)V(q,q')\varphi_i^*(q')\varphi_j(q')\right)$$
$$=\sum_{i=1}^{n}\varepsilon_{ji}\int \mathrm{d}q\psi_j^*(q)\varphi_i(q) \quad . \tag{4.16}$$

As $\psi_j(q)$ is arbitrary, it is necessary that the $\varphi_j(q)$ obey a system of *non-linear integro-differential equations* which can be written in the form

$$-\frac{\hbar^2}{2m}(\nabla^2\varphi_j)(q)+\left[V^{\mathrm{ext}}(q)+U(q)\right]\varphi_j(q)-\int \mathrm{d}q'W(q,q')\varphi_j(q')$$
$$=\sum_{i=1}^{n}\varepsilon_{ji}\varphi_i(q) \quad . \tag{4.17}$$

The Mean Potential and the Exchange Potential

Equation (4.17) introduces the effective potentials U and W. In reference to (4.16), U *acts by multiplication* with the function

$$U(q)=\int \mathrm{d}q'V(q,q')\left(\sum_{i=1}^{n}|\varphi_i(q')|^2\right) \quad , \tag{4.18}$$

while W is an *integral operator* of which the matrix elements in position space are

$$\langle q|W|q'\rangle = W(q,q') = \sum_{i=1}^{n}\varphi_i(q)V(q,q')\varphi_i^*(q') \quad . \tag{4.19}$$

It is clear from (4.18) and (4.19) that U and W are Hermitian

$$U(q)=U^*(q) \quad , \quad W(q,q')=W^*(q',q) \quad . \tag{4.20}$$

Since $\sum_{i=1}^{n} |\varphi_i(q)|^2 = \langle \Phi | n(q) | \Phi \rangle$ is the particle density, $U(q)$ can be interpreted as the *mean potential* created at q by the collection of particles found in state $|\Phi\rangle$. The term W has a non-local action and results from Fermi statistics; it is called the *exchange potential.*

The Lagrange parameters are determined by taking the scalar product (4.17) with $\varphi_i(q)$ or, equivalently, posing $|\psi_j\rangle = |\varphi_i\rangle$ in (4.15)

$$\begin{aligned}\varepsilon_{ji} &= \langle \varphi_i | H | \varphi_j \rangle + \sum_{m=1}^{n} \left(\langle \varphi_i \otimes \varphi_m | V | \varphi_j \otimes \varphi_m \rangle - \langle \varphi_i \otimes \varphi_m | V | \varphi_m \otimes \varphi_j \rangle \right) \\ &= \langle \varphi_i | H | \varphi_j \rangle + \langle \varphi_i | U | \varphi_j \rangle - \langle \varphi_i | W | \varphi_j \rangle \quad . \end{aligned} \tag{4.21}$$

The Hartree–Fock Equations and the Hartree–Fock Energy

The problem now is to find n functions $\varphi_j(q)$ which satisfies (4.17) and then to calculate the corresponding *Hartree–Fock energy*. To put it simply: the energy $\langle \Phi | \mathtt{H}(n) | \Phi \rangle$, as well as the potentials U and W are invariant under change of basis in the n-dimensional space generated by vectors $|\varphi_j\rangle, j = 1, 2, \ldots, n$. Consider a unitary matrix R of dimension $n \times n$ and the transformed states

$$|\tilde{\varphi}_j\rangle = \sum_{m=1}^{n} R_{jm} |\varphi_m\rangle \quad . \tag{4.22}$$

States $|\Phi\rangle$ and $|\tilde{\Phi}\rangle$ are Slater determinants (2.69): the determinant of the composition of two linear applications is a product of their determinants. So, one has $|\tilde{\Phi}\rangle = (\mathrm{Det}R)\,|\Phi\rangle$ and $\langle \tilde{\Phi} | \mathtt{H}(n) | \tilde{\Phi} \rangle = \langle \Phi | \mathtt{H}(n) | \Phi \rangle$, since $|(\mathrm{Det}R)| = 1$. Unitarity of R also implies that

$$\sum_{j=1}^{n} \tilde{\varphi}_j(q) \tilde{\varphi}_j^*(q') = \sum_{j=1}^{n} \varphi_j(q) \varphi_j^*(q')$$

and thus $\tilde{U} = U$ and $\tilde{W} = W$. In conclusion, the $\tilde{\varphi}_j(q)$ equally satisfy (4.17) with the Lagrange parameters $\tilde{\varepsilon}_{ji} = \sum_{m,l=1}^{n} R_{jm} \varepsilon_{ml} R_{li}^{-1}$. Thus, any solution of (4.17) determines a whole family of solutions of (4.17), linked by transformations (4.22), and all giving rise to the same Hartree–Fock energy.

Formula (4.21) shows that matrix ε_{ji} is Hermitian: it can thus be diagonalized by a unitary transformation. Consequently, there always exists in this family a solution to (4.17) (which is again denoted as $\varphi_j(q), j = 1, 2, \ldots, n$) corresponding to the diagonal matrix $\varepsilon_{ji} = \varepsilon_j^{\mathrm{HF}} \delta_{i,j}$

$$\begin{aligned} &-\frac{\hbar^2}{2m} (\nabla^2 \varphi_j)(q) + \left[V^{\mathrm{ext}}(q) + U(q) \right] \varphi_j(q) - \int \mathrm{d}q' W(q, q') \varphi_j(q') \\ &= \varepsilon_j^{\mathrm{HF}} \varphi_j(q) \quad . \end{aligned} \tag{4.23}$$

There is thus no restriction to the generality of adopting form (4.23) for the variational equations: these are the *Hartree–Fock equations.*

Equations (4.23) present a formal analogy with the Schrödinger eigenvalue equations for one particle. They are different, nevertheless, on two essential points. First of all, they are *non-linear*, since U and W depend quadratically on $\varphi_j(q)$. This non-linearity incorporates in a self-consistent manner the effects of two-body interactions. Secondly, *the "eigenvalues" $\varepsilon_j^{\mathrm{HF}}$ do not represent individual energy states* and their sum does not give the Hartree–Fock energy. In fact, one obtains, after taking the scalar product of (4.23) with $\varphi_j(q)$ and summing over j,

$$\sum_{j=1}^{n} \varepsilon_j^{\mathrm{HF}} = \sum_{j=1}^{n} \langle \varphi_j | H | \varphi_j \rangle + \sum_{j=1}^{n} \left(\langle \varphi_j | U | \varphi_j \rangle - \langle \varphi_j | W | \varphi_j \rangle \right) \quad . \tag{4.24}$$

Definitions (4.18) and (4.19) together with expression (4.10) gives

$$\langle \Phi | \mathtt{H}(n) | \Phi \rangle = \sum_{j=1}^{n} \varepsilon_j^{\mathrm{HF}} - \frac{1}{2} \sum_{j=1}^{n} \left(\langle \varphi_j | U | \varphi_j \rangle - \langle \varphi_j | W | \varphi_j \rangle \right) \quad . \tag{4.25}$$

The Hartree–Fock equations only express the stationarity of the energy function: given a solution, it is necessary in principle to verify that it corresponds to a minimum.

While the many-body problem is thus reduced to an effective non-linear one-body problem, the analytical solution of the Hartree–Fock equations is in general not possible, except for the case of the electron gas. Application to the atom or to nuclear matter requires elaborated numerical calculations.

4.2 Electron Gas in the Hartree–Fock Approximation

4.2.1 Electron Gas and Its Hamiltonian

An electron gas is a collection of n electrons, of density $\rho = n/L^3$, obeying Fermi statistics and interacting with the Coulomb potential

$$V(\boldsymbol{x}) = \frac{1}{4\pi\varepsilon_o |\boldsymbol{x}|} \quad . \tag{4.26}$$

To guarantee global neutrality, the electrons interact equally with a classical positive and uniform charge density ρe (neutralizing background). This charge density creates a potential[2]

$$V^{\mathrm{ext}}(\boldsymbol{x}) = \rho e \int_{\Lambda} \mathrm{d}\boldsymbol{y} V(\boldsymbol{x} - \boldsymbol{y}) \quad . \tag{4.27}$$

[2] In this section, $e > 0$ is the magnitude of the charge of an electron.

The complete Hamiltonian of an electron gas is

$$\mathtt{H}(n) = \sum_{j=1}^{n} \frac{|\boldsymbol{p}_j|^2}{2m} + \frac{e^2}{2} \sum_{i \neq j}^{n} V(\boldsymbol{x}_i - \boldsymbol{x}_j)$$
$$- \rho e^2 \sum_{j=1}^{n} \int_{\Lambda} \mathrm{d}\boldsymbol{y} V(\boldsymbol{x}_j - \boldsymbol{y}) + \frac{\rho^2 e^2}{2} \int_{\Lambda} \mathrm{d}\boldsymbol{x} \int_{\Lambda} \mathrm{d}\boldsymbol{y} V(\boldsymbol{x} - \boldsymbol{y}) \quad . \tag{4.28}$$

The first term is the kinetic energy and the second is the Coulomb repulsion between the electrons. The third term, which is negative, is the potential of the electrons due to the oppositely charged background, while the last term represents the self-energy of the neutralizing background.

This model, defined by Hamiltonian (4.28), is called *jellium*. It is the simplest model of the conduction electrons of a metal, in which the dynamics of the ion lattice (the phonons) is ignored. The ions are static and their charge has been spread out uniformly to create the neutralizing background. *The crystal aspects are thus not taken into account in this model and only the collective properties of the electron gas are studied.* A more general model, in which the dynamics of the elastic waves of the solid (phonons) are taken into account, is introduced in Sect. 8.4.3.

The Characteristic Parameters

High and low density regimes of this system are often characterized by the dimensionless parameter r_s defined by

$$r_s = \frac{r_0}{a_{\mathrm{B}}} = \left(\frac{3}{4\pi}\right)^{1/3} \frac{\rho^{-1/3}}{a_{\mathrm{B}}} \quad , \tag{4.29}$$

where r_0 is the average inter-electron distance

$$\frac{4\pi r_0^3}{3} = \frac{L^3}{n} = \rho^{-1} \tag{4.30}$$

and a_{B} is the Bohr radius

$$a_{\mathrm{B}} = \frac{4\pi\varepsilon_0 \hbar^2}{me^2} \quad . \tag{4.31}$$

The parameter r_s is a ratio measure of the potential energy to the kinetic energy of the electron gas, since one can also write

$$r_s = \frac{e^2}{4\pi\varepsilon_0 r_0} \Big/ \frac{\hbar^2}{m r_0^2} \tag{4.32}$$

from (4.30) and (4.31). The numerator of (4.32) is the potential of two electrons at a distance r_0, whereas the denominator is approximately the kinetic energy of a wavepacket of extension r_0.

The high density regime is characterized by $r_s \ll 1$. In this case, the kinetic energy predominates and the phase is homogeneous. Indeed, from (2.97) the kinetic energy density behaves as $\rho^{5/3}$, whereas from (2.114) the potential energy density behaves as $\rho^{4/3}$. Hence, the Coulomb interaction is much less important as r_s becomes smaller. The low-density regime corresponds to $r_s \gg 10$. For such densities, the potential energy dominates and the electrons form a solid phase called the *Wigner crystal*. In this case, the electrons occupy a lattice in the neutralizing background. The Coulomb repulsion keeps the electrons distant from one another. The resulting localized electronic structure cannot be destabilized by a weak kinetic energy when r_s is large enough. Finally, the intermediate regime, for which $2 \lesssim r_s \lesssim 6$, corresponds to the typical electronic densities of metals. In this case, the kinetic and potential energies are comparable and it becomes difficult to apply approximation methods.

In all situations, one wants to know the energy per electron in the ground state as a function of r_s, that is

$$\mathrm{u}(r_s) = \lim_{n\to\infty} \frac{E_0(n)}{n} \quad , \quad \frac{n}{L^3} = \rho \quad , \tag{4.33}$$

where $E_0(n)$ is the energy of the ground state of the Hamiltonian (4.28) at a fixed density. In the following section, the calculation for the high density regime will be presented using the Hartree–Fock method; thus an upper limit to the exact energy will be obtained. Following these calculations, excitations of the electron gas will be studied, as well as their response to external fields. These questions are addressed in Sects. 4.3.2–4.3.4.

Form of the Hamiltonian

The Hamiltonian (4.28) contains a two-body interaction (second term) and the effect of an external field (third and fourth terms). These give rise in the second quantization to operators of type (3.75) and (3.110). In the study of electron gas, it is commonly convenient for the calculations to rewrite the Hamiltonian (4.28) under a slightly different form, regrouping the different terms of the interaction. For this, it is necessary to introduce the plane-wave basis and the Fourier transform $\tilde{V}_L(\boldsymbol{k})$ of the Coulomb potential $V_L(\boldsymbol{x})$, periodic in a box Λ conforming to definitions (3.111) and (3.112).

With V_L replacing V in (4.28), the interaction energy of the electrons with the background is

$$-\rho e^2 \sum_{j=1}^{n} \frac{1}{L^3} \int_\Lambda \mathrm{d}\boldsymbol{y} \sum_{\boldsymbol{k}} \tilde{V}_L(\boldsymbol{k}) \mathrm{e}^{\mathrm{i}\boldsymbol{k}\cdot(\boldsymbol{x}_j - \boldsymbol{y})} = -n\rho e^2 \tilde{V}_L(0) \quad , \tag{4.34}$$

since $(1/L^3)\int_\Lambda \mathrm{d}\boldsymbol{y}\mathrm{e}^{\mathrm{i}\boldsymbol{k}\cdot\boldsymbol{y}} = \delta_{\boldsymbol{k},0}$. In the same manner, the self-energy of the background is

$$\frac{\rho^2 e^2}{2}\int_\Lambda \mathrm{d}\boldsymbol{x}\int_\Lambda \mathrm{d}\boldsymbol{y} V_L(\boldsymbol{x}-\boldsymbol{y}) = \frac{L^3\rho^2 e^2}{2}\tilde{V}_L(0) = \frac{n\rho e^2}{2}\tilde{V}_L(0) \quad . \tag{4.35}$$

Finally, the electronic repulsion is given by

$$\frac{e^2}{2}\sum_{i\neq j}^{n}\frac{1}{L^3}\sum_{\boldsymbol{k}}\tilde{V}_L(\boldsymbol{k})\mathrm{e}^{\mathrm{i}\boldsymbol{k}\cdot(\boldsymbol{x}_i-\boldsymbol{x}_j)}$$
$$= \frac{e^2}{2}\sum_{i\neq j}^{n}\frac{1}{L^3}\sum_{\boldsymbol{k}\neq 0}\tilde{V}_L(\boldsymbol{k})\mathrm{e}^{\mathrm{i}\boldsymbol{k}\cdot(\boldsymbol{x}_i-\boldsymbol{x}_j)} + \frac{n(n-1)e^2}{2L^3}\tilde{V}_L(0) \quad , \tag{4.36}$$

where the last term of (4.36) is the contribution of $\boldsymbol{k} = 0$ to the sum. For $n \gg 1$, this contribution gives

$$\frac{n(n-1)e^2}{2L^3}\tilde{V}_L(0) \simeq \frac{n\rho e^2}{2}\tilde{V}_L(0) \quad , \tag{4.37}$$

which is exactly compensated by terms (4.34) and (4.35). This compensation results in the global neutrality imposed by the choice of background density n/L^3.

Thus the complete Hamiltonian (4.28) with $n \gg 1$ and V_L replacing V is equivalent to a Hamiltonian where only two-body interactions occur, defined by

$$\mathtt{H}(n) = \sum_{j=1}^{n}\frac{|\boldsymbol{p}_j|^2}{2m} + \frac{1}{2}\sum_{i\neq j}^{n} V_L^0(\boldsymbol{x}_i - \boldsymbol{x}_j) \tag{4.38}$$

with

$$V_L^0(\boldsymbol{x}) = \frac{e^2}{L^3}\sum_{\boldsymbol{k}\neq 0}\tilde{V}_L(\boldsymbol{k})\mathrm{e}^{\mathrm{i}\boldsymbol{k}\cdot\boldsymbol{x}} \quad . \tag{4.39}$$

The only difference between the second terms of the right-hand sides of (4.28) and (4.39) is the omission of the $\boldsymbol{k} = 0$ term in $V_L^0(\boldsymbol{x})$. Finally, one should note that the Fourier transform of the Coulomb potential in an infinite volume is defined for $\boldsymbol{k} \neq 0$ with the result

$$\tilde{V}(\boldsymbol{k}) = \frac{1}{4\pi\varepsilon_0}\int \mathrm{d}\boldsymbol{x}\frac{\mathrm{e}^{\mathrm{i}\boldsymbol{k}\cdot\boldsymbol{x}}}{|\boldsymbol{x}|} = \frac{1}{\varepsilon_0|\boldsymbol{k}|^2} \quad , \quad \boldsymbol{k} \neq 0 \quad . \tag{4.40}$$

For large L, one is allowed to replace $\tilde{V}_L(\boldsymbol{k})$ by $\tilde{V}(\boldsymbol{k})$ in (4.39), so that the interaction Hamiltonian generally used for the study of an electron gas is given by (4.38) with $V_L^0(\boldsymbol{x})$ replaced by the interaction

$$V^0(\boldsymbol{x}) = \frac{e^2}{L^3}\sum_{\boldsymbol{k}\neq 0}\frac{1}{\varepsilon_0|\boldsymbol{k}|^2}\mathrm{e}^{\mathrm{i}\boldsymbol{k}\cdot\boldsymbol{x}} = \frac{e^2}{L^3}\sum_{\boldsymbol{k}}\tilde{V}^0(\boldsymbol{k})\mathrm{e}^{\mathrm{i}\boldsymbol{k}\cdot\boldsymbol{x}} \tag{4.41}$$

setting

$$\tilde{V}^0(\boldsymbol{k}) = \begin{cases} e^2/\varepsilon_0|\boldsymbol{k}|^2 & , \quad \boldsymbol{k} \neq 0 \\ 0 & , \quad \boldsymbol{k} = 0 \end{cases} \quad . \tag{4.42}$$

Its version in the second quantization results by taking the potential to be equal to $e^2/\varepsilon_0|\boldsymbol{k}|^2$ for $\boldsymbol{k} \neq 0$ in (3.116) and omitting the $\boldsymbol{k} = 0$ term from the sum.

Conforming to the discussion found in of Sect. 3.2.7, the fact that interactions (3.109) and (3.116) (in particular the replacement of $\tilde{V}(\boldsymbol{k})$ by $\tilde{V}_L(\boldsymbol{k})$) give the same results at the thermodynamic limit needs verification. For example, the Fourier transform of the Coulomb potential in a spherical region of volume $4\pi R^3/3 = L^3$, equivalent to that of the cube, is

$$\tilde{V}_R(\boldsymbol{k}) = \frac{1}{4\pi\varepsilon_0}\int_{|\boldsymbol{x}|\leq R} \mathrm{d}\boldsymbol{x}\frac{e^{i\boldsymbol{k}\cdot\boldsymbol{x}}}{|\boldsymbol{x}|} = \frac{1}{\varepsilon_0|\boldsymbol{k}|^2}\left(1 - \cos|\boldsymbol{k}|R\right) \quad . \tag{4.43}$$

A spherical region is chosen for the convenience of the calculation. When the result is expressed in the form of a sum over modes with a sufficiently regular function $F(\boldsymbol{k})$, one obtains in the thermodynamic limit

$$\lim_{R\to\infty}\int \mathrm{d}\boldsymbol{k}\tilde{V}_R(\boldsymbol{k})F(\boldsymbol{k}) = \int \mathrm{d}\boldsymbol{k}\frac{F(\boldsymbol{k})}{\varepsilon_0|\boldsymbol{k}|^2} \tag{4.44}$$

because the oscillating term of (4.43) does not contribute in the limit $R \to \infty$. Such verifications do not pose any serious problems for the homogeneous phase being addressed and will be taken for granted in the rest of this study.

4.2.2 The Hartree–Fock Energy

The Hartree–Fock equations (4.23) for an electron gas defined by the Hamiltonian (4.38) have simple plane-wave solutions. A plane wave with wavenumber $\boldsymbol{k}_j$ and spin τ_j is denoted by

$$\langle \boldsymbol{x}\sigma|\boldsymbol{k}_j\tau_j\rangle = \frac{1}{L^{3/2}}\delta_{\tau_j,\sigma}e^{i\boldsymbol{k}_j\cdot\boldsymbol{x}} \quad . \tag{4.45}$$

Any selection of n orthogonal plane waves of type (4.45) for $j = 1, 2, \ldots, n$ is a solution of equation (4.23).

The Mean Potential

As the external potential of the neutralizing background has been incorporated in the formulation (4.38) of the Hamiltonian, it is sufficient to calculate the mean potential (4.18) and the exchange potential (4.19) corresponding to (4.41) for the case where the single particle states are plane waves. Explicitly taking the sum over spin states in (4.18), one obtains (as V^0 and U are independent of the spin)

$$\begin{aligned} U(\boldsymbol{x}) &= \sum_{\sigma'} \int_{\Lambda} \mathrm{d}\boldsymbol{x}' V^0(\boldsymbol{x}-\boldsymbol{x}') \sum_{j=1}^{n} |\langle \boldsymbol{x}'\sigma' | \boldsymbol{k}_j \tau_j \rangle|^2 \\ &= \sum_{\sigma'} \sum_{j=1}^{n} \delta_{\tau_j,\sigma'} \frac{1}{L^3} \int_{\Lambda} \mathrm{d}\boldsymbol{x}' V^0(\boldsymbol{x}-\boldsymbol{x}') = \frac{n}{L^3} \tilde{V}^0(0) = 0 \quad , \end{aligned} \tag{4.46}$$

since, according to (4.42), the Fourier transform $\tilde{V}^0(\boldsymbol{k})$ of $V^0(\boldsymbol{x})$ in Λ vanishes for $\boldsymbol{k} = 0$. *The mean potential is thus equal to zero.* Choosing plane waves for the electronic states distributes the electrons uniformly in volume Λ. The potential due to the electrons in such a state is thus exactly compensated by that of the background.

The Exchange Potential

The exchange potential (4.19)

$$\langle \boldsymbol{x}\sigma | W | \boldsymbol{x}'\sigma' \rangle = \sum_{j=1}^{n} \langle \boldsymbol{x}\sigma | \boldsymbol{k}_j \tau_j \rangle \, V^0(\boldsymbol{x}-\boldsymbol{x}') \, \langle \boldsymbol{k}_j \tau_j | \boldsymbol{x}'\sigma' \rangle \tag{4.47}$$

can be calculated by introducing (4.41) and remarking that

$$\langle \boldsymbol{k}_j \tau_j | \boldsymbol{x}\sigma \rangle \, \mathrm{e}^{\mathrm{i}\boldsymbol{l}\cdot\boldsymbol{x}} = \langle (\boldsymbol{k}_j + \boldsymbol{l}) \tau_j | \boldsymbol{x}\sigma \rangle \quad . \tag{4.48}$$

One thus obtains

$$W = \sum_{j=1}^{n} \frac{1}{L^3} \sum_{\boldsymbol{l}} |(\boldsymbol{k}_j + \boldsymbol{l}) \tau_j \rangle \, \tilde{V}^0(\boldsymbol{l}) \, \langle (\boldsymbol{k}_j + \boldsymbol{l}) \tau_j | \quad . \tag{4.49}$$

Finally, applying W to a plane wave $|\boldsymbol{k}\tau\rangle$ gives

$$\begin{aligned} W|\boldsymbol{k}\tau\rangle &= \frac{1}{L^3} \sum_{j=1}^{n} \sum_{\boldsymbol{l}} \tilde{V}^0(\boldsymbol{l}) \, |(\boldsymbol{k}_j + \boldsymbol{l}) \tau_j \rangle \, \langle (\boldsymbol{k}_j + \boldsymbol{l}) \tau_j | \boldsymbol{k}\tau \rangle \\ &= \left(\frac{1}{L^3} \sum_{j=1}^{n} \tilde{V}^0(\boldsymbol{k} - \boldsymbol{k}_j) \right) |\boldsymbol{k}\tau\rangle \end{aligned} \tag{4.50}$$

because of the orthogonality of the plane waves. *Equation (4.50) shows that each plane wave is an eigenvector of W with eigenvalue $(1/L^3)\sum_{j=1}^{n} \tilde{V}^0(\boldsymbol{k}-\boldsymbol{k}_j)$.*

As the plane waves equally diagonalize the kinetic energy with eigenvalues $\hbar^2|\boldsymbol{k}|^2/2m$, it can be concluded that the plane waves (4.45) obey the Hartree–Fock equations (4.23) with

$$\varepsilon_{\boldsymbol{k}}^{\mathrm{HF}} = \frac{\hbar^2|\boldsymbol{k}|^2}{2m} - \frac{1}{L^3} \sum_{j=1}^{n} \tilde{V}^0(\boldsymbol{k} - \boldsymbol{k}_j) \quad . \tag{4.51}$$

Calculation of the Energy

The solution of the variational problem is not unique (discounting invariance under the change of basis (4.22)) since any system of plane waves (4.45) satisfies (4.23). As this discussion focuses on the high-density regime, dominated by the kinetic energy, one expects that the minimum total energy must occur when the states with the smallest wavenumbers $\boldsymbol{k}$, that is, those within the Fermi sphere, are occupied. One thus chooses for each $\boldsymbol{k}$ two states of opposite spin and $0 \leq |\boldsymbol{k}| \leq k_{\mathrm{F}}$, where k_{F} is the Fermi wavenumber. In this case, the Hartree–Fock energy (4.25) becomes, taking into account that $U = 0$,

$$\begin{aligned} E_0^{\mathrm{HF}} &= 2 \sum_{|\boldsymbol{k}| \leq k_{\mathrm{F}}} \varepsilon_{\boldsymbol{k}}^{\mathrm{HF}} + \frac{1}{2} \sum_{|\boldsymbol{k}| \leq k_{\mathrm{F}}} \sum_{\tau} \langle \boldsymbol{k}\tau | W | \boldsymbol{k}\tau \rangle \\ &= 2 \sum_{|\boldsymbol{k}| \leq k_{\mathrm{F}}} \varepsilon_{\boldsymbol{k}}^{\mathrm{HF}} - \frac{1}{L^3} \sum_{\substack{|\boldsymbol{k}| \leq k_{\mathrm{F}} \\ |\boldsymbol{l}| \leq k_{\mathrm{F}}}} \tilde{V}^0(\boldsymbol{k} - \boldsymbol{l}) \\ &= \frac{\hbar^2}{m} \sum_{|\boldsymbol{k}| \leq k_{\mathrm{F}}} |\boldsymbol{k}|^2 - \frac{1}{L^3} \sum_{\substack{|\boldsymbol{k}| \leq k_{\mathrm{F}} \\ |\boldsymbol{l}| \leq k_{\mathrm{F}}}} \tilde{V}^0(\boldsymbol{k} - \boldsymbol{l}) \quad . \end{aligned} \tag{4.52}$$

The sums are evaluated at the limit $L \to \infty$. For the kinetic energy term, as in (2.96), one has

$$\sum_{|\boldsymbol{k}| \leq k_{\mathrm{F}}} |\boldsymbol{k}|^2 \simeq \left(\frac{L}{2\pi} \right)^3 \frac{4\pi}{5} k_{\mathrm{F}}^5 \quad , \quad L \to \infty \tag{4.53}$$

and it follows from (4.40) and (4.42) that

$$\begin{aligned} \sum_{\substack{|\boldsymbol{k}| \leq k_{\mathrm{F}} \\ |\boldsymbol{l}| \leq k_{\mathrm{F}}}} \tilde{V}^0(\boldsymbol{k} - \boldsymbol{l}) &\simeq \left(\frac{L}{2\pi} \right)^6 \int_{|\boldsymbol{k}| \leq k_{\mathrm{F}}} \mathrm{d}\boldsymbol{k} \int_{|\boldsymbol{l}| \leq k_{\mathrm{F}}} \mathrm{d}\boldsymbol{l} \frac{e^2}{\varepsilon_0 |\boldsymbol{k} - \boldsymbol{l}|^2} \\ &= \left(\frac{L}{2\pi} \right)^6 \frac{4\pi^2 e^2}{\varepsilon_0} k_{\mathrm{F}}^4 \quad , \quad L \to \infty \quad . \end{aligned} \tag{4.54}$$

A detailed calculation of this integral is given at the end of this section.

Introducing (4.54) and (4.53) into (4.52), one obtains the Hartree–Fock energy per particle when $n \to \infty$, $n/L^3 = \rho$

$$\lim_{n \to \infty} \frac{E_0^{\mathrm{HF}}(n)}{n} = \frac{\hbar^2 k_{\mathrm{F}}^5}{10\pi^2 m \rho} - \frac{e^2 k_{\mathrm{F}}^4}{16\pi^4 \varepsilon_0 \rho} \quad . \tag{4.55}$$

Finally, recalling that $k_{\mathrm{F}} = (6\pi^2 \rho/2)^{1/3}$ and using (4.31), one can express this energy in terms of r_s defined by (4.29)

$$\lim_{n\to\infty} \frac{E_0^{\mathrm{HF}}(n)}{n} = \frac{me^4}{2\hbar^2(4\pi\varepsilon_0)^2}\left(\frac{3}{5}\frac{1}{(\alpha r_s)^2} - \frac{3}{2\pi}\frac{1}{\alpha r_s}\right)$$
$$= \left(\frac{2.21}{r_s^2} - \frac{0.916}{r_s}\right)[\text{Rydberg}] \quad , \tag{4.56}$$

where $\alpha = (4/9\pi)^{1/3}$, which is the final result.

This result reproduces relations (2.97) and (2.114), namely that the kinetic and potential energy densities are respectively proportional to $\rho^{5/3}$ and $\rho^{4/3}$. Here, the potential energy is entirely due to the exchange term: it is negative and of magnitude comparable to the kinetic energy when r_s is of the order of one. The Hartree theory, which neglects the exchange effects, is not of interest for the present case.

The Hartree–Fock variational state corresponding to energy (4.56) is identical to the ground state $|\Phi_0\rangle$ of a free Fermi gas. It is thus clear that the Hartree–Fock energy $E_0^{\mathrm{HF}} = \langle\Phi_0|\mathtt{H}(n)|\Phi_0\rangle$ is identical to the energy evaluated to first order of the ordinary perturbation theory of quantum mechanics. This consists of calculating the mean value of the interaction Hamiltonian in the eigenstate $|\Phi_0\rangle$ of the unperturbed Hamiltonian. *Thus, in this problem and at this order, the predictions of Hartree–Fock and of the perturbation theory are the same.*

Since the Hartree–Fock state is that of a free gas, the pair-correlation function $\Gamma(\boldsymbol{x}_1, \boldsymbol{x}_2)$ previously discussed in Sect. 3.2.8 is applicable. One can just as well calculate the Hartree–Fock potential energy by formula (3.144) with the two-body potential (4.41) and $\Gamma(\boldsymbol{x}_1, \boldsymbol{x}_2)$ given by (3.155). *Thus the Hartree–Fock energy only takes into account correlations induced by the exclusion principle and does not incorporate correlation effects due to the Coulomb interaction between electrons.* For this reason, the *Coulomb correlation energy* is the difference between the exact energy and that of Hartree–Fock. The calculation of the Coulomb correlation energy requires diagrammatic perturbative methods which will be elaborated in Chap. 10. The dominant contribution to the Coulomb correlation energy in the limit of small r_s is logarithmic

$$\lim_{n\to\infty}\left(\frac{E_0(n)}{n} - \frac{E_0^{\mathrm{HF}}(n)}{n}\right) = (0.062\ln r_s - 0.096)[\text{Ryd}] \tag{4.57}$$

up to corrections which tend to zero with r_s.

Integral (4.54)

We now calculate the integral introduced in (4.54):

$$J(k_{\mathrm{F}}) = \int_{|\boldsymbol{k}_1|\le k_{\mathrm{F}}} \mathrm{d}\boldsymbol{k}_1 \int_{|\boldsymbol{k}_2|\le k_{\mathrm{F}}} \mathrm{d}\boldsymbol{k}_2 \frac{1}{|\boldsymbol{k}_1 - \boldsymbol{k}_2|^2} = k_{\mathrm{F}}^4 J(1) \quad . \tag{4.58}$$

Substituting $k_1 = |\boldsymbol{k}_1|,\ k_2 = |\boldsymbol{k}_2|,\ \mu = \boldsymbol{k}_1 \cdot \boldsymbol{k}_2/k_1k_2 = \cos\theta,\ s = k_2/k_1$, one has

$$\frac{1}{|\boldsymbol{k}_1 - \boldsymbol{k}_2|^2} = \frac{1}{k_1^2 + k_2^2 - 2\mu k_1 k_2} = \frac{1}{k_1^2}\frac{1}{1+s^2-2\mu s} \tag{4.59}$$

and for $s < 1$,

$$\frac{1}{1+s^2-2\mu s} = \left(\sum_{l=0}^{\infty} s^l P_l(\mu)\right)^2 \quad . \tag{4.60}$$

Here, P_l is the Legendre polynomial of order l. Thus, noting the integrand is symmetric in k_1 and k_2, one obtains

$$J(1) = 2\int_{|\boldsymbol{k}_1|\le 1} \mathrm{d}\boldsymbol{k}_1 \frac{1}{k_1^2}\int_0^{k_1} \mathrm{d}k_2 2\pi k_2^2 \int_{-1}^{1}\mathrm{d}\mu \sum_{l,m=0}^{\infty}\left(\frac{k_2}{k_1}\right)^{l+m} P_l(\mu)P_m(\mu) \quad . \tag{4.61}$$

The orthogonality of the Legendre polynomials,

$$\int_{-1}^{1} \mathrm{d}\mu P_l(\mu)P_m(\mu) = \frac{2}{2l+1}\delta_{l,m} \quad , \tag{4.62}$$

implies

$$\begin{aligned} J(1) &= 2\int_{|\boldsymbol{k}_1|\le 1} \mathrm{d}\boldsymbol{k}_1 \int_0^{k_1}\mathrm{d}k_2 2\pi \sum_{l=0}^{\infty}\frac{2}{2l+1}\left(\frac{k_2}{k_1}\right)^{2l+2} \\ &= 8\pi^2 \sum_{l=0}^{\infty}\frac{1}{(2l+1)(2l+3)} \quad . \end{aligned} \tag{4.63}$$

Since

$$\sum_{l=0}^{\infty}\frac{1}{(2l+1)(2l+3)} = \lim_{n\to\infty}\frac{1}{2}\sum_{l=0}^{n}\left(\frac{1}{2l+1} - \frac{1}{2l+3}\right) = \frac{1}{2} \quad , \tag{4.64}$$

one finally obtains $J(1) = 4\pi^2$, which leads to (4.54).

4.3 The Dielectric Function

4.3.1 Screening and the Plasmon

Perfect Screening

Screening and *charge oscillations* are phenomena characteristic of a charged fluid. A qualitative description of screening is presented first. A static charge e_0 is placed in an homogeneous phase of an electron gas at equilibrium. In vacuum, this charge (placed at the origin $\boldsymbol{x} = 0$) creates a potential $V(\boldsymbol{x}) = e_0/(4\pi\varepsilon_0|\boldsymbol{x}|)$ at a distance $|\boldsymbol{x}|$. In the gas, however, the external charge e_0 attracts or repels, according to its sign, the mobile electrons: the electron density readjusts itself in the neighbourhood of the external charge. This creates a local excess or deficiency of electron charge called the polarization cloud or the static *induced charge* $\rho_e^{\text{ind}}(\boldsymbol{x})$

$$\rho_e^{\text{ind}}(\boldsymbol{x}) = e\,[\rho(\boldsymbol{x}) - \rho] \quad . \tag{4.65}$$

In this expression, $\rho(\boldsymbol{x})$ is the electron density modified by the presence of the external charge, and ρ is the background density. In the absence of an external charge, the state is homogeneous and $\rho(\boldsymbol{x}) = \rho$ at any point. This situation is schematically represented in Fig. 4.1.

It is called *perfect screening* when the induced density carries a charge of the same magnitude and opposite in sign as the external charge

$$\int \mathrm{d}\boldsymbol{x}\rho_e^{\text{ind}}(\boldsymbol{x}) = -e_0 \quad . \tag{4.66}$$

At this point it is necessary to make an important remark which applies to the discussions to follow. An infinitely extended system constitutes an infinite reservoir of particles: the formation of an excess (or deficiency) of local charge is thus not in contradiction with conservation of the number of particles.

In a finite system, the creation of a polarization cloud around e_0 is accompanied by an accumulation of opposite charges at the surface of the system (surface polarization) such that the total charge remains constant. If one designates the charge density induced to a finite volume by $\rho_{e\Lambda}^{\text{ind}}(\boldsymbol{x})$, the latter

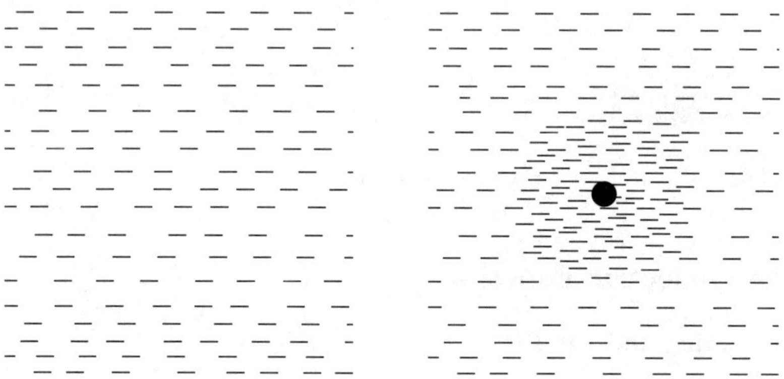

Homogeneous state: $\rho_e^{\text{ind}}(\boldsymbol{x}) = 0$ Polarization cloud: $\rho_e^{\text{ind}}(\boldsymbol{x}) \neq 0$

Fig. 4.1. Homogeneous state of an electron gas (*left*) with density $\rho_e^{\text{ind}}(\boldsymbol{x}) = 0$ and the same gas in the presence of an external charge with $\rho_e^{\text{ind}}(\boldsymbol{x}) \neq 0$ (*right*)

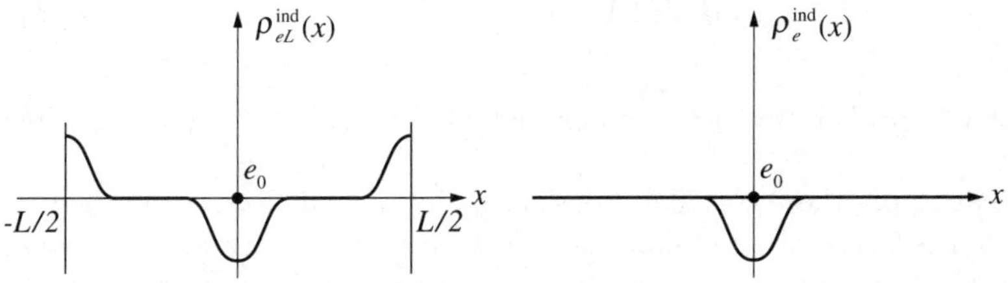

Fig. 4.2. One-dimensional representations of the charge density in a finite volume (*left*) and an infinite volume in the presence of an external charge e_0 (*right*)

describes a modification of the local density neighbouring e_0 as well as the polarization of the surface, in such a manner that for all cases $\int_\Lambda \mathrm{d}\boldsymbol{x} \rho_{e\Lambda}^{\mathrm{ind}}(\boldsymbol{x}) = 0$. In Fig. 4.2, one-dimensional representations of $\rho_{e\Lambda}^{\mathrm{ind}}(\boldsymbol{x})$ and $\rho_e^{\mathrm{ind}}(\boldsymbol{x})$ are illustrated. It is necessary to understand that in (4.66) and all the formulae which follow that $\rho_e^{\mathrm{ind}}(\boldsymbol{x})$ is the thermodynamic limit of $\rho_{e\Lambda}^{\mathrm{ind}}(\boldsymbol{x})$ with the surface polarization effects sent to infinity. Thus in $\rho_e^{\mathrm{ind}}(\boldsymbol{x})$, only the local polarization in the neighbourhood of e_0 exists, for which the relation of perfect screening (4.66) is valid.

The Effective Potential

An *effective potential* can be associated to the external charge, equipped with its polarization cloud

$$V^{\mathrm{eff}}(\boldsymbol{x}) = \frac{1}{4\pi\varepsilon_0} \int \mathrm{d}\boldsymbol{y} \frac{\rho_e^{\mathrm{tot}}(\boldsymbol{y})}{|\boldsymbol{x} - \boldsymbol{y}|} \quad , \tag{4.67}$$

where $\rho_e^{\mathrm{tot}}(\boldsymbol{x})$ is the density due to all charges present

$$\rho_e^{\mathrm{tot}}(\boldsymbol{x}) = \rho_e^{\mathrm{ind}}(\boldsymbol{x}) + e_0 \delta(\boldsymbol{x}) \quad . \tag{4.68}$$

Relation (4.66) can be rewritten as

$$\int \mathrm{d}\boldsymbol{x} \rho_e^{\mathrm{tot}}(\boldsymbol{x}) = 0 \quad . \tag{4.69}$$

Since $V^{\mathrm{eff}}(\boldsymbol{x})$ behaves asymptotically as

$$V^{\mathrm{eff}}(\boldsymbol{x}) \simeq \frac{1}{4\pi\varepsilon_0 |\boldsymbol{x}|} \int \mathrm{d}\boldsymbol{y} \rho_e^{\mathrm{tot}}(\boldsymbol{y}) \quad , \quad |\boldsymbol{x}| \to \infty \quad , \tag{4.70}$$

perfect screening (4.69) is equivalent to $V^{\mathrm{eff}}(\boldsymbol{x})$ having a range shorter than that of the Coulomb potential. In general, screening takes place over a microscopic distance λ, the *screening length*, which is of order k_{F}^{-1}. A form commonly used for the effective potential is

$$V^{\text{eff}}(\boldsymbol{x}) = \frac{e_0}{4\pi\varepsilon_0|\boldsymbol{x}|} \exp\left(-\frac{|\boldsymbol{x}|}{\lambda}\right) \quad , \tag{4.71}$$

such that $V^{\text{eff}}(\boldsymbol{x})$ behaves as a Coulomb potential for $|\boldsymbol{x}| \ll \lambda$ and is negligible for $|\boldsymbol{x}| \gg \lambda$.

One can apply these considerations to the individual electrons in the gas: because of the Coulomb repulsion, each electron creates in its neighbourhood a region which is void of other electrons and thus positively charged due to the background which tends to compensate its own charge. Thus the effective potential between two electrons equipped with their polarization clouds is also of short range. From the macroscopic point of view, *the property of perfect screening signifies that the electron gas behaves as a perfect conductor: the electric field, on a scale of distances greater than* λ, *is zero everywhere.*

The Plasmon

A second important phenomenon tied to screening is the collective movement of the electrons, the plasma oscillation or *plasmon*. When an inhomogeneity is created in an electron gas, the electrons put themselves into motion in order to reestablish neutrality under the effect of the electrostatic potential which has been produced. As the electrons possess inertia, they acquire an oscillatory motion around the perfectly neutral equilibrium state. This can be shown by the following elementary argument. At equilibrium, the electron density neutralizes the background charge at each point. Now suppose that a two-dimensional layer of electrons of thickness d is displaced by a distance $x > 0$ where $x \ll d$, as indicated in Fig. 4.3. This would create an excess of charge to the left and a deficit of charge to the right. The superficial charge density of the two charged planes is $\sigma = \pm e\rho x$ and it gives rise (as in a capacitor) to an electric field $E = \sigma/\varepsilon_0 = -\rho e x/\varepsilon_0$. Thus, each electron of the displaced layer obeys the force law

$$m\frac{\mathrm{d}^2 x}{\mathrm{d}t^2} = eE = -\frac{1}{\varepsilon_0}\rho e^2 x \quad . \tag{4.72}$$

This is the equation of motion of an oscillator of frequency

$$\omega_{\text{pl}} = \left(\frac{\rho e^2}{m\varepsilon_0}\right)^{1/2} \quad . \tag{4.73}$$

The electron layer carries out a collective oscillation of a well-determined frequency ω_{pl}, *called a plasmon.* In this argument, collisions between the electrons and their mutual Coulomb interaction are neglected. In the following sections, screening and plasma oscillations are analyzed on a microscopic level.

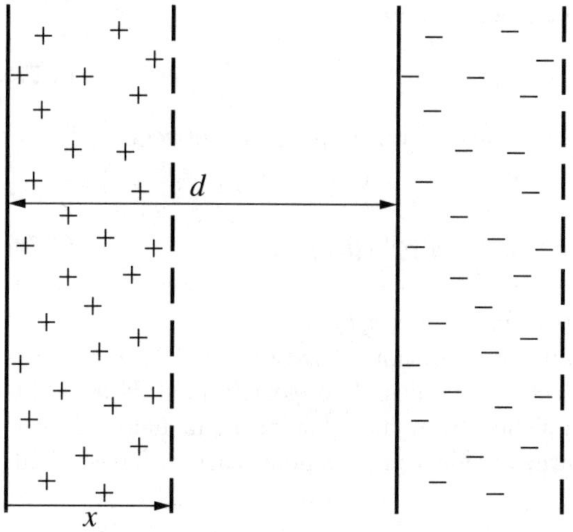

Fig. 4.3. The displacement of a two-dimensional layer of electrons in the gas causes an oscillation of those electrons with a well-determined frequency ω_{pl}, called a plasmon

4.3.2 Response to an External Charge

The dielectric function formalism allows the simultaneous treatment of the static screening mechanisms and the dynamical phenomena taking place in the electron gas. It consists of studying the response of an electron gas to the perturbation caused by an external charge density $\rho_e^{\mathrm{ext}}(\boldsymbol{x}, t)$ with prescribed spatial and temporal variations. *The fundamental hypothesis of the response theory is the existance of a linear relation between an induced charge and the perturbation when the latter is sufficiently weak.*

The induced dynamical charge is defined as in (4.65)

$$\rho_e^{\mathrm{ind}}(\boldsymbol{x}, t) = e\left[\rho(\boldsymbol{x}, t) - \rho\right] \quad , \tag{4.74}$$

where $\rho(\boldsymbol{x}, t)$ is the electron density at time t in the presence of the external charges. When the medium is homogeneous, the most general linear relation is

$$\rho_e^{\mathrm{ind}}(\boldsymbol{x}, t) = \int \mathrm{d}s \int \mathrm{d}\boldsymbol{y} \chi(\boldsymbol{x} - \boldsymbol{y}, t - s) \rho_e^{\mathrm{ext}}(\boldsymbol{y}, s) \quad , \tag{4.75}$$

where $\chi(\boldsymbol{x}, t)$ is the *response function.* Modification of the medium cannot occur later than the perturbation it causes, imposing the *time-retarded* character of the response function

$$\chi(\boldsymbol{x}, t) = 0 \quad , \quad t < 0 \quad . \tag{4.76}$$

The total charge density at time t is given by

$$\rho_e^{\text{tot}}(\boldsymbol{x},t) = \rho_e^{\text{ind}}(\boldsymbol{x},t) + \rho_e^{\text{ext}}(\boldsymbol{x},t) \quad . \tag{4.77}$$

Relation (4.75) shows that there is also a linear relation between $\rho_e^{\text{tot}}(\boldsymbol{x},t)$ and $\rho_e^{\text{ext}}(\boldsymbol{x},t)$. Commonly written as

$$\varepsilon_0 \rho_e^{\text{ext}}(\boldsymbol{x},t) = \int \mathrm{d}s \int \mathrm{d}\boldsymbol{y} \varepsilon(\boldsymbol{x}-\boldsymbol{y}, t-s) \rho_e^{\text{tot}}(\boldsymbol{y},s) \quad , \tag{4.78}$$

this relation defines the *dielectric function* $\varepsilon(\boldsymbol{x},t)$.

The terminology comes from the definition (4.78) of $\varepsilon(\boldsymbol{x},t)$ being analogous to that of macroscopic electrodynamics. The electric field $\boldsymbol{E}(\boldsymbol{x},t)$ due to all charges present is distinguished from the "displacement field" $\boldsymbol{D}(\boldsymbol{x},t)$ due only to the additional charges neglecting the polarization. These fields are determined by

$$\varepsilon_0 \nabla \cdot \boldsymbol{E}(\boldsymbol{x},t) = \rho_e^{\text{tot}}(\boldsymbol{x},t) \tag{4.79}$$

$$\nabla \cdot \boldsymbol{D}(\boldsymbol{x},t) = \rho_e^{\text{ext}}(\boldsymbol{x},t) \quad . \tag{4.80}$$

In an isotropic and homogeneous medium, relation (4.78) is thus equivalent to the linear relation between the longitudinal components $\boldsymbol{D}_{\text{L}}$ and $\boldsymbol{E}_{\text{L}}$ of the fields

$$\boldsymbol{D}_{\text{L}}(\boldsymbol{x},t) = \int \mathrm{d}s \int \mathrm{d}\boldsymbol{y} \varepsilon(\boldsymbol{x}-\boldsymbol{y}, t-s) \boldsymbol{E}_{\text{L}}(\boldsymbol{y},s) \quad . \tag{4.81}$$

This generalizes the usual electrostatic relation $\boldsymbol{D} = \varepsilon \boldsymbol{E}$ to fields which are variable in space and time. One should nevertheless be aware that in view of the remark after (4.66), the *definitions (4.75) and (4.78) do not take into account the effects of surface polarization.*

By exhibiting some properties of the dielectric function, we now show it provides the essential information of the behaviour of electron gas. Through application of the convolution theorem, relations (4.75) and (4.78) take on simple algebraic forms in terms of their Fourier transforms

$$\rho_e^{\text{ind}}(\boldsymbol{k},\omega) = \chi(\boldsymbol{k},\omega) \rho_e^{\text{ext}}(\boldsymbol{k},\omega) \tag{4.82}$$

$$\begin{aligned} \varepsilon_0 \rho_e^{\text{ext}}(\boldsymbol{k},\omega) &= \varepsilon(\boldsymbol{k},\omega) \rho_e^{\text{tot}}(\boldsymbol{k},\omega) \\ &= \varepsilon(\boldsymbol{k},\omega) \left[\rho_e^{\text{ind}}(\boldsymbol{k},\omega) + \rho_e^{\text{ext}}(\boldsymbol{k},\omega)\right] \end{aligned} \tag{4.83}$$

with

$$\rho_e(\boldsymbol{k},\omega) = \int \mathrm{d}t \int \mathrm{d}\boldsymbol{x} \rho_e(\boldsymbol{x},t) \mathrm{e}^{-\mathrm{i}(\boldsymbol{k}\cdot\boldsymbol{x} - \omega t)} \tag{4.84}$$

and similar transforms for χ and ε. If there is no possible ambiguity, the Fourier transforms are denoted by the same symbols $\rho_e, \chi, \varepsilon$. The relation between ε and χ can be deduced from (4.82) and (4.83) to be

$$\frac{\varepsilon(\boldsymbol{k},\omega)}{\varepsilon_0} = \frac{1}{\chi(\boldsymbol{k},\omega)+1} \quad . \tag{4.85}$$

The time-retarded character (4.76) of the response function has an important implication on $\varepsilon(\boldsymbol{k},\omega)$. The temporal Fourier transform of $\chi(\boldsymbol{x},t)$ only extends to the half-axis $t > 0$

$$\chi(\boldsymbol{k},\omega) = \int_0^\infty \mathrm{d}t \int \mathrm{d}\boldsymbol{x}\chi(\boldsymbol{x},t)\mathrm{e}^{-\mathrm{i}(\boldsymbol{k}\cdot\boldsymbol{x}-\omega t)} \quad . \tag{4.86}$$

By consequence, $\chi(\boldsymbol{k},\omega)$ can be continued analytically in the half-plane $\omega + \mathrm{i}\eta, \eta > 0$, since the factor $\mathrm{e}^{\mathrm{i}(\omega+\mathrm{i}\eta t)} = \mathrm{e}^{\mathrm{i}\omega t}\mathrm{e}^{-\eta t}$ converges for $\eta > 0$ and $t \to \infty$. Relation (4.85) shows that the same holds for $\varepsilon(\boldsymbol{k},\omega)$. This implies that $\varepsilon(\boldsymbol{k},\omega)$, *for real* ω, *must be understood as the limit of* $\varepsilon(\boldsymbol{k},\omega+\mathrm{i}\eta)$ *as* η *goes to zero from positive values*

$$\varepsilon(\boldsymbol{k},\omega) = \lim_{\eta\to 0} \varepsilon(\boldsymbol{k},\omega+\mathrm{i}\eta) \quad , \quad \eta > 0 \quad . \tag{4.87}$$

Static Screening

The static perturbation $\rho_e^{\mathrm{ext}}(\boldsymbol{x},t) = e_0\delta(\boldsymbol{x})$ caused by a charge e_0 at the origin can be expressed in terms of the Fourier transform as $\rho_e^{\mathrm{ext}}(\boldsymbol{k},\omega) = e_0 2\pi\delta(\omega)$. According to (4.83), the total corresponding charge is $\rho_e^{\mathrm{tot}}(\boldsymbol{k},\omega) = \rho_e^{\mathrm{tot}}(\boldsymbol{k})2\pi\delta(\omega)$, with

$$\rho_e^{\mathrm{tot}}(\boldsymbol{k}) = \frac{\varepsilon_0}{\varepsilon(\boldsymbol{k},0)}e_0 \quad . \tag{4.88}$$

The associated effective potential (4.67) is thus given by

$$V^{\mathrm{eff}}(\boldsymbol{k}) = \frac{\rho_e^{\mathrm{tot}}(\boldsymbol{k})}{\varepsilon_0|\boldsymbol{k}|^2} = \frac{e_0}{\varepsilon(\boldsymbol{k},0)|\boldsymbol{k}|^2} \quad . \tag{4.89}$$

Knowledge of the dielectric function $\varepsilon(\boldsymbol{k},0) = \varepsilon(\boldsymbol{k},\omega=0)$ at zero frequency allows one to calculate the total and induced charges and the effective potential. *Perfect screening (4.69),* $\int \mathrm{d}\boldsymbol{x}\rho_e^{\mathrm{tot}}(\boldsymbol{x}) = \rho_e^{\mathrm{tot}}(\boldsymbol{k}=0) = 0$, *is equivalent, from (4.88), to the divergence of the dielectric function for small wavenumbers*

$$\lim_{|\boldsymbol{k}|\to 0} \varepsilon(\boldsymbol{k},0) = \infty \quad . \tag{4.90}$$

Charge Oscillations

Knowledge of $\varepsilon(\boldsymbol{k},\omega)$ for $\omega \neq 0$ allows one to determine the possible excitations of an electron gas. In the absence of external charges, (4.83) reduces to

$$\varepsilon(\boldsymbol{k},\omega)\rho_e^{\text{ind}}(\boldsymbol{k},\omega) = 0 \quad . \tag{4.91}$$

For all values of $\boldsymbol{k}$ and ω such that $\varepsilon(\boldsymbol{k},\omega) \neq 0$, one must have $\rho_e^{\text{ind}}(\boldsymbol{k},\omega) = 0$. When this condition is met, the system cannot allow for the free oscillation of charge. However, for frequencies $\omega(\boldsymbol{k})$ which are solutions of equation

$$\varepsilon\left(\boldsymbol{k},\omega(\boldsymbol{k})\right) = 0 \quad , \tag{4.92}$$

relation (4.91) allows for a non-zero amplitude $\rho_e^{\text{ind}}(\boldsymbol{k},\omega)$. *This corresponds to the possibility of a free oscillation of the charge density in the system (since there are no external fields), and $\omega(\boldsymbol{k})$ gives the excitation spectrum.* From this method, the frequency of the plasmon must appear as a zero of the dielectric function. This will be shown in the following section.

4.3.3 Evolution of a Charge Fluctuation

Microscopic Dynamics

In a quantum microscopic treatment, the charge density is identified as the mean value of the electron density operator

$$\rho_e(\boldsymbol{x},t) = e\left\langle n(\boldsymbol{x},t)\right\rangle \tag{4.93}$$

and the induced charge density as

$$\rho_e^{\text{ind}}(\boldsymbol{x},t) = e\left(\left\langle n(\boldsymbol{x},t)\right\rangle - \rho\right) \quad . \tag{4.94}$$

The operator $n(\boldsymbol{x},t) = \sum_\sigma a^*(\boldsymbol{x},\sigma,t)a(\boldsymbol{x},\sigma,t)$ is the electron density (3.70) at time t. It obeys the Heisenberg equation of motion

$$\mathrm{i}\hbar\frac{\partial}{\partial t}n(\boldsymbol{x},t) = \left[n(\boldsymbol{x},t),\mathtt{H}(t)\right] \quad . \tag{4.95}$$

In (4.95), $\mathtt{H}(t)$ is the Hamiltonian of the electron gas under the perturbation of an external charge density $\rho_e^{\text{ext}}(\boldsymbol{x},t)$. The averages in (4.93) and (4.94) are taken in the ground state of the electron gas without the perturbation. One then supposes that the effect of the external charge density is adiabatically switched on. To find the dielectric function, solve the equation of motion (4.95) to first order in the external perturbation to obtain $\varepsilon(\boldsymbol{k},\omega)$ from its definition (4.83).

The total Hamiltonian is the sum of the three contributions

$$\mathtt{H}(t) = \mathtt{H}_0 + \mathtt{H}_\mathrm{I} + \mathtt{H}^{\text{ext}}(t) \quad , \tag{4.96}$$

where $\mathtt{H}_0$ and $\mathtt{H}_\mathrm{I}$ are the kinetic (3.72) and Coulomb (3.117) energies of the electron gas in the second quantization

$$\mathrm{H}_0 = \sum_{\boldsymbol{k},\sigma} \frac{\hbar^2|\boldsymbol{k}|^2}{2m} a^*_{\boldsymbol{k}\sigma} a_{\boldsymbol{k}\sigma} \quad , \tag{4.97}$$

$$\mathrm{H}_\mathrm{I} = \frac{e^2}{2L^3} \sum_{\boldsymbol{k}\neq 0} \tilde{V}^0(\boldsymbol{k}) {:}\tilde{n}^*(\boldsymbol{k})\tilde{n}(\boldsymbol{k}){:} \tag{4.98}$$

with $\tilde{V}^0$ given by (4.42). The interaction Hamiltonian of the electron with the external charge is given by the coupling of the electron density with the external potential $V^\mathrm{ext}(\boldsymbol{x},t)$

$$\mathrm{H}^\mathrm{ext}(t) = e \int_\Lambda \mathrm{d}\boldsymbol{x} n(\boldsymbol{x}) V^\mathrm{ext}(\boldsymbol{x},t) \quad , \tag{4.99}$$

$$V^\mathrm{ext}(\boldsymbol{x},t) = \frac{1}{4\pi\varepsilon_0} \int_\Lambda \mathrm{d}\boldsymbol{y} \frac{1}{|\boldsymbol{x}-\boldsymbol{y}|} \rho_\mathrm{e}^\mathrm{ext}(\boldsymbol{y},t) \quad . \tag{4.100}$$

As in Sects. 4.2.1 and 4.2.2, the Coulomb potential in (4.100) can be replaced by the periodic potential $V_\mathrm{L}(\boldsymbol{x})$ for simplification. So $\mathrm{H}^\mathrm{ext}(t)$ has the form (3.78) in Fourier representation

$$\mathrm{H}^\mathrm{ext}(t) = \frac{e}{L^3} \sum_{\boldsymbol{k}} \tilde{n}^*(\boldsymbol{k}) \tilde{V}^\mathrm{ext}(\boldsymbol{k},t) \tag{4.101}$$

with

$$\tilde{V}^\mathrm{ext}(\boldsymbol{k},t) = \int_\Lambda \mathrm{d}\boldsymbol{x} V^\mathrm{ext}(\boldsymbol{x},t) \mathrm{e}^{-\mathrm{i}\boldsymbol{k}\cdot\boldsymbol{x}} = \tilde{V}^0(\boldsymbol{k}) \tilde{\rho}_\mathrm{e}^\mathrm{ext}(\boldsymbol{k},t) \quad . \tag{4.102}$$

In (4.102), $\tilde{\rho}_\mathrm{e}^\mathrm{ext}(\boldsymbol{k},t)$ is the Fourier transform over Λ of the external charge density

$$\tilde{\rho}_\mathrm{e}^\mathrm{ext}(\boldsymbol{k},t) = \int_\Lambda \mathrm{d}\boldsymbol{x} \rho_\mathrm{e}^\mathrm{ext}(\boldsymbol{x},t) \mathrm{e}^{-\mathrm{i}\boldsymbol{k}\cdot\boldsymbol{x}} = \left[\tilde{\rho}_\mathrm{e}^\mathrm{ext}(-\boldsymbol{k},t)\right]^* \tag{4.103}$$

and one has replaced $\tilde{V}_\mathrm{L}(\boldsymbol{k})$, $\boldsymbol{k} \neq 0$, by $\tilde{V}^0(\boldsymbol{k})$ as in (4.42). *To study the behaviour of the system vis-a-vis the entire spectrum of external perturbations, it is sufficient, in virtue of the linearity of the equations of motion, to excite only one mode* $\boldsymbol{k}$ *at a time*, that is to take $\tilde{\rho}_\mathrm{e}^\mathrm{ext}(\boldsymbol{l},t) = 0, \boldsymbol{l} \neq \pm\boldsymbol{k}$ with $\boldsymbol{k} \neq 0$. In this case, formulae (4.101), (4.102) and (4.103) give simply

$$\mathrm{H}^\mathrm{ext}(t) = \frac{e}{L^3} \tilde{V}^0(\boldsymbol{k}) \left\{ \tilde{n}^*(\boldsymbol{k}) \tilde{\rho}_\mathrm{e}^\mathrm{ext}(\boldsymbol{k},t) + \tilde{n}(\boldsymbol{k}) \left[\tilde{\rho}_\mathrm{e}^\mathrm{ext}(\boldsymbol{k},t)\right]^* \right\} \quad . \tag{4.104}$$

To specify the contents of the Heisenberg equations (4.95), it is convenient to represent $\tilde{n}(\boldsymbol{k},t)$ by the sum (3.79)

$$\tilde{n}(\boldsymbol{k},t) = \sum_{\boldsymbol{l}} c(\boldsymbol{k},\boldsymbol{l},t) \quad , \tag{4.105}$$

where

$$c(\boldsymbol{k},\boldsymbol{l},t) = \sum_{\sigma} a^*_{\boldsymbol{l}\sigma}(t) a_{(\boldsymbol{l}+\boldsymbol{k})\sigma}(t) \tag{4.106}$$

and to study the temporal evolution of operators $c(\boldsymbol{k},\boldsymbol{l},t)$. By annihilating an electron of momentum $\hbar(\boldsymbol{k}+\boldsymbol{l})$ and recreating it in the state $\hbar\boldsymbol{l}$, the momentum of the total system is decreased by $\hbar\boldsymbol{k}$. The equation of motion of $c(\boldsymbol{k},\boldsymbol{l},t)$ introduces three contributions

$$\begin{aligned} i\hbar\frac{\partial}{\partial t}c(\boldsymbol{k},\boldsymbol{l},t) &= [c(\boldsymbol{k},\boldsymbol{l},t),\mathtt{H}(t)] \\ &= [c(\boldsymbol{k},\boldsymbol{l},t),\mathtt{H}_0] + [c(\boldsymbol{k},\boldsymbol{l},t),\mathtt{H}_\mathrm{I}] + [c(\boldsymbol{k},\boldsymbol{l},t),\mathtt{H}^{\mathrm{ext}}(t)] \quad . \end{aligned} \tag{4.107}$$

Calculation of the commutators can be performed by applying repeated use of the anticommutation rules. Details can be found at the end of this section. The final result is

$$i\hbar\frac{\partial}{\partial t}c(\boldsymbol{k},\boldsymbol{l},t) = (\varepsilon_{\boldsymbol{k}+\boldsymbol{l}} - \varepsilon_{\boldsymbol{l}})c(\boldsymbol{k},\boldsymbol{l},t) \tag{4.108}$$

$$+\left(\frac{e^2}{2L^3}\right)\sum_{\boldsymbol{k}'\neq 0}\tilde{V}^0(\boldsymbol{k}')\big\{\left[c(\boldsymbol{k}-\boldsymbol{k}',\boldsymbol{l},t) - c(\boldsymbol{k}-\boldsymbol{k}',\boldsymbol{l}+\boldsymbol{k}',t)\right]\tilde{n}(\boldsymbol{k}',t)$$

$$+\,\tilde{n}^*(\boldsymbol{k}',t)\left[c(\boldsymbol{k}+\boldsymbol{k}',\boldsymbol{l},t) - c(\boldsymbol{k}+\boldsymbol{k}',\boldsymbol{l}-\boldsymbol{k}',t)\right]\big\} \tag{4.109}$$

$$+\left(\frac{e^2}{2L^3}\right)\tilde{V}^0(\boldsymbol{k})\big\{\tilde{\rho}^{\mathrm{ext}}_{e\boldsymbol{k}}(t)\left[c(0,\boldsymbol{l},t) - c(0,\boldsymbol{k}+\boldsymbol{l},t)\right]$$

$$+\left(\tilde{\rho}^{\mathrm{ext}}_{e\boldsymbol{k}}(t)\right)^*\left[c(2\boldsymbol{k},\boldsymbol{l},t) - c(2\boldsymbol{k},\boldsymbol{l}-\boldsymbol{k},t)\right]\big\} \quad . \tag{4.110}$$

The RPA Approximation

Because of (4.105), terms (4.109), which come from the Coulomb interaction of the electrons, are quadratic in $c(\boldsymbol{k},\boldsymbol{l},t)$. Equation (4.108) is a non-linear differential equation for $c(\boldsymbol{k},\boldsymbol{l},t)$ which cannot be solved. For this reason, it is useful to introduce *random phase approximation or RPA*. This is formulated in two steps. *First, take the average of terms (4.108)–(4.110) in the ground state of the electron gas, then assume that the averages of the quadratic terms are factorizable*, that is, that one can write in (4.109)

$$\langle c(\boldsymbol{k},\boldsymbol{l},t)\tilde{n}(\boldsymbol{k}',t)\rangle \simeq \langle c(\boldsymbol{k},\boldsymbol{l},t)\rangle\,\langle\tilde{n}(\boldsymbol{k}',t)\rangle = \langle c(\boldsymbol{k},\boldsymbol{l},t)\rangle\,\tilde{\rho}(\boldsymbol{k}',t) \quad . \tag{4.111}$$

Next, approximate the quantities $\langle c(\boldsymbol{k},\boldsymbol{l},t)\rangle$ which appear in factors of the densities $\tilde{\rho}(\boldsymbol{k},t)$ and $\tilde{\rho}^{\mathrm{ext}}_e(\boldsymbol{k},t)$ in (4.109) and (4.110) by the values that they take for a gas of free electrons

$$\sum_{\sigma}\langle\Phi_0|a^*_{\boldsymbol{l}\sigma}(t)a_{(\boldsymbol{l}+\boldsymbol{k})\sigma}(t)|\Phi_0\rangle = 2\delta_{\boldsymbol{k},0}n_{\boldsymbol{l}} \quad , \quad n_{\boldsymbol{l}} = \begin{cases} 1 & |\boldsymbol{l}| \le k_\mathrm{F} \\ 0 & |\boldsymbol{l}| > k_\mathrm{F} \end{cases} \quad . \tag{4.112}$$

Taking into account that $\tilde{\rho}^*(\boldsymbol{k},t) = \tilde{\rho}(-\boldsymbol{k},t)$, (4.108) reduces to

$$\begin{aligned} \mathrm{i}\hbar\frac{\partial}{\partial t}\langle c(\boldsymbol{k},\boldsymbol{l},t)\rangle &= (\varepsilon_{\boldsymbol{k}+\boldsymbol{l}} - \varepsilon_{\boldsymbol{l}})\langle c(\boldsymbol{k},\boldsymbol{l},t)\rangle \\ &\quad + 2\frac{\tilde{V}^0(\boldsymbol{k})}{L^3}\left[e^2\tilde{\rho}(\boldsymbol{k},t) + e\tilde{\rho}_e^{\mathrm{ext}}(\boldsymbol{k},t)\right] \\ &\quad \cdot (n_{\boldsymbol{l}} - n_{\boldsymbol{k}+\boldsymbol{l}}) \quad . \end{aligned} \tag{4.113}$$

These equations combined with

$$\tilde{\rho}(\boldsymbol{k},t) = \sum_{\boldsymbol{l}} \langle c(\boldsymbol{k},\boldsymbol{l},t)\rangle \tag{4.114}$$

form a *system of linear differential equations* which is the starting point for electron gas theory in the RPA approximation. It is not easy to give a rational justification for the RPA approximation, which can be formulated in a number of different manners. In particular, it will be shown in Sect. 10.2.2 that in perturbation theory the RPA approximation is equivalent to summing an infinite class of diagrams. In any case, its merit is that it provides a simple and physically correct description of screening and of fundamental excitations, as will be seen in Sect. 4.3.4.

Establishment of the Equation of Motion

To establish (4.108), it is first necessary to calculate the commutator

$$\begin{aligned} [c(\boldsymbol{k},\boldsymbol{l}), c(\boldsymbol{k}',\boldsymbol{l}')] &= \sum_\sigma [a^*_{\boldsymbol{l}\sigma}a_{(\boldsymbol{k}+\boldsymbol{l})\sigma}, a^*_{\boldsymbol{l}'\sigma}a_{(\boldsymbol{k}'+\boldsymbol{l}')\sigma}] \\ &= \delta_{\boldsymbol{k}+\boldsymbol{l},\boldsymbol{l}'}\sum_\sigma a^*_{\boldsymbol{l}\sigma}a_{(\boldsymbol{k}'+\boldsymbol{l}')\sigma} - \delta_{\boldsymbol{k}'+\boldsymbol{l}',\boldsymbol{l}}\sum_\sigma a^*_{\boldsymbol{l}'\sigma}a_{(\boldsymbol{k}+\boldsymbol{l})\sigma} \\ &= \delta_{\boldsymbol{k}+\boldsymbol{l},\boldsymbol{l}'}c(\boldsymbol{k}+\boldsymbol{k}',\boldsymbol{l}) - \delta_{\boldsymbol{k}'+\boldsymbol{l}',\boldsymbol{l}}c(\boldsymbol{k}+\boldsymbol{k}',\boldsymbol{l}') \quad . \end{aligned} \tag{4.115}$$

Inserting this into

$$[c(\boldsymbol{k},\boldsymbol{l}), \mathtt{H}_0] = \sum_{\boldsymbol{k}'} \varepsilon_{\boldsymbol{k}'}[c(\boldsymbol{k},\boldsymbol{l}), c(0,\boldsymbol{k}')] \quad , \tag{4.116}$$

one obtains (4.108). It follows from (4.105) and (4.115) that

$$\begin{aligned} [c(\boldsymbol{k},\boldsymbol{l}), \tilde{n}(\boldsymbol{k}')] &= \sum_{\boldsymbol{l}'}[c(\boldsymbol{k},\boldsymbol{l}), c(\boldsymbol{k}',\boldsymbol{l}')] \\ &= c(\boldsymbol{k}+\boldsymbol{k}',\boldsymbol{l}) - c(\boldsymbol{k}+\boldsymbol{k}',\boldsymbol{l}-\boldsymbol{k}') \end{aligned} \tag{4.117}$$

and

$$[c(\boldsymbol{k},\boldsymbol{l}), \tilde{n}^*(\boldsymbol{k}')] = c(\boldsymbol{k}-\boldsymbol{k}',\boldsymbol{l}) - c(\boldsymbol{k}-\boldsymbol{k}',\boldsymbol{l}+\boldsymbol{k}') \quad . \tag{4.118}$$

Finally, one gets

$$[c(\boldsymbol{k},\boldsymbol{l}),\mathrm{H}_1] = \left(\frac{e^2}{2L^3}\right)\sum_{\boldsymbol{k}'\neq 0}\tilde{V}^0(\boldsymbol{k}')\{[c(\boldsymbol{k},\boldsymbol{l}),\tilde{n}^*(\boldsymbol{k}')]\tilde{n}(\boldsymbol{k}') \\ +\tilde{n}^*(\boldsymbol{k}')[c(\boldsymbol{k},\boldsymbol{l}),\tilde{n}(\boldsymbol{k}')]\} \quad , \tag{4.119}$$

which, with the help of (4.117) and (4.118), leads to (4.109). Term (4.110) is obtained in the same manner.

4.3.4 The RPA Dielectric Function

It is possible to obtain an expression for the dielectric function $\varepsilon(\boldsymbol{k},\omega)$ directly from (4.113) and (4.114) without explicitly solving the differential equations. To do this, it is useful to introduce the frequency decompositions

$$\langle c(\boldsymbol{k},\boldsymbol{l},\omega)\rangle = \int \mathrm{d}t\mathrm{e}^{\mathrm{i}\omega t}\,\langle c(\boldsymbol{k},\boldsymbol{l},t)\rangle \tag{4.120}$$

and

$$\tilde{\rho}_e^{\mathrm{ext}}(\boldsymbol{k},\omega) = \int \mathrm{d}t\mathrm{e}^{\mathrm{i}\omega t}\,\tilde{\rho}_e^{\mathrm{ext}}(\boldsymbol{k},t) \quad . \tag{4.121}$$

Equation (4.113) is equivalent to

$$(\hbar\omega - \varepsilon_{\boldsymbol{k}+\boldsymbol{l}} + \varepsilon_{\boldsymbol{l}})\,\langle c(\boldsymbol{k},\boldsymbol{l},\omega)\rangle \\ = 2\frac{\tilde{V}^0(\boldsymbol{k})}{L^3}\left[e^2\tilde{\rho}(\boldsymbol{k},\omega) + e\tilde{\rho}_e^{\mathrm{ext}}(\boldsymbol{k},\omega)\right](n_{\boldsymbol{l}} - n_{\boldsymbol{k}+\boldsymbol{l}}) \quad . \tag{4.122}$$

After summing over $\boldsymbol{l}$ and using (4.114), one finds

$$\tilde{\rho}(\boldsymbol{k},\omega) = \left[\tilde{\rho}_e^{\mathrm{ext}}(\boldsymbol{k},\omega) + e\tilde{\rho}(\boldsymbol{k},\omega)\right]\frac{2e\tilde{V}^0(\boldsymbol{k})}{L^3}\left(\sum_{\boldsymbol{l}}\frac{n_{\boldsymbol{l}} - n_{\boldsymbol{k}+\boldsymbol{l}}}{\hbar\omega - \varepsilon_{\boldsymbol{k}+\boldsymbol{l}} + \varepsilon_{\boldsymbol{l}}}\right) \quad . \tag{4.123}$$

On the other hand, one can express the relation (4.83) for the dielectric function as

$$\rho_e^{\mathrm{ind}}(\boldsymbol{k},\omega) = \left(1 - \frac{\varepsilon(\boldsymbol{k},\omega)}{\varepsilon_0}\right)\left[\rho_e^{\mathrm{ext}}(\boldsymbol{k},\omega) + \rho_e^{\mathrm{ind}}(\boldsymbol{k},\omega)\right] \quad . \tag{4.124}$$

Performing this calculation to linear order of the external charge density, the electron density $e\tilde{\rho}(\boldsymbol{k},\omega)$, can be identified as the induced charge density $\rho_e^{\mathrm{ind}}(\boldsymbol{k},\omega)$. Thus, a comparison between (4.123) and (4.124) gives

$$\frac{\varepsilon(\boldsymbol{k},\omega)}{\varepsilon_0} = 1 - 2e^2\frac{\tilde{V}^0(\boldsymbol{k})}{L^3}\left(\sum_{\boldsymbol{l}}\frac{n_{\boldsymbol{l}} - n_{\boldsymbol{l}+\boldsymbol{k}}}{\hbar\omega - \varepsilon_{\boldsymbol{k}+\boldsymbol{l}} + \varepsilon_{\boldsymbol{l}}}\right) \quad , \tag{4.125}$$

which is the *dielectric function in the RPA approximation.*

To calculate $\varepsilon(\boldsymbol{k},\omega)$ in the thermodynamic limit, recall condition (4.87) which prescribes the method to correctly treat the singularity due to the vanishing of the denominator in (4.125). One thus obtains, in the limit of infinite volume,

$$\frac{\varepsilon(\boldsymbol{k},\omega)}{\varepsilon_0} = 1 - \frac{e^2}{\varepsilon_0|\boldsymbol{k}|^2}\lim_{\eta\to 0}\frac{2}{(2\pi)^3}\int \mathrm{d}\boldsymbol{l}\frac{n_{\boldsymbol{l}} - n_{\boldsymbol{l}+\boldsymbol{k}}}{\hbar(\omega+\mathrm{i}\eta) - \varepsilon_{\boldsymbol{k}+\boldsymbol{l}} + \varepsilon_{\boldsymbol{l}}} \quad . \tag{4.126}$$

It is possible to explicitly evaluate integral (4.126) for all values of $\boldsymbol{k}$ and ω.

Static Screening

Screening properties are determined by the behaviour of the static dielectric function $\varepsilon(\boldsymbol{k},0)$. Calculation of integral (4.126) for $\omega=0$ can be found at the end of this section. It yields the result

$$\frac{\varepsilon(\boldsymbol{k},0)}{\varepsilon_0} = 1 + \frac{1}{\lambda^2 k^2}\left\{\frac{1}{2} + \frac{k_{\mathrm{F}}}{2k}\left[1-\left(\frac{k}{2k_{\mathrm{F}}}\right)^2\right]\ln\left|\frac{2k_{\mathrm{F}}+k}{2k_{\mathrm{F}}-k}\right|\right\} \tag{4.127}$$

with

$$\frac{1}{\lambda^2} = \frac{e^2 m}{\pi^2\hbar^2\varepsilon_0}k_{\mathrm{F}} \quad , \quad |\boldsymbol{k}| = k \quad . \tag{4.128}$$

It is easy to determine the behaviour of $\varepsilon(\boldsymbol{k},0)$ for small $\boldsymbol{k}$ since the bracketed $\{\ldots\}$ part of (4.127) tends to 1 as $k\to 0$

$$\frac{\varepsilon(\boldsymbol{k},0)}{\varepsilon_0} \simeq 1 + \frac{1}{\lambda^2 k^2} \quad , \quad k\to 0 \quad . \tag{4.129}$$

Perfect screening (4.90) is recovered, since $\varepsilon(\boldsymbol{k},0)$ diverges when $|\boldsymbol{k}|\to 0$. The effective potential (4.89) thus behaves as

$$\tilde{V}^{\mathrm{eff}}(\boldsymbol{k}) \simeq \frac{e_0}{\varepsilon_0(|\boldsymbol{k}|^2+\lambda^{-2})} \quad , \quad |\boldsymbol{k}|\to 0 \quad . \tag{4.130}$$

If one allows $\tilde{V}^{\mathrm{eff}}(\boldsymbol{k})$ to be given by (4.130) for all $\boldsymbol{k}$, one finds that, by Fourier transformation, $V^{\mathrm{eff}}(\boldsymbol{x})$ has the form (4.71). The screening length λ (4.128) is of the order of a few angstroms. In reality, the logarithmic singularity of $\varepsilon(\boldsymbol{k},0)$ at $k=2k_{\mathrm{F}}$ creates a non-exponential, oscillating decrease of $V^{\mathrm{eff}}(\boldsymbol{x})$ at large distances

$$V^{\mathrm{eff}}(\boldsymbol{x}) \simeq \frac{\cos(2k_{\mathrm{F}}|\boldsymbol{x}|)}{|\boldsymbol{x}|^3} \quad , \quad |\boldsymbol{x}|\to\infty \quad . \tag{4.131}$$

The corresponding spatial oscillations, called *Friedel's oscillations*, can be observed in magnetic nuclear resonance experiments.

Plasma Oscillations

The dielectric function $\varepsilon(\boldsymbol{k},\omega)$ is studied here at the limit of small wavenumber and for any non-zero ω. Note that integral (4.126) can be written under the form

$$\begin{aligned} &\frac{2}{(2\pi)^3}\int \mathrm{d}\boldsymbol{l}\frac{n_{\boldsymbol{l}}-n_{\boldsymbol{l}+\boldsymbol{k}}}{\hbar(\omega+\mathrm{i}\eta)-\varepsilon_{\boldsymbol{k}+\boldsymbol{l}}+\varepsilon_{\boldsymbol{l}}} \\ &= \frac{2}{(2\pi)^3}\int \mathrm{d}\boldsymbol{l}n_{\boldsymbol{l}}\left(\frac{1}{\hbar(\omega+\mathrm{i}\eta)-\varepsilon_{\boldsymbol{k}+\boldsymbol{l}}+\varepsilon_{\boldsymbol{l}}}-\frac{1}{\hbar(\omega+\mathrm{i}\eta)+\varepsilon_{\boldsymbol{k}-\boldsymbol{l}}-\varepsilon_{\boldsymbol{l}}}\right) \\ &= \frac{|\boldsymbol{k}|^2}{m}\frac{2}{(2\pi)^3}\int \mathrm{d}\boldsymbol{l}n_{\boldsymbol{l}}\frac{1}{[(\omega+\mathrm{i}\eta)-(\hbar/m)\boldsymbol{k}\cdot\boldsymbol{l}]^2-(\hbar^2|\boldsymbol{k}|^2/2m)^2} \quad . \end{aligned} \tag{4.132}$$

This identity results from substituting $\boldsymbol{l}\to\boldsymbol{l}-\boldsymbol{k}$ in the portion of integral (4.126) which is carried out over $n_{\boldsymbol{l}+\boldsymbol{k}}$ and by taking into account that $\varepsilon_{\boldsymbol{k}+\boldsymbol{l}}-\varepsilon_{\boldsymbol{l}}=(\hbar^2/m)(\boldsymbol{k}\cdot\boldsymbol{l}+|\boldsymbol{k}|^2/2)$. When $|\boldsymbol{k}|\to 0$, it is clear that (4.132) behaves as

$$\frac{|\boldsymbol{k}|^2}{m(\omega+\mathrm{i}\eta)^2}\frac{2}{(2\pi)^3}\int \mathrm{d}\boldsymbol{l}n_{\boldsymbol{l}} = \frac{|\boldsymbol{k}|^2}{m(\omega+\mathrm{i}\eta)^2}\rho \tag{4.133}$$

since, according to (2.95), $[2/(2\pi)^3]\int \mathrm{d}\boldsymbol{l}n_{\boldsymbol{l}}$ is equal to the electron gas density. Inserting (4.133) into (4.126), one finally obtains

$$\lim_{|\boldsymbol{k}|\to 0}\frac{\varepsilon(\boldsymbol{k},\omega)}{\varepsilon_0} = 1-\left(\frac{\omega_{\mathrm{pl}}}{\omega}\right)^2 \quad , \tag{4.134}$$

where $\omega_{\mathrm{pl}}=(e^2\rho/\varepsilon_0 m)^{1/2}$ is the plasmon frequency (4.73).

As seen in Sect. 4.3.2, possible excitations of the gas are given by solutions $\omega(\boldsymbol{k})$ of $\varepsilon(\boldsymbol{k},\omega(\boldsymbol{k}))=0$. The result (4.134) shows that such a solution has the property

$$\lim_{|\boldsymbol{k}|\to 0}\omega(\boldsymbol{k}) = \omega_{\mathrm{pl}} \quad . \tag{4.135}$$

The plasmon frequency is thus an oscillation of the gas at the limit of small wavenumber, hence confirming the macroscopic analysis of Sect. 4.3.1. It is not possible to give an explicit form of $\omega(\boldsymbol{k})$ for $\boldsymbol{k}\neq 0$. However, solving $\varepsilon(\boldsymbol{k},\omega(\boldsymbol{k}))=0$ to second order in $\boldsymbol{k}$ gives

$$\omega(\boldsymbol{k}) = \omega_{\mathrm{pl}}+\frac{3}{10}\frac{v_{\mathrm{F}}^2}{\omega_{\mathrm{pl}}}|\boldsymbol{k}|^2+\ldots \quad , \quad v_{\mathrm{F}}=\frac{\hbar k_{\mathrm{F}}}{m} \quad , \tag{4.136}$$

which shows that the plasmon spectrum is parabolic for small $\boldsymbol{k}$.

It is instructive to compare the plasmon energy to that of the individual excitations of the electrons. An individual excitation is that which would take an electron of momentum $\hbar|\boldsymbol{l}|<\hbar k_{\mathrm{F}}$ outside of the Fermi sphere, thus

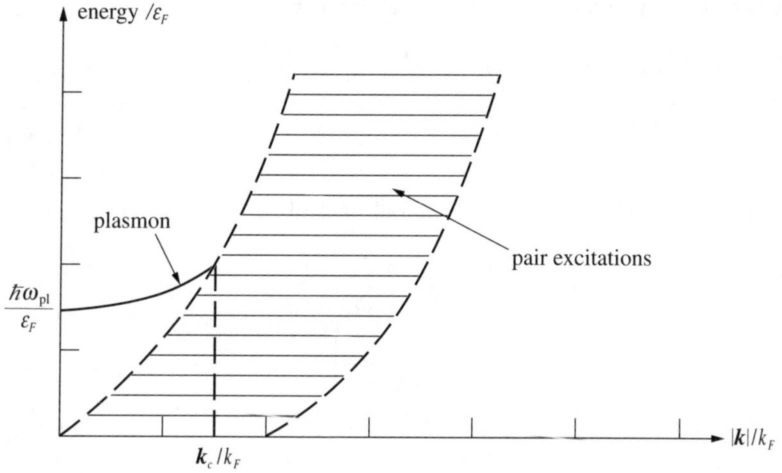

Fig. 4.4. Comparison of plasmon energy $\hbar\omega_{\mathrm{pl}}/\varepsilon_{\mathrm{F}}$ to individual electron excitation energy for sodium ($r_s = 4$)

forming an electron–hole pair (Sect. 2.2.2). The creation operator of such a pair is $a^*_{\boldsymbol{l}+\boldsymbol{k}}a_{\boldsymbol{l}}$ with $|\boldsymbol{l}+\boldsymbol{k}| > k_{\mathrm{F}}$ and its energy, relative to the Fermi energy, is given by

$$\varepsilon_{\boldsymbol{l}+\boldsymbol{k}} - \varepsilon_{\boldsymbol{l}} = \frac{\hbar^2}{m}\left(\boldsymbol{l}\cdot\boldsymbol{k} + \frac{|\boldsymbol{k}|^2}{2}\right) \quad , \quad |\boldsymbol{l}+\boldsymbol{k}| > k_{\mathrm{F}} \quad , \quad |\boldsymbol{l}| < k_{\mathrm{F}} \quad . \tag{4.137}$$

In Fig. 4.4, the plasmon energy is compared to the energies of electron–hole pairs for sodium ($r_s = 4$). The pair excitation region (4.137) is represented by the hatched zone and the plasmon energy by the solid line.

The typical energy of a plasmon in a metal ($r_s \simeq 4$) is of the order of 6 eV. This is much higher than the thermal energy $k_{\mathrm{B}}T \simeq 0.025$ eV at ambient temperature. The plasmons are not commonly excited thermally, but rather by other mechanisms, such as the passage of a high momentum charged particle through the metal. Figure 4.4 shows that, at a value k_{c}, the plasmon energy becomes comparable to the pair energy. Thus for $|\boldsymbol{k}| > k_{\mathrm{c}}$ plasmon energy can be transformed to pair energy and, from this point on, the plasmon is no longer a well defined mode of excitation. In fact, when $\boldsymbol{k} \neq 0$, the solution of $\varepsilon\left(\boldsymbol{k}, \omega(\boldsymbol{k})\right) = 0$ becomes complex

$$\omega(\boldsymbol{k}) = \omega_1(\boldsymbol{k}) - \mathrm{i}\omega_2(\boldsymbol{k}) \quad , \quad \omega_2(\boldsymbol{k}) > 0 \quad . \tag{4.138}$$

The inverse of the imaginary part $\omega_2(\boldsymbol{k})$, which measures the damping of the oscillation, is interpreted as the lifetime of the plasmon. For $|\boldsymbol{k}| \ll k_{\mathrm{c}}$, $\omega_2(\boldsymbol{k})$ is small enough for the plasmon to be considered stable vis-a-vis other excitations of the system.

Calculation of the Static Dielectric Function

In (4.126), the formula

$$\frac{1}{x \pm \mathrm{i}\eta} = P\left(\frac{1}{x}\right) \mp \mathrm{i}\pi\delta(x) \quad , \quad \eta \to 0 \tag{4.139}$$

will be used, where P denotes the principal part of the integral. Thus

$$\begin{aligned}\frac{\varepsilon(\boldsymbol{k},0)}{\varepsilon_0} &= 1 + \frac{e^2}{\varepsilon_0|\boldsymbol{k}|^2}\frac{2}{(2\pi)^3}\left[\mathcal{P}\!\!\int \mathrm{d}\boldsymbol{l}\frac{n_{\boldsymbol{l}} - n_{\boldsymbol{l}+\boldsymbol{k}}}{\varepsilon_{\boldsymbol{k}+\boldsymbol{l}} - \varepsilon_{\boldsymbol{l}}}\right.\\ &\quad \left. + \mathrm{i}\pi\int \mathrm{d}\boldsymbol{l}(n_{\boldsymbol{l}} - n_{\boldsymbol{l}+\boldsymbol{k}})\delta(\varepsilon_{\boldsymbol{k}+\boldsymbol{l}} - \varepsilon_{\boldsymbol{l}})\right]\\ &= 1 + \frac{2e^2}{\varepsilon_0|\boldsymbol{k}|^2}\frac{2}{(2\pi)^3}\mathcal{P}\!\!\int \mathrm{d}\boldsymbol{l}\frac{n_{\boldsymbol{l}}}{\varepsilon_{\boldsymbol{k}+\boldsymbol{l}} - \varepsilon_{\boldsymbol{l}}} \quad .\end{aligned} \tag{4.140}$$

One obtains (4.140) by making the substitution $\boldsymbol{l} \to -\boldsymbol{l} - \boldsymbol{k}$ in the integrals which are carried out over $n_{\boldsymbol{l}+\boldsymbol{k}}$. The imaginary part does not contribute because of the symmetry $\delta(x) = \delta(-x)$ and

$$\begin{aligned}\mathcal{P}\!\!\int \mathrm{d}\boldsymbol{l}\frac{n_{\boldsymbol{l}}}{\varepsilon_{\boldsymbol{k}+\boldsymbol{l}} - \varepsilon_{\boldsymbol{l}}} &= \frac{4\pi m}{\hbar^2}\int_0^{k_\mathrm{F}} \mathrm{d}|\boldsymbol{l}||\boldsymbol{l}|^2\,\mathcal{P}\!\!\int_{-1}^{+1} \mathrm{d}u\frac{1}{2|\boldsymbol{k}||\boldsymbol{l}|u + |\boldsymbol{k}|^2}\\ &= \frac{2\pi m}{\hbar^2|\boldsymbol{k}|}\int_0^{k_\mathrm{F}} \mathrm{d}|\boldsymbol{l}||\boldsymbol{l}|\ln\left|\frac{2|\boldsymbol{l}| + |\boldsymbol{k}|}{2|\boldsymbol{l}| - |\boldsymbol{k}|}\right| \quad .\end{aligned} \tag{4.141}$$

Calculation of the latter integral is elementary and, once (4.141) is substituted into (4.140), one arrives at (4.127).

Exercises

1. Electron–Hole Excitations in the Hartree–Fock Theory

Consider the excited Hartree–Fock state obtained from Φ_0 by removing one electron with wavenumber $\boldsymbol{k}_1$, $|\boldsymbol{k}_1| < k_\mathrm{F}$ and putting it in the state $\boldsymbol{k}_2$, $|\boldsymbol{k}_2| > k_\mathrm{F}$ outside of the Fermi sphere, keeping the wavenumbers of all the other electrons unchanged. If $E_\mathrm{exc}^\mathrm{HF}$ denotes the Hartree–Fock energy of this excited state, show that

$$E_\mathrm{exc}^\mathrm{HF} - E_0^\mathrm{HF} = \varepsilon_{\boldsymbol{k}_2}^\mathrm{HF} - \varepsilon_{\boldsymbol{k}_1}^\mathrm{HF} \quad .$$

Comment: The differences of the Hartree–Fock eigenvalues can be interpreted as electron–hole excitations in this theory.

2. Energy of a Ferromagnetic Hartree–Fock State

In the case for which all electrons have spin up, show that the Hartree–Fock energy per electron is

$$\lim_{n\to\infty} \frac{E_{\uparrow}^{\text{HF}}}{n} = \left(2.21\frac{2^{2/3}}{r_s^2} - 0.916\frac{2^{1/3}}{r_s}\right) [\text{Ryd}] \quad .$$

Conclude that the Hartree–Fock energy (4.56) (with pairs of up-down spins) has a lower value as soon as the density is sufficiently high:

$$\rho > \frac{3}{4\pi\bar{r}}\frac{1}{a_{\text{B}}^3} \quad , \quad \bar{r} \simeq 5.45 \quad .$$

Comment: This exercise confirms that the electronic configuration of the free gas with two spin states per mode has a lower energy at high density. However one should not conclude that for densities smaller than $\bar{r}$, the Hartree–Fock theory predicts ferromagnetism: the theory neglects the Coulomb correlations that become important precisely at lower densities.

3. Dielectric Function and Dynamics of the Ions

One wants to find a simple expression for the dielectric function taking into account the dynamics of a collection of classical ions of mass M and charge $-Ze$. The electronic and ionic charge densities are $\rho_e(\boldsymbol{x}, t)$ and $\rho_i(\boldsymbol{x}, t)$. When there are no external perturbations, the system is uniform and neutral, i.e. $\rho_e = -\rho_i = e\rho$ where ρ is the electronic number density. The introduction of an external charge $\rho^{\text{ext}}(\boldsymbol{x}, t)$ density gives rise to induced electronic and ionic densities

$$\rho_e(\boldsymbol{x}, t) = e\rho + \rho_e^{\text{ind}}(\boldsymbol{x}, t) \quad , \quad \rho_i(\boldsymbol{x}, t) = -e\rho + \rho_i^{\text{ind}}(\boldsymbol{x}, t) \quad ,$$

with total induced charge density $\rho^{\text{ind}}(\boldsymbol{x}, t) = \rho_e^{\text{ind}}(\boldsymbol{x}, t) + \rho_i^{\text{ind}}(\boldsymbol{x}, t)$. Then the dielectric function is defined as in (4.83) and the (frequency-dependent) effective potential is given by $\tilde{V}^{\text{eff}}(\boldsymbol{k}, \omega) = \rho^{\text{ext}}(\boldsymbol{k}, \omega)/\varepsilon(\boldsymbol{k}, \omega)|\boldsymbol{k}|^2$. Since the ions are classical, one can introduce a purely electronic dielectric function $\varepsilon_e(\boldsymbol{x}, t)$ by considering $\rho_i^{\text{ind}}(\boldsymbol{x}, t)$ as part of an external charge density acting on the electrons, namely in the Fourier representation

$$\frac{\varepsilon_e(\boldsymbol{k}\omega)}{\varepsilon_0} = \frac{\rho_i^{\text{ind}}(\boldsymbol{k}, \omega) + \rho^{\text{ext}}(\boldsymbol{k}, \omega)}{\rho_e^{\text{ind}}(\boldsymbol{k}, \omega) + \rho_i^{\text{ind}}(\boldsymbol{k}, \omega) + \rho^{\text{ext}}(\boldsymbol{k}, \omega)} \quad .$$

The classical dynamics of the ions will be treated in the mean-field approximation: each ion moves independently under the action of the total electric field. Moreover the motion of ions is slow, allowing one to neglect terms which are quadratic in their velocities.

(i) Establish the linear response relation

$$\omega^2 \rho_i^{\text{ind}}(\boldsymbol{k}, \omega) = \Omega_i^2 \left[\rho_e^{\text{ind}}(\boldsymbol{k}, \omega) + \rho_i^{\text{ind}}(\boldsymbol{k}, \omega) + \rho^{\text{ext}}(\boldsymbol{k}, \omega)\right] \quad ,$$

where

$$\Omega_i = \left(\frac{Ze^2\rho}{M\varepsilon_0}\right)^{1/2} \quad .$$

Hint: Use the continuity equation for conservation of the ionic charge and the equation of motion of an ion submitted to the electric field generated by the total charge density.

(ii) Approximating $\varepsilon_e(\boldsymbol{k}, \omega) \sim \varepsilon_e^{\text{stat}}(\boldsymbol{k})$ by the static RPA formula (4.129), verify that the excitation spectrum $\omega_i(\boldsymbol{k})$ of the ions behaves in a phononic way

$$\omega_i(\boldsymbol{k}) \simeq \lambda \Omega_i |\boldsymbol{k}| \quad , \quad \boldsymbol{k} \to 0 \quad .$$

Hint: The ionic excitation spectrum is defined as the solution of the equation in (i) when $\rho^{\text{ext}}(\boldsymbol{k}, \omega) = 0$.

(iii) Calculate the dielectric function

$$\frac{\varepsilon(\boldsymbol{k}, \omega)}{\varepsilon_0} = \frac{\varepsilon_e^{\text{stat}}(\boldsymbol{k})}{\varepsilon_0} \left(1 - \frac{\omega_i^2(\boldsymbol{k})}{\omega^2}\right)$$

and the effective potential

$$\tilde{V}^{\text{eff}}(\boldsymbol{k}, \omega) = \frac{\rho^{\text{ext}}(\boldsymbol{k}, \omega)}{\varepsilon_e^{\text{stat}}(\boldsymbol{k})|\boldsymbol{k}|^2} + \frac{\rho^{\text{ext}}(\boldsymbol{k}, \omega)}{\varepsilon_e^{\text{stat}}(\boldsymbol{k})|\boldsymbol{k}|^2} \frac{\omega_i^2(\boldsymbol{k})}{\omega^2 - \omega_i^2(\boldsymbol{k})} \quad .$$

Comment: We recover in the first term of $\tilde{V}^{\text{eff}}(\boldsymbol{k}, \omega)$ the screened Coulomb potential by a static uniform ionic distribution. The second term is due to the dynamics of ions and can become negative in the frequency range $0 < \omega < \omega_i(\boldsymbol{k})$, thus showing here the source of a possible attraction for the electrons. The same phenomenon will occur in the full many-body problem (Sect. 9.4.4): the phonon-mediated effective interaction between the electrons is attractive at low frequency and this attraction is at the origin of superconductivity.

5. Fermion Pairing and Superconductivity

5.1 Does There Exist an Analogue of the Bose Condensation for Fermions?

Chapter 2 examined the essential difference between bosons and fermions in the ground state. The distinction between them was shown to extend to low temperature. Free massive bosons have a phase transition at a finite temperature T_0 (2.86) called Bose–Einstein condensation. *No such phenomenon exists for a system of free fermions, regardless of whether the fermions are massive or not.* While the problem is simple at all temperatures for free particles, things are different when the particles interact.

The case for bosons is illustrated by the superfluidity of ^{4}He. Interpreting the liquid phase as a Bose–Einstein condensation has proven to be very fruitful (Chap. 7). Nevertheless, the existence of an inter-atomic potential complicates all quantitative property predictions of superfluid helium. As will be seen in Chap. 7, involved arguments only roughly explain the qualitative properties of this spectacular phase.

Concerning fermions, particle interaction can manifest itself in two different manners. For very strong interactions, the particles group in multiplets with even numbers of identical components, forming new highly-localized units which obey Bose–Einstein statistics. For this case, one can apply the preceding arguments as for the bosons. This is what occurs, for example, when nucleons form nuclei of integer spin (Sect. 2.1.6).

Equally interesting are fermion systems in the presence of weak interactions. Although they cannot group into localized units, they can still undergo a phase transition when they are subjected to very low temperature. This presents an undeniable analogy to the Bose–Einstein condensation. The most famous example is that of *superconductivity* discovered in 1911 by K. Onnes. It was not until 1957 that a satisfactory microscopic theory of Bardeen, Cooper and Schrieffer succeeded in explaining the nature of this phenomenon. This theory, universally known as the *BCS theory*, describes superconductivity as a phase transition affecting the electron gas of a number of metals at very low temperature.

In BCS theory, the electrons of a metal at temperature very close to absolute zero take advantage of an effective attractive interaction due to the presence of positive ions. They redistribute themselves in wavenumber space so that their total energy is less than a normal electron gas in its ground state. The individual states of this normal electron gas all strictly occupy the Fermi sphere $|\boldsymbol{k}| < k_{\mathrm{F}}$. Presented as such, BCS theory has no surprises. When redistributing the electrons in wavenumber space, the kinetic energy of the gas must necessarily be increased since, in the ground state, it is only minimized when the Fermi sphere is occupied. If an effective attractive interaction between the electrons exists, a redistribution of the occupation rate could compensate for the increase in kinetic energy. Thus the total energy of the gas, including the interaction potential, is reduced.

The originality of BCS theory becomes apparent when one realizes that the authors overcame a number of major difficulties. This also explains why it took many years after the discovery of quantum mechanics to arrive at a microscopic understanding of superconductivity. The theory needed to explain a phenomenological analogy, first noticed by F. London, between superconductivity and the superfluidity of helium. This similarity is absolutely astonishing if one considers only the properties of a free gas of bosons and those of a free gas of fermions.

Application of the variational method to the determination of the ground state energy of interacting fermions meets with a specific difficulty. Imagine that the variational states are not judiciously chosen. Sign changes brought about by the permutations of electrons could completely cancel the benefit of an attractive potential: this is the problem of sign ambiguity. The solution of this problem by BCS is the pairing of fermions, a pairing which permits the gas to take advantage of the attractive interaction.

The formation of pairs of electrons was suggested by Cooper in his analysis of a two-electron problem. As seen in Sect. 9.2.4, Cooper analyzed the interaction of two electrons whose wavenumbers are constrained to remain in the neighbourhood of k_{F} while the other electrons, considered as non-interacting fermions, occupy the Fermi sphere. These two electrons can form a bound state in the presence of an attractive interaction, however weak it may be (Sect. 9.2.4). Cooper pairs *are not* "e–e molecules" as the two electron partners are separated by a distance which surpasses the mean distance between the electrons by two to three orders of magnitude. In fact, in the volume of each pair, around 10^6 other pairs overlap (Sect. 5.3.8).

As suggestive as Cooper's analysis is, superconductivity remains an authentic many-body problem. Sections 5.3.2–5.3.4 show how one can remove the sign ambiguity by means of a fermion pairing mechanism. This procedure underlines both the importance and generality of the BCS theory and helps one to understand why these concepts can be used to describe at least three other systems. In chronological order, these systems are the structure of certain nuclei, the superfluidity of ^{3}He, and superconductivity due to heavy fermions. Other similar applications have been made in astrophysics, particularly in the study of neutron stars.

5.2 The Phenomenology of Superconductivity

5.2.1 Experimental Facts

Pure metals which do not become superconductors are exceptional (they include some ferromagnetic metals, light alkaline metals, and the precious metals Cu, Ag and Au). Many alloys also undergo this transition. Among these is the metal Nb_3Ge which held, until 1985, the record for the highest temperature of transition T_c at 23 K. Since then, a veritable revolution has taken place in the discovery of *high temperature (high-T_c) superconductors*. This revolution has stimulated the field of material physics by popularizing the study of a class of new composites containing principally oxygen, copper and two or three other metals. The compound Tl-Ca-Ba-Cu-O displays superconducting properties up to 120 K! In these new superconductors, the origin of the pairing mechanism is no longer clear, but it appears that the attraction between electrons has a different origin than that of the more conventional superconductors. On the other hand, the existence of electron pairing, the back-bone of the BCS theory, is not questioned by the new superconductors. Some of the new features of high-T_c superconductivity are shortly described in Sect. 5.5. The only pairing mechanism analyzed in this book (Sect. 9.2.4) is the traditional electron coupling due to the dynamics of the ions.

Aside from the observable absence of resistance, the principal characteristics of the superconducting phase, limiting oneself to the elementary properties of type-I superconductors,[1] are presented briefly in the following sections.

Discontinuity of the Specific Heat at the Transition Temperature

At very low temperature, the specific heat C_{tot} of a normal metal can be broken down into two terms

$$C_{\text{tot}} = C_e + C_{\text{lat}} = \gamma_e T + aT^3 \quad , \tag{5.1}$$

where γ_e and a are constants. Contributions C_e and C_{lat} are due to the electrons and to the ionic lattice, respectively. If C_e is plotted as a function of T, one obtains the behaviour shown in Fig. 5.1. It should be emphasized, first of all, that *the transition from a normal state to a superconducting state (called an n-s transition) is a phase transition which, in the absence of a magnetic field, is of second order*. Comparing a low-temperature superconductor with specific heat $C_e^s(T)$ to a non-superconductor with specific heat $C_e^n(T)$, it follows that

[1] Type-I superconductors present all the characteristic properties of superconductivity, in particular complete Meissner effect. Among these are the pure superconducting metals (Hg, Sn, Pb, ...).

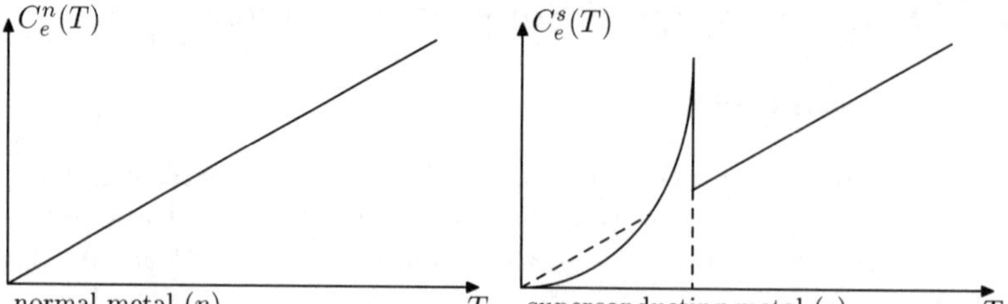

Fig. 5.1. Behaviour of the electron component of the specific heat (*left*) for a normal metal $C_e^n(T)$ and (*right*) for a superconducting metal $C_e^s(T)$ as functions of temperature

$$\int_0^T \mathrm{d}T' C_e^s(T') < \int_0^T \mathrm{d}T' C_e^n(T') \quad , \quad T < T_c \quad , \tag{5.2}$$

provided the two metals have equal values of γ_e. Inequality (5.2) implies that *the energy of the superconducting state is inferior to that of the normal state.*

The Meissner Effect

Once the resistance of a body is zero, the magnetic field cannot change on the interior of that body. In effect, Ohm's law $\boldsymbol{j} = \sigma \boldsymbol{E}$ imposes that the electric field must go to zero when the conductivity σ is infinite. As a consequence, the magnetic field cannot vary

$$\frac{\partial \boldsymbol{B}}{\partial t} = -\nabla \times \boldsymbol{E} = 0 \quad . \tag{5.3}$$

There are two different manners to submit a metallic sample to an external field at a temperature inferior to the transition temperature. The metal may be cooled first and then placed in the field. In this case, the field on the interior of the sample vanishes because of (5.3). *Remarkable, the field also vanishes if one proceeds in the opposite manner, that is, if the metal is submitted to a magnetic field and then cooled below the transition temperature.* In this case the field is expelled outside of the superconductor (Fig. 5.2). This is called the *Meissner effect* and is expressed by the relation

$$\boldsymbol{B} = 0 \quad . \tag{5.4}$$

A superconductor is sometimes called a perfect diamagnetic metal. If the field surpasses a certain critical value $H_c(T)$, the expulsion does not take place and the body remains in a normal state.

Even when the expulsion takes place, one observes an exponential penetration of the field at the surface of the sample down to a *penetration depth* δ of the order of 1000 Å. London gave a microscopic description in 1935 of the Meissner effect which was the first step towards the theory of superconductivity.

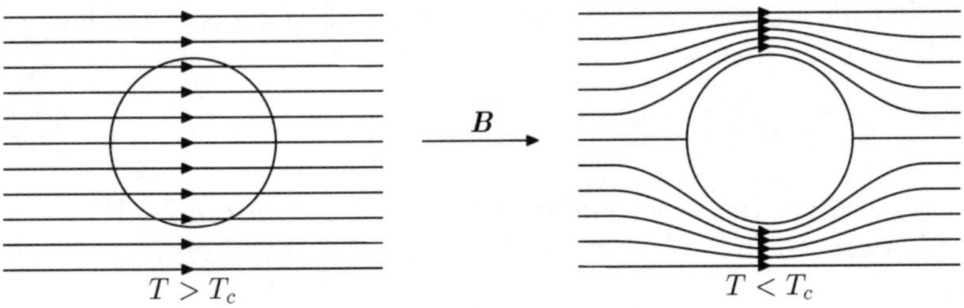

Fig. 5.2. Expulsion of the field outside of the superconductor cooled below the transition temperature

5.2.2 The Phenomenological Approach

Consider the electrons to form a frictionless fluid described by a velocity field $\boldsymbol{v}(\boldsymbol{x}, t)$. In the presence of an electromagnetic field, the electrons obey the Lorentz law of force

$$\frac{\mathrm{d}\boldsymbol{v}}{\mathrm{d}t} = \frac{e}{m}(\boldsymbol{E} + \boldsymbol{v} \times \boldsymbol{B}) \quad . \tag{5.5}$$

The total derivative $\mathrm{d}\boldsymbol{v}/\mathrm{d}t$ can be expressed in terms of the local acceleration $\partial\boldsymbol{v}/\partial t$

$$\frac{\mathrm{d}\boldsymbol{v}}{\mathrm{d}t} = \frac{\partial \boldsymbol{v}}{\partial t} + (\boldsymbol{v} \cdot \nabla)\boldsymbol{v} = \frac{\partial \boldsymbol{v}}{\partial t} + \frac{1}{2}\nabla|\boldsymbol{v}|^2 - \boldsymbol{v} \times (\nabla \times \boldsymbol{v}) \quad . \tag{5.6}$$

The combination of (5.5) and (5.6) gives

$$\frac{\partial \boldsymbol{v}}{\partial t} - \frac{e}{m}\boldsymbol{E} + \nabla\left(\frac{1}{2}|\boldsymbol{v}|^2\right) = \boldsymbol{v} \times \left(\nabla \times \boldsymbol{v} + \frac{e}{m}\boldsymbol{B}\right) \quad . \tag{5.7}$$

Taking into account Maxwell's equation $\partial \boldsymbol{B}/\partial t = -\nabla \times \boldsymbol{E}$ and taking the curl of expression (5.7) gives

$$\frac{\partial \boldsymbol{w}}{\partial t} = \nabla \times (\boldsymbol{v} \times \boldsymbol{w}) \quad , \tag{5.8}$$

where $\boldsymbol{w} = \nabla \times \boldsymbol{v} + (e/m)\boldsymbol{B}$.

This equation is well known in hydrodynamics as the Helmholtz equation. It describes the evolution of a rotational element of a fluid $\nabla \times \boldsymbol{v}$ when viscosity is negligible. The differential equation (5.8), linear in $\boldsymbol{w}$ and of first order in t, has the following property: the solution $\boldsymbol{w}(t)$ corresponding to the initial condition $\boldsymbol{w}(t = 0) = 0$ is the solution $\boldsymbol{w}(t) = 0$ for all t.

The London Equations

Consider a superconducting metal at equilibrium with no magnetic field: $\boldsymbol{v} = 0$, $\boldsymbol{B} = 0$, and $\boldsymbol{w} = 0$. It can be concluded from (5.8) that $\boldsymbol{w}$ will remain zero even after a magnetic field is turned on. This reasoning is at the base of London's hypothesis that *the equation*

$$\boldsymbol{w} = \nabla \times \boldsymbol{v} + \frac{e}{m}\boldsymbol{B} = 0 \tag{5.9}$$

describes the state of the superconductor under all circumstances. It can be immediately deduced from (5.7) and (5.9) that, if one keeps terms quadratic in the velocity,

$$\frac{\partial \boldsymbol{v}}{\partial t} + \nabla\left(\frac{1}{2}|\boldsymbol{v}|^2\right) = \frac{e}{m}\boldsymbol{E} \quad . \tag{5.10}$$

Equations (5.9) and (5.10) constitute the *London equations of superconductivity.* Under the form presented here, they represent a phenomenological model of a superconductor at zero temperature.

If one supposes that the density ρ_s of the conducting electrons is uniform throughout the superconductor, then the charge density must be zero at each point in order to maintain local neutrality. The current density $\boldsymbol{j}(\boldsymbol{x}, t)$ due to electron motion

$$\boldsymbol{j}(\boldsymbol{x}, t) = e\rho_s \boldsymbol{v}(\boldsymbol{x}, t) \tag{5.11}$$

obeys London's equation (5.9)

$$\nabla \times \boldsymbol{j} = -\frac{e^2}{m}\rho_s \boldsymbol{B} \tag{5.12}$$

and the equation of continuity

$$\nabla \cdot \boldsymbol{j} = 0 \quad . \tag{5.13}$$

These last two equations suggest a particularly simple relationship between the current $\boldsymbol{j}$ and the vector potential $\boldsymbol{A}$ when the latter is expressed in the Coulomb gauge. With $\boldsymbol{B} = \nabla \times \boldsymbol{A}$ and $\nabla \cdot \boldsymbol{A} = 0$, equations (5.12) and (5.13) are equivalent to

$$\nabla \times \left(\boldsymbol{j} + \frac{e^2 \rho_s}{m}\boldsymbol{A}\right) = 0 \quad ,$$
$$\nabla \cdot \left(\boldsymbol{j} + \frac{e^2 \rho_s}{m}\boldsymbol{A}\right) = 0 \quad . \tag{5.14}$$

Suppose that a superconducting region, limited by frontier Σ, is simply connected. Imposing $\hat{\boldsymbol{n}} \cdot \boldsymbol{A} = 0$ at all points on Σ (where $\hat{\boldsymbol{n}}$ is a unit-normal vector to Σ), uniquely determines the field $\boldsymbol{A}$ and the general solution to (5.14) is written

$$\boldsymbol{j} + \frac{e^2 \rho_s}{m}\boldsymbol{A} = \nabla \chi(\boldsymbol{x}, t) \quad . \tag{5.15}$$

The function χ, which satisfies

$$\nabla^2\chi = 0 \quad \text{and} \quad \hat{\boldsymbol{n}} \cdot \nabla\chi = \hat{\boldsymbol{n}} \cdot \boldsymbol{j} \quad \text{at all points on } \Sigma, \tag{5.16}$$

is determined up to a time-dependent function. When no current traverses the enclosure Σ (that is $\hat{\boldsymbol{n}} \cdot \boldsymbol{j} = 0$), one can set $\chi = 0$.

Relation (5.15) plays a remarkable role:

- It provides a *link between the microscopic theory and the macroscopic electromagnetic description of the superconductor*. BCS theory allows one to derive (5.15) in the limit of small wavenumbers (not covered here).
- As London suggests, a relation such as (5.15) reveals a quantum origin. Furthermore, it explains a *deep phenomenological relationship between superconductivity and superfluidity* (detailed in Sect. 5.2.3).
- It is the source of interesting analogies to the mechanism for the generation of mass in field theory (Sect. 8.3.7).

Penetration of the Magnetic Field

Equation (5.12) describes a Meissner effect characterized by a penetration depth which is of the same order of magnitude as that measured by experiment. In a stationary regime, taking into account (5.13) and Ampere's relation $\nabla \times \boldsymbol{B} = \mu_0 \boldsymbol{j}$, (5.12) yields

$$\nabla^2 \boldsymbol{j} = -\nabla \times (\nabla \times \boldsymbol{j}) = \frac{e^2}{m}\rho_s \nabla \times \boldsymbol{B} = \frac{e^2}{m}\rho_s \mu_0 \boldsymbol{j} = \frac{e^2 \rho_s}{\varepsilon_0 m c^2}\boldsymbol{j} \quad . \tag{5.17}$$

Taking the curl of (5.17) gives a similar equation for $\boldsymbol{B}$

$$\nabla^2 \boldsymbol{B} = \frac{e^2 \rho_s}{\varepsilon_0 m c^2}\boldsymbol{B} = \delta^{-2}\boldsymbol{B} \quad , \tag{5.18}$$

where δ is defined by

$$\delta = \left(\frac{\varepsilon_0 m c^2}{\rho_s e^2}\right)^{1/2} \quad . \tag{5.19}$$

Numerically, δ is of the order of 1000 Å, comparable to the measured penetration depth of the field.

Imagine a superconducting plate occupying the region $-a \leq x \leq a$, as shown in Fig. 5.3. A stationary induction field $\boldsymbol{B}$ with magnitude less than the critical field B_c is parallel to the plate. Outside of the plate, the field is uniform and equal to $\boldsymbol{B}_0$. By symmetry, both $\boldsymbol{B}$ and $\boldsymbol{j}$ depend only on the x-coordinate inside the metal. Vectors $\boldsymbol{B}(\boldsymbol{x}) = (0, B(x), 0)$ and $\boldsymbol{j}(\boldsymbol{x}) = (0, 0, j(x))$ are orthogonal to each other and both parallel to the plate. Equation (5.18) and Ampere's relation then reduce to

$$\frac{\mathrm{d}^2 B(x)}{\mathrm{d}x^2} = \frac{1}{\delta^2}B(x) \quad , \quad j(x) = \frac{1}{\mu_0}\frac{\mathrm{d}B(x)}{\mathrm{d}x} \tag{5.20}$$

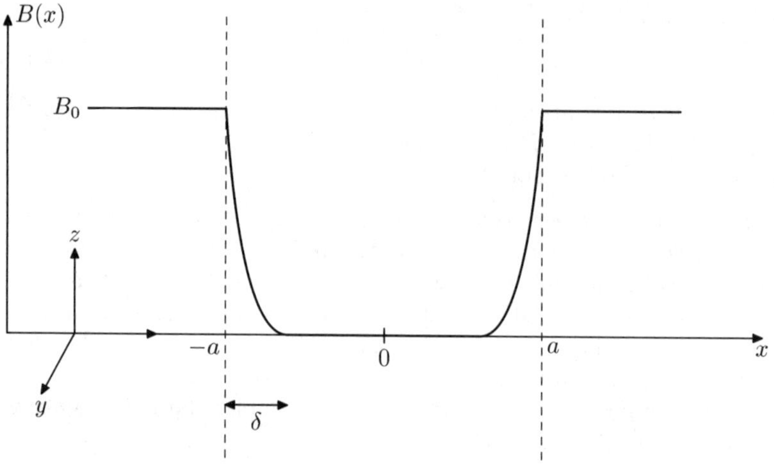

Fig. 5.3. For a superconducting plate submitted to a parallel external field $\boldsymbol{B}$, the field and the current inside the plate decrease exponentially over a distance of the order of δ, the penetration depth

which can both be integrated to give

$$B(x) = B_0 \frac{\cosh(x/\delta)}{\cosh(a/\delta)} \quad , \quad j(x) = \frac{B_0}{\mu_0 \delta} \frac{\sinh(x/\delta)}{\cosh(a/\delta)} \quad . \tag{5.21}$$

The field and the current thus decrease exponentially inside the plate over a distance of the order of δ, justifying the interpretation of δ. This is a general property, regardless of the form of the superconductor, provided the external field to which it is submitted does not reach the value B_c at any point on the surface and under the condition that all of the dimensions of the superconductor exceed δ.

Equations (5.17) and (5.18) are interpreted as follows: *Submitted to an induction field* $\boldsymbol{B}$*, either external or produced by its own current, the superconductor gives rise to surface currents which create a field* $-\boldsymbol{B}$ *exactly compensating the contribution of* $\boldsymbol{B}$ *at all points inside the metal situated sufficiently far from the surface.* This phenomenon is the magnetic analogue to the electrostatic screening effects described in Sect. 4.3.1. The charge distribution inside a normal conductor at equilibrium exactly compensates the field of external charges, with the screening length λ playing an analogous role to that of the penetration depth δ. On the other hand, the analogy between superconductivity and superfluidity will be treated in the following section.

5.2.3 Macroscopic Quantum Fluids

To make relation (5.12) more plausible, London proceeded as follows. He attributed to the electron fluid a complex scalar function $\varphi(\boldsymbol{x}, t)$ possessing several properties inherent to a typical wavefunction. Such a description can nevertheless be considered macroscopic, because $\varphi(\boldsymbol{x},t)$ varies appreciably only over the distances which are characteristic to variations of the electromagnetic fields themselves. This idea had been successfully applied to other systems, grouped under the name *quantum fluids*. It will be presented here with a generality which extends beyond superconductivity but which will be restricted to the case where $\varphi(\boldsymbol{x},t)$ is a scalar.

A quantum fluid obeys certain hydrodynamic relations. Beside the purely hydrodynamic variables, the pseudo wavefunction enters into the description via the particle density $\rho(\boldsymbol{x},t)$ identified as

$$\rho(\boldsymbol{x},t) = |\varphi(\boldsymbol{x},t)|^2 \quad . \tag{5.22}$$

Moreover, φ appears in the definition of the velocity (superfluid ^{4}He) or of the current density (superconductivity), as will be shown in (5.24) and (5.30).

In the two cases considered here, $\rho(\boldsymbol{x},t) = \rho$ is assumed to be constant: ^{4}He is an incompressible superfluid. As for the superconductor, electrical neutrality forces ρ to be uniform. Hence,

$$\varphi(\boldsymbol{x},t) = \rho^{1/2}\mathrm{e}^{\mathrm{i}\alpha(\boldsymbol{x},t)} \quad . \tag{5.23}$$

Thus the phase $\alpha(\boldsymbol{x},t)$ contains the essential information on the function $\varphi(\boldsymbol{x},t)$ and gives the fluid its quantum characteristic.

Neutral Quantum Fluid

If the quantum fluid is not charged, the model (to be further developed in Chap. 7) is of superfluid ^{4}He. Starting with $\varphi(\boldsymbol{x},t)$, a vector field is defined as in quantum mechanics

$$\boldsymbol{j} = \frac{1}{m}\mathrm{Re}(\varphi^*\boldsymbol{p}\varphi) = \frac{1}{2m}\left(\varphi^*\frac{\hbar}{\mathrm{i}}\nabla\varphi - \varphi\frac{\hbar}{\mathrm{i}}\nabla\varphi^*\right) \quad , \quad \boldsymbol{p} = \frac{\hbar}{\mathrm{i}}\nabla \quad . \tag{5.24}$$

Here, $\boldsymbol{j}(\boldsymbol{x},t)$ is interpreted as the particle current, produced from ρ by the velocity field $\boldsymbol{v}(\boldsymbol{x},t)$

$$\boldsymbol{j}(\boldsymbol{x},t) = \rho\boldsymbol{v}(\boldsymbol{x},t) \quad . \tag{5.25}$$

From (5.23) and (5.24), with m being the mass of an atom, one has

$$\boldsymbol{v}(\boldsymbol{x},t) = \frac{\hbar}{m}\nabla\alpha(\boldsymbol{x},t) \quad . \tag{5.26}$$

By consequence, *a neutral and incompressible quantum fluid is irrotational*

$$\nabla\times\boldsymbol{v} = \frac{\hbar}{m}\nabla\times(\nabla\alpha) = 0 \quad . \tag{5.27}$$

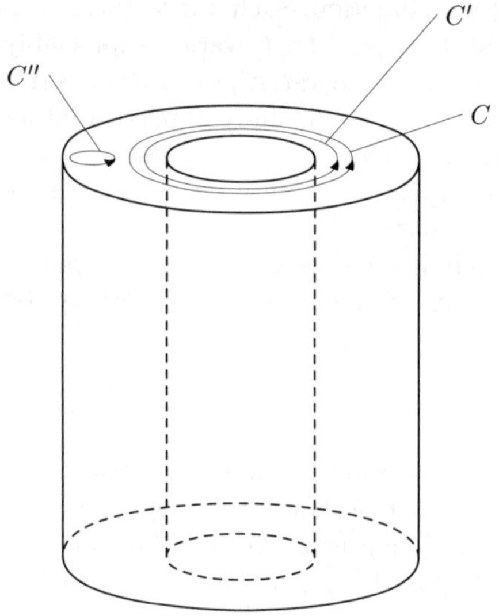

Fig. 5.4. Phase variations of velocity for closed curves have the same value if the curves, such as C and C' are homotopic

The dynamics of superfluid ^{4}He obeys precisely relation (5.27).

One can benefit further from (5.26). If the fluid is enclosed in *a region which is not simply connected* (Fig. 5.4), one can consider a closed curve C. Circulation of the velocity field $\boldsymbol{v}$ along C is expressed as

$$\oint_{\mathrm{c}} \mathrm{d}\boldsymbol{l} \cdot \boldsymbol{v} = \frac{\hbar}{m} \Delta_{\mathrm{c}} \alpha \quad , \tag{5.28}$$

where $\Delta_{\mathrm{c}}\alpha$ is the variation of the phase when one revolution along C is completed. This variation does not depend on C itself, but takes the same value for all curves C' which are homotopic to C.[2] If the properties of $\varphi(\boldsymbol{x}, t)$ are compared to those of a quantum wavefunction, $\varphi(\boldsymbol{x}, t)$ is single valued and, after one complete revolution, the variation of its phase is equal to a multiple of 2π. The result is *the circulation of the velocity is quantified in superfluid* ^{4}He

$$\oint_{\mathrm{c}} \mathrm{d}\boldsymbol{l} \cdot \boldsymbol{v} = \frac{\hbar}{m} 2\pi n = n \frac{h}{m} \quad , \quad n \text{ integer} \quad . \tag{5.29}$$

[2] Two closed curves are homotopic if it is possible to continuously deform one into the other. In Fig. 5.4, C and C' are homotopic, while C'' is homotopic to a point.

Charged Quantum Fluid

The function $\varphi(\boldsymbol{x},t)$ is used to describe the fluid of charged particles in a superconductor. In this case, each particle carries charge e_0 and, in the presence of an electromagnetic field of vector potential $\boldsymbol{A}$, the velocity operator $\boldsymbol{v} = (\hbar/m\mathrm{i})\nabla$ must be replaced by $\boldsymbol{v} = (1/m)[(\hbar/\mathrm{i})\nabla - e_0\boldsymbol{A}]$. Here, the quantity of interest is the electrical current density

$$\begin{aligned} j &= \frac{e_0}{m}\,\mathrm{Re}\,[\varphi^*(\boldsymbol{p} - e_0\boldsymbol{A})\varphi] \\ &= \frac{e_0}{2m}\left[\varphi^*\left(\frac{\hbar}{\mathrm{i}}\nabla - e_0\boldsymbol{A}\right)\varphi - \varphi\left(\frac{\hbar}{\mathrm{i}}\nabla + e_0\boldsymbol{A}\right)\varphi^*\right] \quad . \end{aligned} \tag{5.30}$$

Defining $\boldsymbol{j}$ once again as the field of the hydrodynamic velocity (as in (5.11)) and taking into account that the density $\rho_s = |\varphi|^2$ is constant, one obtains

$$\boldsymbol{j} + \frac{e_0^2\rho_s}{m}\boldsymbol{A} = \frac{\hbar e_0\rho_s}{m}\nabla\alpha \quad . \tag{5.31}$$

Up to a constant factor $\hbar e_0\rho_s/m$, this is relation (5.15). In a simply connected domain, it is possible to choose α to be zero over the entire domain with boundary surface Σ and the condition that the Coulomb gauge is taken with boundary conditions

$$\nabla\cdot\boldsymbol{A} = 0 \quad , \quad \hat{\boldsymbol{n}}\cdot\boldsymbol{A} = 0 \quad \text{for all points on } \Sigma. \tag{5.32}$$

Consider a superconducting volume which is not simply connected. The volume represented in Fig. 5.5 will be chosen here for the sake of discussion, although generalization to a more complex topology would not pose any particular problem. In such a situation, the solution of (5.32) is not unique. One must still specify the flux Φ of the field $\boldsymbol{B}$ on the interior of the cavity. The following supplementary condition is thus added to (5.32)

$$\oint_{C'} \mathrm{d}\boldsymbol{l}\cdot\boldsymbol{A} = \Phi \quad , \tag{5.33}$$

where the integration is carried out over a simple closed curve C' drawn on the internal surface and encircling the cavity (in order to keep Fig. 5.5 from becoming too complicated, only curve C introduced below is drawn).

Concerning relations (5.15) and (5.31), even if they are formally identical, they differ in regard to the interpretations given to functions χ in (5.15) and $\hbar e_0\rho_s\alpha/m$ in (5.31). There is no a priori prescription for the behaviour of χ if the volume is not simply connected. On the other hand, for (5.31) the function $\varphi(\boldsymbol{x},t)$ must be single valued. This implies the same constraints on the determination of the phase α of φ as in the case of a neutral quantum fluid.

Finally, to exploit the conditions which must satisfy the phase α, consider a curve C traced inside of the volume of the superconductor (Fig. 5.5). This

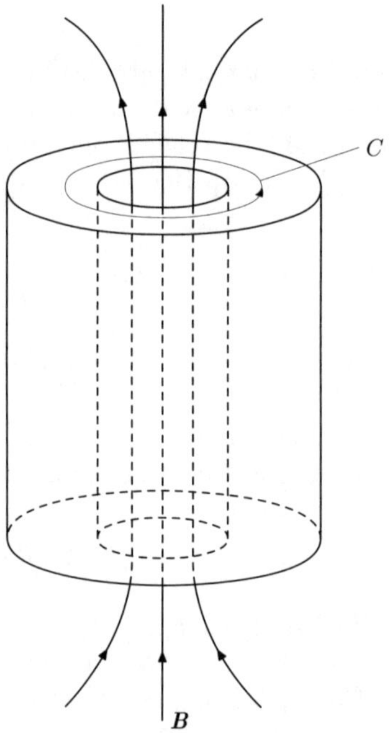

Fig. 5.5. A variation of phase $\Delta_c\alpha = 2\pi n$ corresponds to one revolution along curve C inside the volume of the superconductor

curve encircles a cavity which could be either a hole or a normal metal. When one revolution is completed along C, the corresponding variation of the phase $\Delta_c\alpha$ must be written as

$$\Delta_c\alpha = 2\pi n \quad , \quad n \text{ integer} \tag{5.34}$$

and the quantization relation analogous to (5.29) must be

$$\oint_c \mathrm{d}\boldsymbol{l} \cdot \left(\frac{m}{\rho_s e_0^2} \boldsymbol{j} + \boldsymbol{A} \right) = \frac{\hbar}{e_0} \Delta_c\alpha = n \frac{h}{e_0} \quad . \tag{5.35}$$

The left-hand side of (5.35) is known as a *fluxoid* and *relation (5.35) expresses the quantization of a fluxoid.*

In fact, one speaks in general of the *quantization of flux*. If Σ_c is a surface resting on the closed curve C, then

$$\oint_c \mathrm{d}\boldsymbol{l} \cdot \boldsymbol{A} = \int_{\Sigma_c} \mathrm{d}\boldsymbol{s} \cdot (\nabla \times \boldsymbol{A}) = \int_{\Sigma_c} \mathrm{d}\boldsymbol{s} \cdot \boldsymbol{B} \quad , \tag{5.36}$$

which is exactly the flux traversing Σ_c. If the distance between C and the surface of the sample is larger than the penetration depth δ, then $\boldsymbol{j}$ is zero on C and (5.35) takes the form

$$\int_{\Sigma_c} \mathrm{d}\boldsymbol{s} \cdot \boldsymbol{B} = n\frac{h}{e_0} \quad . \tag{5.37}$$

Experimental verification of flux quantization dates to 1961.[3] It confirmed London's prediction except for the factor e_0 which is *twice the charge of an electron*

$$e_0 = 2e \tag{5.38}$$

and not e as London had thought. Justification for the doubling of the charge (5.38) is found in the BCS theory where, as already mentioned, the electrons are associated into pairs.

The analogy between the quantization of circulation inside a fluid of bosons (superfluid ^{4}He) and the quantization of flux inside a fermion fluid is remarkable.

5.2.4 Existence of the Energy Gap

The electronic specific heat C_e^s of a superconductor decreases exponentially as $T \to 0$

$$C_e^s \simeq \exp\left(-\frac{\Delta_0}{k_\mathrm{B}T}\right) \quad . \tag{5.39}$$

This indicates that it is necessary to have a finite energy quantity Δ_0 to create an excitation above the ground state. This is similar to Einstein's model of lattice vibrations, in which each mode is characterized by an energy $\hbar\omega_0$ which is the same for all modes. The specific heat in this case also decreases exponentially in the neighbourhood of absolute zero.

One refers to Δ_0 as the *energy gap*. In effect, all of the first excited levels of the electron spectrum are separated from the ground level by the quantity Δ_0. From the point of view of orders of magnitude, the energy involved in the superconducting phenomenon

$$\Delta_0 \simeq k_\mathrm{B}T_\mathrm{c} \simeq 10^{-3}\ \mathrm{eV} \tag{5.40}$$

is much smaller than the Coulomb energy per electron (see (2.114) or (4.32)), which is of order

$$|\mathrm{u_{pot}}|\rho^{-1} \simeq \frac{e^2}{\varepsilon_0 a_\mathrm{B}} \simeq 1\ \mathrm{eV} \quad . \tag{5.41}$$

To quote Feynman, "The Coulomb correlation energy is too big, and can therefore be neglected. If this reason for neglecting Coulomb effects seems odd, remember that we are not trying to explain and predict everything

[3] R. Doll and M. Näbauer, Phys. Rev. Letters 7, 51 (1961); B. S. Deaver and W. M. Fairbank, Phys. Rev. Letters 7, 43 (1961).

about the solid. We are just trying to understand superconductivity." In other words, it is possible to neglect the Coulomb energy insofar as one can convince oneself that the mechanism which gives rise to superconductivity has its origin elsewhere. As a result, however, one must abandon making any quantitative predictions for the absolute value of the energies. Only the difference between the energy of a normal state and that of a superconducting state can yield numerical predictions.

Experimental evidence for the energy gap is not limited to the study of specific heat. More direct methods exist, such as the tunnelling effect between a normal metal and a superconductor or the absorption of electromagnetic waves of frequency $\nu \simeq 10^{11}\ \mathrm{s}^{-1}$, that is, satisfying

$$h\nu \simeq \Delta_0 \quad . \tag{5.42}$$

The existence of an energy gap must therefore be considered as a fact and must necessarily appear in any reasonable theory of superconductivity.

5.3 BCS Theory

5.3.1 The Effective Interaction Between Electrons

Electrons repel each other because of the Coulomb potential. The existence of a positive neutralizing charge due to the ion lattice, however, renders their dynamics a little more subtle. To understand the mechanism giving rise to superconductivity, it is necessary to take into account an experimental result which was decisive for its discovery. This is the *isotopic effect* (1950): if one replaces a superconducting element by one of its isotopes, the transition temperature can vary. For Sn and Th, if A is the atomic mass, one finds

$$T_\mathrm{c} \simeq A^{-1/2} \quad . \tag{5.43}$$

It is thus possible to conclude that the electron–electron interaction is composed of two terms, the Coulomb interaction and an interaction which is transmitted by the ions (an electron in motion displaces the ions which then modify the movement of the neighbouring electrons). This interaction is evidently delayed when transmitted by ions which are relatively slow themselves. For now, assume this force is attractive without bothering to determine its exact value; this point will be returned to in Sect. 9.4.4.

The goal of this section is to study how one can benefit from such an attraction in order to lower the energy of the ground state. *The word attractive means, in this context, that a certain collection of interaction matrix elements in a given basis has a determined sign, negative.* Since the specific form of the interaction is not relevant to this study, it will be assumed here that the electrons interact by means of a two-body interaction V, independent of spin. These matrix elements in the plane waves basis are

$$(\boldsymbol{k}_1\sigma_1, \boldsymbol{k}_2\sigma_2|V|\boldsymbol{k}'_1\sigma'_1, \boldsymbol{k}'_2\sigma'_2) = V_{\boldsymbol{k}_1,\boldsymbol{k}_2;\boldsymbol{k}'_1,\boldsymbol{k}'_2}\delta_{\sigma_1,\sigma'_1}\delta_{\sigma_2,\sigma'_2} \quad . \tag{5.44}$$

The dynamics of the ions only appear indirectly through their influence on the form of the coefficient $V_{\boldsymbol{k}_1,\boldsymbol{k}_2;\boldsymbol{k}'_1,\boldsymbol{k}'_2}$. Spin independence is a natural hypothesis if one recalls that the interaction transmitted by the ions is of Coulombic origin. The coefficients $V_{\boldsymbol{k}_1,\boldsymbol{k}_2;\boldsymbol{k}'_1,\boldsymbol{k}'_2}$ satisfy the following relations:

$$V_{\boldsymbol{k}_1,\boldsymbol{k}_2;\boldsymbol{k}'_1,\boldsymbol{k}'_2} = V_{\boldsymbol{k}_2,\boldsymbol{k}_1;\boldsymbol{k}'_2,\boldsymbol{k}'_1} \quad , \tag{5.45}$$

$$V_{\boldsymbol{k}_1,\boldsymbol{k}_2;\boldsymbol{k}'_1,\boldsymbol{k}'_2} = V_{-\boldsymbol{k}_1,-\boldsymbol{k}_2;-\boldsymbol{k}'_1,-\boldsymbol{k}'_2} \quad , \tag{5.46}$$

$$V_{\boldsymbol{k}_1,\boldsymbol{k}_2;\boldsymbol{k}'_1,\boldsymbol{k}'_2} = 0 \quad \text{if} \quad \boldsymbol{k}_1 + \boldsymbol{k}_2 \neq \boldsymbol{k}'_1 + \boldsymbol{k}'_2 \quad . \tag{5.47}$$

Relation (5.45) is the symmetry property (3.103). Relations (5.46) and (5.47) express invariance of V under inversion and translation in space.

Under these conditions, the complete Hamiltonian of the system is

$$\mathtt{H} = \mathtt{H}_0 + \mathtt{V} \quad , \tag{5.48}$$

where $\mathtt{H}_0$ is the kinetic energy of the electrons (3.72).

$$\mathtt{H}_0 = \sum_{\boldsymbol{k}\sigma} \frac{\hbar^2|\boldsymbol{k}|^2}{2m} a^*_{\boldsymbol{k}\sigma} a_{\boldsymbol{k}\sigma} = \sum_{\boldsymbol{k}\sigma} \varepsilon_{\boldsymbol{k}} a^*_{\boldsymbol{k}\sigma} a_{\boldsymbol{k}\sigma} \quad , \quad \varepsilon_{\boldsymbol{k}} = \frac{\hbar^2|\boldsymbol{k}|^2}{2m} \quad . \tag{5.49}$$

With (5.44) and (3.101), the two-body interaction takes the form

$$\mathtt{V} = \frac{1}{2} \sum_{\substack{\boldsymbol{k}_1\boldsymbol{k}_2\boldsymbol{k}'_1\boldsymbol{k}'_2 \\ \sigma_1\sigma_2}} V_{\boldsymbol{k}_1,\boldsymbol{k}_2;\boldsymbol{k}'_1,\boldsymbol{k}'_2} a^*_{\boldsymbol{k}_1\sigma_1} a^*_{\boldsymbol{k}_2\sigma_2} a_{\boldsymbol{k}'_2\sigma_2} a_{\boldsymbol{k}'_1\sigma_1} \quad . \tag{5.50}$$

5.3.2 Application of the Variational Method to Superconductivity

Normal State

In the absence of interaction, the ground state of the system governed by the Hamiltonian $\mathtt{H}_0$ is obtained by filling all of the individual states of the Fermi sphere (Sect. 2.2.2). This is written as

$$|\Phi_0\rangle = \prod_{|\boldsymbol{k}|\leq k_{\mathrm{F}}} (a^*_{\boldsymbol{k}\uparrow} a^*_{-\boldsymbol{k}\downarrow})\,|0\rangle \quad . \tag{5.51}$$

The state $|\Phi_0\rangle$ has energy E_0 (2.94). Convenient notation here replaces $\sigma = 1/2$ $(-1/2)$ with the symbols $\uparrow$ $(\downarrow)$. This notation will be maintained in the future, as long as there is no ambiguity.

The fact that the one-particle states have been grouped into pairs $(\boldsymbol{k}\uparrow; -\boldsymbol{k}\downarrow)$ has no significance, since any change of the order of the occupied states only has the effect of multiplying $|\Phi_0\rangle$ by a factor ± 1.

Formulation of the Problem

In the presence of an interaction V, the appearance of a new ground state $|\Psi_s\rangle$ of energy E_s, possibly separated by a gap from the first excited states of the new system is expected. The variational method is the most appropriate for determining the gain in energy obtained when passing from $|\Phi_0\rangle$ to $|\Psi_s\rangle$. Similar to an atom, the ground state is also separated from the first excited state by a finite difference in energy and for which the variational method is frequently used.

It is thus a matter of determining a class of variational states which are adequate for the problem: the central point rests in making a judicious choice, allowing one to benefit from a lowering of the energy due to the attractive potential.

Consider that the new ground state $|\Psi_s\rangle$ can be written as a linear combination of certain excited states $|\Phi_m\rangle\,, m = 0, 1, 2, \ldots, M$, of the normal state. The nature of these excited states will not be explicitly specified. The goal here is to discuss the variational approach in a qualitative manner in order to make evident the importance of a good choice of these states. It will be shown, as one expects, *that the only states which play a role are the* $|\Phi_m\rangle$ *of energy* E_m *close to* E_0*, otherwise the contribution to the kinetic energy is sharply increased.* Such states can be easily constructed: one creates a few holes inside the Fermi sphere, which is completely filled in the ground state $|\Phi_0\rangle$, by moving each corresponding electron into a one-particle state located just beyond the surface of the sphere. These are electron–hole excitations of type (2.100). It is important to remark that the energy of the excited states $|\Phi_m\rangle$ can be as close as desired to that of the ground state.

Once the choice of the states $|\Phi_m\rangle$ is made, the variational states are of the form

$$|\Psi\rangle = \sum_m c_m\,|\Phi_m\rangle \quad . \tag{5.52}$$

The c_m are complex variational parameters to be determined under the constraint

$$\langle\Psi|\Psi\rangle = \sum_m |c_m|^2 = 1 \quad , \tag{5.53}$$

assuring that $|\Psi\rangle$ is normalized.

Variational Calculation

To determine the c_m, one forms the function $E_\lambda(c_0, c_1, \ldots, c_m)$,

$$E_\lambda = \langle\Psi|(\mathtt{H}_0 + \mathtt{V})|\Psi\rangle - \lambda(\langle\Psi|\Psi\rangle - 1) \quad , \tag{5.54}$$

which takes into account constraint (5.53) by the introduction of the Lagrange parameter λ. Because of (5.52), one has

$$E_\lambda = \sum_m |c_m|^2 E_m + \sum_{m,m'} c_m^* c_{m'} W_{mm'} - \lambda \left(\sum_m |c_m|^2 - 1 \right) \quad , \tag{5.55}$$

where

$$E_m = \langle \Phi_m | \mathtt{H}_0 | \Phi_m \rangle = E_m^* \tag{5.56}$$

and

$$W_{mm'} = \langle \Phi_m | \mathtt{V} | \Phi_{m'} \rangle = W_{m'm}^* \quad . \tag{5.57}$$

It is now a matter of finding the complex parameters c_m for which E_λ is stationary

$$\frac{\partial E_\lambda}{\partial c_m^*} = c_m E_m + \sum_{m'} c_{m'} W_{mm'} - \lambda c_m = 0 \quad . \tag{5.58}$$

If one multiplies (5.58) by c_m^* and sums over m then, taking into account (5.56) and (5.57),

$$\lambda = \sum_m |c_m|^2 E_m + \sum_{m,m'} c_{m'} c_m^* W_{mm'} = \langle \Psi_s | (\mathtt{H}_0 + \mathtt{V}) | \Psi_s \rangle = E_s \quad . \tag{5.59}$$

The Lagrange parameter identifies itself with the energy of the superconductor ground state.

One can rewrite (5.58) under the form

$$c_m = -\frac{\sum_{m'} c_{m'} W_{mm'}}{E_m - E_s} \quad . \tag{5.60}$$

Suppose that, for a choice of excited states $|\Phi_m\rangle$, the matrix elements $W_{mm'}$ (5.57) do not vary much with m and m'. In this case, it is legitimate to discuss the nature of the solution, to replace $W_{mm'}$ by its average

$$\overline{W} = \frac{1}{M^2} \sum_{m,m'=0}^{M} W_{mm'} \quad . \tag{5.61}$$

Because of (5.57), $\overline{W}$ is necessarily real and (5.60) takes the simple form

$$c_m = -\overline{W} \frac{\sum_{m'} c_{m'}}{E_m - E_s} \quad . \tag{5.62}$$

After summing over m, one obtains the equation for E_s

$$-\overline{W} \sum_{m=0}^{M} \frac{1}{E_m - E_s} = 1 \quad . \tag{5.63}$$

This study is concerned with those solutions which satisfy

$$E_s < E_0 \leq E_m \quad , \tag{5.64}$$

that is, for which energy is gained by the redistribution of the electrons in the space $\boldsymbol{k}$. One sees from (5.63) that this is only possible if

$$\overline{W} < 0 \quad . \tag{5.65}$$

To obtain an energy E_s which is less than E_0, it is thus necessary to satisfy two conditions. *The average (5.61) must be meaningful* (it will be shown in the following section that this is not always the case) and $\overline{W}$ *must be negative*. The trick is to find the states $|\Phi_m\rangle$ which allow these conditions to be satisfied.

5.3.3 Sign Ambiguity

Matrix element (5.44) describes the effective interaction of a pair of electrons scattered from states $(\boldsymbol{k}'_1\sigma'_1)$ and $(\boldsymbol{k}'_2\sigma'_2)$ into states $(\boldsymbol{k}_1\sigma_1)$ and $(\boldsymbol{k}_2\sigma_2)$. Its attractive or repulsive character is reflected in the sign of a certain number of matrix elements. On the other hand, (5.65) shows that it is the sign of matrix elements $W_{mm'}$ which is decisive for the search of a superconductor ground state $|\Psi_s\rangle$. *In general, however, there is no simple relation between the signs of matrix elements (5.44) and (5.57).*

To illustrate this difficulty, construct states $|\Phi_m\rangle$ of the form

$$|\Phi_m\rangle = a^*_{\boldsymbol{k}_1\sigma_1} a^*_{\boldsymbol{k}_2\sigma_2} a^*_{\boldsymbol{k}_3\sigma_3} a^*_{\boldsymbol{k}_4\sigma_4} \cdots |0\rangle \quad . \tag{5.66}$$

In order for the specification of an ensemble of wavenumbers and spins $(\boldsymbol{k}_1\sigma_1, \boldsymbol{k}_2\sigma_2, \ldots)$ to determine state (5.66) in a unique manner, it is necessary to decide how the creation operators are ordered. This can be done by defining an order of the $(\boldsymbol{k}\sigma)$ ensemble and by agreeing to write the creators in (5.66) according to the same order: $(\boldsymbol{k}_1\sigma_1) > (\boldsymbol{k}_2\sigma_2) > (\boldsymbol{k}_3\sigma_3) > (\boldsymbol{k}_4\sigma_4) \ldots$. The choice of the relation of the $(\boldsymbol{k}\sigma)$ order is arbitrary, but as with all conventions, it must not be changed during the course of the study.

Consider another state

$$|\Phi_{m'}\rangle = a^*_{\boldsymbol{k}'_1\sigma'_1} a^*_{\boldsymbol{k}'_2\sigma'_2} a^*_{\boldsymbol{k}_3\sigma_3} a^*_{\boldsymbol{k}_4\sigma_4} \cdots |0\rangle \tag{5.67}$$

with $(\boldsymbol{k}'_1\sigma'_1) > (\boldsymbol{k}'_2\sigma'_2) > (\boldsymbol{k}_3\sigma_3) > (\boldsymbol{k}_4\sigma_4) > \ldots$. Up to two pairs of states, $(\boldsymbol{k}'_1\sigma'_1)$, $(\boldsymbol{k}'_2\sigma'_2)$ and $(\boldsymbol{k}_1\sigma_1)$, $(\boldsymbol{k}_2\sigma_2)$, states $|\Phi_{m'}\rangle$ and $|\Phi_m\rangle$ correspond to the occupation of the same ensemble of one-particle states $(\boldsymbol{k}_3\sigma_3), (\boldsymbol{k}_4\sigma_4), \ldots$, otherwise $W_{mm'}$ would vanish. By an analogous calculation to (3.122) it is not difficult to verify that, for $\{(\boldsymbol{k}_1\sigma_1), (\boldsymbol{k}_2, \sigma_2)\} \neq \{(\boldsymbol{k}'_1\sigma'_1), (\boldsymbol{k}'_2\sigma'_2)\}$,

$$W_{mm'} = \frac{1}{2}[(\boldsymbol{k}_1\sigma_1, \boldsymbol{k}_2\sigma_2|V|\boldsymbol{k}'_1\sigma'_1, \boldsymbol{k}'_2\sigma'_2) - (\boldsymbol{k}_1\sigma_1, \boldsymbol{k}_2\sigma_2|V|\boldsymbol{k}'_2\sigma'_2, \boldsymbol{k}'_1\sigma'_1)] \quad . \tag{5.68}$$

If, in place of $|\Phi_m\rangle$ and $|\Phi_{m'}\rangle$, one takes the states

$$\begin{aligned} |\overline{\Phi}_m\rangle &= a^*_{\boldsymbol{k}_1\sigma_1} a^*_{\boldsymbol{k}_2\sigma_2} a^*_{\overline{\boldsymbol{k}}_3\overline{\sigma}_3} a^*_{\boldsymbol{k}_4\sigma_4} |0\rangle \quad , \\ |\overline{\Phi}_{m'}\rangle &= a^*_{\boldsymbol{k}'_1\sigma'_1} a^*_{\overline{\boldsymbol{k}}_3\overline{\sigma}_3} a^*_{\boldsymbol{k}'_2\sigma'_2} a^*_{\boldsymbol{k}_4\sigma_4} |0\rangle \quad , \end{aligned} \tag{5.69}$$

corresponding respectively to orders $(\boldsymbol{k}_1\sigma_1) > (\boldsymbol{k}_2\sigma_2) > (\overline{\boldsymbol{k}}_3\overline{\sigma}_3) > (\boldsymbol{k}_4\sigma_4) > \ldots$ and $(\boldsymbol{k}'_1\sigma'_1) > (\overline{\boldsymbol{k}}_3\overline{\sigma}_3) > (\boldsymbol{k}'_2\sigma'_2) > (\boldsymbol{k}_4\sigma_4) > \ldots$, it is clear that, because of the anticommutation of the creators, the corresponding matrix element $W_{mm'}$ is given by the same expression (5.68), but with opposite sign. As a consequence, *the matrix elements $W_{mm'}$ depend on the wavenumbers $\boldsymbol{k}_1, \boldsymbol{k}_2, \boldsymbol{k}'_1$ and $\boldsymbol{k}'_2$ which index $V_{\boldsymbol{k}_1,\boldsymbol{k}_2;\boldsymbol{k}'_1,\boldsymbol{k}'_2}$, and on the other states which figure into $|\Phi_m\rangle$ and $|\Phi_{m'}\rangle$, that is $\boldsymbol{k}_3, \boldsymbol{k}_4, \ldots$.* Even if the coefficient $V_{\boldsymbol{k}_1,\boldsymbol{k}_2;\boldsymbol{k}'_1,\boldsymbol{k}'_2}$ varies in a regular manner, the sign of $W_{mm'}$ will change each time that one of the other states permutes (relative to the order of convention) with one of the states which indexes $V_{\boldsymbol{k}_1,\boldsymbol{k}_2;\boldsymbol{k}'_1,\boldsymbol{k}'_2}$. *If one allows states of (5.66) without restrictions, the sign of $W_{mm'}$ will continually oscillate in sums (5.60) and (5.61).* The mean value $\overline{W}$ of the $W_{mm'}$ risks to be zero: in any case, its sign has nothing to do with the coefficients $V_{\boldsymbol{k}_1,\boldsymbol{k}_2;\boldsymbol{k}'_1,\boldsymbol{k}'_2}$.

This is the manifestation of the *sign ambiguity*, a source of difficulties which occurs in the search for the mechanism of superconductivity. BCS succeeded in resolving this ambiguity by selecting a family of states $|\Phi_m\rangle$ which does not suffer from this handicap.

5.3.4 Variational Class of BCS States

BCS defines the family of states

$$|\Psi\rangle = \prod_{\boldsymbol{k}} (u_{\boldsymbol{k}} + v_{\boldsymbol{k}} a^*_{\boldsymbol{k}\uparrow} a^*_{-\boldsymbol{k}\downarrow}) |0\rangle \quad , \tag{5.70}$$

where the product is carried out over all values of $\boldsymbol{k}$. *The complex parameters $u_{\boldsymbol{k}}$ and $v_{\boldsymbol{k}}$ are determined by the variational principle.* They satisfy the condition

$$|u_{\boldsymbol{k}}|^2 + |v_{\boldsymbol{k}}|^2 = 1 \tag{5.71}$$

to assure the normalization of $|\Psi\rangle$.

It is important to note that *the order of the factors in (5.70) is indifferent* since pairs of creators commute for all $\boldsymbol{k}_1$ and $\boldsymbol{k}_2$

$$[a^*_{\boldsymbol{k}_1\uparrow}a^*_{-\boldsymbol{k}_1\downarrow}, a^*_{\boldsymbol{k}_2\uparrow}a^*_{-\boldsymbol{k}_2\downarrow}] = 0 \quad . \tag{5.72}$$

Equally, for $\boldsymbol{k}_1 \neq \boldsymbol{k}_2$,

$$[a_{-\boldsymbol{k}_1\downarrow}a_{\boldsymbol{k}_1\uparrow}, a^*_{\boldsymbol{k}_2\uparrow}a^*_{-\boldsymbol{k}_2\downarrow}] = 0 \quad . \tag{5.73}$$

The properties of commutation (5.72) and (5.73) allow one to easily calculate the norm of $|\Psi\rangle$ by a suitable regrouping of the pair operators. Letting $\mathtt{A}^*_{\boldsymbol{k}} = u_{\boldsymbol{k}} + v_{\boldsymbol{k}}a^*_{\boldsymbol{k}\uparrow}a^*_{-\boldsymbol{k}\downarrow}$ one can write

$$\begin{aligned}
\langle\Psi|\Psi\rangle &= \left\langle 0 \left| \left(\prod_{\boldsymbol{k}} \mathtt{A}_{\boldsymbol{k}}\right) \left(\prod_{\boldsymbol{k}'} \mathtt{A}^*_{\boldsymbol{k}'}\right) \right| 0 \right\rangle \\
&= \left\langle 0 \left| \left(\prod_{\boldsymbol{k}'\neq\boldsymbol{k}} \mathtt{A}_{\boldsymbol{k}'}\mathtt{A}^*_{\boldsymbol{k}'}\right) \mathtt{A}_{\boldsymbol{k}}\mathtt{A}^*_{\boldsymbol{k}} \right| 0 \right\rangle \\
&= \left\langle 0 \left| \left(\prod_{\boldsymbol{k}'\neq\boldsymbol{k}} \mathtt{A}_{\boldsymbol{k}'}\mathtt{A}^*_{\boldsymbol{k}'}\right) \right[|u_{\boldsymbol{k}}|^2 + |v_{\boldsymbol{k}}|^2 a_{-\boldsymbol{k}\downarrow}a_{\boldsymbol{k}\uparrow}a^*_{\boldsymbol{k}\uparrow}a^*_{-\boldsymbol{k}\downarrow} \right. \\
&\qquad \left. + u^*_{\boldsymbol{k}}v_{\boldsymbol{k}}a^*_{\boldsymbol{k}\uparrow}a^*_{-\boldsymbol{k}\downarrow} + u_{\boldsymbol{k}}v^*_{\boldsymbol{k}}a_{-\boldsymbol{k}\downarrow}a_{\boldsymbol{k}\uparrow} \Big] \right| 0 \right\rangle \\
&= \left(|u_{\boldsymbol{k}}|^2 + |v_{\boldsymbol{k}}|^2\right) \left\langle 0 \left| \prod_{\boldsymbol{k}'\neq\boldsymbol{k}} \mathtt{A}_{\boldsymbol{k}'}\mathtt{A}^*_{\boldsymbol{k}'} \right| 0 \right\rangle \quad .
\end{aligned} \tag{5.74}$$

To obtain (5.74), relation (3.16) has been used ($a^*_{\boldsymbol{k}\uparrow}a^*_{-\boldsymbol{k}\downarrow}$ commutes with the product $\prod_{\boldsymbol{k}'\neq\boldsymbol{k}} \mathtt{A}_{\boldsymbol{k}'}\mathtt{A}^*_{\boldsymbol{k}'}$ and can act on the left of the latter), as well as $a_{-\boldsymbol{k}\downarrow}a_{\boldsymbol{k}\uparrow}a^*_{\boldsymbol{k}\uparrow}a^*_{-\boldsymbol{k}\downarrow}|0\rangle = |0\rangle$, which follows immediately from the anticommutation rules (3.33). If one repeats the same calculation for each term $\mathtt{A}_{\boldsymbol{k}'}\mathtt{A}^*_{\boldsymbol{k}'}$ of the product in (5.74) and takes into account (5.71), one verifies that $|\Psi\rangle$ is normalized to 1.

Resolution of the Sign Ambiguity

After the expansion of the product (5.70), $|\Psi\rangle$ is clearly a linear combination similar to (5.52) with excited states $|\Phi_m\rangle$ of the form

$$|\Phi_m\rangle = (a^*_{\boldsymbol{k}_1\uparrow}a^*_{-\boldsymbol{k}_1\downarrow})(a^*_{\boldsymbol{k}_2\uparrow}a^*_{-\boldsymbol{k}_2\downarrow})(a^*_{\boldsymbol{k}_3\uparrow}a^*_{-\boldsymbol{k}_3\downarrow})\dots|0\rangle \tag{5.75}$$

and with coefficients formed from products of $u_{\boldsymbol{k}}$ and $v_{\boldsymbol{k}}$. The remarkable property of the BCS choice (5.70) is that the one-particle states are *grouped into pairs*: $(\boldsymbol{k}\uparrow)$ and $(-\boldsymbol{k}\downarrow)$ which are simultaneously either occupied or unoccupied. For a given $\boldsymbol{k}$ one always writes the creation operators of a pair in the order $a^*_{\boldsymbol{k}\uparrow}a^*_{-\boldsymbol{k}\downarrow}$. Note that *these states do not suffer from a sign ambiguity*. If one poses

$$|\Phi_{m'}\rangle = (a^*_{\boldsymbol{k}'_1\uparrow}a^*_{-\boldsymbol{k}'_1\downarrow})(a^*_{\boldsymbol{k}_2\uparrow}a^*_{-\boldsymbol{k}_2\downarrow})(a^*_{\boldsymbol{k}_3\uparrow}a^*_{-\boldsymbol{k}_3\downarrow})\dots|0\rangle \quad , \tag{5.76}$$

then, using (5.44) in (5.68), for $\boldsymbol{k}_1 \neq \boldsymbol{k}'_1$, one obtains

$$W_{mm'} = \frac{1}{2}V_{\boldsymbol{k}_1,-\boldsymbol{k}_1;\boldsymbol{k}'_1,-\boldsymbol{k}'_1} \quad . \tag{5.77}$$

Relation (5.77) is not changed when $\boldsymbol{k}_2, \boldsymbol{k}_3, \dots$ are varied since, according to (5.72), the order of pairs is unimportant in (5.75) and (5.76). Contrary to what happens in (5.68), the sign of $W_{mm'}$ depends only on $\boldsymbol{k}_1$ and $\boldsymbol{k}'_1$ and not on the other one-particle states which constitute $|\Phi_m\rangle$ and $|\Phi_{m'}\rangle$.

Grouping of the fermions into pairs thus removes the sign ambiguity. It therefore allows one, by application of the variational method discussed in Sect. 5.3.2, to obtain the anticipated gain in energy: *this is the phenomenon of fermion pairing.* The choice (5.70) of BCS is not the only one which can remove the sign ambiguity. The formation of pairs $(\boldsymbol{k}\uparrow; -\boldsymbol{k}\downarrow)$ is motivated by physical arguments. In order for the interaction matrix elements to be non-zero, relation (5.47) requires that $\boldsymbol{k}_1 + \boldsymbol{k}_2 = \boldsymbol{k}'_1 + \boldsymbol{k}'_2$. As the state being described has zero total current, this requirement is evidently satisfied by imposing $\boldsymbol{k}_1 + \boldsymbol{k}_2 = 0$ for each pair of individual states. Likewise, a simple manner to obtain a state with zero total spin is to chose pairs of electrons with opposite spin. One should equally note that the existence of pairs $(\boldsymbol{k}\uparrow; -\boldsymbol{k}\downarrow)$ had been suggested by Cooper from an analysis of the two-electron problem (Sect. 9.2.4).[4] Note again that BCS can also describe states of non-zero total current by choosing a non-zero $\boldsymbol{k}_1 + \boldsymbol{k}_2$.

While the BCS states do resolve the sign ambiguity, they also present a characteristic which can pose some difficulties: *the number of particles is not fixed.* Sum (5.70) includes a zero-particle state ($\prod_{\boldsymbol{k}} u_{\boldsymbol{k}}|0\rangle$) and states of $2, 4, \dots, 2n, \dots$ particles. *It is this property, in fact, which plays a fundamental role in the description of superconductivity* and which will later reappear as the object of more profound discussion (Sect. 5.4).

5.3.5 How to Calculate with a BCS State

The Bogoliubov Transformation

In the following, it will be necessary to calculate the average of a certain number of products of creators and annihilators in the states (5.70). These calculations are greatly facilitated by the introduction of new operators $\alpha_{\boldsymbol{k}}$ and $\beta_{\boldsymbol{k}}$, defined by the *Bogoliubov transformation*

$$\alpha_{\boldsymbol{k}} = u_{\boldsymbol{k}}a_{\boldsymbol{k}\uparrow} - v_{\boldsymbol{k}}a^*_{-\boldsymbol{k}\downarrow} \quad , \tag{5.78}$$

$$\beta_{\boldsymbol{k}} = u_{\boldsymbol{k}}a_{-\boldsymbol{k}\downarrow} + v_{\boldsymbol{k}}a^*_{\boldsymbol{k}\uparrow} \quad . \tag{5.79}$$

4 It is common to refer to the pairs $(\boldsymbol{k}\uparrow; -\boldsymbol{k}\downarrow)$ as *Cooper pairs*; this term, however, more properly refers to the two electrons of the Cooper problem.

This transformation has two remarkable properties. *It is canonical*

$$\begin{aligned}
&[\alpha_{\boldsymbol{k}_1}, \alpha^*_{\boldsymbol{k}_2}]_+ = [\beta_{\boldsymbol{k}_1}, \beta^*_{\boldsymbol{k}_2}]_+ = \delta_{\boldsymbol{k}_1,\boldsymbol{k}_2} \quad , \\
&[\alpha_{\boldsymbol{k}_1}, \alpha_{\boldsymbol{k}_2}]_+ = [\beta_{\boldsymbol{k}_1}, \beta_{\boldsymbol{k}_2}]_+ = 0 \quad , \\
&[\alpha_{\boldsymbol{k}_1}, \beta_{\boldsymbol{k}_2}]_+ = [\alpha^*_{\boldsymbol{k}_1}, \beta^*_{\boldsymbol{k}_2}]_+ = 0 \quad ,
\end{aligned} \tag{5.80}$$

which can be verified easily using the normalization condition (5.71). In addition, *the state (5.70) plays the role of "vacuum" for the* $\alpha_{\boldsymbol{k}}$ *and the* $\beta_{\boldsymbol{k}}$

$$\alpha_{\boldsymbol{k}} |\Psi\rangle = \beta_{\boldsymbol{k}} |\Psi\rangle = 0 \quad . \tag{5.81}$$

As $a_{\boldsymbol{k}\uparrow}$ and $a^*_{-\boldsymbol{k}\downarrow}$ commute with all of the factors $\boldsymbol{k}' \neq \boldsymbol{k}$ of the product (5.70), one has

$$\begin{aligned}
\alpha_{\boldsymbol{k}} |\Psi\rangle = &\left[\prod_{\boldsymbol{k}' \neq \boldsymbol{k}} (u_{\boldsymbol{k}'} + v_{\boldsymbol{k}'} a^*_{\boldsymbol{k}'\uparrow} a^*_{-\boldsymbol{k}'\downarrow})\right] (u_{\boldsymbol{k}} a_{\boldsymbol{k}\uparrow} - v_{\boldsymbol{k}} a^*_{-\boldsymbol{k}\downarrow}) \\
&\times (u_{\boldsymbol{k}} + v_{\boldsymbol{k}} a^*_{\boldsymbol{k}\uparrow} a^*_{-\boldsymbol{k}\downarrow}) |0\rangle \quad .
\end{aligned} \tag{5.82}$$

The anticommutation rules (3.33) and (3.16) imply that the factor depending on $\boldsymbol{k}$ in (5.82) gives $u_{\boldsymbol{k}} v_{\boldsymbol{k}} a^*_{-\boldsymbol{k}\downarrow} (a_{\boldsymbol{k}\uparrow} a^*_{\boldsymbol{k}\uparrow} - 1) |0\rangle = 0$, hence (5.81). The same procedure is followed to establish $\beta_{\boldsymbol{k}} |\Psi\rangle = 0$.

It is possible to invert (5.78) and (5.79). Multiplying (5.78) by $u^*_{\boldsymbol{k}}$ and the complex conjugate of (5.79) by $v_{\boldsymbol{k}}$ then summing, one obtains, taking into account (5.72),

$$a_{\boldsymbol{k}\uparrow} = u^*_{\boldsymbol{k}} \alpha_{\boldsymbol{k}} + v_{\boldsymbol{k}} \beta^*_{\boldsymbol{k}} \tag{5.83}$$

and

$$a^*_{\boldsymbol{k}\uparrow} = u_{\boldsymbol{k}} \alpha^*_{\boldsymbol{k}} + v^*_{\boldsymbol{k}} \beta_{\boldsymbol{k}} \quad . \tag{5.84}$$

Similarly,

$$a_{\boldsymbol{k}\downarrow} = u^*_{-\boldsymbol{k}} \beta_{-\boldsymbol{k}} - v_{-\boldsymbol{k}} \alpha^*_{-\boldsymbol{k}} \tag{5.85}$$

and

$$a^*_{\boldsymbol{k}\downarrow} = u_{-\boldsymbol{k}} \beta^*_{-\boldsymbol{k}} - v^*_{-\boldsymbol{k}} \alpha_{-\boldsymbol{k}} \quad . \tag{5.86}$$

Calculation of Average Values

As the $a_{\boldsymbol{k}\sigma}$ are linear combinations of the $\alpha_{\boldsymbol{k}}$ and the $\beta_{\boldsymbol{k}}$ and their complex conjugates, it is clear that *calculations of the average value in the state* $|\Psi\rangle$ *of monomials of* $a_{\boldsymbol{k}\sigma}$ *and* $a^*_{\boldsymbol{k}\sigma}$ *are reduced to applications of the Wick theorem relative to* $\alpha_{\boldsymbol{k}}$ *and* $\beta_{\boldsymbol{k}}$. In effect, the validity of the rules in Sect. 3.2.4 depends on the anticommutation rules (5.80) and on the existence of a "vacuum"

(5.81). Starting with (5.83), (5.84) and (5.81), one immediately obtains the value of contractions relative to the state $|\Psi\rangle$

$$\begin{aligned}\underbracket{a^*_{\boldsymbol{k}_1\uparrow}a_{\boldsymbol{k}_2\uparrow}} &= \langle\Psi|a^*_{\boldsymbol{k}_1\uparrow}a_{\boldsymbol{k}_2\uparrow}|\Psi\rangle \\ &= \langle\Psi|\,(u_{\boldsymbol{k}_1}u^*_{\boldsymbol{k}_2}\alpha^*_{\boldsymbol{k}_1}\alpha_{\boldsymbol{k}_2} + u_{\boldsymbol{k}_1}v_{\boldsymbol{k}_2}\alpha^*_{\boldsymbol{k}_1}\beta^*_{\boldsymbol{k}_2} \\ &\quad + v^*_{\boldsymbol{k}_1}u^*_{\boldsymbol{k}_2}\beta_{\boldsymbol{k}_1}\alpha_{\boldsymbol{k}_2} + v^*_{\boldsymbol{k}_1}v_{\boldsymbol{k}_2}\beta_{\boldsymbol{k}_1}\beta^*_{\boldsymbol{k}_2})\,|\Psi\rangle = |v_{\boldsymbol{k}_1}|^2\delta_{\boldsymbol{k}_1,\boldsymbol{k}_2}\end{aligned} \tag{5.87}$$

and

$$\begin{aligned}\underbracket{a^*_{\boldsymbol{k}_1\downarrow}a_{\boldsymbol{k}_2\downarrow}} &= |v_{-\boldsymbol{k}_1}|^2\delta_{\boldsymbol{k}_1,\boldsymbol{k}_2} \quad , \\ \underbracket{a^*_{\boldsymbol{k}_1\uparrow}a_{\boldsymbol{k}_2\downarrow}} &= \underbracket{a^*_{\boldsymbol{k}_1\downarrow}a_{\boldsymbol{k}_2\uparrow}} = 0 \quad .\end{aligned} \tag{5.88}$$

Similarly, one finds

$$\underbracket{a^*_{\boldsymbol{k}_1\uparrow}a^*_{\boldsymbol{k}_2\downarrow}} = (\underbracket{a_{\boldsymbol{k}_2\downarrow}a_{\boldsymbol{k}_1\uparrow}})^* = u_{\boldsymbol{k}_1}v^*_{\boldsymbol{k}_1}\delta_{\boldsymbol{k}_1,-\boldsymbol{k}_2} \quad , \tag{5.89}$$

$$\underbracket{a^*_{\boldsymbol{k}_1\uparrow}a^*_{\boldsymbol{k}_2\uparrow}} = (\underbracket{a_{\boldsymbol{k}_2\uparrow}a_{\boldsymbol{k}_1\uparrow}})^* = 0 \quad . \tag{5.90}$$

Relation (5.87) (with $\boldsymbol{k}_1 = \boldsymbol{k}_2 = \boldsymbol{k}$) shows that $|v_{\boldsymbol{k}}|^2(|u_{\boldsymbol{k}}|^2 = 1 - |v_{\boldsymbol{k}}|^2)$ is nothing other than the probability that the one-particle states ($\boldsymbol{k}\uparrow$) and ($-\boldsymbol{k}\downarrow$) be simultaneously occupied (empty). Moreover, the second part of (5.89) must be zero unless $u_{\boldsymbol{k}_1}$ and $v_{\boldsymbol{k}_1}$ are simultaneously non-zero. Because $|\Psi\rangle$ is a linear combination of states corresponding to different numbers of particles, it is possible for contraction (5.89) to be non-zero.

The average number of particles and the average kinetic energy are

$$N(\{u_{\boldsymbol{k}}, v_{\boldsymbol{k}}\}) = \langle\Psi|\sum_{\boldsymbol{k}\sigma} a^*_{\boldsymbol{k}\sigma}a_{\boldsymbol{k}\sigma}\,|\Psi\rangle = \sum_{\boldsymbol{k}}(|v_{\boldsymbol{k}}|^2 + |v_{-\boldsymbol{k}}|^2) = 2\sum_{\boldsymbol{k}}|v_{\boldsymbol{k}}|^2 \tag{5.91}$$

and

$$E^{\text{kin}}(\{u_{\boldsymbol{k}}, v_{\boldsymbol{k}}\}) = \langle\Psi|\sum_{\boldsymbol{k}\sigma}\varepsilon_{\boldsymbol{k}}a^*_{\boldsymbol{k}\sigma}a_{\boldsymbol{k}\sigma}\,|\Psi\rangle = 2\sum_{\boldsymbol{k}}\varepsilon_{\boldsymbol{k}}|v_{\boldsymbol{k}}|^2 \quad . \tag{5.92}$$

To calculate the potential energy, it is useful to rewrite interaction (5.50), explicitly summing over the spins

$$\begin{aligned}\mathsf{V} = \frac{1}{2}\sum_{\boldsymbol{k}_1\boldsymbol{k}_2\boldsymbol{k}'_1\boldsymbol{k}'_2} & V_{\boldsymbol{k}_1,\boldsymbol{k}_2;\boldsymbol{k}'_1,\boldsymbol{k}'_2}(a^*_{\boldsymbol{k}_1\uparrow}a^*_{\boldsymbol{k}_2\uparrow}a_{\boldsymbol{k}'_2\uparrow}a_{\boldsymbol{k}'_1\uparrow} + a^*_{\boldsymbol{k}_1\downarrow}a^*_{\boldsymbol{k}_2\downarrow}a_{\boldsymbol{k}'_2\downarrow}a_{\boldsymbol{k}'_1\downarrow}) \\ + \sum_{\boldsymbol{k}_1\boldsymbol{k}_2\boldsymbol{k}'_1\boldsymbol{k}'_2} & V_{\boldsymbol{k}_1,\boldsymbol{k}_2;\boldsymbol{k}'_1,\boldsymbol{k}'_2}a^*_{\boldsymbol{k}_1\uparrow}a^*_{\boldsymbol{k}_2\downarrow}a_{\boldsymbol{k}'_2\downarrow}a_{\boldsymbol{k}'_1\uparrow} \quad .\end{aligned} \tag{5.93}$$

The last term of (5.93) appears with a supplementary factor of 2 due to symmetry (5.45). Application of the Wick theorem (see Sect. 3.2.4) with contractions (5.87)–(5.90) yields

$$\langle\Psi|\, a^*_{\boldsymbol{k}_1\uparrow}a^*_{\boldsymbol{k}_2\downarrow}a_{\boldsymbol{k}'_2\downarrow}a_{\boldsymbol{k}'_1\uparrow}\,|\Psi\rangle$$
$$= a^*_{\boldsymbol{k}_1\uparrow}a^*_{\boldsymbol{k}_2\downarrow}a_{\boldsymbol{k}'_2\downarrow}a_{\boldsymbol{k}'_1\uparrow} + a^*_{\boldsymbol{k}_1\uparrow}a^*_{\boldsymbol{k}_2\downarrow}a_{\boldsymbol{k}'_2\downarrow}a_{\boldsymbol{k}'_1\uparrow} + a^*_{\boldsymbol{k}_1\uparrow}a^*_{\boldsymbol{k}_2\downarrow}a_{\boldsymbol{k}'_2\downarrow}a_{\boldsymbol{k}'_1\uparrow}$$
$$= \delta_{\boldsymbol{k}_1,-\boldsymbol{k}_2}\delta_{\boldsymbol{k}'_1,-\boldsymbol{k}'_2}u_{\boldsymbol{k}_1}v^*_{\boldsymbol{k}_1}u^*_{\boldsymbol{k}'_1}v_{\boldsymbol{k}'_1} + \delta_{\boldsymbol{k}_1,\boldsymbol{k}'_1}\delta_{\boldsymbol{k}_2,\boldsymbol{k}'_2}|v_{\boldsymbol{k}_1}|^2|v_{-\boldsymbol{k}_2}|^2 \quad . \tag{5.94}$$

Similarly,

$$\langle\Psi|\, a^*_{\boldsymbol{k}_1\uparrow}a^*_{\boldsymbol{k}_2\uparrow}a_{\boldsymbol{k}'_2\uparrow}a_{\boldsymbol{k}'_1\uparrow}\,|\Psi\rangle = (\delta_{\boldsymbol{k}_1,\boldsymbol{k}'_1}\delta_{\boldsymbol{k}_2,\boldsymbol{k}'_2} - \delta_{\boldsymbol{k}_1,\boldsymbol{k}'_2}\delta_{\boldsymbol{k}_2,\boldsymbol{k}'_1})|v_{\boldsymbol{k}_1}|^2|v_{\boldsymbol{k}_2}|^2 \tag{5.95}$$

and

$$\langle\Psi|\, a^*_{\boldsymbol{k}_1\downarrow}a^*_{\boldsymbol{k}_2\downarrow}a_{\boldsymbol{k}'_2\downarrow}a_{\boldsymbol{k}'_1\downarrow}\,|\Psi\rangle$$
$$= (\delta_{\boldsymbol{k}_1,\boldsymbol{k}'_1}\delta_{\boldsymbol{k}_2,\boldsymbol{k}'_2} - \delta_{\boldsymbol{k}_1,\boldsymbol{k}'_2}\delta_{\boldsymbol{k}_2,\boldsymbol{k}'_1})|v_{-\boldsymbol{k}_1}|^2|v_{-\boldsymbol{k}_2}|^2 \quad . \tag{5.96}$$

The latter two relations allow one finally to obtain the BCS energy functional (employing symmetry (5.46))

$$\langle\Psi|\,(\mathtt{H}_0 + \mathtt{V})\,|\Psi\rangle \tag{5.97}$$
$$= 2\sum_{\boldsymbol{k}}\varepsilon_{\boldsymbol{k}}|v_{\boldsymbol{k}}|^2 \tag{5.97a}$$
$$+\sum_{\boldsymbol{k}_1\boldsymbol{k}_2}(V_{\boldsymbol{k}_1,\boldsymbol{k}_2;\boldsymbol{k}_1,\boldsymbol{k}_2} - V_{\boldsymbol{k}_1,\boldsymbol{k}_2;\boldsymbol{k}_2,\boldsymbol{k}_1} + V_{\boldsymbol{k}_1,-\boldsymbol{k}_2;\boldsymbol{k}_1,-\boldsymbol{k}_2})|v_{\boldsymbol{k}_1}|^2|v_{\boldsymbol{k}_2}|^2 \tag{5.97b}$$
$$+\sum_{\boldsymbol{k}_1\boldsymbol{k}_2}V_{\boldsymbol{k}_1,-\boldsymbol{k}_1;\boldsymbol{k}_2,-\boldsymbol{k}_2}u_{\boldsymbol{k}_1}v^*_{\boldsymbol{k}_1}u^*_{\boldsymbol{k}_2}v_{\boldsymbol{k}_2} \quad . \tag{5.97c}$$

Reduced Hamiltonian

Note that the normal ground state (5.51) is equivalent to the particular choice

$$\begin{aligned} v^0_{\boldsymbol{k}} &= 1 \quad u^0_{\boldsymbol{k}} = 0 \quad , \quad |\boldsymbol{k}| < k_{\mathrm{F}} \quad , \\ v^0_{\boldsymbol{k}} &= 0 \quad u^0_{\boldsymbol{k}} = 1 \quad , \quad |\boldsymbol{k}| > k_{\mathrm{F}} \quad . \end{aligned} \tag{5.98}$$

With this choice, term (5.97b) is the interaction energy to first order of the perturbation calculation. Moreover, the superconducting state cannot correspond to a dramatic redistribution of the occupation rate of the $\boldsymbol{k}$ states without causing the kinetic energy to grow too large. *The result is that $|v_{\boldsymbol{k}}|^2$ can only differ appreciably from $|v^0_{\boldsymbol{k}}|^2$ in the neighbourhood of the Fermi surface* (this will be confirmed in the following section). As a result, term (5.97b), proportional to $|v_{\boldsymbol{k}_1}|^2|v_{\boldsymbol{k}_2}|^2$, cannot sensibly be modified by the passage from the normal state to the superconducting state. One may then conclude that the mechanism of superconductivity results from term (5.97c).

The absolute values of the energies (see Sect. 5.2.4) are not being sought here. To simplify the analysis and to make evident the determining mechanism, Hamiltonian (5.48) can be replaced by a *reduced Hamiltonian*

$$\begin{aligned} \mathtt{H}^{\mathrm{red}} &= \sum_{\boldsymbol{k}\sigma} \varepsilon_{\boldsymbol{k}} a^*_{\boldsymbol{k}\sigma} a_{\boldsymbol{k}\sigma} + \sum_{\boldsymbol{k}_1 \neq \boldsymbol{k}_2} V_{\boldsymbol{k}_1,\boldsymbol{k}_2} a^*_{-\boldsymbol{k}_1\downarrow} a^*_{\boldsymbol{k}_1\uparrow} a_{\boldsymbol{k}_2\uparrow} a_{-\boldsymbol{k}_2\downarrow} \\ &= \mathtt{H}_0 + \mathtt{V}^{\mathrm{red}} \quad , \end{aligned} \tag{5.99}$$

which gives energy (5.97) excluding, however, all terms involving $|v_{\boldsymbol{k}_1}|^2$ and $|v_{\boldsymbol{k}_2}|^2$ in the interaction

$$\begin{aligned} E(\{u_{\boldsymbol{k}}, v_{\boldsymbol{k}}\}) &= \langle \Psi | \, \mathtt{H}^{\mathrm{red}} \, | \Psi \rangle \\ &= 2 \sum_{\boldsymbol{k}} \varepsilon_{\boldsymbol{k}} |v_{\boldsymbol{k}}|^2 + \sum_{\boldsymbol{k}_1 \neq \boldsymbol{k}_2} V_{\boldsymbol{k}_1,\boldsymbol{k}_2} u_{\boldsymbol{k}_1} v^*_{\boldsymbol{k}_1} u^*_{\boldsymbol{k}_2} v_{\boldsymbol{k}_2} \\ &= E^{\mathrm{kin}}(\{u_{\boldsymbol{k}}, v_{\boldsymbol{k}}\}) + U^{\mathrm{pot}}(\{u_{\boldsymbol{k}}, v_{\boldsymbol{k}}\}) \end{aligned} \tag{5.100}$$

with

$$V_{\boldsymbol{k}_1,\boldsymbol{k}_2} = V_{-\boldsymbol{k}_1,\boldsymbol{k}_1;-\boldsymbol{k}_2,\boldsymbol{k}_2} = V_{\boldsymbol{k}_1,-\boldsymbol{k}_1;\boldsymbol{k}_2,-\boldsymbol{k}_2} = V^*_{\boldsymbol{k}_2,\boldsymbol{k}_1} \quad . \tag{5.101}$$

One sees that the interaction term in $\mathtt{H}^{\mathrm{red}}$ precisely incorporates the matrix elements (5.77) which are free from sign ambiguity.

5.3.6 Search for a Minimum-Energy State

It remains to find parameters $u_{\boldsymbol{k}}$ and $v_{\boldsymbol{k}}$ which will minimize $E(\{u_{\boldsymbol{k}}, v_{\boldsymbol{k}}\})$. Since the number of particles is variable, it is necessary to minimize the energy with the constraint that the mean density $N(\{u_{\boldsymbol{k}}, v_{\boldsymbol{k}}\})/L^3$ is fixed at the value ρ. The Lagrange parameter μ is introduced as well as the functional

$$F(\{u_{\boldsymbol{k}}, v_{\boldsymbol{k}}\}) = E(\{u_{\boldsymbol{k}}, v_{\boldsymbol{k}}\}) - \mu N(\{u_{\boldsymbol{k}}, v_{\boldsymbol{k}}\}) \quad . \tag{5.102}$$

Parameter μ is the chemical potential and the treatment is analogous to a grand canonical ensemble in statistical mechanics: once the minimum of (5.102) is found, $\mu = \mu(\rho)$ is determined as a function of the density. For a free Fermi gas at zero temperature, $\mu(\rho)$ is identical to the Fermi energy (2.19). The distribution of the BCS occupied states must not vary by much compared to the normal state (Sect. 5.3.5). One thus expects that μ *still coincides to first approximation with the Fermi energy.*

The Gap Equation

The variational problem is resolved without great difficulty. Taking into account normalization condition (5.71), one can introduce new real parameters $h_{\boldsymbol{k}}$ and $\alpha(\boldsymbol{k})$, such that

$$u_{\boldsymbol{k}} = (1 - h_{\boldsymbol{k}})^{1/2} \quad , \quad v_{\boldsymbol{k}} = h_{\boldsymbol{k}}^{1/2} \mathrm{e}^{\mathrm{i}\alpha(\boldsymbol{k})} \quad . \tag{5.103}$$

Referring to (5.87), $h_{\boldsymbol{k}}$ is the probability that the pair $(\boldsymbol{k}\uparrow, -\boldsymbol{k}\downarrow)$ is present; $\alpha(\boldsymbol{k})$ is the phase difference between $v_{\boldsymbol{k}}$ and $u_{\boldsymbol{k}}$. One can take $u_{\boldsymbol{k}}$ to be real and positive, since multiplication of $u_{\boldsymbol{k}}$ and $v_{\boldsymbol{k}}$ by an identical factor of modulus 1 only multiplies $|\Psi\rangle$ by a constant phase factor.

With the definition

$$\xi_{\boldsymbol{k}} = \varepsilon_{\boldsymbol{k}} - \mu \quad , \tag{5.104}$$

F takes the form

$$F = 2\sum_{\boldsymbol{k}} \xi_{\boldsymbol{k}} h_{\boldsymbol{k}} + \sum_{\boldsymbol{k}_1 \neq \boldsymbol{k}_2} V_{\boldsymbol{k}_1,\boldsymbol{k}_2} [(1 - h_{\boldsymbol{k}_1}) h_{\boldsymbol{k}_1}]^{1/2} \times [(1 - h_{\boldsymbol{k}_2}) h_{\boldsymbol{k}_2}]^{1/2} \mathrm{e}^{\mathrm{i}(\alpha(\boldsymbol{k}_2) - \alpha(\boldsymbol{k}_1))} \quad . \tag{5.105}$$

F is required to be stationary with respect to $\alpha(\boldsymbol{k})$ and $h_{\boldsymbol{k}}$. Taking into account $\partial F / \partial \alpha(\boldsymbol{k}) = 0$, this is expressed as

$$\frac{\partial F}{\partial h_{\boldsymbol{k}}} = 2\xi_{\boldsymbol{k}} + \sum_{\boldsymbol{k}' \neq \boldsymbol{k}} V_{\boldsymbol{k},\boldsymbol{k}'} [(1 - h_{\boldsymbol{k}'}) h_{\boldsymbol{k}'}]^{1/2} \frac{1 - 2h_{\boldsymbol{k}}}{[(1 - h_{\boldsymbol{k}}) h_{\boldsymbol{k}}]^{1/2}} \mathrm{e}^{\mathrm{i}(\alpha(\boldsymbol{k}') - \alpha(\boldsymbol{k}))} = 0 \quad . \tag{5.106}$$

This equation is notably simplified by an adequate redefinition of variables

$$h_{\boldsymbol{k}} = \frac{1}{2}\left(1 - \frac{\xi_{\boldsymbol{k}}}{E_{\boldsymbol{k}}}\right) \quad , \quad E_{\boldsymbol{k}} = \left(\xi_{\boldsymbol{k}}^2 + |\Delta_{\boldsymbol{k}}|^2\right)^{1/2} \quad ,$$
$$\Delta_{\boldsymbol{k}} = |\Delta_{\boldsymbol{k}}| \mathrm{e}^{\mathrm{i}\alpha(\boldsymbol{k})} \quad , \quad u_{\boldsymbol{k}}^* v_{\boldsymbol{k}} = \frac{\Delta_{\boldsymbol{k}}}{2E_{\boldsymbol{k}}} \quad . \tag{5.107}$$

Equation (5.106) can thus be rewritten as

$$\Delta_{\boldsymbol{k}} = -\frac{1}{2} \sum_{\boldsymbol{k}' \neq \boldsymbol{k}} V_{\boldsymbol{k},\boldsymbol{k}'} \frac{\Delta_{\boldsymbol{k}'}}{E_{\boldsymbol{k}'}} \quad . \tag{5.108}$$

It is an integral equation which determines $\Delta_{\boldsymbol{k}}$ (the "gap" equation).

The Solution

It is clear that (5.108) allows the trivial solution $\Delta_{\boldsymbol{k}} = 0$ for all $\boldsymbol{k}$

$$\mu = \varepsilon_{\mathrm{F}} \quad , \quad h_{\boldsymbol{k}}^0 = \frac{1}{2}\left(1 - \frac{\xi_{\boldsymbol{k}}}{|\xi_{\boldsymbol{k}}|}\right) = \begin{cases} 1 & , \quad |\boldsymbol{k}| < k_{\mathrm{F}} \\ 0 & , \quad |\boldsymbol{k}| > k_{\mathrm{F}} \end{cases} \quad .$$

This corresponds to the normal state (5.98).

A second non-zero solution exists when the matrix elements $V_{\boldsymbol{k},\boldsymbol{k}'}$ are negative. In order to be able to resolve (5.108) in an analytical manner, BCS propose a simplified form for $V_{\boldsymbol{k},\boldsymbol{k}'}$.

$$V_{\boldsymbol{k},\boldsymbol{k}'} = \begin{cases} -V_0/L^3 < 0 & \text{for } |\xi_{\boldsymbol{k}}|, |\xi_{\boldsymbol{k}'}| \leq \hbar\omega_{\mathrm{D}} \\ 0 & \text{for all other cases} \end{cases} \quad , \tag{5.109}$$

where ω_{D} is the Debye frequency

$$\omega_{\mathrm{D}} \simeq c_{\mathrm{s}} k_{\mathrm{F}} \tag{5.110}$$

and c_{s} is the speed of sound in the metal ($c_{\mathrm{s}} \simeq 10^3$ ms^{-1}) such that ω_{D} is the maximum frequency of a phonon. The result is that $\hbar\omega_{\mathrm{D}}$ is the maximum energy of a phonon. Since the attraction between the electrons responsible for superconductivity is transmitted by the ions in motion (see Sect. 9.4.4), *it is reasonable to limit its action to the band of energy* $-\hbar\omega_{\mathrm{D}} \leq \xi_{\boldsymbol{k}} = \varepsilon_{\boldsymbol{k}} - \mu \leq \hbar\omega_{\mathrm{D}}$. Assuming that $V_{\boldsymbol{k},\boldsymbol{k}'}$ is constant in this domain additionally simplifies the model.

Introducing form (5.109) into (5.108) implies that

$$\Delta_{\boldsymbol{k}} = \begin{cases} \Delta = |\Delta| \mathrm{e}^{\mathrm{i}\alpha} & , \quad |\xi_{\boldsymbol{k}}| < \hbar\omega_{\mathrm{D}} \\ 0 & , \quad |\xi_{\boldsymbol{k}}| > \hbar\omega_{\mathrm{D}} \end{cases} \quad . \tag{5.111}$$

The parameter Δ is thus a constant complex number in the domain of the definition of $V_{\boldsymbol{k},\boldsymbol{k}'}$. *As for its phase, it disappears completely from (5.108)*, which is written in the infinite volume limit

$$\frac{V_0}{2} \int_{-\hbar\omega_{\mathrm{D}}}^{\hbar\omega_{\mathrm{D}}} \mathrm{d}\xi \frac{g(\xi)}{(\xi^2 + |\Delta|^2)^{1/2}} = 1 \quad . \tag{5.112}$$

The quantity $g(\xi)$ is the *density of states* which appears when one replaces the sum over the wave vectors by an integral over the energy ξ

$$\frac{1}{L^3} \sum_{\boldsymbol{k}} \to \frac{1}{(2\pi)^3} \int \mathrm{d}\boldsymbol{k} \ldots = \int \mathrm{d}\xi g(\xi) \ldots \quad ,$$

$$g(\xi) = \frac{1}{2\pi^2} |\boldsymbol{k}|^2 \left(\frac{\mathrm{d}\varepsilon_{\boldsymbol{k}}}{\mathrm{d}|\boldsymbol{k}|} \right)^{-1} \quad , \quad \xi_{\boldsymbol{k}} = \varepsilon_{\boldsymbol{k}} - \mu \quad . \tag{5.113}$$

In the energy band considered, $g(\xi)$ varies only slightly. One can thus replace it by the value that it takes for $\xi = 0$ ($\varepsilon = \mu$). Integral (5.112) can now be carried out, giving

$$\begin{aligned} \Delta &= \hbar\omega_{\mathrm{D}} \left[\sinh \left(\frac{1}{g(0)V_0} \right) \right]^{-1} \\ &\simeq 2\hbar\omega_{\mathrm{D}} \exp \left(-\frac{1}{g(0)V_0} \right) \simeq 10^{-3} \text{ eV} \ll \varepsilon_{\mathrm{F}} \simeq 1 \text{ eV} \quad , \end{aligned} \tag{5.114}$$

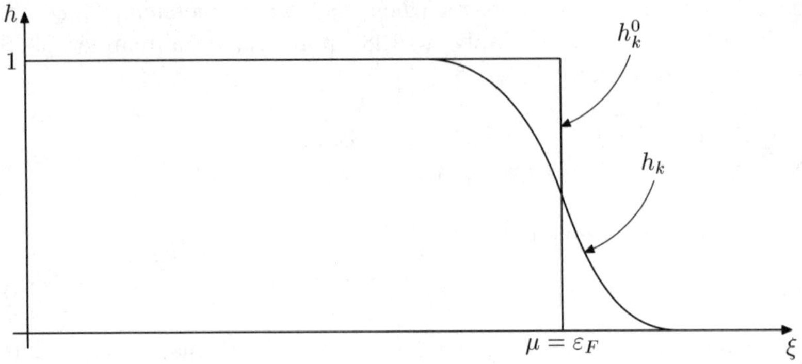

Fig. 5.6. The function $h_{\boldsymbol{k}}$ corresponds to a redistribution of electrons in $\boldsymbol{k}$ space neighbouring the Fermi surface

where the fact that $g(0)V_0 \ll 1$ has been taken into account.

In this relation and in the two sections which follow, Δ is taken to be real and positive. Section 5.4 will return to the question of the indetermination of the phase.

Electron Distribution in the BCS State

Relations (5.87), (5.88) and (5.107) allow one to write the distribution function of the superconductor ground state $|\Psi_{\mathrm{BCS}}\rangle$

$$\begin{aligned}\langle\Psi_{\mathrm{BCS}}|\, a^*_{\boldsymbol{k}\sigma}a_{\boldsymbol{k}\sigma}\,|\Psi_{\mathrm{BCS}}\rangle &= |v_{\pm\boldsymbol{k}}|^2 = h_{\boldsymbol{k}} \\ &= \frac{1}{2}\left(1 - \frac{\xi}{(\xi^2+\Delta^2)^{1/2}}\right) \quad , \quad \sigma = \pm\frac{1}{2} \quad . \end{aligned} \tag{5.115}$$

The function $h_{\boldsymbol{k}}$ is represented in Fig. 5.6. *It closely corresponds to a redistribution of electrons in* **k** *space closely neighbouring the Fermi surface.* Its form suggests the identification of the chemical potential with the Fermi level. In fact, the calculation shows that $\mu = \varepsilon_{\mathrm{F}}$ up to a term of order $\Delta^2/\varepsilon_{\mathrm{F}}$, which is negligible according to (5.114).

Energy of the BCS State

Finally, it is necessary to verify that the augmentation of the kinetic energy due to the appearance of occupied states with $|\boldsymbol{k}| > k_{\mathrm{F}}$ is compensated by a decrease in the potential energy. With (5.92), (5.103), (5.108), (5.115) and the same approximations $\mu \simeq \varepsilon_{\mathrm{F}}$ and $g(\xi) \simeq g(0)$, the increase in the kinetic energy density is given by

$$
\begin{aligned}
u_{\mathrm{BCS}}^{\mathrm{kin}} - u_0^{\mathrm{kin}} &= \frac{E_{\mathrm{BCS}}^{\mathrm{kin}} - E_0^{\mathrm{kin}}}{L^3} \\
&= \frac{2}{L^3} \sum_{\boldsymbol{k}} \varepsilon_{\boldsymbol{k}} \left[v_{\boldsymbol{k}}^2 - (v_{\boldsymbol{k}}^0)^2 \right] = \frac{1}{L^3} \sum_{\boldsymbol{k}} \varepsilon_{\boldsymbol{k}} \left(\frac{\xi_{\boldsymbol{k}}}{|\xi_{\boldsymbol{k}}|} - \frac{\xi_{\boldsymbol{k}}}{E_{\boldsymbol{k}}} \right) \\
&\simeq g(0) \int_{-\hbar\omega_{\mathrm{D}}}^{\hbar\omega_{\mathrm{D}}} \mathrm{d}\xi \xi^2 \left(\frac{1}{|\xi|} - \frac{1}{(\xi^2 + \Delta^2)^{1/2}} \right) > 0 \quad . \qquad (5.116)
\end{aligned}
$$

The kinetic energy E_0^{kin} is obtained by formally replacing $v_{\boldsymbol{k}}$ by $v_{\boldsymbol{k}}^0$ in (5.97a). It represents the kinetic energy in the ground state. Due to (5.111), the only contribution to integral comes from the band $|\xi| \leq \hbar\omega_{\mathrm{D}}$. The symmetry $\xi \to -\xi$ permits ε to be replaced by $\xi = \varepsilon - \mu$. Integral (5.116) can be evaluated explicitly, giving the result

$$
u_{\mathrm{BCS}}^{\mathrm{kin}} - u_0^{\mathrm{kin}} = -\frac{g(0)\Delta^2}{2} + \frac{\Delta^2}{V_0} \quad , \qquad (5.117)
$$

which is valid when $\Delta \ll \hbar\omega_{\mathrm{D}}$.

With $u_{\boldsymbol{k}}^* v_{\boldsymbol{k}} = \Delta_{\boldsymbol{k}}/2E_{\boldsymbol{k}}$ and (5.108), the potential energy density (5.100) $u_{\mathrm{BCS}}^{\mathrm{pot}} = U_{\mathrm{BCS}}^{\mathrm{pot}}/L^3$ takes the form

$$
\begin{aligned}
u_{\mathrm{BCS}}^{\mathrm{pot}} &= \frac{1}{L^3} \sum_{\boldsymbol{k}_1 \neq \boldsymbol{k}_2} V_{\boldsymbol{k}_1,\boldsymbol{k}_2} u_{\boldsymbol{k}_1} v_{\boldsymbol{k}_2}^* u_{\boldsymbol{k}_2}^* v_{\boldsymbol{k}_2} = -\frac{1}{2L^3} \sum_{\boldsymbol{k}} \frac{|\Delta_{\boldsymbol{k}}|^2}{E_{\boldsymbol{k}}} \\
&\simeq -\frac{g(0)}{2} \int_{-\hbar\omega_{\mathrm{D}}}^{\hbar\omega_{\mathrm{D}}} \mathrm{d}\xi \frac{\Delta^2}{(\xi^2 + \Delta^2)^{1/2}} \quad . \qquad (5.118)
\end{aligned}
$$

It can be concluded from (5.112) that $u_{\mathrm{BCS}}^{\mathrm{pot}} = -\Delta^2/V_0$ and that

$$
u_{\mathrm{BCS}}^{\mathrm{kin}} + u_{\mathrm{BCS}}^{\mathrm{pot}} - u_0^{\mathrm{kin}} = \frac{E_{\mathrm{BCS}}^{\mathrm{tot}} - E_0^{\mathrm{tot}}}{L^3} \simeq -\frac{g(0)\Delta^2}{2} \quad . \qquad (5.119)
$$

The BCS state leads to an extensive decrease (proportional to the volume) of the total energy. Since, according to (5.114), Δ has an essential singularity at $V_0 = 0$, it is evident that *this result cannot be obtained by means of a perturbative method.*

5.3.7 The Energy Gap

Excited States

It is important to determine the first excited states and the energy spectrum in the vicinity of the ground state in order to identify the energy gap. The state $|\Psi_{\mathrm{BCS}}\rangle$, obtained by the variational method, is only an approximation of the ground state of $\mathtt{H}^{\mathrm{red}}$. It is not possible to determine the exact form of the excited states, but use of the Bogoliubov operators (5.78) and (5.79) allows one to give an approximate description.

Anticommutation relations (5.80) and property (5.81) show that operators $\alpha^*_{\boldsymbol{k}}, \beta^*_{\boldsymbol{k}}$ $(\alpha_{\boldsymbol{k}}, \beta_{\boldsymbol{k}})$ behave as creators (annihilators) of independent particles. The collection of states $\alpha^*_{\boldsymbol{k}_1}\beta^*_{\boldsymbol{k}_2}\cdots\alpha^*_{\boldsymbol{k}_n}|\Psi_{\mathrm{BCS}}\rangle$ obtained by the action of some products of $\alpha^*_{\boldsymbol{k}}$ and $\beta^*_{\boldsymbol{k}}$ constitute an orthonormal basis of Fock space. $\alpha^*_{\boldsymbol{k}}$ and $\beta^*_{\boldsymbol{k}}$ create *quasi-particles* and these states contain n individual quasi-particles. Although this basis does not exactly diagonalize $\mathtt{H}^{\mathrm{red}}$, *it is reasonable to admit that states* $\alpha^*_{\boldsymbol{k}}|\Psi_{\mathrm{BCS}}\rangle$ *and* $\beta^*_{\boldsymbol{k}}|\Psi_{\mathrm{BCS}}\rangle$ *and their linear combinations approximate the first excited states of the spectrum.* Consider, for example

$$\begin{aligned}\alpha^*_{\boldsymbol{k}}|\Psi_{\mathrm{BCS}}\rangle &= \prod_{\boldsymbol{k}'\neq\boldsymbol{k}}(u_{\boldsymbol{k}'}+v_{\boldsymbol{k}'}a^*_{\boldsymbol{k}'\uparrow}a^*_{-\boldsymbol{k}'\downarrow})\\ &\quad\times(u^*_{\boldsymbol{k}}a^*_{\boldsymbol{k}\uparrow}-v^*_{\boldsymbol{k}}a_{-\boldsymbol{k}\downarrow})(u_{\boldsymbol{k}}+v_{\boldsymbol{k}}a^*_{\boldsymbol{k}\uparrow}a^*_{-\boldsymbol{k}\downarrow})|0\rangle\\ &= \prod_{\boldsymbol{k}'\neq\boldsymbol{k}}(u_{\boldsymbol{k}'}+v_{\boldsymbol{k}'}a^*_{\boldsymbol{k}'\uparrow}a^*_{-\boldsymbol{k}'\downarrow})a^*_{\boldsymbol{k}\uparrow}|0\rangle \quad . \end{aligned} \tag{5.120}$$

The significance of this state is clear: the pair $(\boldsymbol{k}\uparrow; -\boldsymbol{k}\downarrow)$ has been replaced in $|\Psi_{\mathrm{BCS}}\rangle$ by a non-paired electron $(\boldsymbol{k}\uparrow)$. That is, *a pair was broken.*

Energy of the Excited States

To calculate the energy of state (5.120), one evaluates

$$E^{\mathrm{exc}}_{\boldsymbol{k}} = \langle\Psi_{\mathrm{BCS}}|\,\alpha_{\boldsymbol{k}}\mathtt{H}^{\mathrm{red}}\alpha^*_{\boldsymbol{k}}\,|\Psi_{\mathrm{BCS}}\rangle - \langle\Psi_{\mathrm{BCS}}|\,\mathtt{H}^{\mathrm{red}}\,|\Psi_{\mathrm{BCS}}\rangle \quad , \tag{5.121}$$

where $E^{\mathrm{exc}}_{\boldsymbol{k}}$ represents the energy of state $\alpha^*_{\boldsymbol{k}}|\Psi_{\mathrm{BCS}}\rangle$ measured with respect to the energy of the ground state. This difference is easily calculated with the help of the Wick theorem for $\alpha_{\boldsymbol{k}}$ and $\alpha^*_{\boldsymbol{k}}$. The average $\langle\Psi_{\mathrm{BCS}}|\,\alpha_{\boldsymbol{k}}\mathtt{H}^{\mathrm{red}}\alpha^*_{\boldsymbol{k}}\,|\Psi_{\mathrm{BCS}}\rangle$ can be evaluated as a sum of products of contractions

$$\begin{aligned}\langle\Psi_{\mathrm{BCS}}|\,\alpha_{\boldsymbol{k}}\mathtt{H}^{\mathrm{red}}\alpha^*_{\boldsymbol{k}}\,|\Psi_{\mathrm{BCS}}\rangle &= \langle\Psi_{\mathrm{BCS}}|\,\underbrace{\alpha_{\boldsymbol{k}}\mathtt{H}^{\mathrm{red}}\alpha^*_{\boldsymbol{k}}}\,|\Psi_{\mathrm{BCS}}\rangle\\ &\quad+\langle\Psi_{\mathrm{BCS}}|\,\underbrace{\alpha_{\boldsymbol{k}}\mathtt{H}}\underbrace{{}^{\mathrm{red}}\alpha^*_{\boldsymbol{k}}}\,|\Psi_{\mathrm{BCS}}\rangle \quad . \end{aligned}\tag{5.122}$$

The notation used here is the following. Among all these products, the first term of the right hand side of (5.122) designates all those for which $\alpha_{\boldsymbol{k}}$ is contracted with $\alpha^*_{\boldsymbol{k}}$. Since $\underbrace{\alpha_{\boldsymbol{k}}\alpha^*_{\boldsymbol{k}}}=1$, it is evident that

$$\langle\Psi_{\mathrm{BCS}}|\,\underbrace{\alpha_{\boldsymbol{k}}\mathtt{H}^{\mathrm{red}}\alpha^*_{\boldsymbol{k}}}\,|\Psi_{\mathrm{BCS}}\rangle = \langle\Psi_{\mathrm{BCS}}|\,\mathtt{H}^{\mathrm{red}}\,|\Psi_{\mathrm{BCS}}\rangle \quad . \tag{5.123}$$

The difference $E^{\mathrm{exc}}_{\boldsymbol{k}}$ (5.121) is thus equal to the second term of (5.122) which represents all the products where $\alpha_{\boldsymbol{k}}$ and $\alpha^*_{\boldsymbol{k}}$ are separately contracted with the creators and annihilators of $\mathtt{H}^{\mathrm{red}}$.

Referring back to the calculation methods developed in Sect. 5.3.5, one arrives at the following expression for the kinetic energy,

$$
\begin{aligned}
&\langle \Psi_{\mathrm{BCS}} |\, \alpha_{\boldsymbol{k}} \mathrm{H}_0 \alpha^*_{\boldsymbol{k}} \,| \Psi_{\mathrm{BCS}} \rangle \\
&= \sum_{\boldsymbol{k}'} \xi_{\boldsymbol{k}'} \left(\alpha_{\boldsymbol{k}} a^*_{\boldsymbol{k}'\uparrow} a_{\boldsymbol{k}'\uparrow} \alpha^*_{\boldsymbol{k}} + \alpha_{\boldsymbol{k}} a^*_{-\boldsymbol{k}'\downarrow} a_{-\boldsymbol{k}'\downarrow} \alpha^*_{\boldsymbol{k}} \right) \\
&= \sum_{\boldsymbol{k}'} \xi_{\boldsymbol{k}} \delta_{\boldsymbol{k},\boldsymbol{k}'} \left(|u_{\boldsymbol{k}}|^2 - |v_{\boldsymbol{k}}|^2 \right) = \xi_{\boldsymbol{k}} (1 - 2h_{\boldsymbol{k}}) = \frac{\xi^2_{\boldsymbol{k}}}{E^2_{\boldsymbol{k}}} \quad .
\end{aligned}
\tag{5.124}
$$

Proceeding in the same manner for the potential energy in (5.99), one obtains

$$
\begin{aligned}
&\langle \Psi_{\mathrm{BCS}} |\, \alpha_{\boldsymbol{k}} \mathrm{V}^{\mathrm{red}} \alpha^*_{\boldsymbol{k}} \,| \Psi_{\mathrm{BCS}} \rangle \\
&= \sum_{\boldsymbol{k}_1 \neq \boldsymbol{k}_2} V_{\boldsymbol{k}_1,\boldsymbol{k}_2} \big[\alpha_{\boldsymbol{k}} a^*_{-\boldsymbol{k}_1\downarrow} a^*_{\boldsymbol{k}_1\uparrow} a_{\boldsymbol{k}_2\uparrow} a_{-\boldsymbol{k}_2\downarrow} \alpha^*_{\boldsymbol{k}} \\
&\qquad + \alpha_{\boldsymbol{k}} a^*_{-\boldsymbol{k}_1\downarrow} a^*_{\boldsymbol{k}_1\uparrow} a_{\boldsymbol{k}_2\uparrow} a_{-\boldsymbol{k}_2\downarrow} \alpha^*_{\boldsymbol{k}} \big] \\
&= - \left(u_{\boldsymbol{k}} v^*_{\boldsymbol{k}} \sum_{\boldsymbol{k}' \neq \boldsymbol{k}} V_{\boldsymbol{k},\boldsymbol{k}'} u^*_{\boldsymbol{k}'} v_{\boldsymbol{k}'} + v_{\boldsymbol{k}} u^*_{\boldsymbol{k}} \sum_{\boldsymbol{k}' \neq \boldsymbol{k}} V_{\boldsymbol{k}',\boldsymbol{k}} u_{\boldsymbol{k}'} v^*_{\boldsymbol{k}'} \right) \quad .
\end{aligned}
\tag{5.125}
$$

Contractions which vanish are not included in (5.125). Making use of (5.107) and (5.108), one finally obtains

$$
E^{\mathrm{exc}}_{\boldsymbol{k}} = \frac{\xi^2_{\boldsymbol{k}}}{E_{\boldsymbol{k}}} + \frac{|\Delta_{\boldsymbol{k}}|^2}{E_{\boldsymbol{k}}} = E_{\boldsymbol{k}} \geq \Delta \quad . \tag{5.126}
$$

An analogous calculation shows that the deviation between the energy of $\beta^*_{\boldsymbol{k}} |\Psi_{\mathrm{BCS}}\rangle$ and that of the ground state is also equal to $E_{\boldsymbol{k}}$. *All the excited states $\alpha^*_{\boldsymbol{k}} |\Psi_{\mathrm{BCS}}\rangle$ and $\beta^*_{\boldsymbol{k}} |\Psi_{\mathrm{BCS}}\rangle$ correspond to an energy exceeding that of the ground state by a quantity at least equal to Δ which, assuming the validity of the hypotheses presented, can be identified with an energy gap in the energy spectrum of the superconductors.*

One should note that when an external perturbation acts on a superconductor, breaking a pair, it produces an excited state of the form $\beta^*_{\boldsymbol{k}} \alpha^*_{\boldsymbol{k}} |\Psi_{\mathrm{BCS}}\rangle$ for which the energy, to first approximation, is equal to $2E_{\boldsymbol{k}}$. These two-quasi-particle states are separated from the ground state by the quantity 2Δ. This originates from the fact that an external perturbation which conserves the number of particles is bilinear in the electron operators. Since the Bogoliubov transformation is linear, this perturbation is equally bilinear in the operators $\alpha_{\boldsymbol{k}}$ and $\beta_{\boldsymbol{k}}$ and only induces transitions with states comprising at least two quasi-particles.

5.3.8 Spatial Extension of a Cooper Pair

Suppose, that the concept of a "pair wavefunction" $\varphi_{\mathrm{pair}}(\boldsymbol{x})$ could be defined. The order of magnitude of its spatial extension can be determined with the help of a few simple considerations. The function $\varphi_{\text{pair}}(\boldsymbol{x})$ can be expressed as the transform of an equivalent object $\tilde{\varphi}(\boldsymbol{k})$

$$\varphi_{\text{pair}}(\boldsymbol{x}) \simeq \int_{\delta k} \mathrm{d}\boldsymbol{k}\tilde{\varphi}(\boldsymbol{k})\mathrm{e}^{\mathrm{i}\boldsymbol{k}\cdot\boldsymbol{x}} \quad . \tag{5.127}$$

The sum over $\boldsymbol{k}$ is limited to the domain δk such that $|\xi_{\boldsymbol{k}}| \leq \hbar\omega_{\text{D}}$, for which

$$|\xi_{\boldsymbol{k}}| \simeq \left|\frac{\hbar^2|\boldsymbol{k}|^2}{2m} - \varepsilon_{\text{F}}\right| \simeq \left|\frac{\hbar^2(k_{\text{F}}+\delta k)^2}{2m} - \frac{\hbar^2 k_{\text{F}}^2}{2m}\right| \simeq \frac{\hbar^2 k_{\text{F}}\delta k}{m} \quad . \tag{5.128}$$

The uncertainty principle gives the magnitude of the spatial extension δx for a pair

$$\delta x \geq (\delta k)^{-1} \simeq \frac{\hbar k_{\text{F}}}{m\omega_{\text{D}}} \simeq \frac{\hbar}{mc_{\text{s}}} \simeq 1000\ \text{Å} \quad . \tag{5.129}$$

The ensemble of paired electrons (the "condensate") is characterized by a specific number ρ_s which is much smaller than the total number of electrons per unit volume. In the ground state, the electrons are considered as free

$$\rho = \int_0^{\varepsilon_{\text{F}}} \mathrm{d}\varepsilon g(\varepsilon) = A\varepsilon_{\text{F}}^{3/2} \simeq k_{\text{F}}^3 \quad . \tag{5.130}$$

The constant A has been determined by relation (2.98). As for the density of states (5.113), it becomes

$$g(\varepsilon) = \frac{3A}{2}\varepsilon^{1/2} \quad . \tag{5.131}$$

On the other hand, the paired electrons belong to a band of energy of width $\hbar\omega_{\text{D}}$ neighbouring the Fermi energy. Thus

$$\rho_s = \int_{\varepsilon_{\text{F}}}^{\varepsilon_{\text{F}}+\hbar\omega_{\text{D}}} \mathrm{d}\varepsilon g(\varepsilon) \simeq \hbar\omega_{\text{D}} g(\varepsilon_{\text{F}}) \simeq \hbar\omega_{\text{D}}\frac{\rho}{\varepsilon_{\text{F}}} \quad . \tag{5.132}$$

Under these conditions

$$\frac{\rho_s}{\rho} \simeq \frac{\hbar\omega_{\text{D}}}{\varepsilon_{\text{F}}} \simeq \frac{c_{\text{s}}m}{\hbar k_{\text{F}}} \simeq \frac{c_{\text{s}}}{v_{\text{F}}} \quad , \tag{5.133}$$

where v_{F} is the speed of an electron at the Fermi level. Thus

$$\rho_s \simeq \frac{c_{\text{s}} m k_{\text{F}}^2}{\hbar} \quad . \tag{5.134}$$

Moreover, $\rho_s \simeq d_s^{-3}$ where d_s is the mean distance between two electrons of the condensate. If one wished to evaluate the number of pairs which occupy the region of space of volume $(\delta x)^3$ corresponding to the extension of a determined pair, it would be given by the ratio

$$\left(\frac{\delta x}{d_s}\right)^3 \simeq \left(\frac{\hbar k_{\mathrm{F}}}{mc_{\mathrm{s}}}\right)^2 \simeq \left(\frac{v_{\mathrm{F}}}{c_{\mathrm{s}}}\right)^2 \simeq 10^6 \quad . \tag{5.135}$$

This numerical result demonstrates that a pair cannot be considered as an object spatially separated from the others. That is, a Cooper pair is not an *e–e* pseudo-molecule satisfying Bose statistics as in the case of an ^{4}He atom acting as the constituent of a gas, liquid or solid (Chap. 7).

5.4 Particle Number and Phase in Superconductivity

5.4.1 Is it Necessary to Fix the Particle Number or the Phase?

As shown earlier, the ground state $|\Psi_{\mathrm{BCS}}\rangle$ does not correspond to a fixed number of particles (Sect. 5.3.4). This could be considered as a defect of the theory. Taking a closer look, however, this particularity conceals a profound physical significance. If one rewrites (5.103) taking into account (5.111), then

$$\begin{aligned} u_{\boldsymbol{k}} &= (1 - h_{\boldsymbol{k}})^{1/2} \quad , \\ v_{\boldsymbol{k}} &= h_{\boldsymbol{k}}^{1/2} \mathrm{e}^{\mathrm{i}\alpha} \quad . \end{aligned} \tag{5.136}$$

The phase α is independent of $\boldsymbol{k}$, but otherwise arbitrary. One can make use of this degree of freedom to build a ground state $|\Psi_{\mathrm{BCS}}\rangle_n$ corresponding to a fixed number $2n$ of particles. One writes

$$\begin{aligned} |\Psi_{\mathrm{BCS}}\rangle_n &= \int_0^{2\pi} \frac{\mathrm{d}\alpha}{2\pi} \mathrm{e}^{-in\alpha} \prod_{\boldsymbol{k}} (u_{\boldsymbol{k}} + |v_{\boldsymbol{k}}| \mathrm{e}^{\mathrm{i}\alpha} a^*_{\boldsymbol{k}\uparrow} a^*_{-\boldsymbol{k}\downarrow})\, |0\rangle \\ &= \int_0^{2\pi} \frac{\mathrm{d}\alpha}{2\pi} \mathrm{e}^{-in\alpha}\, |\Psi_{\mathrm{BCS}}\rangle_\alpha \quad . \end{aligned} \tag{5.137}$$

In (5.137), the state $|\Psi_{\mathrm{BCS}}\rangle_\alpha$ is defined as the member of the family of states $|\Psi_{\mathrm{BCS}}\rangle$ which is formed with coefficients (5.136) and all having a well-defined phase, α. As for $|\Psi_{\mathrm{BCS}}\rangle_n$, its expression shows the integral of a sum of terms having $\mathrm{e}^{\mathrm{i}(l-n)\alpha}$ as a factor (where l is an integer). The integration causes all the terms for which $l \neq n$ to disappear. All that remains is a sum of contributions of the form $c(\{u_{\boldsymbol{k}}, v_{\boldsymbol{k}}\}) \prod_{\boldsymbol{k}}^{(n)} a^*_{\boldsymbol{k}\uparrow} a^*_{-\boldsymbol{k}\downarrow} |0\rangle$ where the product is carried over exactly n pairs each characterized by a different value of $\boldsymbol{k}$. The coefficient $c(\{u_{\boldsymbol{k}}, v_{\boldsymbol{k}}\})$ formed with the help of the $u_{\boldsymbol{k}}$ and $v_{\boldsymbol{k}}$ depends on these n values. Finally, the state $|\Psi_{\mathrm{BCS}}\rangle_n$ is the sum of all possible contributions of n different wave vectors.

Relation (5.137) can be inverted to give

$$|\Psi_{\mathrm{BCS}}\rangle_\alpha = \sum_{n=0}^{\infty} \mathrm{e}^{in\alpha}\, |\psi_{\mathrm{BCS}}\rangle_n \quad . \tag{5.138}$$

In light of (5.137) and (5.138), the state $|\Psi_{\mathrm{BCS}}\rangle_\alpha$, obtained by the variational method of BCS, has a fixed phase, while the state $|\Psi_{\mathrm{BCS}}\rangle_n$ has a fixed particle number. *It is not possible to simultaneously fix the phase and the number of particles.*

Relations (5.137) and (5.138) call to mind the Fourier transformation formulae which, in Schrödinger's theory, allows one to pass from the representation of position q to that of momentum p (see (2.6), (2.9) and (2.10)). These observables are subject to the uncertainty principle $\Delta p \Delta q \geq \hbar/2$. One is thus tempted to deduce from (5.137) and (5.138) an analogous inequality $\Delta n \Delta \alpha \geq 1/2$ satisfying the intuition, since it confirms that one cannot simultaneously fix the phase and the number of particles. Note, however, that the establishment of an uncertainty relation between the phase and the number of particles runs into some difficulties because the spectrum of the operator $\mathtt{N}$ is necessarily discrete. Strictly speaking, it is not possible to define a self-adjoint phase operator that obeys a canonical commutation rule with $\mathtt{N}$.

Having made these remarks, one can legitimately pose the question: in a given situation, must one provide $|\Psi_{\mathrm{BCS}}\rangle_\alpha$ or $|\Psi_{\mathrm{BCS}}\rangle_n$ to represent the ground state of a superconductor? Before answering this question, one should note two points:

- It is in general easier to calculate with $|\Psi_{\mathrm{BCS}}\rangle_\alpha$ than with $|\Psi_{\mathrm{BCS}}\rangle_n$.
- *The relative fluctuation $\langle\Delta\mathtt{N}\rangle / \langle\mathtt{N}\rangle$ of the mean number $\langle\mathtt{N}\rangle$ of particles in the state $|\Psi_{\mathrm{BCS}}\rangle_\alpha$ decreases when* $\langle\mathtt{N}\rangle = {}_\alpha\langle\Psi_{\mathrm{BCS}}|\,\mathtt{N}\,|\Psi_{\mathrm{BCS}}\rangle_\alpha = 2\sum_{\boldsymbol{k}}|v_{\boldsymbol{k}}|^2$ *increases:*

$$\frac{\langle\Delta\mathtt{N}\rangle}{\langle\mathtt{N}\rangle} = \frac{\left[{}_\alpha\langle\Psi_{\mathrm{BCS}}|\,\mathtt{N}^2\,|\Psi_{\mathrm{BCS}}\rangle_\alpha - \left({}_\alpha\langle\Psi_{\mathrm{BCS}}|\,\mathtt{N}\,|\Psi_{\mathrm{BCS}}\rangle_\alpha\right)^2\right]^{1/2}}{{}_\alpha\langle\Psi_{\mathrm{BCS}}|\,\mathtt{N}\,|\Psi_{\mathrm{BCS}}\rangle_\alpha} \quad . \tag{5.139}$$

In effect, one can write

$$\begin{aligned}
&{}_\alpha\langle\Psi_{\mathrm{BCS}}|\,\mathtt{N}^2\,|\Psi_{\mathrm{BCS}}\rangle_\alpha - \left({}_\alpha\langle\Psi_{\mathrm{BCS}}|\,\mathtt{N}\,|\Psi_{\mathrm{BCS}}\rangle_\alpha\right)^2 \\
&\quad = \sum_{\boldsymbol{k}_1\sigma_1\,\boldsymbol{k}_2\sigma_2} {}_\alpha\langle\Psi_{\mathrm{BCS}}|\, a^*_{\boldsymbol{k}_1\sigma_1} a_{\boldsymbol{k}_1\sigma_1} a^*_{\boldsymbol{k}_2\sigma_2} a_{\boldsymbol{k}_2\sigma_2}\,|\Psi_{\mathrm{BCS}}\rangle_\alpha \\
&\qquad - \sum_{\boldsymbol{k}\sigma}\left({}_\alpha\langle\Psi_{\mathrm{BCS}}|\, a^*_{\boldsymbol{k}\sigma} a_{\boldsymbol{k}\sigma}\,|\Psi_{\mathrm{BCS}}\rangle_\alpha\right)^2 \\
&\quad = 2\left(\sum_{\boldsymbol{k}}|v_{\boldsymbol{k}}|^2 - \sum_{\boldsymbol{k}}|v_{\boldsymbol{k}}|^4\right) \quad .
\end{aligned} \tag{5.140}$$

Thus

$$\begin{aligned}
\frac{\langle\Delta\mathtt{N}\rangle}{\langle\mathtt{N}\rangle} &= \frac{\left(2\sum_{\boldsymbol{k}}(|v_{\boldsymbol{k}}|^2 - |v_{\boldsymbol{k}}|^4)\right)^{1/2}}{\langle\mathtt{N}\rangle} \\
&= \frac{\left(2\sum_{\boldsymbol{k}}|v_{\boldsymbol{k}}|^2|u_{\boldsymbol{k}}|^2\right)^{1/2}}{\langle\mathtt{N}\rangle} \leq \left(\frac{1}{\langle\mathtt{N}\rangle}\right)^{1/2} \quad .
\end{aligned} \tag{5.141}$$

With a large number of particles present ($\langle \mathtt{N} \rangle \gg 1$), the choice between $|\Psi_{\mathrm{BCS}}\rangle_\alpha$ and $|\Psi_{\mathrm{BCS}}\rangle_n$ is of little importance. Convenience favors the choice of $|\Psi_{\mathrm{BCS}}\rangle_\alpha$. The question of principle remains, however: must one work with $|\Psi_{\mathrm{BCS}}\rangle_\alpha$ or with $|\Psi_{\mathrm{BCS}}\rangle_n$? An analogy with statistical physics provides an appropriate response.

5.4.2 Analogy with Statistical Physics

Envision a physical system Σ comprising a large number of particles n and for which statistical mechanics is applicable. One frequently encounters into two important cases:

- If Σ is isolated, its energy E is conserved. Here, in principle, the statistics of a micro-canonical ensemble apply.
- If Σ is only part of a much larger system Σ', it can exchange energy with its environment; in doing so, it generally evolves toward an equilibrium characterized by a uniformity of the temperature T in Σ with respect to its neighbourhood. When the temperature is uniform and equilibrium has been attained, the canonical ensemble can be used to describe the statistics of Σ, an ensemble where the energy of Σ fluctuates, with its relative fluctuation $\langle \Delta E \rangle / \langle E \rangle$ varying equally as $\langle \mathtt{N} \rangle^{-1/2}$.

In an inhomogeneous situation, but close to equilibrium, one assumes that the system can be described by a local temperature of equilibrium $T(\boldsymbol{x})$. This temperature varies slowly at the molecular scale and the exchanges of energy depend linearly on its gradient

$$\boldsymbol{j}_Q(\boldsymbol{x}) = -\kappa \nabla T(\boldsymbol{x}) \quad , \tag{5.142}$$

where $\boldsymbol{j}_Q(\boldsymbol{x})$ is the heat current density and κ the thermal conductivity. The statistical duality between E and T allows one to better understand the analogous duality occurring on the quantum level between n and α (in the case of the superconductor, only the ground state is discussed). One can formulate an analogy in the following terms:

- An isolated superconductor Σ has a fixed number of electrons n. Its ground state is thus identified with $|\Psi_{\mathrm{BCS}}\rangle_n$.
- *As part of a larger system Σ', the system Σ evolves toward a state where the number of particles is not fixed. It is another quantity, the phase α, which guarantees by its uniformity that Σ is indeed in its ground state* $|\Psi_{\mathrm{BCS}}\rangle_\alpha$. The fluctuation of the number of particles is given by (5.141).

In an inhomogeneous situation, where macroscopic currents are present, one supposes that the phase $\alpha(\boldsymbol{x})$ varies from point to point, but that the superconductor can still be described locally by a state $|\Psi_{\mathrm{BCS}}\rangle_{\alpha(\boldsymbol{x})}$. At the microscopic scale, the phase varies sufficiently slowly enough that, in the neighbourhood of a point $\boldsymbol{x}$, the definition of the state $|\Psi_{\mathrm{BCS}}\rangle_{\alpha(\boldsymbol{x})}$ maintains its sense. This is in analogy with the assumption in statistical physics of a local thermal equilibrium. In analogy with (5.142) one equally conceives that the macroscopic field of the velocities inside a superconductor must be bound to the gradient of the phase $\nabla\alpha(\boldsymbol{x})$. This renders relation (5.31) plausible and natural. The goal of these very qualitative considerations is to show that the BCS theory can also justify the phenomenology of superconductors in the presence of currents and electromagnetic fields.

A complete deduction of this phenomenology was originally made by Gork'hov,[5] who limited himself to a study of the static case and to the region of temperatures close to T_c.

5.5 High-T_c Superconductivity

The BCS theory of electrons interacting via the phonon-exchange mechanism describes adequately the superconductors of type I (mostly pure metals, e.g. Al, $\delta = 500$ Å, $\xi = 16000$ Å) and the superconductors of type II (mostly metallic alloys, e.g. Nb-Ti, $\delta = 600$ Å, $\xi = 450$ Å). The first class is characterized by $\delta \leq (1/\sqrt{2})\xi$ and the second class by the inverse inequality $(1/\sqrt{2})\xi \leq \delta$ where δ is the penetration length (5.19) and ξ is the coherence length. The coherence length can be identified with the extension of a Cooper pair discussed in Sect. 5.3.8. Type-I superconductors show a complete Meissner effect whereas type-II superconductors exhibit an incomplete Meissner effect with flux penetration (vortex lines) above a certain critical magnetic field. In either case, the critical temperatures $k_B T_c \sim \hbar\omega_D \exp\left(-1/g(0)V_0\right)$ predicted by the BCS model for these metals or alloys (see Exercise 1, part (iv)) give values of the order of 1 K to 10 K. There has of course been an active search over the years for increasing critical temperatures with the ultimate dream of discovering a room-temperature superconductor opening the door for dramatic changes in the technology of electricity. The highest T_c obtained until 1986 was about 20 K for the intermetallic compound Nb_3–Si when G. Bednorz and A. Müller discovered a superconducting transition at 30 K in the cuprate oxide La-Ba-Cu-O. Since then, superconductors with $T_c \sim 100$ K have been found in the family of cuprate oxides (e.g. Y-Ba-Cu-O, $T_c = 92$ K). These compounds show the following characteristic properties.

- The structure of cuprate superconductors is highly anisotropic. It can be viewed as stacks of closely spaced quasi two-dimensional CuO_2 planes separated by atomic layers acting as charge reservoirs. Superconductivity essentially takes place within these CuO_2 planes.
- The cuprate oxide has to be doped to become superconducting. This is achieved by adding oxygen ions that attracts electrons from the CuO_2 planes into the reservoirs to maintain charge balance. The remaining holes

[5] L.P. Gork'hov, Soviet Phys. JETP *7*, 505 (1958); ibid. *9*, 1364 (1959).

in these planes are conducting and are subject to fermion pairing below T_c. The existence of pairing is confirmed by observation of flux quantization with effective charge $2e$.

- A variety of experiments indicate that an energy gap $\Delta(\boldsymbol{k}) \neq 0$ exists everywhere on the Fermi surface. $\Delta(\boldsymbol{k})$ may be anisotropic.
- High-T_c superconductors have very short coherence lengths: there are only a few pairs in the coherence volume (in contrast with the conventional Cooper pairs, see Sect. 5.3.8). Consequently they are extremely type-II, e.g. for Y-Ba-Cu-O, $\delta = 4000$ Å, $\xi = 10$ Å(δ and ξ can depend on the axis directions). Lattices of flux vortices are also observed.

The general BCS concept of pairing with a resulting macroscopic quantum state of condensed pairs remains an essential ingredient of the presently proposed theories for high-T_c cuprates, at least for the optimally doped materials. The origin of pairing is controversial. An important issue to discriminate between various theories is the symmetry of the energy gap $\Delta(\boldsymbol{k})$, namely if it behaves as s-wave, a p-wave or a d-wave. For conventional metal superconductors, the electron–phonon interaction creates the pairs, and the resulting energy gap has an isotropic s-wave symmetry, as illustrated in the simple BCS calculation of Sect. 5.3.6. For cuprates, several arguments plead against a phonon mediated interaction. In particular, the low density of carriers in the CuO_2 planes give vanishingly small T_c in the BCS formula $k_B T_c \sim \hbar\omega_D \exp(-1/g(0)V_0)$ because of the correspondingly small value of the density of states $g(0)$. This could be compensated by a particularly large electron–phonon coupling constant V_0 (due to peculiar phonon modes) but then the compound could deform and lose its structural stability under such strong couplings. More conclusivly the energy gap $\Delta(\boldsymbol{k})$ changes its sign as a function of the momentum $\boldsymbol{k}$, so excluding the isotropic s-wave symmetry typical for a pairing due to lattice vibrations. Recent experimental evidence seems to favor a d-symmetry $\Delta(\boldsymbol{k}) \sim k_x^2 - k_y^2$. In this case, the change of sign $\Delta(\boldsymbol{k})$ along different momentum directions allows for the occurence of half integer flux quanta: these have been observed in the elegant experiment of Kirtley et al. The d-wave symmetry requires a non-phonon mechanism for superconductivity. Among the new proposals is the theory of spin fluctuations: the motion of a charge carrier creates a spin wave by modifying locally the spin orientations of the ions in the crystal, and this spin wave attracts a second carrier. This is analogous to the pairing mechanism leading to the superfluidity of ^{3}He which is mediated by magnetic excitations, not by phonons (Sect. 7.5.2). The question is not yet settled and some physicists think that a strongly anisotropic and extended s-wave is possible.

Ideally, one wants a theory that explains the already observed facts and discriminates between the different pairing mechanisms by experimentally verifiable predictions. Currently, more theoretical and experimental work is needed to reach agreement on the origin of high-T_c superconductivity. Since its discovery by K. Onnes in 1911, understanding superconductivity has been a continous challenge. The advent of quantum mechanics, in contrast to many other open problems, did not provide any clues of effective one-body theories. One had to wait for the BCS theory (1957) to elucidate the subtle many-body coherence effects that are at the onset of conventional superconductivity. The observation of high-T_c superconductivity in 1986 will most likely keep the subject alive for the next few decades.[6]

Exercises

1. BCS Theory at Non-Zero Temperature

The variational principle of statistical mechanics is applied to derive the thermodynamical properties of the superconducting phase. Let $F_\rho = E_\rho - TS_\rho - \mu N_\rho$ be the grand-canonical potential with E_ρ, S_ρ, N_ρ the energy, the entropy and the particle number evaluated for a variational class of states ρ. To define this class, one assumes that at low temperature the superconducting state is well approximated by a free Fermi gas of the elementary excitations discussed in Sect. 5.3.7, corresponding to the Hamiltonian

$$\mathrm{H}^{\mathrm{exc}} = \sum_{\boldsymbol{k}} E_{\boldsymbol{k}}(\alpha_{\boldsymbol{k}}^* \alpha_{\boldsymbol{k}} + \beta_{\boldsymbol{k}}^* \beta_{\boldsymbol{k}}) + E_{\mathrm{BCS}}^{\mathrm{tot}} \quad .$$

The class is parametrized by the same variational parameters $\{u_{\boldsymbol{k}}, v_{\boldsymbol{k}}\}$ as before, occurring in the Bogoliubov transformation (5.78), (5.79). Then for $\rho = \exp(-\beta \mathrm{H}^{\mathrm{exc}})/\mathrm{Tr}[\exp(-\beta \mathrm{H}^{\mathrm{exc}})]$, the BCS thermodynamical potential F_{BCS} is obtained by minimizing F_ρ on the $\{u_{\boldsymbol{k}}, v_{\boldsymbol{k}}\}$.

(i) The formula (5.100) generalizes to

$$\begin{aligned} E_\rho = {} & 2\sum_{\boldsymbol{k}} \varepsilon_{\boldsymbol{k}} \left[|u_{\boldsymbol{k}}|^2 f_{\boldsymbol{k}} + |v_{\boldsymbol{k}}|^2 (1 - f_{\boldsymbol{k}}) \right] \\ & + \sum_{\boldsymbol{k}_1 \neq \boldsymbol{k}_2} V_{\boldsymbol{k}_1,\boldsymbol{k}_2} u_{\boldsymbol{k}_1} v_{\boldsymbol{k}_1}^* u_{\boldsymbol{k}_2}^* v_{\boldsymbol{k}_2} (1 - 2f_{\boldsymbol{k}_1})(1 - 2f_{\boldsymbol{k}_2}) \quad , \end{aligned}$$

where $f_{\boldsymbol{k}}$ is the average occupation number of an excitation of wave number $\boldsymbol{k}$ at temperature T. Moreover, one has

$$S_\rho = -2k_{\mathrm{B}} \sum_{\boldsymbol{k}} [f_{\boldsymbol{k}} \ln f_{\boldsymbol{k}} + (1 - f_{\boldsymbol{k}}) \ln(1 - f_{\boldsymbol{k}})] \quad .$$

---

[6] J. Rowell (editor), *High-T_c superconductivity*, Physics Today, *44* (1991); P. Fulde, P. Horsch, *Theoretical Models for High-T_c Superconducting Materials*, Eur. Phys. News *24*, 73 (1993); J. P. Kirtley et al., Phys. Rev. Lett., *76*, 1336 (1996) and books quoted in the bibliography.

Hint: Generalize the calculations (5.87)–(5.94) by evaluating the contractions in the above thermal state ρ of independent excitations.

From now on we look for the extremum of F_ρ treating the $\{u_{\boldsymbol{k}}, v_{\boldsymbol{k}}\}$ and the $\{f_{\boldsymbol{k}}\}$ as independent variational parameters.

(ii) Show that the extremum of F_ρ with respect to the variation of the $\{u_{\boldsymbol{k}}, v_{\boldsymbol{k}}\}$ (keeping the $\{f_{\boldsymbol{k}}\}$ fixed) gives the temperature-dependent gap equation

$$\Delta_{\boldsymbol{k}} = -\sum_{\boldsymbol{k}'\neq\boldsymbol{k}} V_{\boldsymbol{k},\boldsymbol{k}'} \frac{\Delta_{\boldsymbol{k}'}}{2E_{\boldsymbol{k}'}}(1-2f_{\boldsymbol{k}'}) \quad .$$

Hint: Introduce the same changes of variables as in (5.103) and (5.107).

(iii) By varying now the $\{f_{\boldsymbol{k}}\}$, confirm that $f_{\boldsymbol{k}} = [\exp(\beta E_{\boldsymbol{k}})+1]^{-1}$ is the Fermi–Dirac thermal distribution of excitations.

(iv) For the simplified BCS interaction (5.109), the gap equation in the infinite volume limit becomes

$$\frac{V_0}{2}\int_{-\hbar\omega_{\mathrm{D}}}^{\hbar\omega_{\mathrm{D}}} \mathrm{d}\xi \, \frac{g(\xi)}{\sqrt{\xi^2+|\Delta|^2}} \tanh\left(\beta\frac{\sqrt{\xi^2+|\Delta|^2}}{2}\right) = 1$$

with $g(\xi)$ the density of states (5.113). Show that there exists a critical temperature T_{c} for which Δ vanishes and, within the approximation $g(\xi) \simeq g(0)$ and $k_{\mathrm{B}}T_{\mathrm{c}} \ll \hbar\omega_{\mathrm{D}}$, the order of magnitude of T_{c} is $k_{\mathrm{B}}T_{\mathrm{c}} \simeq \hbar\omega_{\mathrm{D}} \exp[-1/(g(0)V_0)]$.

Comment: The solution of the gap equation is now a temperature-dependent quantity $|\Delta(T)|$ that reduces to the value (5.114) at $T=0$ and vanishes at $T=T_{\mathrm{c}}$. It is natural to interpret $\Delta(T)$ as the (complex) order parameter of the superconducting phase transition.

6. Nucleon Pairing and the Structure of the Nucleus

6.1 Introduction

At an early stage of nuclear physics, a systematic difference was noted between nuclei depending on whether their proton and neutron numbers were even or odd. After the discovery of BCS theory, A. Bohr, B. Mottelson and D. Pines were first to recognize that the concepts useful for treating the electronic correlations in superconductors could provide a basis for analyzing the pair correlations in nuclei (1958).

The physical consequences of fermion pairing in nuclei are nowhere near as spectacular as in superconductors. The two main effects of nuclei pairing are concerned with the structure of the nuclear spectra as well as a tendency of nuclei with strong pairing effects to assume a spherical shape. Nevertheless, the nuclear case underlines the fundamental character of fermion pairing in the many-body problem. It is important to note that methods developed to describe one system can be transferred to another case, even if the results are not the same because the parameters of the second system are much different from those of the first. While superconducting electrons are very numerous, have low mass and obey the Coulomb force, paired nucleons are few in number, they are much more massive and are essentially under the influence of the strong interaction. As a consequence, a description of the nucleus through a pairing concept is very sensitive to the fact that the BCS theory describes states which are not eigenvectors of the particle-number operator. Such a description is particularly interesting for a qualitative understanding of the nuclei. In fact what the two systems have essentially in common is Fermi statistics and the possibility of realizing a gain in energy by pairing.

Here we tackle the structure of the nucleus from the viewpoint of many-body systems. It is important to note that there have been extensive developments of powerful field-theoretical tools in order to investigate the properties of the nuclei. Nevertheless, applying the BCS theory to nuclei, as will be presented here, provides a relatively simple way of understanding why spectra of odd and even–even nuclei differ so strongly from each other (Sect. 6.4.1) as well as providing additional interesting information about the shell theory of nuclei.

6.2 A Broad Outline of the Nuclear Structure

6.2.1 The Short Range of Nuclear Forces

Nucleons, although formed of three strongly bound quarks, can still be considered the structural units of the nucleus. It is reasonable to consider *the nucleus as an ensemble of nucleons and not of quarks.*

The proton and the neutron are sensitive to four types of forces known today: *the strong, electromagnetic, weak and gravitational forces.* Neither the gravitational force (which is much too weak) nor the weak force (which only appears in certain radioactive transitions) play any role in the architecture of the nucleus. The electromagnetic interaction expresses itself in the Coulombic repulsion between the protons. It is this force which makes massive nuclei comprising of too many nuclei unstable. Thus *the existence of stable nuclei originates from the strong force alone.* But what are the characteristics of the forces which bind the nucleons in the nucleus? Experimental facts point to short range forces.

First of all, nuclei are very small. Their diameters range from 1 to 10 F (1 Fermi = 1 F = 10^{-15} m). Moreover, the nuclear density is approximately constant. It makes sense to compare the nucleus with a liquid drop. The properties of liquid drops stem from the nature of intermolecular forces which behave in various ways. However, these forces share two important properties. They are short range and attractive, apart from a very short range repulsive force. We can therefore expect the nuclear forces to exhibit the same general features.

In fact, the range of the attractive part of the nuclear forces is much smaller than the radius of most nuclei. If their range would cover the whole region occupied by the nucleus, then the nuclear binding energy would roughly scale as $A(A-1)/2$, where A is its atomic number, namely its number of nucleons. As a consequence, the binding energy per nucleon B/A would increase with A and therefore so would the nuclear density. Now Fig. 6.1 shows that this is no longer true for $A \gtrsim 10$. For heavier nuclei, the force saturates. It is easy to estimate the number of bindings n_b each nucleon shares with its neighbours. Inspection of the curve representing B/A as a function of A allows one to estimate the approximate value of n_b because it strongly depends on the intersection of the two main parts of the curve, namely the steep line going through the light isotopes and the nearly horizontal segment going through the heavy elements. The marked bend between the two segments gives some information on the way the nuclear force saturates. It turns out to happen for n_b between 6 and 10.

6.2.2 The Liquid Drop Model

The short range of nuclear forces allows to understand the success of an early semiphenomenological model which emphasizes the similarities between the nucleus and a liquid drop.

For a nucleus (A, Z) containing A nucleons and Z protons, the missing mass $B(A, Z)$ is defined as

$$B(A,Z) = Zm_{\rm p}c^2 + (A-Z)m_{\rm n}c^2 - M(A,Z)c^2 \quad , \tag{6.1}$$

where $m_{\rm p}, m_{\rm n}$ and $M(A, Z)$ are the masses of the proton, the neutron and the nucleus, respectively. The quantity $B(A, Z)$, which is none other than the *binding energy of the nucleus* (A, Z), is reasonably well described by the *Bethe–Weizsäcker semi-empirical formula*

$$\begin{aligned} B(A,Z) &= B_v - B_s - B_{\rm c} - B_a \\ &= a_v A - a_s A^{2/3} - a_{\rm c}\frac{Z^2}{A^{1/3}} - a_a\frac{(N-Z)^2}{A} \quad , \end{aligned} \tag{6.2}$$

where $N = A - Z$ is the number of neutrons. Experiments can precisely determine the positive parameters $a_v, a_s, a_{\rm c}$, and a_a.

As stressed above, the binding energy per nucleon B/A is approximately constant for stable nuclei with $A > 10$. Its value is roughly equal to 8 MeV as can be seen from Fig. 6.1. The various contributions to B/A are outlined as a function of A in the same figure. Notice that the curve labelled "binding energy" is not given by expression (6.2). It corresponds to the true observed value of the binding energy per nucleon. Its marked oscillations in the region of light nuclei cannot be understood from the simple model (6.2).

- The *volume contribution* B_v is proportional to A, the total number of nuclei. It is a direct consequence of the short range of the nuclear force as discussed above.
- The *surface contribution* B_s is due to the fact that the nucleons which reach the surface of the nucleus are less bound than the others (a smaller number of bindings need to be broken to remove a surface nucleon than one which is located inside the volume). This leads to a correction proportional to $A^{2/3}$ since the radius r_n of a nucleus grows as $A^{1/3}$. One can remark that the difference $B_v - B_v$ is analogous to that which occurs for a drop of liquid for which the free energy is composed of a term proportional to the volume and a surface contribution which represents the presence of a surface tension.
- The *Coulomb contribution* $B_{\rm c}$ is due to Coulomb repulsion between the protons. As $B_{\rm c} \sim (eZ)^2/r_n$, we have $B_{\rm c} \sim Z^2/A^{1/3}$.
- The *asymmetry energy* B_a originates from the exclusion principle. The nucleons are fermions, so that the neutrons and the protons must separately fill the one-particle states, thus adopting the Fermi distribution at zero temperature. The nucleons, in this simplified model, are similar to free particles moving in a spherical box with the dimension of the nucleus. On the other hand, the excitation energies are too large to be thermally activated for all temperatures $T < 10^7$ K.

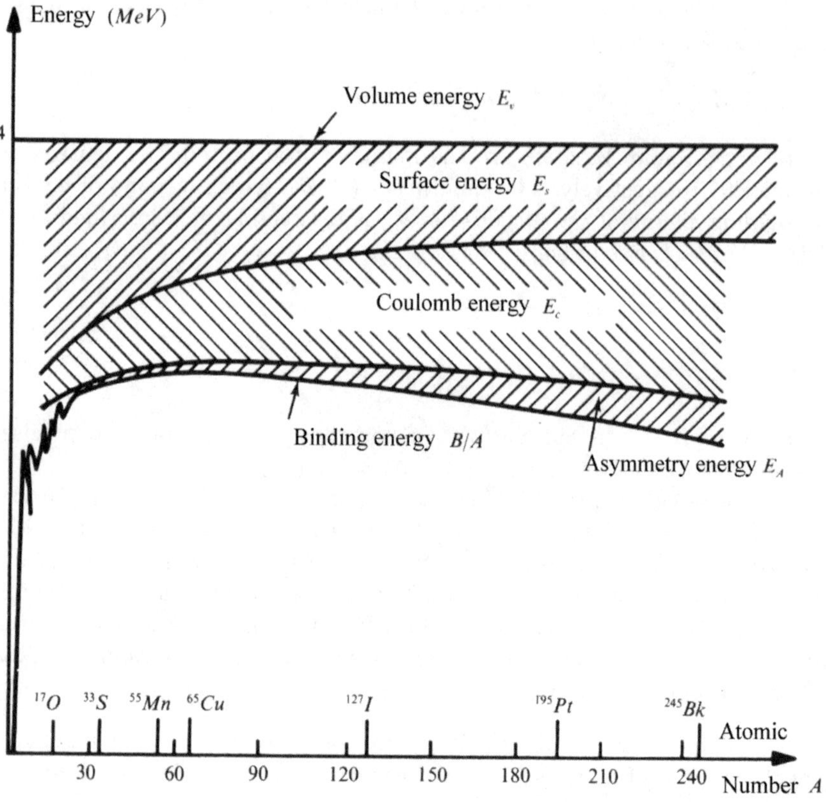

Fig. 6.1. Binding energy per nucleon presented as a function of atomic number A for stable nuclei

To evaluate the form of B_a as a function of A and Z, one only needs recall (2.97). The kinetic energy of the neutrons E_n^{kin} and of the protons E_p^{kin} contained in a nucleus of volume $V \simeq A$ is written, neglecting a numerical factor,

$$E_n^{\text{kin}} \simeq A\frac{\hbar^2}{m_{\text{n}}}\left(\frac{N}{A}\right)^{5/3} = \frac{\hbar^2}{m_{\text{n}}}A\left(1-\frac{Z}{A}\right)^{5/3} ,$$

$$E_p^{\text{kin}} \simeq \frac{\hbar^2}{m_{\text{p}}}A\left(\frac{Z}{A}\right)^{5/3} . \tag{6.3}$$

In this simple model, as in the rest of the chapter, the mass of the proton and the mass of the neutron are taken to be equal ($m_{\text{p}} \simeq m_{\text{n}}$). Therefore, one recognizes immediately that for fixed A, $E^{\text{kin}} = E_n^{\text{kin}} + E_p^{\text{kin}}$ is minimized for $Z = A - Z = N$. If one writes

$$Z = \frac{A}{2} - \frac{N-Z}{2} \quad , \tag{6.4}$$

the kinetic energy becomes

$$\begin{aligned} E^{\mathrm{kin}} &= E_n^{\mathrm{kin}} + E_p^{\mathrm{kin}} \\ &\simeq \frac{\hbar^2}{m_{\mathrm{n}}} A \left[\left(1 + \frac{N-Z}{A} \right)^{5/3} + \left(1 - \frac{N-Z}{A} \right)^{5/3} \right] \\ &\simeq \frac{\hbar^2}{m_{\mathrm{n}}} A + \frac{10}{9} \frac{\hbar^2}{m_{\mathrm{n}}} \frac{(N-Z)^2}{A} + O\left(\frac{(N-Z)^4}{A^3} \right) \quad . \end{aligned} \tag{6.5}$$

The first term is just the contribution of B_v. The term in $(N-Z)^2/A$ is called the energy asymmetry, as it gives rise to an asymmetry in the distribution of the protons and the neutrons.

6.2.3 More About Nuclear Forces

For any nucleus, the atomic number A is the sum of Z, the proton number, and of N, the neutron number. For fixed A, the two nuclei for which $Z = N \pm 1$ are called *mirror nuclei*. They make up pairs, with examples being ($^3_1$H, $^3_2$He), ($^5_2$He, $^5_3$Li) or ($^9_4$Be, $^9_5$B). Nuclear specroscopy shows that their spectra is very similar. This observation strongly points to the fact that the nuclear force is *charge symmetric* meaning that it does not change by replacing protons by neutrons and vice-versa. The Coulomb repulsion between protons only weakly breaks the charge symmetry, which shows that the nuclear force is much stronger that the Coulomb force.

Actually, the symmetry displayed by the nuclear force is even greater than exhibited by mirror nuclei. The nuclear force is *charge independent*, which means that the p–p, n–n, and p–n forces are equal. This can be proven by comparing the spectra of nuclei with fixed A for various values of Z and N. However, even if one still ignores the Coulomb interaction, charge independence is not a very simple concept, because p–n systems are not identical to p–p and n–n systems. Thus of the charge independence of the nuclear force, nucleons are assigned an *isospin* $T = 1/2$. The projection T_z of T along some arbitrary axis is $+1/2$ for neutrons and $-1/2$ for protons[1]. Thus, p–p and n–n systems must have $T = 1$. On the other hand, T is equal to either 0 or 1 in p–n systems. Due to charge independence, one expects that the proton–neutron interaction in a $T = 1$ state is the same as the p–p or n–n interactions. However, charge independence does not say anything about the p–n interaction in a $T = 0$ state. Nuclear spectra tend to show that the interaction is actually stronger in the $T = 0$ state. For instance, $^{27}_{12}$Mg is less bound than its isobar $^{27}_{13}$Al even though it has fewer protons and might be

[1] The choice of the signs is arbitrary. As a consequence, it is far from beeing universal.

expected to be more tightly bound. In fact, the Coulomb repulsion between 12 protons should be weaker than the repulsion between 13. Actually $^{27}_{12}$Mg is more weakly bound because in addition it has fewer p–n interactions in the $T = 0$ state. For these purposes, it is also instructive to study the case of deuterium.

The nucleus of the *deuterium* $^2_1$H is often called *deuteron*. Its properties confirm the tendency of the nucleon-nucleon interactions to be stronger in the $T = 0$ state. Particles made up of only two identical nucleons, therefore in a $T = 1$ state, do not exist, since they do not build a stable state. Because of the Pauli principle, the spins of this hypothetical particles would be antiparallel. The deuteron, on the other hand, is a stable particle with a relatively weak binding energy equal to 2.23 MeV. Actually, the bound state of the deuteron has a spin equal to 1 and a even parity ($J^{\Pi} = 1^+$). In this case, the Pauli principle demands that the spin of both nucleons are parallel. Moreover, the proton and the neutron are both in a $1s_{1/2}$ state so that, in the first approximation, there is no orbital contribution to the total angular momentum of the deuteron. The low value of the deuteron binding energy points to an important fact, specifically that the so-called strong nuclear force is only strong in comparison with the Coulomb repulsion.

The case of the deuteron sheds light on the role of quantum effects in the structure of the nucleus. These effects can be classified into different categories, the shell model of the nucleus structure and the existence of a pairing energy. We shall introduce two types of quantum effects which are important for our purposes, namely the shell model of the nucleus structure and the existence of a pairing energy. We shall discuss their respective meanings in later sections.

6.3 The Shell Model of the Nucleus

6.3.1 The One-Particle Potential Inside the Nucleus

In principle, the Hartree–Fock method introduced in Sect. 4.1 can be used in order to better understand the structure of the nucleus. For the purposes of studying a system of particles interacting through a two-body potential, it may be useful to replace the two-particle interaction by an effective potential $U(q)$ interpreted as a mean one-particle potential seen by a particle at q (4.18). The two-particle potential $W(q, q')$ given by (4.19) is then the exchange potential, which results from the Fermi statistics. In the case of the atom, the Coulomb field due to the nucleus is included from the start as a one-particle potential, and the Hartree–Fock method is only applied to the electron-electron interaction. The merit of this method is to assign to each electron an individual wave function characterized by a collection of quantum numbers (the principal quantum numbers, as well as the numbers related to the angular momentum, the magnetic moment and the spin). The wavefunctions are determined by the solution of a non-linear system (the Hartree–Fock equations) but, despite this fact, the advantages of an atomic description in terms of individual states should not be underestimated. Even if one does not know the energy spectrum precisely, the ability to find a collection of one-body observables which commute with the Hamiltonian allows one to establish the principal characteristics of the periodic table of the elements.

The situation is different when the Hartree–Fock method is used in order to determine the stationary states of the nucleons inside the nucleus. The nuclear forces are still largely unknown so that it is not possible to write down the explicit form of the nucleon-nucleon interaction. Instead, experimental data are used in order to make an initial guess of the best form for $U(q)$ and $W(q, q')$. In this section, we will begin by determining the empirical form of $U(q)$, the mean potential seen by nucleons inside the nucleus.

The atomic electrons feel the nuclear Coulomb potential, which determines the structure of the energy spectrum. An important parameter of this spectrum is the ionization energy of the atom. Even if we did not know the form of the electrostatic force experienced by the orbiting electrons, knowledge of the ionization energies of all elements would give us important clues on the form of the force that the nuclei exerts on them. In a similar way, the structure of the nucleus can be partially inferred from the proton and the neutron separation energies of the various nuclei.

The *neutron separation energy* for a given nucleus is defiened as the energy required to remove one neutron from this nucleus

$$S_n(A, Z) = B(A, Z) - B(A-1, Z) \quad . \tag{6.6}$$

The *proton separation energy* is defined in a similar way:

$$S_p(A, Z) = B(A, Z) - B(A-1, Z-1) \quad . \tag{6.7}$$

Comparing the experimental value of $S_n(A, Z)$, denoted by $(S_n)_{exp}$, with the value $(S_n)_{cal}$ obtained from the help of the Bethe–Weizsäcker formula (6.2), one notices some maxima of $(S_n)_{exp}-(S_n)_{calc}$ for a series of values of the neutron number N (Fig. 6.2). These peaks correspond to abnormally stable nuclear states. Moreover, the same peaks occur at the same places if one plots the proton separation energy as a function of the proton number Z. Numbers of the series $2, 8, 20, 28, 50, 82, 126, \ldots$ corresponding to these stability peaks are called *magic numbers*. Comparing with the atomic case, we may conclude that *each time the number of protons or neutrons reaches a number of the magic series, a shell is filled: an energy level with a multiple degeneracy is entirely occupied.*

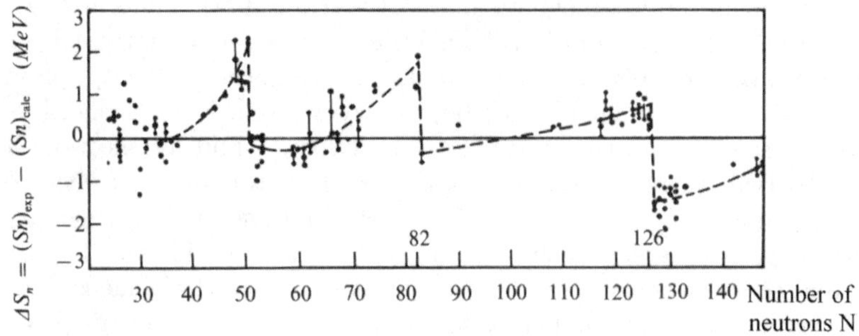

Fig. 6.2. Plot of the difference between two values of the neutron separation energy as a function of the neutron number N. The first value comes from experimental measurements while the second one is given by the semi-empirical formula

6.3.2 Interpretation of the Magic Numbers

Knowledge of a finite set of numbers corresponding to filled shells of one-dimensional states does not uniquely determine the potential responsible for this set. There is necessarily a whole family of potentials yielding the same numbers. Those who were looking for a one-particle potential inside the nucleus had to find further physical arguments in order to choose the best candidate U among the family of acceptable potentials.

To give an expression for U implies that we must return to a single-body problem where the nucleon obeys the Schrödinger equation

$$\left(\frac{1}{2\mu}|\boldsymbol{p}|^2 + U\right)|\varphi_\alpha\rangle = \varepsilon_\alpha\,|\varphi_\alpha\rangle \quad , \quad \alpha = 1, 2, \ldots, A \quad , \tag{6.8}$$

where $\mu = (A-1)m_n/Am_n$ is the *reduced mass* of the nucleon. U itself takes the form

$$U = U(\boldsymbol{x}, \boldsymbol{l}, \boldsymbol{s}) = \left(\frac{1}{2}\right)\mu\omega^2|\boldsymbol{x}^2| - f\boldsymbol{l}\cdot\boldsymbol{s} \quad , \tag{6.9}$$

The first term is a highly degenerate harmonic potential; $\boldsymbol{x}$ is the radial vector originating at the center-of-mass of the nucleus. It expresses the fact that each nucleon remains confined to the interior of the nucleus, provided that it is stable, due to the strong attractive interaction between the nucleons. But from the perspective of finding a useful potential, the *spin-orbit* term U^{SO} can be considered the greatest achievement. It was suggested by Maria Goeppert–Mayer and independently by Haxel, Jensen and Suess. It takes the form

$$U^{\mathrm{SO}} = -f\boldsymbol{l}\cdot\boldsymbol{s} \quad . \tag{6.10}$$

In order to simplify the discussion, we will take f as a constant, even though it is customary to write f as a increasing function of the distance to the center of the nucleus.

The quantities $\boldsymbol{l}$ and $\boldsymbol{s}$ are the operators of the orbital angular momentum and the spin, respectively, so that f is the energy of the spin–orbit interaction. The operators $|\boldsymbol{l}|^2$ and $|\boldsymbol{s}|^2$ have eigenvalues $\hbar^2 l(l+1)$ and $\hbar^2 s(s+1) = \hbar^2 3/4$. The number l takes on non-negative integer values, whereas $s = 1/2$. The total angular momentum operator $\boldsymbol{j}$ is given by

$$\boldsymbol{j} = \boldsymbol{l} + \boldsymbol{s} \quad . \tag{6.11}$$

The operator $|\boldsymbol{j}|^2$ has eigenvalues $\hbar^2 j(j+1)$ where

$$|l - s| \leq j \leq |l + s| \quad , \tag{6.12}$$

which gives

$$\begin{aligned} j &= 1/2 \quad , \quad l = 0 \quad , \\ j &= l \pm 1/2 \quad , \quad l > 0 \quad . \end{aligned} \tag{6.13}$$

As will be explained in the next section, the potential (6.9) satisfies the required condition (provided ω and f take on suitable values), specifically, that it should lead to degenerate energy levels which are filled every time that a magic number is reached, for both nucleon species. It is worth noticing that the form of U and especially the choice of a spin-orbit interaction can be justified on phycical grounds.[2]

6.3.3 Distribution of the Energy Levels in the Shell Model

A three-dimensional harmonic oscillator with Hamiltonian

$$H^{\text{harm}} = \frac{|\boldsymbol{p}|^2}{2\mu} + \frac{\mu}{2}\omega^2 |\boldsymbol{x}|^2 \tag{6.14}$$

has eigenvalues

$$E_K = (K + 3/2)\hbar\omega \quad , \tag{6.15}$$

where

$$K = 2(n-1) + l = 0, 1, 2, \ldots \quad , \tag{6.16}$$

where K is the principal quantum number and n is the radial quantum number. For fixed K, $n \geq 1$ and $l \geq 0$, n and l can have all values compatible with relation (6.16). This determines the degeneracy of the harmonic-oscillator levels if one takes into account the spin degeneracy.

[2] Richard Casten discusses this point in the reference given at the end of this book.

Among the different angular momentum operators, the *spin–orbit potential* U^{SO}, given by

$$\begin{aligned} U^{\mathrm{SO}} &= -\, f \boldsymbol{l} \cdot \boldsymbol{s} \\ &= -\frac{f}{2}\left[|\boldsymbol{l}+\boldsymbol{s}|^2 - |\boldsymbol{l}|^2 - |\boldsymbol{s}|^2\right] \\ &= -\frac{f}{2}\left[|\boldsymbol{j}|^2 - |\boldsymbol{l}|^2 - |\boldsymbol{s}|^2\right] \quad , \end{aligned} \tag{6.17}$$

only commutes with $|\boldsymbol{l}|^2$, $|\boldsymbol{s}|^2$ and the components j_x, j_y and j_z of the total angular momentum, which can easily be demonstrated from the commutation relations of angular momentum.

As these components do not commute with themselves, one chooses a basis formed from the eigenfunctions $|\boldsymbol{j}|^2$, j_z, $|\boldsymbol{l}|^2$ and $|\boldsymbol{s}|^2$. In such a basis, U^{SO} is necessarily diagonal since $f > 0$ is a constant. The potential U^{SO} only removes one part of the degeneracy of the levels of H^{harm}

$$\begin{aligned} \langle j, j_z, l, s|U^{\mathrm{SO}}|j, j_z, l, s\rangle &= -\frac{1}{2} f \, \langle j, j_z, l, s| \left(|\boldsymbol{j}|^2 - |\boldsymbol{l}|^2 - |\boldsymbol{s}|^2\right) |j, j_z, l, s\rangle \\ &= -\frac{\hbar^2}{2} f \left[j(j+1) - l(l+1) - 3/4\right] \quad . \end{aligned} \tag{6.18}$$

Because of (6.13) and (6.15), the value of the total energy of the state $|n, j, j_z, l, s\rangle$, for the case where $l \geq 1$, can be written as

$$\begin{aligned} E_{n,l,j} &= \langle n, j, j_z, l, s|(H^{\mathrm{harm}} + U^{\mathrm{SO}})|n, j, j_z, l, s\rangle \\ &= (K + 3/2)\hbar\omega \begin{cases} -\hbar^2 (l/2) f & \text{if} \quad j = l + 1/2 \\ +\hbar^2((l+1)/2) f & \text{if} \quad j = l - 1/2 \end{cases} \quad . \end{aligned} \tag{6.19}$$

If $l = 0$, one has

$$E_{n,0,j} = (K + 3/2)\hbar\omega \quad . \tag{6.20}$$

Thus, for fixed n and $l \geq 0$, K is also fixed and the spin–orbit coupling gives rise to a separation in the levels

$$E_{n,l,j=l+1/2} - E_{n,l,j=l-1/2} = -\frac{\hbar^2}{2} f(2l+1) \quad , \quad l > 0 \quad . \tag{6.21}$$

Each *shell* of the harmonic oscillator characterized by the number K is divided into subshells of energy $E_{n,l,j}$ indexed by the values of n, l and j. One *subshell* is designated by the triplet nlj which is symbolized by the letters $s, p, d, f, g, h, \ldots$ and by $j = l \pm 1/2$ $(l > 0)$. Each state of the subshell nlj is also characterized by the eigenvalue $\hbar m$ of the operator j_z which takes on $2j+1$ values $(-j \leq m \leq j)$. The degeneracy of the subshell is thus $2j+1$. Taking into account the fact that $f > 0$, (6.21) shows that the energy of a subshell decreases as j and l increase.

Figure 6.3 shows the distribution of the different levels of the Hamiltonian $H^{\mathrm{harm}} + U^{\mathrm{SO}}$ as well as the degrees of degeneracy of each of the subshells. Some of these levels regroup into new shells which are different than those of the harmonic oscillator, except for the 3 lowest: one notes in particular that the subshell $1f_{7/2}$ ($n = 1, l = 3, j = 7/2$) makes up an entire shell of its own. In what follows, the notion of a level will refer to the new classification, stemming from the study of the complete Hamiltonian, and in accordance with the description of Fig. 6.3. The magic number of rank k corresponds to a level completely filled up to the k lowest energy shells, all other levels being empty. A nucleus which only has shells completely filled by one (or two) types of nucleons is particularly stable: this is known as a *magic (or doubly magic) nucleus.*

The effect of the other nucleons on a proton or a neutron arises through the Pauli exclusion principle. Once the energy levels are determined, the method of filling starts with the lower levels and proceeds such that each state is occupied by a number of neutrons or protons equal to the degeneracy of the level.

It is important to note that protons and neutrons are different particles. It is advisable to treat them separately, as far as statistics are concerned. In what follows, it is understood that all which is said for one type of nucleon (filling method, pairing) must be repeated in an identical manner for the other type of nucleon: the differences between the two populations (different mass, presence or absence of Coulomb interaction) will never be mentioned.

6.4 Pairing of the Nucleons

6.4.1 Nature of the Residual Interaction

As discussed above, the form of the nuclear potential has been inferred from general arguments so that the two-body nucleon-nucleon interaction can be considered as a *residual interaction* to be deduced from physical data and not from first principles, as in the case of the atom. We are especially interested in the part of the residual interaction which leads to the pairing of nuclei, so that we will focus on two types of experimental data: the difference between *even–even* and *odd–odd* nuclei and some features of separation energies which do not appear in Fig. 6.2

According to (6.1) and (6.2), on can write the nuclear mass $M(A, Z)$ in the form

$$M(A, Z) = aZ^2 + bZ + c \quad , \tag{6.22}$$

where a, b and c depend only on A. For fixed A, (6.6) represents the equation of a parabola giving the mass $M(A, Z)$ of the *isobars*, or nuclei having the same number of nucleons.

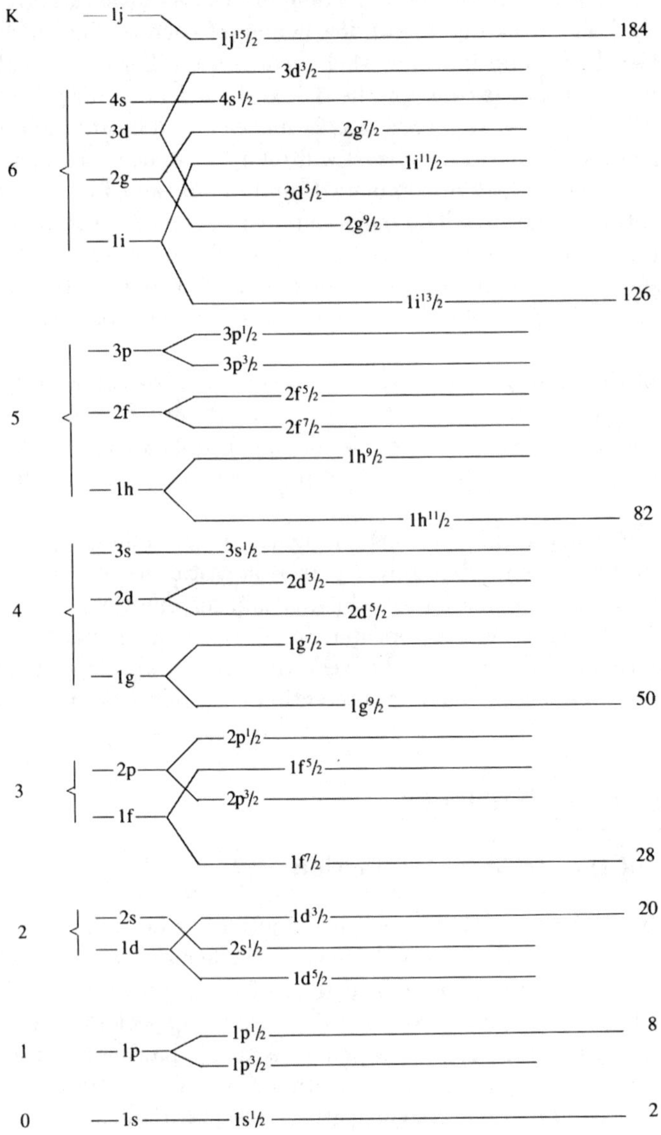

Fig. 6.3. Distribution of energy levels of the Hamiltonian $H^{\mathrm{harm}} + U^{\mathrm{SO}}$ including degeneracy of the subshells

Figure 6.4 shows that this is verified for *odd isobars (o)* – odd A – but that for the case of even isobars (e) – even A – one observes two distinct mass parabolae. The lowest in binding energy are the even–even $(e$–$e)$ isobars, for which N and Z are both even. The highest binding energy corresponds to odd–odd (o–o) nuclei.

Of course, this effect is not found in the Bethe–Weizsäcker formula (6.2). Numerically, the deviation between the two parabolae is of the order of only 0.2 % of the binding energy, but it is very revealing.

First of all, *stable e–e nuclei are much more numerous than stable o or o–o nuclei.* There exist only four stable o–o, all light: ${}^{1}_{2}$H, ${}^{6}_{3}$Li, ${}^{10}_{5}$B and ${}^{14}_{7}$N. In addition, elements having odd Z have one or, exceptionally, two stable isotopes, while elements with even Z have several.

To take this into account, one adds to $B(A, Z)$ a corrective term $\Delta(A, Z)$ which is positive, zero or negative for nuclei e–e, o and o–o respectively. Approximately, $\Delta(A, Z) \simeq 12\ \mathrm{MeV}/A^{1/2}$ and this is known as the *pairing energy* for reasons which will become apparent below.

6.4.2 Pairing Interaction: Further Experimental Facts

The existence of a pairing energy $\Delta(A, Z)$ can be best understood if a significant part of the residual interaction involves a special attraction between pairs of protons or neutrons, the *pairing interaction.* At this stage, we simply list a series of experimental facts that strengthen the case for such a force.

- The ground state of all even-even nuclei has $J^{+} = 0^{+}$. As will be discussed below, this fact gives an interesting hint about the way protons or neutrons pair.
- The spectra of odd and even-even nuclei are different.

Spectrum of an Odd Nucleus

The extreme case of this type is the addition or removal of one nucleon to a nucleus which is doubly magic. To focus on this idea, we will qualitatively discuss the former case, which is analogous to that of the alkaline metals of atomic physics.

The total angular momentum of the nucleus is a conserved quantity: the energy levels are indexed by the quantum numbers $J = 1/2, 3/2, 5/2, \ldots$ and $\Pi = \pm 1$ which correspond to the angular momentum and the parity of the nucleus as a whole. The spectrum of an odd nucleus has the following characteristics:

- *The ground state has total angular momentum J and parity Π equal to that of the additional nucleon occupying the state immediately above the doubly magic core.* The angular momentum of the core is always zero and its parity is always even, so that $\Pi = (-1)^{l}$, where l is the orbital angular momentum of the additional nucleon.
- *The first excited states of the nucleus have a one-to-one correspondence with the different individual states which can be occupied by the additional nucleon* (Fig. 6.5). This correspondence remains valid as long as the

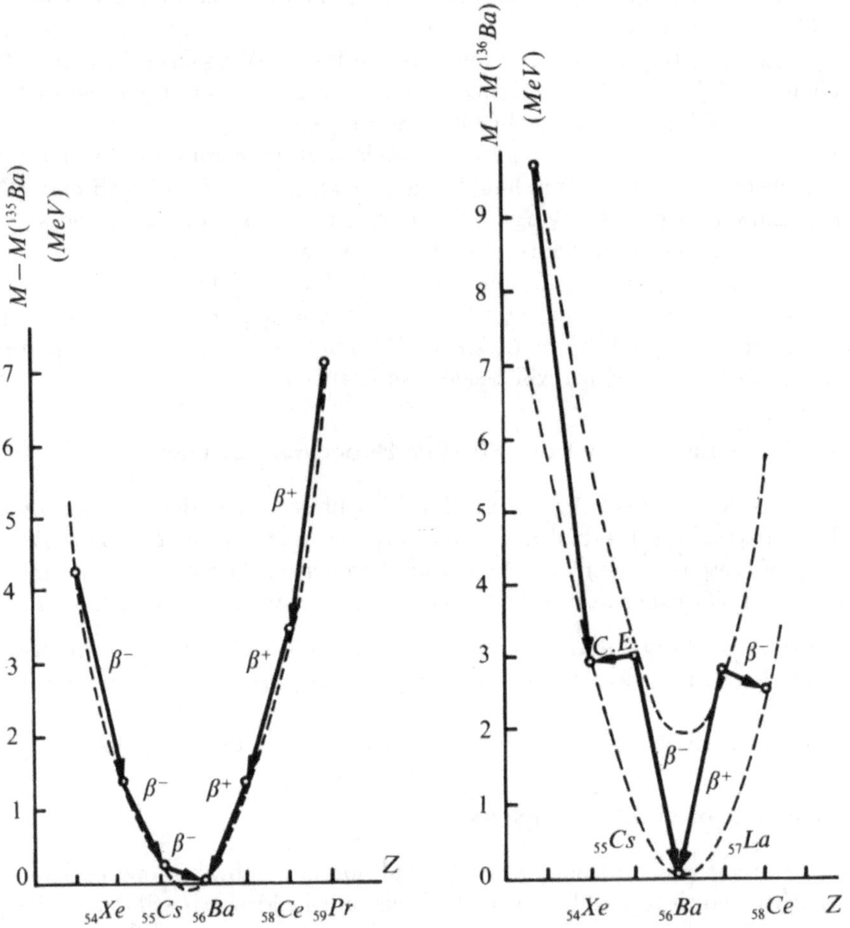

Fig. 6.4. Mass M versus proton number Z for odd (*left*) and even (*right*) isobars. In the left-hand plot, arrows marked $\beta^{\pm}$ indicate a decay characterized by the emission of an electron (β^-) or a positron (β^+). In the right-hand plot, the arrow marked $C.E.$ indicates the capture of an electron of the atomic shell by the nucleus

excitation energy does not reach the separation energy ΔE between the last shell occupied by the core and the shell immediately above.

An example of such a nucleus is given by $^{209}_{82}$Pb for which $\Delta E \simeq 3$ MeV. In the case of a doubly magic nucleus where one nucleon is missing, the behaviour is completely symmetric: the hole replaces the additional nucleon in the description given above.

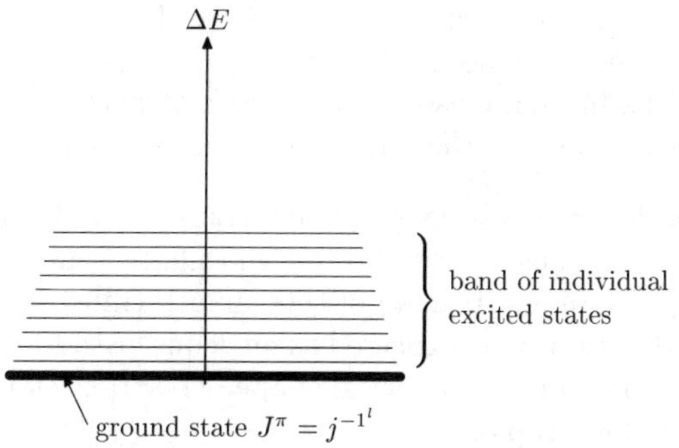

Fig. 6.5. Energy spectrum of an odd nucleus in the case for which addition or removal of a nucleon leads to a doubly magic core. The excited states are in a one-to-one correspondence with the individual states of the extra nucleon outside the core. The excitation energy never reaches that of the separation energy ΔE between the core and the next shell

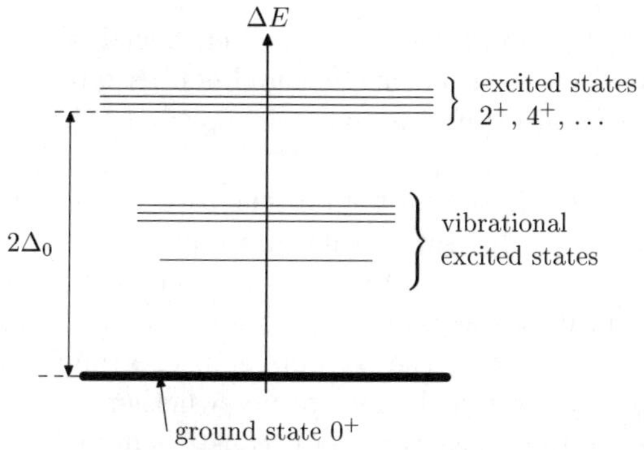

Fig. 6.6. Typical features of the energy spectrum of an even–even nucleus. The lowest excited states bunch together and have a vibrational character

Spectrum of an Even–Even Nucleus

The spectrum of a typical *e–e* nucleus has the following characteristics (Fig. 6.6):

- *The ground state has total angular momentum 0 and even parity* ($J^\Pi = 0^+$).
- *The first excited states are collective states of a vibrational nature.*
- *A band of excited states bunched together close to one another appears at a distance* $2\Delta_0$ *from the ground state.*

If one disregards the vibrational states, which do not have an analogy in the electron gas, one sees how the great stability of the ground state and the existence of a gap between the ground state and the first quasi-continuous excited states, encourages an analogy with BCS theory. The small number of nucleons allows for a distinction between even and odd nuclei. As a consequence, *one expects that pairing has the largest effect for e–e nuclei.*

- Doubly magic nuclei are spherical because they involve completely filled shells. But this is also the case for nuclei whose proton and neutron numbers lie in the vicinity of magic numbers. Instead of the smooth transition toward deformation that would normally be expected as nucleons are added or removed, nuclei retain their more or less spherical shape. The transition toward deformed nuclei then follows rapidly.
- Empirical moments of inertia extracted from the spectra of deformed nuclei are systematically lower than they should be according the shell model.

We shall come to these last two points later.

6.4.3 The Interaction Responsible for the Pairing

In the following discussion, we restrict ourselves to even-even nuclei whose proton and neutron numbers are not far from magic numbers. as stressed in the preceding section, such nuclei are spherical and are expected to show important effects due to pairing.

In the independent particle approximation of the nucleus each individual nucleon state could be described by a large set of quantum numbers. These include the principal quantum number n, the orbital angular momentum l, the total angular momentom j and its projection $j_z = m$ along some axis z. In the absence of any radioactive processes, the isotopic spin of a nucleon remains invariant so that, in general, we shall not specify whether it is a proton or a neutron. The reader is reminded that both constituents of the pairs we will be discussing are similar with respect to the isotopic spin.

The experimental facts listed in the preceding sections hint towards the existence of an attraction between nucleons of the same species. This attraction, the *pairing interaction*, is only one component of the overall residual interaction which includes any attraction or repulsion not taken into account by the potential $U(\boldsymbol{x}, \boldsymbol{l}, \boldsymbol{s})$ defined by (6.9). We will now introduce some properties of the pairing interaction which can be deduced from the fact that the ground state of e–e nuclei is always $J^+ = 0^+$.

The spectra of nuclei display features characteristic of collective motions of nucleons. This means that the residual interaction involves fairly long range components. However, as mentioned at the beginning of this chapter, the most important component of the forces between nucleons is at very short range. It is this component of the residual interaction which is responsible for the pairing of nucleons with identical isotopic spin. As a consequence, one should look for those properties of the two members of a pair which most allow to take the advantage of the short range part of the residual interaction.

One can answer this question in a qualitative way: the spatial component of their wave functions should overlap as much as possible. Now, if we disregard the lightest nuclei, the paired nucleons are likely to occupy levels with a relatively high energy. This is a consequence of the fact that the low energy levels are already filled. Comparing with the case of the electron gas, one could say that the nucleons subjected to pairing lie very close to the Fermi level. Such nucleons are called *valence nucleons* in analogy with the atomic case.

Let us now consider a particular valence nucleon. We first disregard the orientation of its spin. Its orbital angular momentum l is well defined because, as we saw above, the nucleus is spherically symmetric. The orientation of its angular momentum is given, say, by l_z, so that the best candidate for pairing is a nucleon with the same quantum numbers, except that l_z must now be replaced by $-l_z$. Using classical language, we could say that the pairing partner of a nucleon is the nucleon which orbits at the same distance from the nucleus center, with the same speed, but in the opposite direction. On average, the members of this pair will remain closer that a pair of valence electrons chosen at random. As a consequence, the pair $(l_z, -l_z)$ consists of nucleons whose spatial components of the wave function maximize their overlap. Including the spin into this description does not change these qualitative results markedly. Because of the spin-orbit coupling, all states of a subshell are characterized by the same value of $j = l + 1/2$ or $l - 1/2$. Replacing l_z by $j_z = l_z + s_z$ does not change the conclusion of the above discussion. In other words, the overlap is maximized between two identical nucleons with the same values of n, l, j but with opposite values of j_z.

6.4.4 Applying the BCS Theory to the Nucleus

We now consider the simple case of a hypothetical nucleus where the number, n, of one type of nucleons – say the protons – is strictly between 20 and 28. According to Fig. 6.3, this means that subshell $1f_{7/2}$ is only partially filled with protons. In this case, one passes from $_{21}$Sc to $_{28}$Ni. This subshell contains $2j + 1 = 8$ states. In the shell model, it is characterized by the fact that it is more or less isolated from the other shells, so that $1f_{7/2}$ almost makes up a single shell of its own. In the following, we shall restrict ourselves to the states belonging to this subshell, but it is important to note that this choice has been made in order to simplify the notation rather than for any physical meaning. When pairing occurs, states from distinct subshells near the Fermi level superpose, a fact which does not imply restricting our discussion to states originating from a unique subshell. In fact, we are only interested in showing how BCS theory and methods can be used in nuclei, and so we will not discuss the complex details involved in a realistic investigation of the nuclear structure.

In the model considered here, where each nucleon is under the influence of a mean potential $U(\boldsymbol{x}, \boldsymbol{l}, \boldsymbol{s})$, all the individual states $|n = 1, j = 7/2, m, l = 3,$

$s = 1/2\rangle = |1f_{7/2}, m\rangle$ have the same energy. As complete shells do not contribute to the value of the angular momentum, a nucleus containing 20 protons in complete low-energy shells and two supplementary protons p_1 and p_2 occupying the states $|1f_{7/2}, m_1\rangle$ and $|1f_{7/2}, m_2\rangle$ has an angular momentum J determined by p_1 and p_2. As the two protons have the same total angular momentum $j = 7/2$, the value of J satisfies the inequalities

$$|j - j| = 0 \leq J < j + j = 7 \quad . \tag{6.23}$$

The value $J = 7$ is not allowed, as it would impose $m_1 = m_2$, which is forbidden by the Pauli exclusion principle. Analogous considerations based on the rules of composition of the angular momentum show that J can only have the even values: 0, 2, 4 or 6.

The fact that the ground state 0^+ has a net energy lower than the quasi-degenerate excited states $2^+, 4^+, 6^+$ shows that there is a residual interaction between the nucleons. If one writes the Hamiltonian $\mathtt{H}$ which incorporates this residual interaction between the members of the same shell, one can limit $\mathtt{H}$ to the subspace $\{|\alpha, m\rangle\}$ (fixed α, $-j \leq m \leq j$) and it becomes

$$\begin{aligned} \mathtt{H}_{(\alpha)} &= \mathtt{H}^{\text{kin}}_{(\alpha)} + \mathtt{V}_{(\alpha)} \\ &= \sum_{m=-j}^{j} \varepsilon_{\alpha m} a^*_{\alpha m} a_{\alpha m} \\ &\quad + \frac{1}{2} \sum_{\substack{m_1, m_2 \\ m'_1, m'_2}} V_{(\alpha) m_1, m_2; m'_1, m'_2} a^*_{\alpha m_1} a^*_{\alpha m_2} a_{\alpha m'_2} a_{\alpha m'_1} \quad , \end{aligned} \tag{6.24}$$

where

$$\varepsilon_{\alpha m} = \left\langle \alpha, m \left| \frac{|\boldsymbol{p}|^2}{2\mu} \right| \alpha, m \right\rangle \quad .$$

Note again that the situation here is restricted to the case of an incomplete shell, composed of one single subshell for which the individual states only differ by the eigenvalue m of $j_z/\hbar$. This simple case allows one to understand the principle of the method. In (6.24), the coefficients $V_{(\alpha) m_1, m_2; m'_1, m'_2}$ satisfy symmetry relations analogous to those presented in Chap. 5 (they have the same origin as (5.45) and (5.46))

$$\begin{aligned} V_{(\alpha) m_1, m_2; m'_1, m'_2} &= V_{(\alpha) m_2, m_1; m'_2, m'_1} \quad , \\ V_{(\alpha) m_1, m_2; m'_1, m'_2} &= V_{(\alpha) -m_1, -m_2; -m'_1, -m'_2} \quad . \end{aligned} \tag{6.25}$$

In what follows we will no longer write the index α, as there can be no confusion.

Now we have to calculate the ground state of the nucleus. For this purpose, we choose a variational state $|\Psi\rangle$ of the form

$$|\Psi\rangle = \prod_{m>0} \left[u + v(-1)^{j-m} a_m^* a_{-m}^*\right] |\Psi_0\rangle \quad , \tag{6.26}$$

where u and v are two real parameters. The state $|\Psi_0\rangle$ represents an ensemble of nucleons occupying a complete shell and which, because of this, are not submissive to the variational process. As j and m are half-integers, $j+m$ and $j-m$ are necessarily integers and of opposite parity

$$(-1)^{j+m} = -(-1)^{j-m} \quad . \tag{6.27}$$

Concerning the choice of the variational state, one notes several differences between the case of superconductivity and the case of nucleus. The phase factor chosen here, namely $(-1)^{j-m}$, corresponds to a ground state (6.26) whose total angular is zero.

At this point, the BCS method applies without making any changes to the original method used for the electronic case and discussed in Chap. 5. One looks for values of u and v minimizing the energy of the ground state. This extremum calculations must satisfy the condition

$$u^2 + v^2 = 1 \quad , \tag{6.28}$$

which ensures the correct normalization of the state (6.26). The whole calculation involves simplifications which are analogous to those made in Chap. 5. Note that one has to be careful when using of the BCS state (6.26) because of the phase factor (6.27). The final result confirms the existence of a gap 2Δ between the ground state and the first excited states whose total angular momentum and parity are equal in the general case to 2^+, 4^+, $6^+, \ldots$. They correspond to states with a broken pair. In the case where $J = 2$, one nucleon characterized by the quantum number m is present with probability 1, while the nucleon $-m$ is absent with certainty. There are several excited states corresponding to this picture leading to a degeneracy. Notice the similarity of such nuclear states to the first excited states of the BCS theory of superconductivity. The case where J takes on a larger value but remains even is similar. The value of the gap Δ is proportional to the matrix element (6.25), chosen to be proportional to a unique constant G in this simple model (Fig. 6.7). Calculations are left as exercises.

Notice that the state (6.26) has a major inconvenience: the number of nucleons is not fixed. This characteristic is more questionable than in superconductivity since the number of nucleons under consideration is relatively small (in the case of the $1f_{7/2}$ shell, it is even less than 9). One can nevertheless work with the state (6.26) rather than with a state where the number of nucleons would be fixed, at least while one is qualitatively discussing the problem of pairing.

In the next section, we look for a general physical property which stems from the overall pairing scheme of the nucleus in the same way that superconductivity arises at low temperature as a consequence of an indirect attraction between electrons.

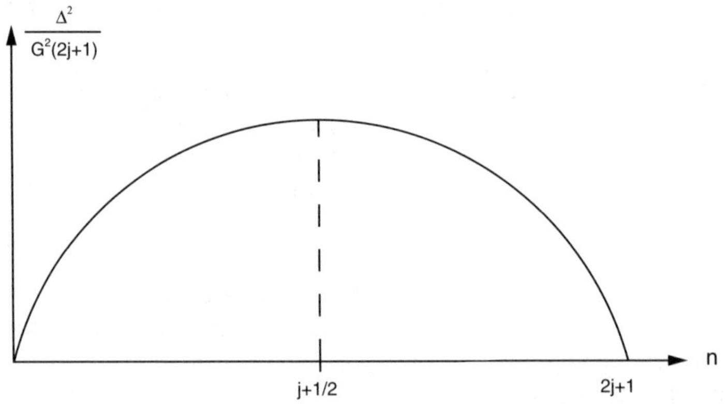

Fig. 6.7. Δ^2, the square of the energy gap in the spectrum of a even-even nucleus due to the pairing of a species of nucleons. This gap is expressed as a function of the pairing constant G and of n, the number of nucleons partially filling the j-subshell involved in the pairing scheme

6.5 Which Nuclear Properties are Affected by Nucleon Pairing?

6.5.1 Superconductivity, Superfluidity and Nuclei

Superconductivity is an important phenomenon which was first observed in metals at very low temperature (Chap. 5). It took physicists more than fifty years before they understood the relation between superconductivity and electron pairing in the vicinity of the Fermi level. In the case of the nucleus, the special properties of even–even nuclei near magic numbers and the existence of a gap in their spectrum were associated with the likely existence of pairing between similar nucleons. At this point, a question nesessarily arose: do these nuclei exhibit some dynamical property analogous to superconductivity? The answer is yes. The rotational properties of nuclei can be related to the amount of pairing involved.

Actually, the similarity between these two systems can be even better understood if one extends the discussion to the superfluidity of helium 4 which will be discussed in Chap. 7[3]. Both superconductivity and superfluidity are macroscopic quantum phenomena, i.e. they are quantum phenomena leading to macroscopic effects, appearing at low temperatures when there are too few thermal excitations to be disrupted by. Superconductivity is due to the pairing of electrons in solids while superfluidity arises in a liquid made of atoms satisfying Bose–Einstein statistics. Although the two systems are different, they share a common property: they lack a dissipation mechanism.

[3] See the discussion A. Bohr and B. Mottelson elaborate on this point in the work quoted in the bibliography at the end of this book.

On the order hand, nuclei belong exclusively to the microscopic quantum world. As a result, they seem to have nothing in common with either superconductivity or superfluidity. This is not true, however. A heavy nucleus consists of more than two hundred nucleons, so that statistical effects begin to become relevant. This is the reason why nucleon pairing can be described through a BCS model state whose particle number is not fixed. This is also the reason why the BCS method is so fruitful for even - even nuclei (the results of the method can be extendet to odd nuclei as well as to odd - odd nuclei, even though such an extension is less fruitful). As mentioned above, one expects nuclei for which pairing is important to share some properties in common.

As superconductivity involves Fermi pairing, one would expect the dynamical properties of nuclei associated with pairing to be closer related to superconductivity rather than superfluidity, because ^{4}He atoms are bosons. This is not true, simply because of the fact that the predominant role of the electrodynamical properties in superconductivity has no counterpart within nuclei. Because of the relative weakness of the electrodynamical interaction compared to the strong interaction, the former hardly affects the rotation of deformed nuclei. As a consequence, if one asks for some special property of nuclei associated to pairing, one should better look for neutral quantum fluids and especially to (5.26) relating velocity to the global phase α:

$$v(x,t) = (\hbar/m_0)\nabla a \quad .$$

In this simple model, the parameter m_0 is identified with the mass of an atom of the hypothetical quantum fluid under consideration. Equation (5.26) is satisfied whenever the whole system is characteriezed by the existence of a *macroscopic quantum state*, i. e. a one-particle state whose occupation number is a non-negligible fraction of the total paricle number of the system. The phenomenology of both superconductivity and superfluidity point towards the existence of such a state, even though Chapters 5 and 7 show that our knowledge of the microscopic properties of both phenomena cannot be related in a simple way to a macroscopic quantum state. In the case of nuclei, the existence of a phase α can be understood provided we first return to the variational state (6.26). The parameters u and v which enter into this state were taken to be real, which ensures that the angular momentum of the ground state vanishes. Actually, the value of the angular momentum depends on the relative phase of the pair creator $a^*_m a^*_{-m}$ in (6.26) *of* $|\Psi\rangle$. In (6.26), we have assumed straightaway that u and v are real by simplicity, but we could take a complex v and a state of the form

$$|\Psi\rangle = \prod_{m>0} \left[u + v\mathrm{e}^{\mathrm{i}\alpha}(-1)^{j-m} a^*_m a^*_{-m}\right] |0\rangle \quad . \tag{6.29}$$

The results of the variational calculation would be the same: they do not determine the phase factor $\mathrm{e}^{\mathrm{i}\alpha}$. This can be compared with the ground state $|\Psi_{\mathrm{BCS}}\rangle$ used in superconductivity,

$$|\Psi_{\mathrm{BCS}}\rangle = \prod_{\boldsymbol{k}} (u_{\boldsymbol{k}} + v_{\boldsymbol{k}} \mathrm{e}^{\mathrm{i}\alpha_{\boldsymbol{k}}} a^*_{\boldsymbol{k}\uparrow} a^*_{-\boldsymbol{k}\downarrow}) \, |0\rangle \quad , \tag{6.30}$$

and it is useful to specify how the similarities and differences arise. In the two cases, the variational process reveals the presence of an arbitrary phase factor $\mathrm{e}^{\mathrm{i}\alpha}$ common to all pairs. On the other hand, the particular phase factor $(-1)^{j-m}$ appearing in (6.29) has been chosen so that the angular momentum vanishes (see exercise 2).

If we restrict ourselves to the simplest dynamical behavior, rotating deformed nuclei can exhibit two limiting cases: rigid rotation, i. e. rotation as a solid body, and irrotational flow. In fact the case of irrotational flow is a direct consequence of (5.26). If the dynamical state of a nucleus is controlled by nucleon pairing alone, the nuclear rotation has to be described by an irrotational flow of the "nuclear fluid". It is therefore interesting to compare a nucleus excited to a rotational state with an uncharged quantum fluid as discussed in Sect. 5.2.3. We leave this until Sect. 6.5.3.

Before going on, let us give a definition of *spherical* and of *deformed nuclei*. This definition only makes sense if we restrict ourselves to nuclei excited just above the ground state. In this region, the excited states can be divided into simple groups. For instance, one can easily make a distinction between vibrational or rotational levels. Rotational states can be identified because of their analogy with the rotational sprectrum of diatomic molecules. They obey simple rules (see for instance relation (6.32)) and the corresponding excitation energy is low compared with other types of excited levels.

There is an important point concerning rotational states of even–even nuclei: they only occur for nuclei far from magic number states. This fact can be interpreted in the following way. When the nucleus is deformed (the opposite of a spherical nucleus), it can be excited into a rotational state along an axis perpendicular to its symmetry axis, if such an axis exists. It is much easier to rotate it than to excite it in any other way. The same is true for a diatomic molecule. If the nucleus is spherically symmetric, there is no sequence of rotational states. All axes passing through the nucleus are equivalent. The nucleus behaves as a rigid sphere, strictly invariant with respect to a rotation.

The interpretation given above is strengthened by the following fact. Deformed nuclei can be identified by the presence of an electrical quadrupole moment, a property that enables to characterize them independently of the presence of rotational states in their spectrum. As such nuclei lie relatively far from magic numbers, nucleon pairing is not so strong in the case of deformed nuclei as it is for "almost magic nuclei". In order to clarify and to sum up what has been mentioned above, the next section is devoted to some general properties of the rotational states of nuclei.

6.5.2 An Excited Nucleus in a Rotational State

In quantum physics, one speaks of a *rigid rotator* when the Hamiltonian of the system is written

$$H_I = \frac{|\boldsymbol{L}|^2}{2I} \quad , \tag{6.31}$$

where $\boldsymbol{L}$ is the orbital angular momentum operator. The quantity $I > 0$ is thus interpreted as the moment of inertia of the system in analogy with the classical problem of two masses M maintained at a fixed distance d from each other (the moment of inertia is thus equal to $Md^2/2$). A diatomic molecule can be described by the Hamiltonian (6.31) provided the energy is insufficient for the molecule to be excited in a vibrational state.

The eigenstates of this system depend on the orbital quantum number l (integer $l \geq 0$) and on the magnetic quantum number m (integer m, $-l \leq m \leq l$) and one has

$$H_I \,|l,m\rangle = \frac{\hbar^2}{2I} l(l+1)\,|l,m\rangle = E_I\,|l,m\rangle \quad . \tag{6.32}$$

The form of the spectrum H_I is typical and allows one to distinguish between the vibrational states whose spectrum is similar to the harmonic oscillator.

If one wishes to calculate the moment of inertia I of a nucleus, it is necessary to split the Hamiltonian into two terms

$$H = \frac{|\boldsymbol{L}|^2}{2I} + H^{\text{int}} \quad . \tag{6.33}$$

The term $|\boldsymbol{L}|^2/2I$ describes the rotation of the ensemble of nucleons and H^{int} represents the contribution to the Hamiltonian due only to internal degrees of freedom associated with the relative motion of the nucleons. This decomposition is not rigorously possible (a global rotation gives rise to inertial forces which must occur in the expression H^{int}).

Nevertheless, measurement of the dipolar electric moment associated with the quasi-uniform repartition of the protons gives rise to the notion of the shape of the nucleus. A number of important points ought to be noted here:

- When the shape of the nucleus is spherical, one does not observe the rotational degrees of freedom.
- When the nucleus has an elongated shape along an axis of revolution (a "cigar" shape), one only observes rotation around the directions perpendicular to this axis.

Concerning the subject of this section, it is enough to remember two points:

- Nuclei for which the number of one given species of nucleon differs markedly from one of the magic numbers generally have an elongated shape.
- A nucleus which lacks a quadrupole moment – let us call it spherical – does not contain excited states of energy proportional to $l(l+1)$.

6.5.3 Moment of Inertia of a Deformed Nucleus

When one has a model of a deformed nucleus for which the decomposition (6.33) makes sense, it is possible to calculate the constant I. It is interesting to study this value as a function of the shape of the nucleus. As deformed nuclei are frequently ellipsoids of revolution, one uses the dimensionless parameter β, defined essentially as the difference between the length of the longer and shorter axes of the ellipse. The parameter β thus completely describes the deformation of the nucleus.

For a deformed nucleus, it is useful to compare the constant I, which is measured or calculated, with a reference value I_0, the moment of inertia of a rigid classical nucleus with the same shape, size and density as the nucleus being considered. If one plots the value I/I_0 as a function of the deformation β for a series of e–e rare-earth nuclei (circles in Fig. 6.8), one notes that the values are all significantly less than 1.

In Fig. 6.8, the dashed curve is the result of a calculation which compares the nucleus to a liquid with an irrotational motion typical to a superfluid. One imagines an incompressible fluid for which the vector field $\boldsymbol{v}_f$ satisfies

$$\boldsymbol{v}_f = \nabla f \quad , \quad \nabla \cdot \boldsymbol{v}_f = \nabla^2 f = 0 \quad . \tag{6.34}$$

In the case of the neutral quantum fluids, the function f has to be identified to the product $(\hbar/m_0)\alpha$ occurring in (5.26).

The fluid is contained inside the interior of an imaginary cavity corresponding to the limits of the nucleus. The rotational motion of the cavity defines that of the nucleus and that of a fluid which, at each point of the boundary, must obey the condition

$$\hat{\boldsymbol{n}} \cdot \boldsymbol{v}_f = 0 \quad , \tag{6.35}$$

where $\hat{\boldsymbol{n}}$ is a vector normal to the cavity. Relations (6.34) and (6.35) determine $\boldsymbol{v}_f$ in a unique manner. If the cavity is spherical, the fluid is at rest

$$\boldsymbol{v}_f = 0 \quad . \tag{6.36}$$

The solution $\boldsymbol{v}_0$, corresponding to a uniform rotational motion

$$\boldsymbol{v}_0 = \boldsymbol{\omega} \times \boldsymbol{r}$$

which would describe the motion of a solid, is excluded because it does not correspond to irrotational motion

$$\nabla \times \boldsymbol{v}_0 = 2\boldsymbol{\omega} \neq 0 \quad .$$

If, on the other hand, the cavity takes the shape of an ellipsoid in rotation around an axis normal to the axis of revolution, the velocity field does not vanish and one can calculate the moment of inertia I. Referring to Fig. 6.8,

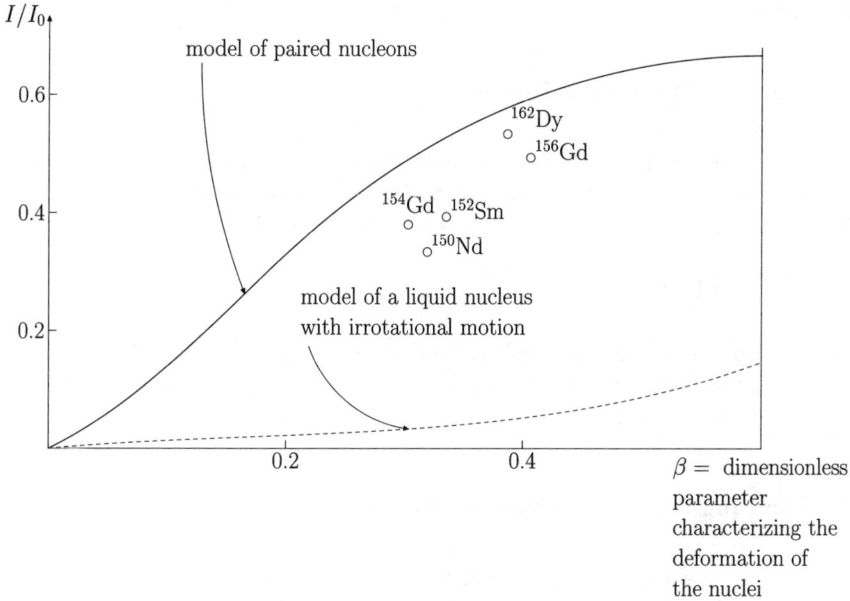

Fig. 6.8. Relative moment of inertia of e–e rare-earth nuclei plotted as a function of the deformation of the nuclei

one notes that the angular momentum calculated in this manner is less than the measured value.

The solid curve of Fig. 6.8 represents the value obtained by calculating the energy $E(\omega)$ of an e–e nucleus moving at a speed of rotation ω using a more sophisticated model of paired nucleons than has been presented in this section. In this model, one defines I by the relation

$$E(\omega) - E(0) = \frac{1}{2} I \omega^2 \quad .$$

If one wishes to summarize the conclusions that can be made regarding a rotating nucleus, one can make the following remarks:

- *A deformed rotating e–e nucleus cannot be compared to a rotating solid.*
- *Furthermore, it cannot be compared to a rotating superfluid, in spite of a closer similarity.*
- *The pairing model of nucleons, which falls between the two preceding models, gives a better description of a rotating nucleus.*

Note that there is a difficulty in terminology: one frequently speaks of the superfluid model of the nucleus to describe predictions given by the pairing model. It is more judicious to use this term to describe the model of a fluid nucleus with irrotational motion.

Exercises

1. Choice of the Phase Factor

In the definition of the BCS state (6.26), parameters u and v are choosen to be real and j has a fixed value equal to 7/2. The product (6.26) involves only positive values of m. Show that the state (6.26) can be written as

$$|\Psi\rangle = \prod_{m \in M_+} \left(u + v a^*_m a^*_{-m}\right) |\Psi_0\rangle \quad .$$

i. e. without involving any explicit mention of a phase factor. Show that for this case, the condition $m > 0$ is replaced by $m \in M_+$, where M_+ is the set $\{7/2, 3/2, -1/2, -5/2\}$.

2. Total angular momentum of the BCS state

Verify that the total angular momentum of BCS state (6.26) vanishes. To this aim, calculate the eigenvalue of the operator J_z when applied to (6.26).

3. The Bogoliubov Transformation

To carry out calculations with the state (6.26), one introduces the canonical transformation of Bogoliubov

$$\alpha^*_m = u a^*_m - (-1)^{j-m} v a_{-m} = \begin{cases} u a^*_m - v a_{-m} & , \quad m \in M_+ \\ u a^*_m + v a_{-m} & , \quad m \in M_- \end{cases} ,$$

where M_- is a set complementary to $M_+ : M_- = \{5/2, 1/2, -3/2, -7/2\}$. Show that the transformation preserves the anticommutation rules, i. e. that

$$[\alpha_m, \alpha^*_{m'}]_+ = \delta_{m,m'} \quad .$$

Show that the state $|\Psi\rangle$ plays the role of "vacuum" with respect to the α_m, i. e. that $\alpha_m |\Psi\rangle = 0$.

4. Particle Number of the Shell $j = 7/2$

Show that the average particle number

$$n = N(u, v) = \sum_{m \in M_+} \langle\Psi| a^*_m a_m |\Psi\rangle$$

of the shell is given by $(2j + 1)v^2$.

5. Determination of the Ground State of a Paired Nucleus

Using methods similar to those developed in Chap. 5, one obtains a reduced hamiltonian

$$\mathtt{H}^{\text{red}} = \varepsilon \sum a_m^* a_m - 2G \sum \sum a_m^* a_{-m}^* a_{-m'} a_{m'}$$

The first term corresponds to the kinetic energy which assumed to be constant. One has to sum up over all values of m belonging to M, union of M_+ and M_-. The second term describes the pair interaction whose strength is given by the value of the positive constant G. It involves a double sum over indices m and m' belonging to M_+.

In this very simplified model, the ground state properties are then obtained by minimizing the function $F(u, v)$ given by the expectation value of $\mathtt{H}^{\text{red}} - \mu N$ in the BCS ground state (6.26):

$$F(u, v) = (2j + 1)(\varepsilon - \mu)v^2 - 2G(j + 1/2)^2 u^2 v^2 \quad .$$

Following the methods of Chap. 5, determine the values of u and v corresponding to the minimum of F. Using the same change of variables then in (5.107), determine the relation between the gap Δ and the constant G,

$$\Delta^2 = nG^2(2j + 1)\left(1 - \frac{n}{2j + 1}\right) \quad ,$$

n is the mean particle number in the shell.

7. The Superfluidity of Liquid Helium

7.1 Experimental Facts

7.1.1 Phase Diagram of $^4_2$He

Helium is a particularly simple, stable and symmetric atom. The K shell is filled by two $1s$ electrons which are closely bound to the nucleus: its ionization potential of 24.5 eV is larger than for any other atom. Transitions between atomic levels caused by thermal excitation have a very low probability for temperatures below 10^3 K.

Two stable isotopes exist, $^3_2$He and $^4_2$He. The nucleus of the lighter isotope, $^3_2$He, contains one neutron, making it a fermion of spin 1/2. The nucleus of the heavier isotope, $^4_2$He, contains two neutrons, making it a boson of spin zero (Sect. 2.1.6). In this chapter, we will study this latter isotope, which we will refer to simply as *helium*. We will return to a discussion of $^3_2$He below.

The helium ground state neither has an electric nor a magnetic dipole and its electric polarizability is extremely low, making the van der Waals forces very weak. This explains why its critical temperature and boiling point are no higher than 5.20 K and 4.21 K respectively, and, above all, why *helium remains a liquid at atmospheric pressure as the temperature goes to zero.* In Sect. 2.2.1, we saw that a perfect Bose gas undergoes a transition, namely Bose–Einstein condensation, at a temperature (2.86)

$$T_0 = 3.31 \frac{\hbar^2}{k_B m} \left(\frac{\rho}{2s+1} \right)^{2/3} \quad . \tag{7.1}$$

Replacing m by the He atom mass, the spin s by zero and ρ by the observed density of liquid He (where each atom occupies a volume of 45 Å^3 on average), one obtains $T_0 = 3.13$ K. While (7.1) is valid only for the case of a perfect gas, it is reasonable to expect the measured value of T_0 to be close by in the presence of such a weak interaction. *The liquefaction temperature is thus found to be slightly above the value predicted for Bose–Einstein condensation.*

For all other integer spin isotopes occurring in nature, freezing takes place at a much higher temperature than a perfect Bose–Einstein condensation. Either the mass m of the atom is too large or, as in the case of hydrogen, the interactions are much stronger than for He and the perfect gas approximation is questionable.

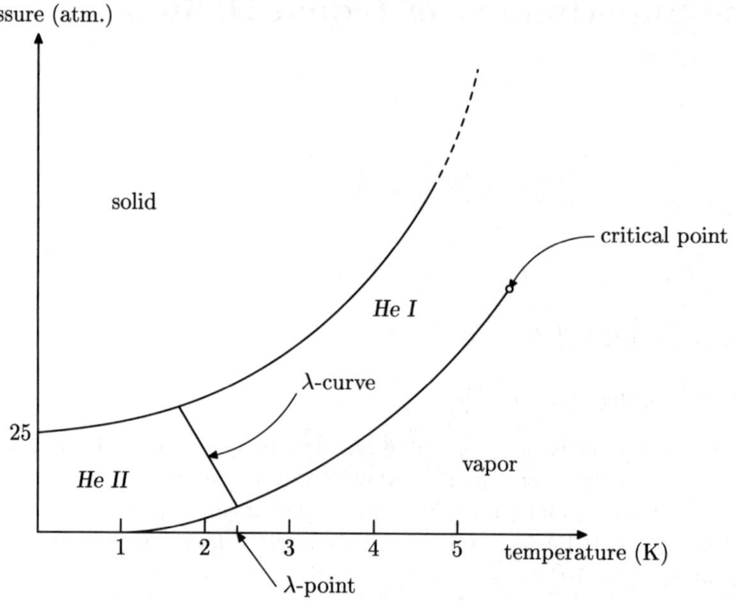

Fig. 7.1. Phase diagram of helium. Note the absence of freezing point below 25 atm and the presence of two different types of liquid helium

Apart from the fact that He does not freeze at ambient pressure, its phase diagram displays another special feature (Fig. 7.1): there are two types of liquid helium, He I and He II, separated by a phase transition. One can hardly distinguish He I from a typical liquid, except for the fact that it possesses a rather weak viscosity of around 3.5×10^{-5} poise (compared to 10^{-2} for water. He II, however, possesses a spectacular property known as *superfluidity*.

The transition between He I and He II is indeed a phase transition as shown by the behaviour of the specific heat as a function of the temperature (Fig. 7.2). The shape of this curve, which resembles the Greek letter lambda, is called the lambda transition, and the temperature is referred to as T_λ. At a pressure of 38 mm *Hg*, $T_\lambda = 2.18$ K.

Note that the numerical value for T_λ is very close to T_0. Furthermore, the phase diagrams of ^{3}He and ^{4}He are very different. In ^{3}He, instead of a lambda transition, one observes two abnormal liquid phases compared to a typical liquid, appearing only below 3 mK. This is the temperature where the interaction between the spins of the ^{3}He atoms plays a non-negligible role.[1] These two comments show that *the origin of the lambda transition lies in the Bose statistics of* ^{4}He *atoms.* While it may be reckless to compare this

[1] If a magnetic field is applied to ^{3}He then one observes three superfluid phases, not just two.

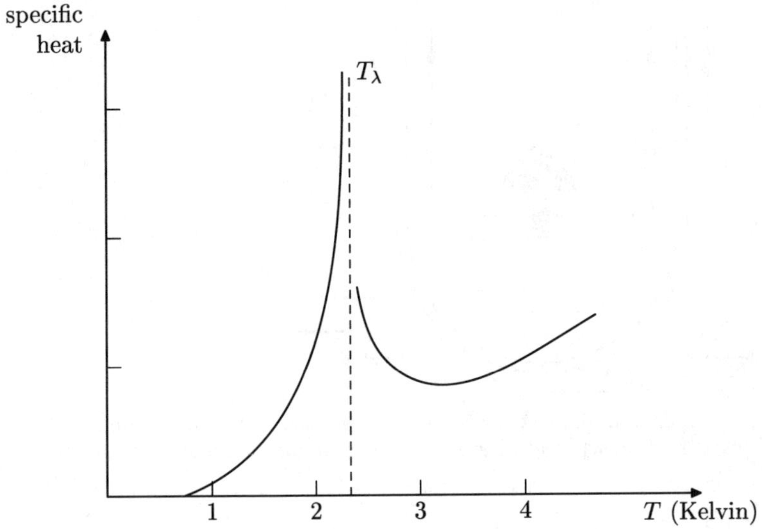

Fig. 7.2. Specific heat versus temperature for liquid helium. The transition at T_λ from He I to He II is called the lambda transition

to Bose–Einstein condensation, it is nevertheless worth keeping this analogy in mind when interpreting the properties of He II.

7.1.2 Properties of the Superfluid Phase of He II

Below, we draw up a list of the most spectacular properties of He II.

Absence of Viscosity

Under certain conditions, the *viscosity of* He II *vanishes* (its upper limit can reach 10^{-11} poises). The liquid is capable of passing through very thin capillaries without dissipation, as long as the speed of the liquid is below a critical speed v_c which increases as the diameter of the capillary decreases. Alternatively, if one measures the viscosity of He II by immersing in it a series of rotating discs, it is found that the viscosity is only slightly weaker than for He I. This point is addressed in Sect. 7.2.2.

Superfluid Film

Consider an open container of He II, itself immersed in a bath of He II. A thin film of He II then forms along the walls of the container, which, like the capillary, is capable of transporting He II without dissipation until the internal level is raised or lowered to match the external level (Fig. 7.3).

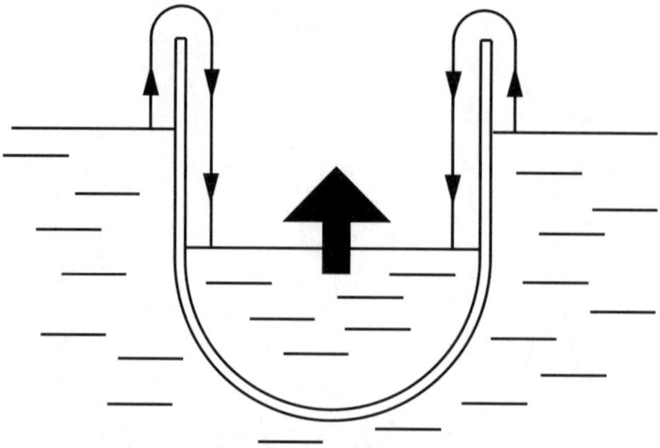

Fig. 7.3. An open container of liquid He II immersed in He II. A thin film forms, transporting the liquid without dissipation until the internal and external levels are equal

Thermal Conductivity

The thermal conductivity of He II *is larger than that of any other known substance by several orders of magnitude.* More specifically, heat propagates in He II by a very special mechanism. In a normal substance, if one creates a temperature difference ΔT with respect to the average temperature, ΔT will obey a diffusion equation. In He II, *ΔT obeys a wave equation* characterized by a propagation velocity of the order of 20 $\mathrm{m\,s^{-1}}$ whereas the speed of sound is around 240 $\mathrm{m\,s^{-1}}$. This mode of heat propagation is referred to as second sound.

Thermomechanical Effect

The He II phase introduces a *thermomechanical effect*, which has many manifestations. Suppose two containers A and B, filled with He II are connected by a very thin capillary (Fig. 7.4). Each container is separately maintained at a fixed temperature (T in A and $T + \Delta T$ in B). One then observes a non-dissipative flow of the liquid through the capillary so that at equilibrium there is a difference in pressure between the two containers

$$\Delta p = s\Delta T \quad , \tag{7.2}$$

where s is the entropy per unit volume.

This effect can be easily explained by supposing that He II does not carry any entropy when passing through the capillary. Neither the temperature nor the pressure difference between the two containers can equalize; the chemical

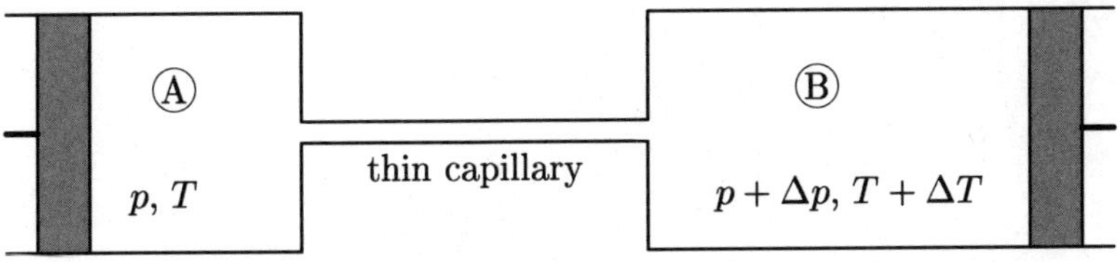

Fig. 7.4. Two containers of He II maintained at slightly different temperatures and attached by a very thin capillary. A pressure difference then arises

potentials alone must be equal due the possible transfer of matter between A and B. If one supposes that Δp and ΔT are small, one has

$$\begin{aligned}\mu(p,T) &= \mu(p+\Delta p, T+\Delta T)\\ &\simeq \mu(p,T) + \left(\frac{\partial \mu}{\partial p}\right)_{\mathrm{T}} \Delta p + \left(\frac{\partial \mu}{\partial T}\right)_p \Delta T \quad .\end{aligned} \tag{7.3}$$

Because the chemical potential μ is the same as the free enthalpy per particle, one can write $\mu = G(p,T)/N$. From the relations $(\partial G/\partial T)_p = -S$ and $(\partial G/\partial p)_{\mathrm{T}} = V$, where S and V are the entropy and the volume respectively, one deduces the thermodynamic identity

$$s = \frac{S}{V} = -\left(\frac{\partial \mu}{\partial T}\right)_p \bigg/ \left(\frac{\partial \mu}{\partial p}\right)_{\mathrm{T}} \quad , \tag{7.4}$$

yielding equality (7.2).

The fountain effect is a spectacular illustration of the same phenomenon (Fig. 7.5). A container R, with two openings blocked by cotton wads, is filled with emery powder and immersed in He II. The upper opening is attached to a tube which points out of the liquid. The porous cotton wad plays the role of the capillary described above. The emery powder inside R is heated up by radiation so that the temperature and pressure rise, and liquid then spurts out. We will return to this thermomechanical effect in Sect. 7.2.2.

7.2 Quantum Liquid and the Two-Fluid Model

7.2.1 The Superfluid Phase and Quantum Liquid

The equation of motion for a typical isotropic fluid is the *Navier–Stokes equation*,

$$\frac{\mathrm{d}\boldsymbol{v}}{\mathrm{d}t} = \frac{\partial \boldsymbol{v}}{\partial t} + \frac{1}{2}\nabla|\boldsymbol{v}|^2 - \boldsymbol{v} \times (\nabla \times \boldsymbol{v}) = -\frac{1}{\rho_{\mathrm{M}}}\nabla p + \nu \nabla^2 \boldsymbol{v} \quad , \tag{7.5}$$

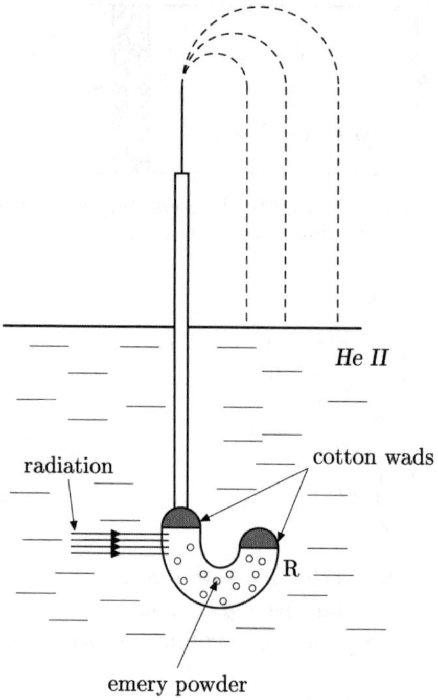

Fig. 7.5. Container of emery powder plugged at both ends by cotton wads and immersed in He II. The fountain of liquid passing through the tube is due to the thermomechanical effect of He II which causes the temperature and pressure inside the container to increase

where p is the pressure, ρ_{M} is the specific mass and ν is the specific viscosity coefficient. Even when ρ_{M} and ν are constant (one refers to this case as a Newtonian incompressible fluid), the Navier–Stokes equation is in general very difficult to solve because of the non-linear character of (7.5).

The situation simplifies considerably if one considers a neutral and incompressible quantum fluid. Because of the phase α (see (5.26) and (5.27)), the velocity is the gradient of a potential and its curl is zero; that is, the flow is irrotational:

$$\boldsymbol{v} = \nabla\Phi = \nabla\left(\frac{\hbar\alpha}{m}\right) \quad , \tag{7.6}$$

$$\nabla \times \boldsymbol{v} = 0 \quad . \tag{7.7}$$

Moreover, the continuity equation for an incompressible fluid is reduced to the condition

$$\nabla \cdot \boldsymbol{v} = 0 \quad . \tag{7.8}$$

Furthermore, the vector identity

$$\nabla^2 \boldsymbol{v} = \nabla(\nabla \cdot \boldsymbol{v}) - \nabla \times (\nabla \times \boldsymbol{v}) \tag{7.9}$$

shows that the term $\nu \nabla^2 \boldsymbol{v}$ vanishes for an irrotational and incompressible system (relations (7.7) and (7.8)). The Navier–Stokes equation thus takes on a much more simple form

$$\frac{\partial \boldsymbol{v}}{\partial t} + \nabla \left(\frac{1}{2} |\boldsymbol{v}|^2 + \frac{p}{\rho_{\mathrm{M}}} \right) = 0 \quad . \tag{7.10}$$

Relations (7.7) and (7.10) are reminiscent of the London equations (5.9) and (5.10). Furthermore, relation (7.6), where α is interpreted as a quantum-mechanical phase, gives rise to the quantization of circulation (see Sect. 5.2.3)

$$\oint_{\mathrm{C}} \mathrm{d}\boldsymbol{l} \cdot \boldsymbol{v}_s = n \frac{h}{m} \quad , \quad \text{integer } n \quad . \tag{7.11}$$

To keep the integer n from vanishing, it is necessary to have the fluid in a non-simply connected domain. The curve C must wind around a region containing no fluid (a vortex).

We have already mentioned in Sect. 7.1 that, depending on the experimental setup, He II may flow with or without viscosity. Therefore it is not possible to completely describe its dynamics with relations such as (7.7) and (7.10), in which all dissipation is neglected (the specific viscosity coefficient ν does not appear). In order to correctly describe He II, one must first of all examine the nature of the dissipation.

7.2.2 Dissipation in a Superfluid

We already mentioned in Sect. 7.1.2 the paradox of oscillating disks immersed in He II measuring a viscosity the same order of magnitude as for He I. This can be related to another paradox, that of the thermomechanical effect. He II passes through a capillary without transporting entropy, whereas the same fluid in a container of macroscopic dimensions is characterized by a well-determined entropy.

The Two-Fluid Model

These two observations, among numerous others, give rise to the *two-fluid model* in which He II is pictured as a mixture of two components. We emphasize straightaway that this does not imply the existence of two different chemical species, but is rather a convenient way to describe the phenomenology of superfluid He. The microscopic counterparts of these two components will be clarified later. One of them, the normal fluid, behaves as an ordinary fluid; in particular, it obeys the usual thermodynamic relations and cannot pass through a capillary in the absence of a sufficiently strong pressure gradient.

The other component, the superfluid, is the origin of the spectacular properties of He II. One might be tempted to identify it, at least partially, with the incompressible quantum fluid described in the last section. Only by analysing diverse fundamental experiments can its dynamics be determined more precisely.

The first problem is concerned with determining the respective concentrations $\rho_n(T)$ and $\rho_s(T)$ of the two fluids, both of which depend on temperature. If one writes the specific mass ρ_{M} of He II in the form

$$\rho_{\mathrm{M}} = \rho_n(T) + \rho_s(T) \quad , \tag{7.12}$$

one can postulate that the temperature dependence of the densities satisfies, respectively

$$\lim_{T \to T_\lambda} \frac{\rho_s(T)}{\rho_{\mathrm{M}}} = 0 \quad , \quad \lim_{T \to 0} \frac{\rho_s(T)}{\rho_{\mathrm{M}}} = 1 \quad . \tag{7.13}$$

The relevance of the decomposition (7.12) was confirmed in 1946 by Andronikashvili with the help of measurements of the viscosity as a function of T (Fig. 7.6). A series of equidistant disks rigidly separated from each other is immersed in He II, and oscillatory twists are applied. The distance between neighboring disks (0.21 mm) is sufficiently small that at a temperature $T > T_\lambda$ all of the liquid between the disks is put into motion because of the viscosity. For $T < T_\lambda$, only the normal fluid oscillates, the superfluid remains still. A measurement of the global dissipation allows one to determine ρ_s/ρ_{M} and to verify (7.13).

The thermomechanical effect and Andronikashvili's experiment confirm the idea that the flow of the superfluid is non-dissipative. To better distinguish between the flows of the two components, it is natural to assign a different velocity to each of them. These will be denoted as $\boldsymbol{v}_n$ and $\boldsymbol{v}_s$. The velocity $\boldsymbol{v}$ associated with the center of mass of the fluid is written as

$$\rho_{\mathrm{M}} \boldsymbol{v} = \rho_n \boldsymbol{v}_n + \rho_s \boldsymbol{v}_s \quad . \tag{7.14}$$

We require that the flow of the superfluid component be irrotational

$$\nabla \times \boldsymbol{v}_s = 0 \quad . \tag{7.15}$$

Indeed, this is the most fundamental characteristic of a quantum fluid. On the other hand, one cannot cancel the divergence of $\boldsymbol{v}_s$ as this would require a complete identification of the superfluid component with an incompressible quantum fluid (Sect. 7.2.1). Indeed, (7.8) is a conservation equation; as the normal and superfluid components can transform from one to the other (as proven by the thermomechanical effect and (7.13)), it is not possible to write

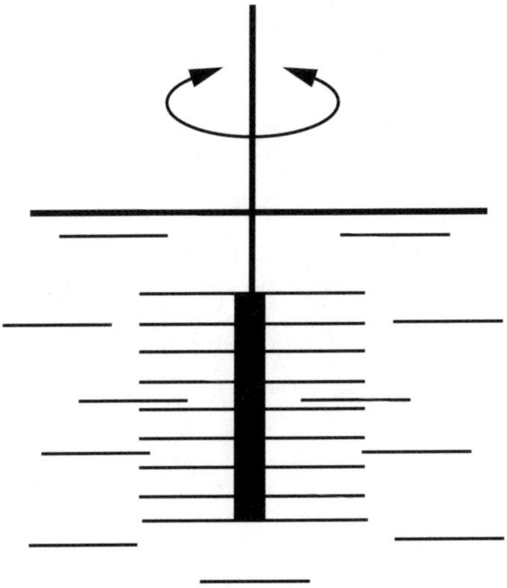

Fig. 7.6. Andronikashvili's experiment measuring the viscosity of He II. Neighbouring disks attached at a fixed distance are immersed in the liquid and rotated with oscillating twists

a continuity equation for the superfluid component alone. In general, the divergence of $\boldsymbol{v}_s$ does not vanish, except for one particular case, a uniform rotation which will be studied below.

Vortices

Imagine applying a uniform rotation to a container of He II of uniform temperature. If the temperature is uniform, it is natural to suppose that the normal and superfluid components would be separately conserved: one can thus successively apply (7.8) and (7.6) to the velocity $\boldsymbol{v}_s$ which leads to

$$\nabla \cdot \boldsymbol{v}_s = \nabla^2 \Phi = 0 \quad , \tag{7.16}$$

with the boundary conditions

$$\hat{\boldsymbol{n}} \cdot \boldsymbol{v}_s = \hat{\boldsymbol{n}} \cdot \nabla \Phi = 0 \tag{7.17}$$

at the walls of the container, with normal vector $\hat{\boldsymbol{n}}$. If the domain is simply connected, the only solution to (7.16) and (7.17) is given by

$$\boldsymbol{v}_s = 0 \quad . \tag{7.18}$$

This analysis predicts that only the normal component should follow the rotation of the container, and in particular that the liquid's moment of inertia should be less than for a typical liquid under the same conditions. However,

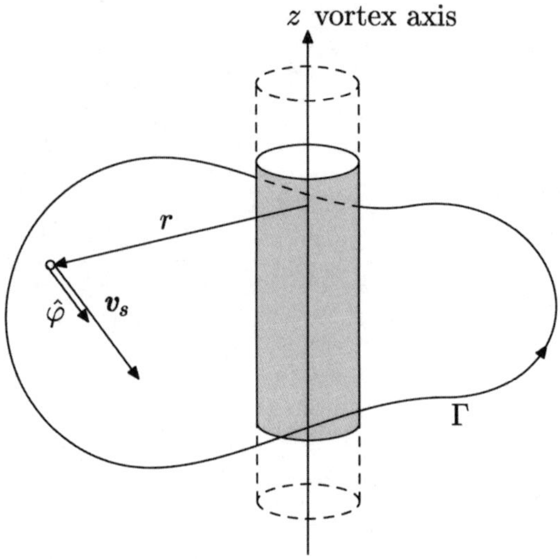

Fig. 7.7. A two dimensional velocity field $\boldsymbol{v}_s$ in a perfect fluid surrounding an infinite cylinder

this is not the case: the superfluid component rotates with the rest of the fluid, even at zero temperature.

Contrary to what one would think, it is not necessary to abandon the concept of irrotational motion (7.15) to explain this fact. But it is necessary to introduce the notion of *vortices*. Imagine a two-dimensional velocity field $\boldsymbol{v}_s$ in a perfect fluid circulating around an infinite cylinder (Fig. 7.7). Outside the cylinder, the velocity field is written as

$$\boldsymbol{v}_s = \frac{\kappa}{2\pi r}\hat{\boldsymbol{\varphi}} \quad , \quad |\hat{\boldsymbol{\varphi}}| = 1 \quad . \tag{7.19}$$

Throughout the fluid $\boldsymbol{v}_s$ is irrotational ($\nabla \times \boldsymbol{v}_s = 0$) and its circulation on a closed curve Γ surrounding the cylinder is given by

$$\oint_\Gamma \mathrm{d}\boldsymbol{l} \cdot \boldsymbol{v}_s = \kappa \quad . \tag{7.20}$$

If the radius of the cylinder was reduced sufficiently, then the liquid could be removed and a cylindrical cavity would remain with the fluid held to the walls of the cavity by centrifugal force. In this way a vortex has been created, which allows a non-uniform rotation to be applied to the fluid. When κ has the form (7.11), the circulation is quantized. Relation (7.20) is reminiscent of Ampere's relation

$$\oint_\Gamma \mathrm{d}\boldsymbol{l} \cdot \boldsymbol{H} = I \quad , \tag{7.21}$$

which allows the determination of the magnetic field $\boldsymbol{H}$ created by a current I; $\boldsymbol{H}$ and I are the respective analogues of $\boldsymbol{v}_s$ and κ.

Suppose now that the fluid, contained in a cylindrical cavity of radius R, is penetrated from one end to the other by a dense lattice of vortices which are straight and parallel to the axis of rotation; each of the lines is characterized by the same circulation κ. The problem of fluid motion is thus a two-dimensional problem, the velocity field $\boldsymbol{v}_s(\boldsymbol{r}, \varphi)$ does not depend on the z coordinate. At a given moment the vortex j is determined by the coordinates $\boldsymbol{r}_j$ and φ_j; it is moving at the velocity $\boldsymbol{v}_s(\boldsymbol{r}_j, \varphi_j)$ due to the effect of all the other lines. At a point in the fluid where there is no vortex, the velocity $\boldsymbol{v}_s(\boldsymbol{x})$ is due to the superposition of velocity vectors created by the circulation around each of the tubes. To evaluate it at the moment t, one can make an approximation that the vortices are replaced by a continuous distribution of circulation density $\boldsymbol{\kappa}(\boldsymbol{x})$ defined so that if $\mathrm{d}\boldsymbol{s}$ is a surface element traced in the fluid and limited by a closed curve Γ, one has

$$\boldsymbol{\kappa} \cdot \mathrm{d}\boldsymbol{s} = \kappa \mathrm{d}N \quad , \tag{7.22}$$

where $\mathrm{d}N$ is the number of vortices which penetrate $\mathrm{d}\boldsymbol{s}$ in the positive direction (the orientation is defined as in the case of Ampere's relation). Under these conditions,

$$\nabla \times \boldsymbol{v}_s = \boldsymbol{\kappa}(\boldsymbol{x}) \quad , \tag{7.23}$$

as in the magnetostatic case, where

$$\nabla \times \boldsymbol{H} = \boldsymbol{j}(\boldsymbol{x}) \quad . \tag{7.24}$$

Just as a uniform current density $\boldsymbol{j} = (0, 0, j_z)$ in a straight cylindrical wire corresponds to a field $\boldsymbol{H}$ whose field lines are circles,

$$\boldsymbol{H} = \frac{\boldsymbol{j} \times \boldsymbol{r}}{2} = \left(0, H_\varphi = \frac{|\boldsymbol{j}||\boldsymbol{r}|}{2}, 0\right) \quad , \tag{7.25}$$

a uniform circulation density $\boldsymbol{\kappa} = (0, 0, \kappa \mathrm{d}N/\mathrm{d}s)$ corresponds to a velocity field

$$\boldsymbol{v}_s = \frac{\boldsymbol{\kappa} \times \boldsymbol{r}}{2} = \left(0, \boldsymbol{v}_{s\varphi} = \frac{1}{2}\kappa|\boldsymbol{r}|\frac{\mathrm{d}N}{\mathrm{d}s}, 0\right) \quad . \tag{7.26}$$

This field describes a global rotation of the field, with angular velocity Ω given by

$$\Omega = \frac{1}{2}\kappa\frac{\mathrm{d}N}{\mathrm{d}s} = \frac{nh}{2m}\frac{\mathrm{d}N}{\mathrm{d}s} \quad , \tag{7.27}$$

where n is the quantum number of circulation per vortex (relation (7.11)). For a given Ω, the velocity field is most uniform for $n = 1$, which is indeed observed experimentally. Returning to the discrete description, we can imagine a distribution of velocities as presented in Fig. 7.8, namely a hexagonal lattice of vortices in a reference frame moving with angular velocity Ω.

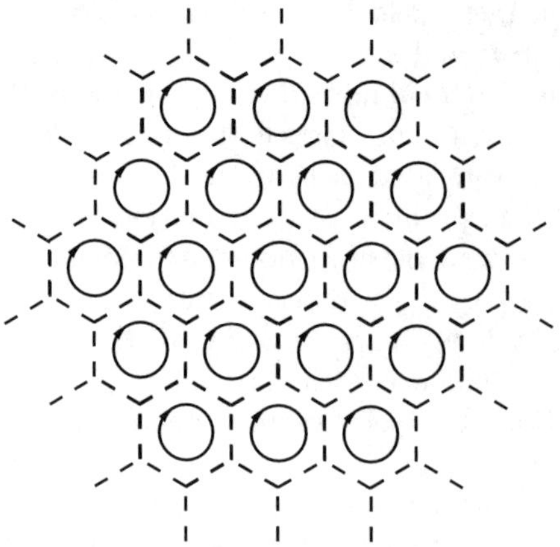

Fig. 7.8. A hexagonal lattice of vortices in a reference frame moving with angular velocity Ω. The solid circles are lines of current in the neighbourhood of a vortex, while the dashed lines represent zones where the velocity is zero

7.2.3 Second Sound

In the previous section, we pointed out that, as a rule, the velocity divergence of the superfluid component cannot be cancelled away. The rapid propagation of heat through He II (one of its characteristic phenomena) will allow us to complete the list of equations of motion in the two-fluid model.

To establish the equation describing *second sound*, one must first write down the description of "first sound", that is the propagation of sound through a normal fluid. In the first approximation one writes down an equation which does not take into account dissipation: this is Euler equation

$$\frac{\mathrm{d}\boldsymbol{v}}{\mathrm{d}t} = \frac{\partial \boldsymbol{v}}{\partial t} + (\boldsymbol{v} \cdot \nabla)\boldsymbol{v} = -\frac{1}{\rho_{\mathrm{M}}} \nabla p \quad , \tag{7.28}$$

obtained from the Navier–Stokes equation (7.5) by neglecting the dissipation term $\nu \nabla^2 \boldsymbol{v}$.

When (7.28) is coupled to the conservation equation (7.34), and keeping only terms linear in the deviation from homogeneity, one derives a wave equation for the density

$$\frac{1}{c_{\mathrm{s}}^2} \frac{\partial^2}{\partial t^2} \rho_{\mathrm{M}}(\boldsymbol{x}, t) = \nabla^2 \rho_{\mathrm{M}}(\boldsymbol{x}, t) \quad , \tag{7.29}$$

which describes the propagation of ordinary, first sound with speed $c_{\mathrm{s}} = c_I$.

In order to establish the existence of second sound, another mode of propagation, one can proceed by neglecting the dissipative phenomena which only contribute to the attenuation of the signal. As first step we will show that the superfluid velocity field $\boldsymbol{v}_s$ satisfies an equation analogous to the Euler equation (7.28).

The field $\boldsymbol{v}$ represents the velocity of the center of mass of a fluid element. In the two-fluid model, $\boldsymbol{v}$ is expressed with the help of the specific masses appearing in (7.12) and the velocities $\boldsymbol{v}_n$ and $\boldsymbol{v}_s$ which characterize respectively the normal component (or normal fluid) and the superfluid component (or superfluid) (7.14). The thermomechanical effect shows that the motions of the two components are not necessarily the same. If one considers He II as an incompressible fluid, it is logical to take ρ_{M} as a constant. On the other hand, it is crucial for what follows to allow the values ρ_n and ρ_s, which are proportional to the concentrations of the two components, to vary in space and in time.

Concerning the equation of motion for the superfluid component, the thermomechanical effect can guide us. Relation (7.3) shows that there is flow in the capillary which appears as a difference of chemical potential between the two ends. Moreover, the fluid flows from the container which has the larger chemical potential μ. To be convinced of this, suppose that the temperature of container B is raised holding the pressure constant. This gives $\Delta T > 0$ and $\Delta p = 0$ using the notation of Sect. 7.1.2. From this, one has the inequality

$$\mu(B) = \mu(p, T + \Delta T) \simeq \mu(p, T) - s\Delta T < \mu(p, T) = \mu(A) \quad . \tag{7.30}$$

Since raising temperature of B leads its superfluid concentration to decrease, there must be a flow from A to B to reestablish equilibrium. Thus one obtains an equation of motion of the form

$$\frac{\mathrm{d}\boldsymbol{v}_s}{\mathrm{d}t} = -k\nabla\mu \quad , \tag{7.31}$$

where k is a positive constant which can easily be determined. Taking into account (7.3) and the remarks that follow, one can put (7.31) into the form

$$\frac{\mathrm{d}\boldsymbol{v}_s}{\mathrm{d}t} = -k\frac{V}{N}(\nabla p - s\nabla T) = -km\frac{1}{\rho_{\mathrm{M}}}(\nabla p - s\nabla T) \quad , \tag{7.32}$$

where m and V/N are the mass and volume per particle. Moreover, it is logical to demand that the superfluid component behave as a normal frictionless fluid in the absence of a temperature gradient and thus that (7.32) reduce to the Euler equation (7.28). This equation is just Newton's second law for fluid dynamics. One can thus set $k = 1/m$ and (7.32) takes on the final form

$$\frac{\mathrm{d}\boldsymbol{v}_s}{\mathrm{d}t} = \frac{\partial \boldsymbol{v}_s}{\partial t} + (\boldsymbol{v}_s \cdot \nabla)\boldsymbol{v}_s = -\frac{1}{\rho_{\mathrm{M}}}(\nabla p - s\nabla T) \quad . \tag{7.33}$$

To obtain the equation of motion of second sound, it is necessary to take into account the continuity equation

$$\frac{\partial \rho_{\mathrm{M}}}{\partial t} + \nabla \cdot (\rho_{\mathrm{M}}\boldsymbol{v}) = 0 \quad . \tag{7.34}$$

Finally, we return to the thermomechanical effect to formulate one last hypothesis. Dissipation is associated with a normal fluid: the superfluid is capable of non-viscous flow through a very fine capillary (Fig. 7.4). During this experiment, the temperatures at each end of the tube do not equalize: no entropy passes from A to B. One can thus write a transport equation for entropy which generalizes the capillary experiment: the convective support of entropy is provided by the normal fluid

$$\frac{\partial s}{\partial t} + \nabla \cdot (s\boldsymbol{v}_n) = 0 \quad . \tag{7.35}$$

As in (7.28), one neglects here the dissipative phenomena as well as heat conduction and local heating due to friction. To establish the propagation equation of second sound, it is sufficient to recall (7.12), (7.28), (7.14), (7.33), (7.34), and (7.35) and to impose two conditions. First of all, the specific mass $\rho_{\mathrm{M}}(\boldsymbol{x}, t) = \rho_{\mathrm{M}}$ of the fluid remains constant throughout. One searches for a mode of propagation different from the usual acoustic mode (7.29) such that there are no local variations in density. Secondly, one limits oneself to oscillations with a small amplitude in the neighbourhood of the equilibrium values and then linearizes the equations. That is, one neglects all terms which are quadratic in deviations from equilibrium. As a result, contributions in $(\boldsymbol{v}\cdot\nabla)\boldsymbol{v}$ are always neglected (at equilibrium all velocities vanish). In addition, for a term such as $\nabla\cdot(s\boldsymbol{v}_n)$, one only retains the equilibrium value of s, since s is multiplied by the value $\boldsymbol{v}_n$, which is zero at equilibrium. One finally arrives at the system of equations

$$\rho_n + \rho_s = \rho_{M0} \quad , \tag{7.36}$$

$$\frac{\partial \boldsymbol{v}}{\partial t} = -\frac{1}{\rho_{M0}} \nabla p \quad , \tag{7.37}$$

$$\boldsymbol{v} = \frac{1}{\rho_{M0}} (\rho_{n0}\boldsymbol{v}_n + \rho_{s0}\boldsymbol{v}_s) \quad , \tag{7.38}$$

$$\frac{\partial \boldsymbol{v}_s}{\partial t} = -\frac{1}{\rho_{M0}} (\nabla p - s_0 \nabla T) \quad , \tag{7.39}$$

$$\nabla \cdot \boldsymbol{v} = 0 \quad , \tag{7.40}$$

$$\frac{\partial s}{\partial t} + s_0 \nabla \cdot \boldsymbol{v}_n = 0 \quad . \tag{7.41}$$

In (7.36)–(7.41), the values at equilibrium (which do not depend on $\boldsymbol{x}$ or on t) are indicated by the subscript 0. If one eliminates ∇p and $\boldsymbol{v}$ between (7.37), (7.38) and (7.39) and takes into account (7.36), one obtains

$$\rho_{n0} \frac{\partial (\boldsymbol{v}_s - \boldsymbol{v}_n)}{\partial t} = s_0 \nabla T \quad . \tag{7.42}$$

Taking the divergence of this equation, one can eliminate $\boldsymbol{v}_s$ because of (7.40)

$$\rho_{n0} \frac{\partial}{\partial t} \left(-\frac{\rho_{M0}}{\rho_{n0}} \nabla \cdot \boldsymbol{v}_n \right) = s_0 \nabla^2 T \quad , \tag{7.43}$$

and with the help of (7.41), one obtains

$$\frac{\rho_{n0} \rho_{M0}}{\rho_{s0}} \frac{1}{s_0} \frac{\partial^2 s}{\partial t^2} = s_0 \nabla^2 T \quad . \tag{7.44}$$

Finally, one considers s as a function of ρ_{M} and $T(\boldsymbol{x}, t)$ and uses the definition of specific heat C_V for a constant volume per unit mass

$$\rho_{\mathrm{M}} C_V = T \left(\frac{\partial s}{\partial T} \right)_{\rho_{\mathrm{M}}} \quad . \tag{7.45}$$

Continuing to neglect terms which are non-linear with respect to deviations from equilibrium, (7.45) allows one to find a relation between $\partial^2 s / \partial t^2$ and $\partial^2 T / \partial t^2$. Then (7.44) takes the form of a wave equation

$$\nabla^2 T - \frac{1}{c_{\mathrm{II}}^2} \frac{\partial^2 T}{\partial t^2} = 0 \quad , \tag{7.46}$$

where the velocity of second sound c_{II} is given by

$$c_{\mathrm{II}} = \left(\frac{\rho_{s0} s_0^2 T_0}{\rho_{n0} \rho_{M0}^2 C_{V0}} \right)^{1/2} \quad . \tag{7.47}$$

Relation (7.46) was originally deduced by Tisza and (7.47) is experimentally verified above 1 K. Below 1 K, the experimental value of c_{II} does not agree with (7.47). Here it is necessary to turn to Landau's theory.

7.3 The Energy Spectrum of He II

7.3.1 Excitations of He II

General Remarks

Superfluid helium is a liquid. This very fact makes it quite complicated to study, since the liquid state, intermediate between solid and gas, is not well understood at the microscopic level. However, an important fact described in Sect. 7.1.1 will set us on the path to the interpretation of this phenomenon. Specifically, the fact that He atoms obey Bose statistics is responsible for the appearance of the He II phase.

Landau and Feynman, adopting different points of view, arrived at descriptions of superfluidity which are admittedly schematic. However, both views have the merit of treating the system as a liquid. Historically, Landau's model preceded Feynman's, though it is useful to examine the latter first for the sake of coherence. In this section, we restrict ourselves to a partial description of these models, which is enough to establish a bridge between the experimental properties described above and the microscopic analysis of an imperfect Bose gas presented in Sect. 7.4.

Phonons: Feynman's Argument

Using a qualitative argument, Feynman suggested that *the excitations of a Bose liquid at very low temperatures are phonons.* This liquid is described by the ground state $\Phi_0(\boldsymbol{x}_1, \ldots, \boldsymbol{x}_n) \in \mathcal{H}_+^{\otimes n}$ where $\boldsymbol{x}_i, i = 1, \ldots, n$, are the positions of n atoms of zero spin. Suppose that the interaction between the atoms can be represented by a potential $V_{ij} = V(\boldsymbol{x}_i - \boldsymbol{x}_j)$, strongly repulsive at short distance. The energy of the ground state is written as

$$\begin{aligned} E_0 &= \langle \Phi_0 | \mathtt{H} | \Phi_0 \rangle \\ &= \sum_{j=1}^{n} \int \mathrm{d}\boldsymbol{x}_1 \ldots \mathrm{d}\boldsymbol{x}_n \Phi_0^*(\boldsymbol{x}_1, \ldots, \boldsymbol{x}_n) \left(-\frac{\hbar^2}{2m} \nabla_j^2 \right) \Phi_0(\boldsymbol{x}_1, \ldots, \boldsymbol{x}_n) \\ &\quad + \frac{1}{2} \sum_{i \neq j}^{n} \int \mathrm{d}\boldsymbol{x}_1 \ldots \mathrm{d}\boldsymbol{x}_n V(\boldsymbol{x}_i - \boldsymbol{x}_j) |\Phi_0(\boldsymbol{x}_1, \ldots, \boldsymbol{x}_n)|^2 \quad . \end{aligned} \tag{7.48}$$

We can apply the result of Sect. 2.2.1 here: Φ_0 can be chosen to be positive, non-degenerate and symmetric with respect to all permutations of the ensemble of coordinates.

To understand better Feynman's argument, one should keep in mind the quantum-mechanical interpretation of $\Phi_0(\boldsymbol{x}_1, \ldots, \boldsymbol{x}_n)$. The complex number Φ_0 is the probability amplitude for the fluid configurations, that is, for the set of positions of the atom centers. Some of these configurations correspond to a value $|\Phi_0|^2 = \Phi_0^2$ which is practically zero. Others, on the contrary, called *typical configurations*, maximize Φ_0^2: these have the largest probability of being realized. In the following discussion, we will use simple physical arguments to describe the typical configurations of the ground state and certain excited states.

With this in mind, it is useful to distinguish between several different distance scales: the microscopic scale, of the order of the average distance a between neighbouring atoms (in liquid He $a \simeq 3.6$ Å, while the diameter of an atom is of the order of 2.7 Å); the macroscopic scale, given by the length L of a side of the container; and the local intermediate scale d, where $a \ll d \ll L$. A fluid cell of linear dimension d contains many atoms, but its volume is much smaller than the total volume of fluid. One can already consider the

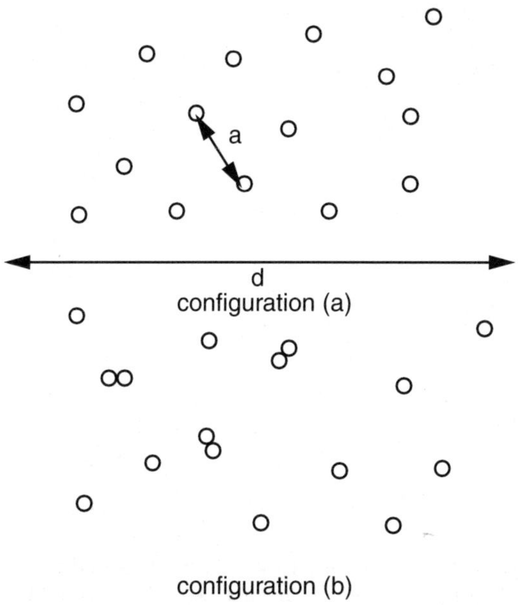

Fig. 7.9. Lower-energy (**a**) and higher-energy (**b**) configurations of liquid He atoms at scale d. The higher potential energy of (**b**) is due to the highly repulsive nature of the potential V_{ij} at very short distances

typical configurations of the ground state and low-energy states to be locally uniform, that is, that the atoms are distributed uniformly throughout any cell. In Fig. 7.9, configuration (a) corresponds to a lower energy than that of configuration (b), in which certain atoms are very close to one another: at short distance, V_{ij} is highly repulsive and the energy potential term in (7.48) will certainly be larger if $\Phi_0(\boldsymbol{x}_1, \dots, \boldsymbol{x}_n)$ takes on maximum values for a family of type (b) configurations than for a family of type (a) configurations. Concerning the kinetic energy term, which appears in the derivatives, one expects that it will take on a minimum value when the probability amplitude Φ_0 varies by a small amount each time one passes from one high probability configuration to a neighbouring one. By neighbouring configuration, we mean a configuration obtained from the previous one by a displacement of atoms less than the average distance a between the neighbouring atoms. As a consequence, the typical ground state configurations are essentially uniform over the entire volume of the liquid.

Consider an excited state just above the ground state. This state is characterized by a function $\Phi_1(\boldsymbol{x}_1, \dots, \boldsymbol{x}_n) \in \mathcal{H}_+^{\otimes n}$. Just like Φ_0, Φ_1 is symmetric and can be taken to be real. For the same reasons specified above, Φ_1 is also locally uniform at the scale of d. Because of the orthogonality condition,

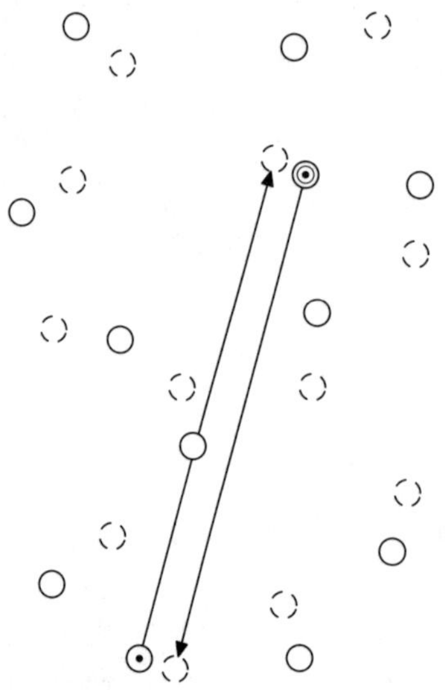

Fig. 7.10. The displacement of an atom (represented by a dot surrounded by one circle) must be compensated by the displacement of another atom (represented by a dot surrounded by two circles) to maintain local uniformity

$$\int \mathrm{d}\boldsymbol{x}_1 \ldots \mathrm{d}\boldsymbol{x}_n \Phi_0 \Phi_1 = 0 \quad , \tag{7.49}$$

Φ_1 must be positive for certain configurations and negative for others, since $\Phi_0 > 0$. Calling upon a purely qualitative argument, we would like to demonstrate the plausibility of the fact that the typical configurations for Φ_1 represent a phonon. The latter excitation essentially corresponds to a locally uniform distribution at the scale d if the phonon is a low-energy excitation (thus having a long wavelength).

With this in mind, we reason by contradiction, by supposing that the typical configurations of Φ_1 do not represent phonons. Consider two typical configurations, C_+ and C_-, such that $\Phi_1(C_+) > 0$ and $\Phi_1(C_-) < 0$, which correspond to the maximum and the minimum of Φ_1 respectively. These configurations are locally uniform for the reasons described above, and since they do not represent a phonon, they must also be uniform at the scale of the entire volume of liquid. In Figs. 7.10 and 7.11, the solid (dashed) circles represent atoms of configuration C_+ (C_-).

One can thus make the hypothesis that, to pass from C_+ to C_-, one displaces an atom (the dot surrounded by a single circle in Fig. 7.10) to another location (dashed circle). This displacement must be made over a large distance or else the term $[\Phi_1, -(\hbar^2\nabla^2/2m)\Phi_1]$ would be too large. For the density to remain locally uniform, one must then compensate this displacement by displacing a second atom (dot surrounded by two circles), which

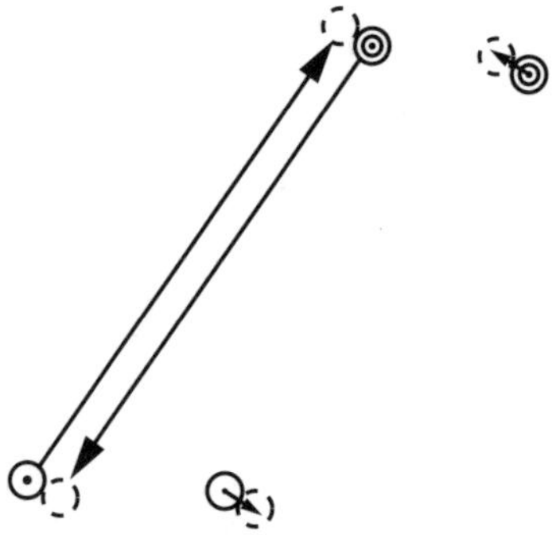

Fig. 7.11. Because of the Bose character of the ^{4}He atoms, the long range rearrangement of atoms (*left*) is equivalent to two short range ones

would fill in the void left by the first atom. This prevents too large a deviation from uniform density from occurring. But, the corresponding rearrangement represented in Fig. 7.10 is exactly equal to that of Fig. 7.11 since the atoms are indistinguishable and Φ_1 is symmetric vis-a-vis any permutation of particles. Passing from C_+ to C_- moving a very small distance would give rise to a large value of $[\Phi_1, -(\hbar^2\nabla^2/2m)\Phi_1]$, contradicting the hypothesis that Φ_1 is an excited state with energy just above that of the ground state.

At higher energies, the situation obviously changes: excitations which give rise to higher kinetic energy of the atoms must be taken into account. One must also allow for the possibility of the formation of vortices. Nevertheless, the excitation spectrum $\varepsilon(\boldsymbol{k})$ of He II, as measured by experiment, has the linear phononic form for small wavenumbers $\boldsymbol{k}$. In Fig. 7.12, note that we indeed have

$$\lim_{|\boldsymbol{k}|\to 0} \varepsilon(\boldsymbol{k}) = \hbar c_{\mathrm{s}} |\boldsymbol{k}| \quad , \tag{7.50}$$

where c_{s}, the speed of sound, is constant when $|\boldsymbol{k}|$ is small. The energy excitations $\varepsilon(\boldsymbol{k})$ when $|\boldsymbol{k}|$ is close to k_0 are called *rotons* ($\varepsilon(k_0)$, and correspond to a local minimum in Fig. 7.12). The roton is associated with a local motion of the atoms which gives rise to an appreciable kinetic energy.

7.3.2 Non-Viscous Flow Through a Capillary

The spectrum displayed in Fig. 7.12 was postulated by Landau who demonstrated how condition (7.50) could explain the non-viscous flow of He II through a capillary.

At zero temperature, He II is in the ground state, and only the superfluid component is present in the two-fluid model. The state Φ_0, introduced above, describes the system in the reference frame R_0, where He II is at rest.

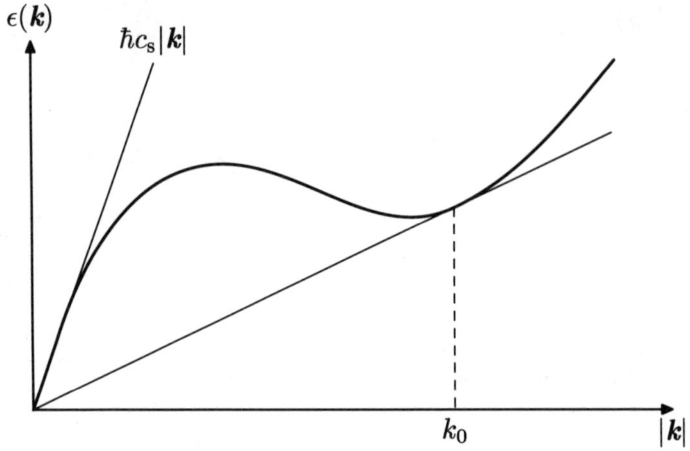

Fig. 7.12. Energy excitation ε of atoms as a function of wavenumber $|\boldsymbol{k}|$. Excitations near $|\boldsymbol{k}| = k_0$ are called *rotons*

Consider a different reference frame R' in which the He II is in motion at a uniform velocity $\boldsymbol{u}$. One moves from R_0 to R' via a Galilean transformation which, classically, is written as

$$\boldsymbol{r}' = \boldsymbol{r}_0 + \boldsymbol{u}t \quad , \tag{7.51}$$

$$\boldsymbol{P}' = \boldsymbol{P}_0 + M\boldsymbol{u} \quad , \tag{7.52}$$

$$E' = E_0 + \boldsymbol{P}_0 \cdot \boldsymbol{u} + \frac{1}{2}M|\boldsymbol{u}|^2 \quad . \tag{7.53}$$

These relations give the transformation laws for the position $\boldsymbol{r}$, the momentum $\boldsymbol{P}$ and the energy E from R_0 to R', where M is the mass of a portion of the fluid. One can verify that the center-of-mass velocities $\boldsymbol{v}'$ and $\boldsymbol{v}_0$ satisfy the expected relation

$$\boldsymbol{v}' = \nabla_{\boldsymbol{P}'} E' = \nabla_{\boldsymbol{P}_0} E' = \nabla_{\boldsymbol{P}_0} E_0 + \boldsymbol{u} = \boldsymbol{v}_0 + \boldsymbol{u} \quad . \tag{7.54}$$

In the rest frame R_0, one has

$$\boldsymbol{P}_0 = (\Phi_0, \mathrm{P}\Phi_0) = 0 \tag{7.55}$$

and we fix the energy of the superfluid at rest to $E_0 = 0$.

If M represents a certain mass of fluid flowing at a uniform velocity in a capillary, one fixes R' such that the tube is at rest. Furthermore, one assumes that the inertia of the tube is very large: no exchange of momentum with the fluid modifies its rest state in R'. Because of (7.52), (7.53) and (7.54),

$$\boldsymbol{P}' = M\boldsymbol{u} \quad , \quad \boldsymbol{P}_0 = 0 \quad , \tag{7.56}$$

$$E' = \frac{M|\boldsymbol{u}|^2}{2} \quad , \quad E_0 = 0 \quad , \tag{7.57}$$

and

$$\boldsymbol{v}' = \boldsymbol{u} \quad . \tag{7.58}$$

If one creates an excitation of wavenumber $\boldsymbol{k}$ in He II one can write, in R_0,

$$\boldsymbol{P}_{\text{exc},0} = \hbar\boldsymbol{k} \quad , \tag{7.59}$$

$$E_{\text{exc},0} = \varepsilon(\boldsymbol{k}) = \hbar c_{\text{s}}|\boldsymbol{k}| \quad , \tag{7.60}$$

keeping the linear part of spectrum (7.50). In the reference frame R' of the capillary,

$$\boldsymbol{P}'_{\text{exc}} = \hbar\boldsymbol{k} + M\boldsymbol{u} \quad , \tag{7.61}$$

$$E'_{\text{exc}} = \varepsilon(\boldsymbol{k}) + \hbar\boldsymbol{k}\cdot\boldsymbol{u} + \frac{1}{2}M|\boldsymbol{u}|^2 = \hbar(c_{\text{s}}|\boldsymbol{k}| + \boldsymbol{u}\cdot\boldsymbol{k}) + \frac{1}{2}M|\boldsymbol{u}|^2 \quad . \tag{7.62}$$

Relation (7.62) is very important: for the case where $|\boldsymbol{u}| < c_{\text{s}}$, one always has

$$c_{\text{s}}|\boldsymbol{k}| + \boldsymbol{u}\cdot\boldsymbol{k} > (c_{\text{s}} - u)|\boldsymbol{k}| > 0 \quad , \tag{7.63}$$

which shows that the creation of such an excitation costs energy from the liquid. As long as its velocity $|\boldsymbol{u}|$ remains below c_{s}, the liquid cannot dissipate its energy by contact with the capillary: superfluidity is possible. If $\varepsilon(\boldsymbol{k})$ goes to zero more rapidly than $|\boldsymbol{k}|$ when $|\boldsymbol{k}|$ itself goes to zero, the creation of an excitation of small $\boldsymbol{k}$ would be energetically favorable each time that $\boldsymbol{u}\cdot\boldsymbol{k} < 0$. A large number of excitations would appear until the fluid stops with respect to the capillary: this is what happens with a typical fluid or, in the case of He II, when the dimension of the capillary grows. Under these circumstances, it becomes easier and easier for vortices with quantized circulation, as described in Sect. 7.2.2, to appear.

We are lead to conclude from (7.63) that, as long as the velocity of He II is below the speed of sound c_{s}, the fluid will flow through the capillary without dissipation. The velocity c_{s} thus represents a critical velocity for the flow of He II through the capillary. We have seen that the actual critical velocities v_{c} are much slower than the speed of sound and, as one would expect, decrease strongly as the diameter of the tube increases: c_{s} is of the order of 220 $\text{m}\,\text{s}^{-1}$, while maximum values for v_{c} are of the order of 50 $\text{cm}\,\text{s}^{-1}$ for films of a few Å in thickness. The discrepancy between the theoretical prediction and the experimentally observed value must be blamed on the crudeness of the model being used: it is unrealistic to exclude all excitations other than the phonons, especially as the diameter of the capillary increases.

Now, suppose that conditions are favorable for the creation of excitations in the tube. Relation (7.63) shows that the most probable momentum is in the opposite direction to $\boldsymbol{u}$. The total momentum $\boldsymbol{P}_{\text{exc}}$ of the excitations can thus be put in the form

$$\boldsymbol{P}_{\text{exc}} = -M'\boldsymbol{u} \quad . \tag{7.64}$$

The coefficient $M' > 0$ has the dimension of mass. Relation (7.64) shows that the excitations can be assigned a mass which depends on their number and thus, in general, on the temperature. In the two-fluid model, one identifies the normal component of He II to the ensemble of excitations of the fluid. Relation (7.64) allows one to identify the density and the velocity field of this normal component. Relations (7.12) and (7.13) thus become plausible.

The considerations presented here, in the form prescribed by Landau, are not purely microscopic (the shape of the spectrum was postulated to conform with experimental observations). If one wishes to start from a completely microscopic picture, the best model to study is the imperfect Bose gas. We will see in the following section that this can be done at the expense of making certain hypotheses whose validity is very difficult to demonstrate. The interesting part is found in the analogies between the imperfect Bose gas and the model developed in this section.

Before closing this section, it is useful to note that we still have not justified the irrotational nature of the velocity field (relation (7.15)). In fact, this fundamental property of a superfluid can not be deduced in a simple way, starting from the hypotheses of Landau and Feynman.

7.4 Imperfect Bose Gas

7.4.1 Bogoliubov's Approximation and Transformation

We noted in Sect. 7.1 that ^{4}He undergoes a transition at the temperature T_λ, which offers an analogy with the Bose–Einstein condensation. Moreover, Feynman's argument tends to explain the excitation spectrum by making direct use of Bose statistics, obeyed by He atoms (Sect. 7.3.1). All the difficulty in studying He II is related to the fact that it is a liquid, for which there exists no simple treatment. In the following, we limit ourselves to a study of an imperfect Bose gas made up of atoms of mass m and of zero spin. In fact, this model is better suited to describe a Bose–Einstein condensation of dilute cold atoms presented in Sect. 2.2.1 (with the important difference that the latter condensation occurs in a confining potential not introduced here).

We start with a Hamiltonian $\mathtt{H}$ given by

$$\mathtt{H} = \sum_{\boldsymbol{k}} \frac{\hbar^2|\boldsymbol{k}|^2}{2m} a^*_{\boldsymbol{k}} a_{\boldsymbol{k}} + \frac{1}{2L^3} \sum_{\boldsymbol{k}_1,\boldsymbol{k}_2,\boldsymbol{k}} \tilde{V}(\boldsymbol{k}) a^*_{\boldsymbol{k}_1} a^*_{\boldsymbol{k}_2} a_{\boldsymbol{k}_2+\boldsymbol{k}} a_{\boldsymbol{k}_1-\boldsymbol{k}} \quad , \tag{7.65}$$

where

$$\tilde{V}(\boldsymbol{k}) = \int \mathrm{d}\boldsymbol{x} \mathrm{e}^{-\mathrm{i}\boldsymbol{k}\cdot\boldsymbol{x}} V(|\boldsymbol{x}|) = \tilde{V}(|\boldsymbol{k}|) = \tilde{V}^*(|\boldsymbol{k}|) \tag{7.66}$$

where we have used relation (3.116) for particles of zero spin. Suppose in the following that the potential $V(\boldsymbol{x})$ is such that its Fourier transform is positive,

$$\tilde{V}(\boldsymbol{k}) \geq 0 \quad , \tag{7.67}$$

corresponding to the repulsive character of the potential. Also, suppose that the effect of the interaction is weak enough for the Bose–Einstein condensation to occur; moreover, consider only the case at zero temperature. If the interaction was not present, only the single-particle ground state would be occupied:

$$n_0 = n \quad , \quad n_{\boldsymbol{k}} = 0 \quad , \quad \boldsymbol{k} \neq 0 \quad . \tag{7.68}$$

Suppose that, in the presence of a weak interaction, (7.68) still holds approximately true, so that the majority of particles remain in the state $\boldsymbol{k} = 0$, so that

$$\frac{n - n_0}{n} \ll 1 \quad . \tag{7.69}$$

This hypothesis has an important consequence. The eigenvalues of $a_0^* a_0$ and $a_0 a_0^*$ are n_0 and $n_0 + 1$. Since n_0 is macroscopic, one can ignore the 1 compared to n_0. This implies that the commutator $[a_0, a_0^*]$, which is of order 1, is negligible with respect to n_0 and that one can treat a_0 and a_0^* as numbers, writing

$$a_0 = a_0^* = \sqrt{n_0} \quad . \tag{7.70}$$

To see how the condensation affects the interaction, one sorts the terms of (7.65) in order of powers of a_0 and a_0^*:

$$\begin{aligned} \mathtt{H} = & \sum_{\boldsymbol{k}} \frac{\hbar^2 |\boldsymbol{k}|^2}{2m} a_{\boldsymbol{k}}^* a_{\boldsymbol{k}} + \frac{1}{2L^3} \tilde{V}(0) a_0^* a_0^* a_0 a_0 \\ & + \frac{1}{2L^3} \tilde{V}(0) \sum_{\boldsymbol{k} \neq 0} (a_0^* a_{\boldsymbol{k}}^* a_{\boldsymbol{k}} a_0 + a_{\boldsymbol{k}}^* a_0^* a_0 a_{\boldsymbol{k}}) \\ & + \frac{1}{2L^3} \sum_{\boldsymbol{k} \neq 0} \tilde{V}(\boldsymbol{k}) (a_{\boldsymbol{k}}^* a_0^* a_{\boldsymbol{k}} a_0 + a_0^* a_{\boldsymbol{k}}^* a_0 a_{\boldsymbol{k}} \\ & + a_0^* a_0^* a_{\boldsymbol{k}} a_{-\boldsymbol{k}} + a_{\boldsymbol{k}}^* a_{-\boldsymbol{k}}^* a_0 a_0) + \mathtt{V}' \quad , \end{aligned} \tag{7.71}$$

where $\mathtt{V}'$ comprises of all other terms containing at most one single a_0 or a_0^* operator. The approximation proposed by Bogoliubov consists of keeping only terms of order n_0^2 and n_0; that is, one neglects V'. Taking into account (7.66) and (7.70), one obtains the approximate Hamiltonian

$$\begin{aligned} \mathtt{H} \simeq & \sum_{\boldsymbol{k} \neq 0} \frac{\hbar^2 |\boldsymbol{k}|^2}{2m} a_{\boldsymbol{k}}^* a_{\boldsymbol{k}} + \tilde{V}(0) \left(\frac{n_0^2}{2L^3} + \frac{n_0}{L^3} \sum_{\boldsymbol{k} \neq 0} a_{\boldsymbol{k}}^* a_{\boldsymbol{k}} \right) \\ & + \frac{n_0}{2L^3} \sum_{\boldsymbol{k} \neq 0} \tilde{V}(\boldsymbol{k}) (a_{\boldsymbol{k}}^* a_{-\boldsymbol{k}}^* + a_{\boldsymbol{k}} a_{-\boldsymbol{k}}) + \frac{n_0}{L^3} \sum_{\boldsymbol{k} \neq 0} \tilde{V}(\boldsymbol{k}) a_{\boldsymbol{k}}^* a_{\boldsymbol{k}} \quad . \end{aligned} \tag{7.72}$$

Noting again that we neglect terms of order $(n - n_0)^2$, we have

$$\begin{aligned} n_0^2 &= [n + (n_0 - n)]^2 \simeq n^2 + 2(n_0 - n)n_0 \\ &= n^2 - 2n_0 \sum_{\boldsymbol{k}\neq 0} a_{\boldsymbol{k}}^* a_{\boldsymbol{k}} \quad . \end{aligned} \tag{7.73}$$

Thus, one can now write

$$\begin{aligned} \mathtt{H} \simeq \frac{\tilde{V}(0)n^2}{2L^3} &+ \sum_{\boldsymbol{k}\neq 0} \left[\left(\frac{\hbar^2|\boldsymbol{k}|^2}{2m} + \frac{n}{L^3}\tilde{V}(\boldsymbol{k}) \right) a_{\boldsymbol{k}}^* a_{\boldsymbol{k}} \right. \\ &\left. + \frac{n}{2L^3}\tilde{V}(\boldsymbol{k})(a_{\boldsymbol{k}}^* a_{-\boldsymbol{k}}^* + a_{\boldsymbol{k}} a_{-\boldsymbol{k}}) \right] \quad . \end{aligned} \tag{7.74}$$

The Hamiltonian (7.74) is a quadratic form of $a_{\boldsymbol{k}}$ and $a_{\boldsymbol{k}}^*$. It can be diagonalized by a canonical Bogoliubov transformation, formally analogous to the one which appears in superconductivity. We thus introduce the new operators

$$\alpha_{\boldsymbol{k}} = u_{\boldsymbol{k}} a_{\boldsymbol{k}} - v_{\boldsymbol{k}} a_{-\boldsymbol{k}}^* \quad , \quad \boldsymbol{k} \neq 0 \quad , \tag{7.75}$$

where the real coefficients $u_{\boldsymbol{k}}$ and $v_{\boldsymbol{k}}$ are to be determined. Recall that the transformation is canonical if it preserves the commutation relations

$$[\alpha_{\boldsymbol{k}}, \alpha_{\boldsymbol{k}'}] = [\alpha_{\boldsymbol{k}}^*, \alpha_{\boldsymbol{k}'}^*] = 0 \quad , \tag{7.76}$$

$$[\alpha_{\boldsymbol{k}}, \alpha_{\boldsymbol{k}'}^*] = \delta_{\boldsymbol{k},\boldsymbol{k}'} \quad . \tag{7.77}$$

This is the case if

$$u_{\boldsymbol{k}}^2 - v_{\boldsymbol{k}}^2 = 1 \quad . \tag{7.78}$$

One can invert (7.75) assuming that $u_{\boldsymbol{k}}$ and $v_{\boldsymbol{k}}$ depend only on $|\boldsymbol{k}|$

$$u_{\boldsymbol{k}} = u_{-\boldsymbol{k}} \quad , \quad v_{\boldsymbol{k}} = v_{-\boldsymbol{k}} \quad . \tag{7.79}$$

Then, one obtains

$$a_{\boldsymbol{k}} = u_{\boldsymbol{k}} \alpha_{\boldsymbol{k}} + v_{\boldsymbol{k}} \alpha_{-\boldsymbol{k}}^* \quad . \tag{7.80}$$

The operators $\alpha_{\boldsymbol{k}}$ and $\alpha_{\boldsymbol{k}}^*$ annihilate or create a quasi-particle of momentum $\hbar\boldsymbol{k}$. If one substitutes (7.80) into (7.74) and defines the density $\rho = n/L^3$, then

$$\begin{aligned} \mathtt{H} \simeq \frac{\tilde{V}(0)\rho^2 L^3}{2} &+ \sum_{\boldsymbol{k}\neq 0} \left[\left(\frac{\hbar^2|\boldsymbol{k}|^2}{2m} + \rho\tilde{V}(\boldsymbol{k}) \right) v_{\boldsymbol{k}}^2 + \rho\tilde{V}(\boldsymbol{k}) u_{\boldsymbol{k}} v_{\boldsymbol{k}} \right] \\ &+ \sum_{\boldsymbol{k}\neq 0} \left[\left(\frac{\hbar^2|\boldsymbol{k}|^2}{2m} + \rho\tilde{V}(\boldsymbol{k}) \right) (u_{\boldsymbol{k}}^2 + v_{\boldsymbol{k}}^2) + 2\rho\tilde{V}(\boldsymbol{k}) u_{\boldsymbol{k}} v_{\boldsymbol{k}} \right] \alpha_{\boldsymbol{k}}^* \alpha_{\boldsymbol{k}} \\ &+ \sum_{\boldsymbol{k}\neq 0} \left[\frac{\rho\tilde{V}(\boldsymbol{k})}{2} (u_{\boldsymbol{k}}^2 + v_{\boldsymbol{k}}^2) \right. \\ &\left. + \left(\frac{\hbar^2|\boldsymbol{k}|^2}{2m} + \rho\tilde{V}(\boldsymbol{k}) \right) u_{\boldsymbol{k}} v_{\boldsymbol{k}} \right] (\alpha_{\boldsymbol{k}}^* \alpha_{-\boldsymbol{k}}^* + \alpha_{\boldsymbol{k}} \alpha_{-\boldsymbol{k}}) \quad . \end{aligned} \tag{7.81}$$

We would like to diagonalize H; that is, to give it the form

$$\mathtt{H} = E_0 + \sum_{\boldsymbol{k}\neq 0} E_{\boldsymbol{k}} \alpha^*_{\boldsymbol{k}} \alpha_{\boldsymbol{k}} \quad , \tag{7.82}$$

where E_0 is a constant. To do this, one needs only set the coefficient of $(\alpha^*_{\boldsymbol{k}}\alpha^*_{-\boldsymbol{k}} + \alpha_{\boldsymbol{k}}\alpha_{-\boldsymbol{k}})$ to zero in (7.81). One can then verify that the equation

$$\frac{\rho\tilde{V}(\boldsymbol{k})}{2}(u^2_{\boldsymbol{k}} + v^2_{\boldsymbol{k}}) + \left(\frac{\hbar^2|\boldsymbol{k}|^2}{2m} + \rho\tilde{V}(\boldsymbol{k})\right) u_{\boldsymbol{k}} v_{\boldsymbol{k}} = 0 \tag{7.83}$$

has the solution

$$u^2_{\boldsymbol{k}} + v^2_{\boldsymbol{k}} = \frac{\rho\tilde{V}(\boldsymbol{k}) + \varepsilon_{\boldsymbol{k}}}{E_{\boldsymbol{k}}} \quad , \tag{7.84}$$

$$2u_{\boldsymbol{k}} v_{\boldsymbol{k}} = -\frac{\rho\tilde{V}(\boldsymbol{k})}{E_{\boldsymbol{k}}} \tag{7.85}$$

with

$$\varepsilon_{\boldsymbol{k}} = \frac{\hbar^2|\boldsymbol{k}|^2}{2m} \tag{7.86}$$

and

$$E_{\boldsymbol{k}} = \left[\left(\rho\tilde{V}(\boldsymbol{k}) + \varepsilon_{\boldsymbol{k}}\right)^2 - \left(\rho\tilde{V}(\boldsymbol{k})\right)^2\right]^{1/2} \quad , \tag{7.87}$$

so that (7.78) is automatically satisfied. The Hamiltonian (7.81) thus takes the form (7.82), where $E_{\boldsymbol{k}}$ is given by (7.87) and

$$E_0 = \frac{\tilde{V}(0)\rho^2 L^3}{2} + \frac{1}{2}\sum_{\boldsymbol{k}\neq 0}\left[E_{\boldsymbol{k}} - \varepsilon_{\boldsymbol{k}} - \rho\tilde{V}(\boldsymbol{k})\right] \quad . \tag{7.88}$$

It is useful to note that $\tilde{V}(\boldsymbol{k})$ is real (relation (7.66)) and hence, because of (7.67), E_0 and $E_{\boldsymbol{k}}$ are also real.

Moreover, (7.87) shows that the excited states are of the form predicted by Landau. Let $|\Psi_0\rangle$ be the state defined by

$$\alpha_{\boldsymbol{k}}\,|\Psi_0\rangle = 0 \quad . \tag{7.89}$$

The ground state $|\Psi_0\rangle$ represents the "vacuum" of quasi-particles (or of excitations). By successive application of the creation operators $\alpha^*_{\boldsymbol{k}}$, one obtains the excited states of the system. Their energies are of the form

$$E(\{n_{\boldsymbol{k}}\}) = E_0 + \sum_{\boldsymbol{k}\neq 0} n_{\boldsymbol{k}} E_{\boldsymbol{k}} \quad , \quad n_{\boldsymbol{k}} = 0, 1, 2, \ldots \quad , \tag{7.90}$$

where E_0 is the energy of the ground state. Note that the form of (7.90) is a consequence of the fact that the quasi-particles are bosons (relations (7.76) and (7.77)). Moreover, $E_{\boldsymbol{k}}$ is linear in $|\boldsymbol{k}|$ in the small wavenumber limit

$$E_{\boldsymbol{k}} \simeq \left(2\varepsilon_{\boldsymbol{k}}\rho\tilde{V}(\boldsymbol{k}) + \varepsilon_{\boldsymbol{k}}^2\right)^{1/2} \simeq |\boldsymbol{k}|\hbar\sqrt{\frac{\rho\tilde{V}(0)}{m}} \quad , \quad |\boldsymbol{k}| \to 0 \quad . \tag{7.91}$$

We thus conclude that the low-energy excitations of a condensed Bose gas are phonons. It is nevertheless reasonable to wonder to what extent these conclusions apply to He II. This point is briefly discussed in the following section.

7.4.2 Bose Gas or Liquid?

We introduced the model in the previous section under the name of an imperfect Bose gas without making clear whether we are indeed limiting ourselves to a gas, or whether the model remains valid when extended to the case of a liquid. One might be tempted to conclude that we have provided a microscopic description of superfluidity. This would be a little bit too hasty, though, as there are a few difficulties which still need to be resolved.

Concerning the calculation of the excitation energy spectrum, inspection of relation (7.87) shows that one must impose certain conditions on the potential for $E_{\boldsymbol{k}}$ to be real. In fact, $E_{\boldsymbol{k}}$ can only be real if the potential satisfies condition (7.67) or at least the following condition:

$$\rho\tilde{V}(\boldsymbol{k}) \geq -\frac{\hbar^2|\boldsymbol{k}|^2}{4m} \quad . \tag{7.92}$$

So that the low-energy excitations are really phonons obeying (7.91), it must be true that

$$\tilde{V}(0) = \int \mathrm{d}\boldsymbol{x} V(|\boldsymbol{x}|) > 0 \quad . \tag{7.93}$$

That is, the repulsive part of the potential must be dominant. It is interesting to note that for a Bose gas, it is the repulsive nature of the potential which is important while, for superconductivity, it is the attractive nature of the potential which is dominant.

We have assumed that Bose–Einstein condensation would occur in the state $\boldsymbol{k} = 0$ relative to the plane wave basis (relation (7.69)). This is correct for a perfect gas and could remain approximately valid for a diluted weakly-interacting gas, provided that we can demonstrate this. For a liquid, on the other hand, one could easily imagine that for a consistent description it is necessary to take into account the strong repulsion of atoms at a short distance. Then, it is not obvious how to formulate the concept of Bose–Einstein condensation in terms of the occupation of single-particle states, since correlations between atoms due to this repulsion cannot be neglected.

At the density of He II, the Bogoliubov approximation which leads to the approximate Hamiltonian (7.74) is questionable. One knows, for example, that in the ground state of interacting bosons the condensed fraction n_0/n does not approach 1, but is around 10 %. This fact, established in an estimation by Penrose and Onsager, is confirmed by observation. Under these conditions, (7.69) is not valid and one can not neglect the $\mathtt{V}'$ terms in (7.71).

In conclusion, one can say that the Bogoliubov model presented in this section does not play the same role in a microscopic theory of superfluidity as the BCS theory does in superconductivity. Rather, it provides an alternative description of He II which is no more fundamental than the semi-phenomenological approaches of Feynman and Landau.

7.5 Superfluidity of the Light Isotope ^{3}He

7.5.1 A Fermi Liquid

The goal of this section is to very briefly present the properties of liquid ^{3}He, since its behavior is very different than the heavier isotope liquid ^{4}He, and because these differences arise due to Fermi statistics (for ^{3}He) and Bose statistics (for ^{4}He).

Like ^{4}He, ^{3}He is a permanent liquid (at all temperatures, it remains liquid at atmospheric pressure, solidifying only at a pressure of 34.4 atm near 0 K). Its critical temperature T_c is less than that of ^{4}He ($T_c(^3\text{He}) = 3.32$ K; $T_c(^4\text{He}) = 5.2$ K). We have seen in this chapter that the properties of He II are similar to those of a boson gas. This suggests that the behaviour of the lighter isotope, under liquid form, could also be similar to a gas which obeys Fermi statistics.

As shown in Sect. 2.2.2, a perfect fermion gas does not undergo a transition. It is nonetheless characterized by a degeneracy temperature T_0 given by (2.102). By definition, for $T \ll T_0$, effects due to Fermi–Dirac statistics become dominant while, for $T \gg T_0$, the properties of the gas are not notably different from a classical perfect gas.

Studies of the specific heat, the magnetic susceptibility, the thermal conductivity and the viscosity show that the properties of liquid ^{3}He approach those of a degenerate gas for $T < 100$ mK. But, if one looks back at relation (2.102), one would expect for this degenerate character to occur at all temperatures lower than $T_c(^3\text{He})$. Indeed, (2.102) establishes a relation between T_0 and the density ρ, with the constants $\hbar$, k_B and m as parameters. If one replaces these constants by their values (m being the mass of an atom of ^{3}He) and ρ by ρ_M/m, ρ_M being the specific mass of the liquid at 1 atmosphere (ρ_M varies only slightly with temperature in this region), one obtains $T_0 = 4.99 \text{ K} > T_c(^3\text{He})$.

Why does ^{3}He resemble a degenerate Fermi gas only under 100 mK? In general, a liquid does not have the same properties as a perfect gas, and it is necessary to examine the reasons for this more closely. Compared to liquid ^{4}He, there is a much better basis for identifying ^{3}He with a perfect gas. The reason is simple: a perfect Fermi gas does not undergo a transition. One can thus follow this identity to lower temperatures, up to the point where the phenomenon of superfluidity, described in the next section, occurs. Landau showed how to realize an identity which becomes even more accurate as the temperature decreases: one needs only establish a correspondence between the fermions of the perfect gas, not with the atoms of the liquid, but with quasi-particles which are analogous to those of the Bogoliubov transformation presented in Sect. 7.4.1. In the presence of an interaction, the atoms in the liquid are not independent: one therefore turns to the quasi-particles which are independent, at least in the first approximation. The quasi-particles have, in general, properties which are close to those of the particles they come from. Their spin does not change; they behave according to the same statistics. Their mass m^*, on the other hand, differs from that of the atoms ($m^*/m \simeq 3$). If in (2.102) one replaces m by m^*, ρ must be replaced by $\rho^* = \rho m/m^*$, giving rise to a new degeneracy temperature:

$$T_0^* = T_0 \left(\frac{m}{m^*}\right)^{5/3} \simeq 0.8 \text{ K} \quad . \tag{7.94}$$

The analogy between liquid ^{3}He and a perfect gas of quasi-particles is thus justified for temperatures $T \ll T_0^*$. One has a better understanding why it is only below 100 mK that the specific heat and the other properties of ^{3}He mentioned above show a behaviour analogous to that of a perfect gas obeying Fermi–Dirac statistics.

7.5.2 Superfluidity of ^{3}He

While the perfect Fermi gas does not undergo a transition analogous to Bose–Einstein condensation, an attraction between its constituents can give rise to a transition via a BCS state, characterized by the existence of pairs. In the case of liquid ^{3}He, such a transition was sought out for quite some time. Before it was observed, it was very difficult to determine its properties, in particular its transition temperature. The interaction between atoms is strongly repulsive at short distance, so that one expects for the formation of pairs to take place in a state of orbital angular momentum $l > 0$ (an s state has a non-zero probability for finding the two members of the pair at vanishing relative distance, a very unfavorable situation for binding because of the repulsive potential of the core).

It was only in 1972 that Osheroff, Gully, Richardson and Lee observed the superfluidity of ^{3}He for the first time.[2] In fact they discovered two transitions rather than just one. At the pressure of solidification of 34.4 atm where, in

[2] D.D. Osheroff, W.J. Gully, R.C. Richardson and D.M. Lee, Phys. Rev. Lett. *29*, 920 (1972).

the neighbourhood of 0 K, ^{3}He is in equilibrium with its solid, one observes the following succession of events:

- at 2.79 mK, a normal transition of a Fermi liquid to a first superfluid phase, A;
- at 2.16 mK, another superfluid phase, B, which takes the place of phase A.

Below 21.5 atm, only phase B appears, and at a transition temperature which depends on the pressure. In the presence of a magnetic field, things become even more complicated, with the appearance of a third superfluid phase.

At the present time, the superfluidity of ^{3}He is well understood, in spite of the difficulty in attaining the extremely low temperatures needed to study it. Phases A and B are anisotropic superfluid phases with the following main characteristics:

- In each of the two phases, two quasi-particles form a pair of orbital angular momentum $l = 1$ and spin $s = 1$. However, the pairing differs in the two phases.
- Apart from the phase factor $\mathrm{e}^{\mathrm{i}\alpha}$ which is common to the ensemble of pairs, they are still characterized by the unit vectors $\hat{l}$ and $\hat{s}$ which determine the direction of the vectors $\boldsymbol{l}$ and $\boldsymbol{s}$, the orbital angular momentum and the spin respectively.
- Because of the existence of the vectors $\hat{l}$ and $\hat{s}$, superfluids A and B are anisotropic (A more so than B).
- When a global rotation is applied, a lattice of vortices forms in the center of superfluid ^{3}He. The structure of this lattice is not the same as in the case of He II. It also differs from phase A to phase B.

We do not propose to analyze these phenomena in detail, but to note that, once again, a collection of fermions benefits from an attractive interaction by pairing according to the BCS mechanism, even if the details of the pairing differ from one system to another. The universality of this mechanism is astonishing since in principle, pairing is only one way among many other possibilities to avoid the sign ambiguity, as described in Chap. 5. It is not obvious a priori that such a mechanism would be energetically favorable.

Exercises

1. The Feynman Variational Function

It is useful to describe the excitations of liquid helium by means of trial wave-functions $\Psi(\boldsymbol{x}_1, \ldots, \boldsymbol{x}_n)$ that are orthogonal to the ground state: $\langle\Psi|\Phi_0\rangle = 0$. One makes the following simple choice, compatible with Bose statistics

$$\Psi(\boldsymbol{x}_1,\ldots,\boldsymbol{x}_n) = F(\boldsymbol{x}_1,\ldots,\boldsymbol{x}_n)\Phi_0(\boldsymbol{x}_1,\ldots,\boldsymbol{x}_n) \quad ,$$

where

$$F(\boldsymbol{x}_1,\ldots,\boldsymbol{x}_n) = \sum_{i=1}^{n} f(\boldsymbol{x}_i)$$

and $f(\boldsymbol{x})$ is to be determined variationally (the fluid is enclosed in a cubic box with periodic boundary conditions).

(i) With $\mathtt{H}$ the Hamiltonian of the Bose gas (as given in (7.48)) and $E = \langle\Psi|H|\Psi\rangle \,/\, \langle\Psi|\Psi\rangle$, express the excitation energy $\varepsilon = E - E_0$ as

$$\varepsilon = \frac{\hbar^2}{2m}\frac{\sum_{i=1}^{n}\int \mathrm{d}\boldsymbol{x}_1\cdots \mathrm{d}\boldsymbol{x}_n\,|\nabla_i F(\boldsymbol{x}_1,\ldots,\boldsymbol{x}_n)|^2\Phi_0^2(\boldsymbol{x}_1,\ldots,\boldsymbol{x}_n)}{\int \mathrm{d}\boldsymbol{x}_1\ldots \mathrm{d}\boldsymbol{x}_n|F(\boldsymbol{x}_1,\ldots,\boldsymbol{x}_n)|^2\Phi_0^2(\boldsymbol{x}_1,\ldots,\boldsymbol{x}_n)} \quad .$$

Hint: Integrate the kinetic energy by parts and use the ground state eigenvalue equation $H\Phi_0 = E_0\Phi_0$.

(ii) Introduce the ground state density $\rho(\boldsymbol{x}) = \langle\Phi_0|\, n(\boldsymbol{x})\,|\Phi_0\rangle$ (3.135) and the ground state density correlations $\Gamma(\boldsymbol{x}_1,\boldsymbol{x}_2) = \langle\Phi_0|\, {:}n(\boldsymbol{x}_1)n(\boldsymbol{x}_2){:}\, |\Phi_0\rangle$ (3.142) to write ε as

$$\varepsilon = \frac{\hbar^2}{2m}\frac{\int \mathrm{d}\boldsymbol{x}\,\rho(\boldsymbol{x})|\nabla f(\boldsymbol{x})|^2}{\int \mathrm{d}\boldsymbol{x}_1\int \mathrm{d}\boldsymbol{x}_2\, S(\boldsymbol{x}_1,\boldsymbol{x}_2)f^*(\boldsymbol{x}_1)f(\boldsymbol{x}_2)}$$

with $S(\boldsymbol{x}_1,\boldsymbol{x}_2) = \Gamma(\boldsymbol{x}_1,\boldsymbol{x}_2) + \delta(\boldsymbol{x}_1-\boldsymbol{x}_2)\rho(\boldsymbol{x}_1)$.

(iii) In an homogeneous state, $\rho(\boldsymbol{x}) = \rho$, $S(\boldsymbol{x}_1,\boldsymbol{x}_2) = S(\boldsymbol{x}_1-\boldsymbol{x}_2,0)$, show that for each wave vector $\boldsymbol{k} \neq 0$, the variational problem has the solution

$$\varepsilon(\boldsymbol{k}) = \frac{\hbar^2\rho}{2m}\frac{|\boldsymbol{k}|^2}{\tilde{S}(\boldsymbol{k})} \quad , \quad f_{\boldsymbol{k}}(\boldsymbol{x}) = \mathrm{e}^{\mathrm{i}\boldsymbol{k}\cdot\boldsymbol{x}} \quad ,$$

where $\tilde{S}(\boldsymbol{k})$ is the Fourier transform of $S(\boldsymbol{x}_1-\boldsymbol{x}_2,0)$.

(iv) Verify that the corresponding excited states $\Psi_{\boldsymbol{k}}(\boldsymbol{x}_1,\ldots,\boldsymbol{x}_n) = \sum_{i=1}^{n}\mathrm{e}^{\mathrm{i}\boldsymbol{k}\cdot\boldsymbol{x}_i}\Phi_0(\boldsymbol{x}_1,\ldots,\boldsymbol{x}_n)$ are pair-wise orthogonal, and orthogonal to Φ_0 for $\boldsymbol{k}\neq 0$.

Hint: The $\Psi_{\boldsymbol{k}}$ are eigenvectors of the total momentum.

Comment: This theory relates the excitation spectrum $\varepsilon(\boldsymbol{k})$ to the structure factor $\tilde{S}(\boldsymbol{k})$ of the liquid, which can be measured by X-ray scattering experiments. $\tilde{S}(\boldsymbol{k})$ is linear as $\boldsymbol{k}\to 0$, leading to the phonon spectrum (7.50), and has a maximum around a^{-1} which gives rise to the roton local minimum of Fig. 7.12. The prediction obtained in this way agrees, at least qualitatively, with the excitation spectrum measured independently by neutron scattering experiments.

8. Quantum Fields

8.1 Introduction

In classical physics, a field is described by one or more space–time functions which satisfy certain partial differential equations. Several important examples were already noted in Chap. 1: the vector potential $\boldsymbol{A}_{\text{cl}}(\boldsymbol{x}, t)$ of the electromagnetic field and the elastic displacement field $\boldsymbol{u}_{\text{cl}}(\boldsymbol{x}, t)$. A classical field can be imagined as an infinite and continuous ensemble of degrees of freedom: a pair of dynamical variables is attached to each point $\boldsymbol{x}$ in space $\mathcal{R}^3$, the amplitude $\boldsymbol{u}_{\text{cl}}(\boldsymbol{x}, t)$ of the field at time t and its time derivative $(\partial/\partial t)\boldsymbol{u}_{\text{cl}}(\boldsymbol{x}, t)$. The coupled evolution of this ensemble of degrees of freedom is governed by the differential equation of the field.

One can conceive of a quantum field in the same way. According to the correspondence principle between classical mechanics and quantum mechanics, the classical variables must be replaced by the operators $\boldsymbol{u}_{\text{op}}(\boldsymbol{x}, t)$ and $(\partial/\partial t)\boldsymbol{u}_{\text{op}}(\boldsymbol{x}, t)$. In this manner, *quantum degrees of freedom are attached to each point in space* $\boldsymbol{x}$*, and their ensemble forms the quantum field*, whose evolution is also governed by a partial differential equation.

To illustrate this concept, we present the example of the quantum linear chain in the continuum limit (Sect. 1.4.6). This example is obtained by letting the lattice constant a tend to zero while keeping the length of the chain $L = na$, the speed of sound $c_{\text{s}} = \omega a$ and the mass density $\rho_{\text{M}} = m/a$ all constant. It follows immediately from (1.250) and (1.236) that the field at a point x in $\mathcal{R}$ such that $ja \to x$ is given by

$$\begin{aligned}\lim_{a\to 0} u_j(t) &= u_{\text{op}}(x, t) \\ &= \left(\frac{\hbar}{\rho_{\text{M}} L}\right)^{1/2} \sum_k \frac{1}{\sqrt{2\omega_k}} \left(a_k^* \mathrm{e}^{-\mathrm{i}(kx-\omega_k t)} + a_k \mathrm{e}^{\mathrm{i}(kx-\omega_k t)}\right) \quad , \end{aligned} \tag{8.1}$$

where $\omega_k = c_s k$ and the sum is taken over the wavenumbers $k = 2\pi r/L$ for integer r (see (1.229) and (1.230)).

Note that the expansion (8.1) is identical to that of the classical elastic field (1.214) and (1.218), with the only difference being the replacement of the Fourier amplitudes β_k^*, β_k with the operators a_k^*, a_k. One can immediately verify from (8.1) that $u_{\text{op}}(x, t)$ still obeys the wave equation

$$\frac{1}{c_s^2}\frac{\partial^2}{\partial t^2}u_{\mathrm{op}}(x,t) = \frac{\partial^2}{\partial x^2}u_{\mathrm{op}}(x,t) \quad . \tag{8.2}$$

The family of operators $u_{\mathrm{op}}(x,t), x \in \mathcal{R}$, is the prototypical quantum field: the longitudinal quantum elastic field.

The associated velocity field is written as

$$\begin{aligned}\lim_{a\to 0}\frac{p_j(t)}{m} &= \frac{\partial}{\partial t}u_{\mathrm{op}}(x,t)\\ &= \mathrm{i}\left(\frac{\hbar}{\rho_{\mathrm{M}}L}\right)^{1/2}\sum_k\sqrt{\frac{\omega_k}{2}}\left(a_k^*\mathrm{e}^{-\mathrm{i}(kx-\omega_k t)} - a_k\mathrm{e}^{\mathrm{i}(kx-\omega_k t)}\right) \quad .\end{aligned} \tag{8.3}$$

Taking into account commutation relations (1.246) and (1.247), one finds that the commutator of the quantum fields of displacement and momentum density $\rho_{\mathrm{M}}(\partial/\partial t)u_{\mathrm{op}}(x,t)$ is equal to

$$\rho_{\mathrm{M}}\left[u_{\mathrm{op}}(x,t), \frac{\partial}{\partial t}u_{\mathrm{op}}(y,t)\right] = \mathrm{i}\hbar\frac{1}{L}\sum_k \mathrm{e}^{\mathrm{i}k(x-y)} = \mathrm{i}\hbar\delta(x-y) \tag{8.4}$$

and, moreover,

$$[u_{\mathrm{op}}(x,t), u_{\mathrm{op}}(y,t)] = \left[\frac{\partial}{\partial t}u_{\mathrm{op}}(x,t), \frac{\partial}{\partial t}u_{\mathrm{op}}(y,t)\right] = 0 \quad . \tag{8.5}$$

These relations are similar, in the continuum, to the canonical commutation relations of quantum variables. The quantized field is indeed equivalent to an infinite ensemble of quantum degrees of freedom. In this example, the quantum nature of the field is of course inherited from the particles which make up the chain. In particular, the creation and annihilation operators necessarily satisfy the same commutation rules as boson operators. This follows from the oscillations of atoms around their equilibrium positions.

On the other hand, the vibrations of a linear chain can only be approximately described by a continuous field (8.1). It is valid for distances $x \gg a$ ($a \simeq 10^{-8}$ cm) where the discrete nature of the atomic medium can be ignored.

The quantum-electromagnetic field, as well as the various other fields which appear in the theory of elementary particles, are of a much more fundamental nature. *Their quantum nature does not follow from any underlying material substrate, but represents the quantum dynamics of the elementary particles themselves.* This description is considered to be valid down to the smallest distances that can investigated (for example, the Compton length associated to W bosons is $\hbar/m_{\mathrm{W}}c \simeq 10^{-15}$ cm). We mention just two reasons which motivate the introduction of quantum fields. It appears that the interactions between particles can be described by the mediation of quantum fields. This fact is already visible in the classical form $-\int \mathrm{d}\boldsymbol{x}\,\boldsymbol{j}_{\mathrm{cl}}(\boldsymbol{x})\cdot\boldsymbol{A}_{\mathrm{cl}}(\boldsymbol{x})$

of the interaction of an electromagnetic field with a current. *Such an interaction, which couples the field to the current at point $\boldsymbol{x}$, is said to be local. The dynamics of the elementary particles results from local interactions of the associated quantum fields.* Finally, using the formalism of quantum fields, the principles of quantum mechanics and special relativity can be reconciled. In this chapter, we will limit ourselves to a study of quantum-electromagnetic and scalar fields. The quantum-electromagnetic field, introduced by Dirac in 1928, is a particulary time-honored example and remained a source of inspiration until recent developments of the gauge theories (Sect. 8.3.6).

The formalism of quantum fields in an expression of the richness of the wave–particle duality. It is already quite visible in the example of a phonon field (8.1). The wave aspect manifests itself by the wave equation (8.2), just as the particle aspect is revealed by the form of the expansion (8.1): a_k^* and a_k are none other than the creators and annihilators of the phonons. It is thus conceivable to introduce the notion of a free quantum field under either of these two complementary aspects. Since we have already become familiar with systems of variable particle number (Chap. 3), we will adopt the second point of view: we begin by defining in Sect. 8.2.1 the quantum-electromagnetic field associated with photons and from there deducing its properties. The other point of view consists of determining the canonical variables of the classical field and then quantizing them by rules analogous to (8.4) and (8.5); this will be briefly addressed in Sect. 8.3.5. The two methods are equivalent, and this is one of the great beauties of the theory of quantum fields.

8.2 The Quantum-Electromagnetic Field

8.2.1 The Free Field

Definition of the Free Field

The photon was introduced in Sect. 1.3.6 as a quantum particle of momentum $\hbar\boldsymbol{k}$ and energy $\hbar\omega_{\boldsymbol{k}} = \hbar c|\boldsymbol{k}|$. According to Planck's black-body law, photons must obey Bose statistics. Moreover this bosonic behavior is in accordance with the properties of the electromagnetic field reviewed in Chap. 1. First of all, the helicity of the photon equals one because particles with integer angular momentum must behave as bosons (Sect. 8.3.4). On the other hand, the decomposition of the electromagnetic field as a collection of harmonic oscillators calls for a quantum field satisfying commutation rather than anticommutation rules. A collection of photons with variable number is thus described, according to the formalism of Chap. 3, by the states of Fock space $\mathcal{F}_+(\mathcal{H}_\gamma)$ where $\mathcal{H}_\gamma$ is the single-particle space of the photon (1.162). According to (3.72), the Hamiltonian for free photons is

$$\mathrm{H}_0 = \sum_{\boldsymbol{k}\lambda} \hbar\omega_{\boldsymbol{k}} a^*_{\boldsymbol{k}\lambda} a_{\boldsymbol{k}\lambda} \quad . \tag{8.6}$$

At each space point $\boldsymbol{x}$, we attach the operator

$$\mathtt{A}(\boldsymbol{x},t) = \left(\frac{\hbar}{\varepsilon_0 L^3}\right)^{1/2} \sum_{\boldsymbol{k}\lambda} \frac{\boldsymbol{e}_\lambda(\boldsymbol{k})}{\sqrt{2\omega_{\boldsymbol{k}}}} \left(a^*_{\boldsymbol{k}\lambda} \mathrm{e}^{-\mathrm{i}(\boldsymbol{k}\cdot\boldsymbol{x}-\omega_{\boldsymbol{k}}t)} + a_{\boldsymbol{k}\lambda} \mathrm{e}^{\mathrm{i}(\boldsymbol{k}\cdot\boldsymbol{x}-\omega_{\boldsymbol{k}}t)}\right) \tag{8.7}$$

$$= \mathtt{A}_+(\boldsymbol{x},t) + \mathtt{A}_-(\boldsymbol{x},t) \quad . \tag{8.8}$$

The family of operators (8.7) defines the *quantum free field of photons*. Definition (8.7) shows that $\mathtt{A}(\boldsymbol{x},t)$ *has an expansion absolutely identical to a classical vector potential (1.154), the only difference being that the complex amplitudes* $\alpha^*_{\boldsymbol{k}\lambda}$ *and* $\alpha_{\boldsymbol{k}\lambda}$ *are replaced by the creation and annihilation operators of the photons*: $\mathtt{A}(\boldsymbol{x},t)$ is thus the quantum vector potential in the Coulomb gauge. From there, the *quantum electric and magnetic fields* are obtained in the usual manner (1.97):

$$\mathtt{E}(\boldsymbol{x},t) = -\frac{\partial}{\partial t}\mathtt{A}(\boldsymbol{x},t) \quad , \quad \mathtt{B}(\boldsymbol{x},t) = \nabla \times \mathtt{A}(\boldsymbol{x},t) \quad . \tag{8.9}$$

We have used symbols $\mathtt{A}(\boldsymbol{x},t)$, $\mathtt{E}(\boldsymbol{x},t)$ and $\mathtt{B}(\boldsymbol{x},t)$ analogous to those of Sect. 1.3; as long as there is no ambiguity, they will henceforth denote the quantum fields (8.7) and (8.9).

In (8.8), $\mathtt{A}_+(\boldsymbol{x},t)$ ($\mathtt{A}_-(\boldsymbol{x},t)$) denotes the contribution of the creators (annihilators) to expansion (8.7). Because of the signs of the arguments of the factors $\mathrm{e}^{\mathrm{i}\omega_{\boldsymbol{k}}t}$ and $\mathrm{e}^{-\mathrm{i}\omega_{\boldsymbol{k}}t}$ which appear in $\mathtt{A}_+(\boldsymbol{x},t)$ and $\mathtt{A}_-(\boldsymbol{x},t)$, these quantities are called the positive and negative frequency field components. Clearly, one has

$$(\mathtt{A}_+(\boldsymbol{x},t))^* = \mathtt{A}_-(\boldsymbol{x},t) \quad , \tag{8.10}$$

from which one concludes that *the photon field is a Hermitian operator.*

Equations of Motion

It follows immediately from (8.7) that the family of operators $\mathtt{A}(\boldsymbol{x},t)$ obeys the wave equation

$$\frac{1}{c^2}\frac{\partial^2}{\partial t^2}\mathtt{A}(\boldsymbol{x},t) - \nabla^2\mathtt{A}(\boldsymbol{x},t) = 0 \tag{8.11}$$

and the transversality condition

$$\nabla \cdot \mathtt{A}(\boldsymbol{x},t) = 0 \quad . \tag{8.12}$$

Moreover, it follows from (3.88) and (3.89) that the free field obeys the usual law of quantum evolution in the Heisenberg representation with respect to the Hamiltonian of the photons (8.6)

$$\mathtt{A}(\boldsymbol{x},t)=\exp\left(\frac{\mathrm{i}\mathtt{H}_0 t}{\hbar}\right)\mathtt{A}(\boldsymbol{x})\exp\left(-\frac{\mathrm{i}\mathtt{H}_0 t}{\hbar}\right)\quad, \tag{8.13}$$

where $\mathtt{A}(\boldsymbol{x})=\mathtt{A}(\boldsymbol{x},t=0)$ is the field at time $t=0$. One thus sees that $\mathtt{A}(\boldsymbol{x},t)$ is the solution of the Heisenberg equations of motion

$$\frac{\partial}{\partial t}\mathtt{A}(\boldsymbol{x},t)=\frac{\mathrm{i}}{\hbar}\left[\mathtt{H}_0,\mathtt{A}(\boldsymbol{x},t)\right]\quad. \tag{8.14}$$

In the same manner, it follows from (8.9) and (8.11) that the quantum fields $\mathtt{E}(\boldsymbol{x},t)$ and $\mathtt{B}(\boldsymbol{x},t)$ obey the free Maxwell equations, and it is not difficult to convince oneself that these equations coincide with Heisenberg's equations of motion for these fields. Thus the law of evolution for the field displays two different, but equivalent, aspects. Firstly wave equation (8.11) is a *partial differential equation* like in classical physics, and secondly (8.14) has the form of a quantum equations of motion.

Hamiltonian of the Free Field

With definition (8.7), the Hamiltonian of the photons (8.6) is expressed with the help of the quantum fields by the same formulae as in the classical theory. To see this, we will calculate an expression analogous to (1.141), which gives the classical energy, but where $\mathtt{E}(\boldsymbol{x},t)$ and $\mathtt{B}(\boldsymbol{x},t)$ are now taken to be quantum operators (8.9). As the relation between the fields and the vector potential is the same as in the classical case and the expansion of the vector potential is the same as (1.140), all of the calculations which lead to relations (1.141)–(1.149) can be reproduced line by line without modification. In all cases, one must be cautious of the fact that the coefficients $f_{\boldsymbol{k}\lambda}=(\hbar/2\varepsilon_0\omega_{\boldsymbol{k}})^{1/2}a_{\boldsymbol{k}\lambda}$ are non-commuting operators and to preserve the order of the factors. On thus obtains

$$\begin{aligned}\frac{1}{2}\int \mathrm{d}\boldsymbol{x}\,\left(\varepsilon_0|\mathtt{E}(\boldsymbol{x},t)|^2+\mu_0^{-1}|\mathtt{B}(\boldsymbol{x},t)|^2\right)&=\sum_{\boldsymbol{k}\lambda}\hbar\omega_{\boldsymbol{k}}\frac{1}{2}(a^*_{\boldsymbol{k}\lambda}a_{\boldsymbol{k}\lambda}+a_{\boldsymbol{k}\lambda}a^*_{\boldsymbol{k}\lambda})\\&=\sum_{\boldsymbol{k}\lambda}\hbar\omega_{\boldsymbol{k}}a^*_{\boldsymbol{k}\lambda}a_{\boldsymbol{k}\lambda}+\frac{1}{2}\sum_{\boldsymbol{k}\lambda}\hbar\omega_{\boldsymbol{k}}\quad.\end{aligned} \tag{8.15}$$

This expression only differs from the photon Hamiltonian (8.6) by the divergent constant $1/2(\sum_{\boldsymbol{k}\lambda}\hbar\omega_{\boldsymbol{k}})$, called the *zero-point energy*. In Sect. 3.2.1, we have conventionally fixed the vacuum energy to zero (see (3.7)). If one maintains this convention, one sees that $\mathtt{H}_0$ is obtained by replacing $a_{\boldsymbol{k}\lambda}a^*_{\boldsymbol{k}\lambda}$ with $a^*_{\boldsymbol{k}\lambda}a_{\boldsymbol{k}\lambda}$, that is, by placing the creators to the left of the annihilators. This is precisely the normal ordering defined in Sect. 3.2.4. We thus have

$$\mathtt{H}_0 = \frac{1}{2}\int \mathrm{d}\boldsymbol{x} :\left(\varepsilon_0|\mathtt{E}(\boldsymbol{x},t)|^2 + \mu_0^{-1}|\mathtt{B}(\boldsymbol{x},t)|^2\right): $$
$$= \frac{1}{2}\sum_{\alpha=1}^{3}\int \mathrm{d}\boldsymbol{x} : \left[\varepsilon_0\left(\frac{\partial \mathtt{A}^\alpha}{\partial t}\right)^2 + \mu_0^{-1}|\nabla \mathtt{A}^\alpha|^2\right]: \quad . \tag{8.16}$$

Thus, placing the operators in the normal ordering in (8.16) is equivalent to fixing the vacuum energy to zero. It is important to realize that the choice of fixing the vacuum energy to zero does not change the equations of motion or their classical limit: this is enough to justify the rewriting of the Hamiltonian (8.16) for free photons.

According to (8.7) and (8.8), one can interpret $\mathtt{A}_+^\alpha(\boldsymbol{x}, t=0) = a^*(|\boldsymbol{x},\alpha\rangle)$ as the creator of a photon in the individual state

$$|\boldsymbol{x},\alpha\rangle = \left(\frac{\hbar}{\varepsilon_0 L^3}\right)^{1/2} \sum_{\boldsymbol{k}\lambda} \frac{\mathrm{e}_\lambda^\alpha(\boldsymbol{k})}{\sqrt{2\omega_{\boldsymbol{k}}}} \mathrm{e}^{-i\boldsymbol{k}\cdot\boldsymbol{x}} \, |\boldsymbol{k},\lambda\rangle \quad . \tag{8.17}$$

It is important to emphasize that this relation differs by a factor $1/\sqrt{2\omega_{\boldsymbol{k}}}$ from the usual Fourier transformation that relates the configuration q and momentum p representations of a non-relativistic quantum particle. This extra factor appears due to the fact that the photon is a relativistic particle, governed by a wave equation of second order in t. As a consequence, the variable $\boldsymbol{x}$ in (8.17) is not a canonically conjugate to the momentum and should not be interpreted as the position of the particle. *Physically, the variable $\boldsymbol{x}$ relates to the localization of the electromagnetic energy density*, as shown in formula (8.16).

Fluctuations

The average values of fields in vacuum are of course zero

$$\langle 0|\mathtt{E}(\boldsymbol{x},t)|0\rangle = \langle 0|\mathtt{B}(\boldsymbol{x},t)|0\rangle = 0 \quad , \tag{8.18}$$

but calculations of the electric and magnetic field fluctuations give infinite values

$$\varepsilon_0 \, \langle 0||\mathtt{E}(\boldsymbol{x},t)|^2|0\rangle = \mu_0^{-1} \, \langle 0||\mathtt{B}(\boldsymbol{x},t)|^2|0\rangle = \frac{1}{2L^3}\sum_{\boldsymbol{k}\lambda} \hbar\omega_{\boldsymbol{k}} = \infty \quad . \tag{8.19}$$

The divergent sum (8.19) has two different interpretations. First of all, it can be a (infinite) constant occurring in the evaluation (8.15) of the field energy in any state, including the vacuum (the integration only gives rise to a constant factor L^3). This constant has no significance, since one only measures energy differences between different states. In the case of the theory of harmonic solids, one can give a physical meaning to the finite sum $1/2(\sum_{\boldsymbol{k}\lambda} \hbar\omega_{\boldsymbol{k}\lambda})$ ($\hbar\omega_{\boldsymbol{k}\lambda}$ is the energy of a phonon). Indeed, dissociating the solid into a collection of individual atoms corresponds to a variation in the energy of the system (the binding energy), whose calculation includes the above sum. However, in the case of the electromagnetic field in the vacuum, this explanation makes no sense: the energy $1/2(\sum_{\boldsymbol{k}\lambda} \hbar\omega_{\boldsymbol{k}\lambda})$ is not measurable.

On the other hand, if considered as *vacuum fluctuations* (consequences of the quantum nature of the fields), the expressions $\langle 0||\mathtt{E}(\boldsymbol{x},t)|^2|0\rangle$ and $\langle 0||\mathtt{B}(\boldsymbol{x},t)|^2|0\rangle$ are observables in the sense that a charged particle will oscillate under the influence of these fluctuations, modifying its energy. In particular, these fluctuations have an influence on the energy of an electron under the influence of the Coulomb field of an atomic nucleus. As a result, the states $2S_{1/2}$ and $2P_{1/2}$, which are degenerate in Dirac's theory, are slightly separated in the atomic spectrum of hydrogen: this is the *Lamb shift* whose numerical value receives a contribution from the vacuum fluctuations of quantum electrodynamics (Sect. 9.4.3).

Concerning the fact that the sum (8.19) is infinite, one can show that this divergence does not have unacceptable consequences on the modification of the energy of an electron in the presence of vacuum fluctuations. A point charge would indeed acquire an infinite energy, but an electron accompanied by photons behaves like a charge extending over a region of dimension $r_0 = \hbar/mc$. Hence the wave vectors $|\boldsymbol{k}| > r_0^{-1}$ appearing in sum (8.19) no longer contribute to the oscillations of the electron, thus giving a finite value to the energy correction.

8.2.2 Canonical Variables

In order to interpret the field $\mathtt{A}(\boldsymbol{x}) = \mathtt{A}(\boldsymbol{x}, t = 0)$ as an ensemble of dynamical quantum variables, it is necessary to determine the canonically conjugate operator to $\mathtt{A}^\alpha(\boldsymbol{x})$. We will show first of all that the operators $\mathtt{A}^\alpha(\boldsymbol{x})$ and $\mathtt{A}^\beta(\boldsymbol{y})$ associated to two different points, $\boldsymbol{x}$ and $\boldsymbol{y}$, are independent. That is,

$$\left[\mathtt{A}^\alpha(\boldsymbol{x}), \mathtt{A}^\beta(\boldsymbol{y})\right] = 0 \quad , \quad \alpha, \beta = 1, 2, 3 \quad . \tag{8.20}$$

Indeed, as the terms with frequencies of the same sign commute, we can conclude from (8.7), (8.8) and $[a_{\boldsymbol{k}\lambda}, a^*_{\boldsymbol{k}',\lambda'}] = \delta_{\boldsymbol{k},\boldsymbol{k}'}\delta_{\lambda,\lambda'}$ that

$$\begin{aligned}\left[\mathtt{A}^\alpha(\boldsymbol{x}), \mathtt{A}^\beta(\boldsymbol{y})\right] &= \left[\mathtt{A}^\alpha_+(\boldsymbol{x}), \mathtt{A}^\beta_-(\boldsymbol{y})\right] + \left[\mathtt{A}^\alpha_-(\boldsymbol{x}), \mathtt{A}^\beta_+(\boldsymbol{y})\right] \\ &= -\frac{\hbar}{\varepsilon_0 L^3} \sum_{\boldsymbol{k}} \frac{1}{2\omega_{\boldsymbol{k}}} \mathrm{e}^{-\mathrm{i}\boldsymbol{k}\cdot(\boldsymbol{x}-\boldsymbol{y})} (\delta_{\alpha,\beta} - \hat{k}^\alpha \hat{k}^\beta) \\ &\quad + \frac{\hbar}{\varepsilon_0 L^3} \sum_{\boldsymbol{k}} \frac{1}{2\omega_{\boldsymbol{k}}} \mathrm{e}^{\mathrm{i}\boldsymbol{k}\cdot(\boldsymbol{x}-\boldsymbol{y})} (\delta_{\alpha,\beta} - \hat{k}^\alpha \hat{k}^\beta) \quad .\end{aligned} \tag{8.21}$$

In (8.21), we have used the fact that the polarization vectors $\boldsymbol{e}_1(\boldsymbol{k}), \boldsymbol{e}_2(\boldsymbol{k})$ and the unit vector $\hat{\boldsymbol{k}} = \boldsymbol{k}/|\boldsymbol{k}|$ form an orthonormal system, which gives rise to the sum

$$\sum_{\lambda=1}^{2} e_\lambda^\alpha(\boldsymbol{k}) e_\lambda^\beta(\boldsymbol{k}) = \delta_{\alpha,\beta} - \frac{k^\alpha k^\beta}{|\boldsymbol{k}|^2} \quad . \tag{8.22}$$

If one changes $\boldsymbol{k}$ with $-\boldsymbol{k}$ in one of the two terms of (8.21), they will become equal up to a sign, verifying (8.20).

As the form (8.16) of the energy suggests, the canonical conjugate of $\mathtt{A}^\alpha(\boldsymbol{x})$ is given by

$$\dot{\mathtt{A}}^\alpha(\boldsymbol{x}) = \frac{\partial}{\partial t}\mathtt{A}^\alpha(\boldsymbol{x}, t)|_{t=0} \tag{8.23}$$

up to a constant factor ε_0 (see (8.25)). To calculate $[\mathtt{A}^\alpha(\boldsymbol{x}), \dot{\mathtt{A}}^\beta(\boldsymbol{y})]$, one notes that differentiating (8.7) with respect to time multiplies the terms of the first sum of (8.21) by $-\mathrm{i}\omega_{\boldsymbol{k}}$ and those of the second sum by $\mathrm{i}\omega_{\boldsymbol{k}}$. Then, after making the substitution $\boldsymbol{k} \to -\boldsymbol{k}$, one obtains

$$\left[\mathtt{A}^\alpha(\boldsymbol{x}), \dot{\mathtt{A}}^\beta(\boldsymbol{y})\right] = \frac{\mathrm{i}\hbar}{\varepsilon_0}\frac{1}{L^3}\sum_{\boldsymbol{k}} \mathrm{e}^{\mathrm{i}\boldsymbol{k}\cdot(\boldsymbol{x}-\boldsymbol{y})}\left(\delta_{\alpha,\beta} - \frac{k^\alpha k^\beta}{|\boldsymbol{k}|^2}\right) \tag{8.24}$$

and taking the infinite volume limit,

$$\left[\mathtt{A}^\alpha(\boldsymbol{x}), \dot{\mathtt{A}}^\beta(\boldsymbol{y})\right] = \frac{\mathrm{i}\hbar}{\varepsilon_0}\delta^{\mathrm{tr}}_{\alpha,\beta}(\boldsymbol{x}-\boldsymbol{y}) \quad . \tag{8.25}$$

We have defined in (8.25) the transverse Dirac function

$$\begin{aligned} \delta^{\mathrm{tr}}_{\alpha,\beta}(\boldsymbol{x}) &= \frac{1}{(2\pi)^3}\int \mathrm{d}\boldsymbol{k}\, \mathrm{e}^{\mathrm{i}\boldsymbol{k}\cdot\boldsymbol{x}}\left(\delta_{\alpha,\beta} - \frac{k^\alpha k^\beta}{|\boldsymbol{k}|^2}\right) \\ &= \delta_{\alpha,\beta}\delta(\boldsymbol{x}) + \frac{1}{4\pi}\frac{\partial^2}{\partial x^\alpha \partial x^\beta}\frac{1}{|\boldsymbol{x}|} \quad . \end{aligned} \tag{8.26}$$

The second term in (8.26) ensures that the commutation relation (8.25) is compatible with the transversality of the field. One can also verify that

$$\left[\dot{\mathtt{A}}^\alpha(\boldsymbol{x}), \dot{\mathtt{A}}^\beta(\boldsymbol{y})\right] = 0 \quad . \tag{8.27}$$

Equations (8.20), (8.25) and (8.27) show that, up to a factor due to transversality, $\{\mathtt{A}^\alpha(\boldsymbol{x}), \dot{\mathtt{A}}^\beta(\boldsymbol{y})\}$ have canonical commutation rules analogous to $\{q_i, p_j; [q_i, p_j] = \mathrm{i}\hbar\delta_{i,j}\}$. Thus the quantum field indeed appears as an ensemble of degrees of freedom attached to points in space (up to the constraint imposed by the transversality factor). We have calculated the commutators of two fields at the same time $t = 0$. As unitarity evolution preserves the commutators, it is clear that these relations continue to hold for fields taken at any time t.

An important conclusion can be drawn from this analysis and from the form of the energy (8.16), which we will elevate to a principle: *The Hamiltonian is expressed, as in classical theory, by an integral over the Hamiltonian density, which can be written in terms of canonical variables of the quantum fields. Up to normal ordering, the Hamiltonian density has the same form as in classical theory.* This principle of quantum field theory is analogous to the correspondence principle of quantum mechanics.

8.2.3 Invariant Commutation Function and Microcausality

Commutation Function

As in classical theory, the observables which are independent of the choice of gauge are the electric field $\mathtt{E}(\boldsymbol{x},t)$ and the magnetic field $\mathtt{B}(\boldsymbol{x},t)$ and expressions formed from them, such as the Hamiltonian (8.16). In this case one speaks of physical observables. As $\mathtt{E}(\boldsymbol{x},t)$ and $\mathtt{B}(\boldsymbol{x},t)$ are operators here, it is interesting to know their commutation relations. As an example, we will calculate the commutator of the components $\mathtt{E}^\alpha(\boldsymbol{x},t)$ and $\mathtt{B}^\beta(\boldsymbol{y},s)$ taken at two different space–time points $\boldsymbol{x},t$ and $\boldsymbol{y},s$. Decomposing $\mathtt{E}(\boldsymbol{x},t)$ and $\mathtt{B}(\boldsymbol{y},s)$ into components of positive and negative frequency, with $\mathtt{E}_\pm = -(\partial/\partial t)\mathtt{A}_\pm$ and $\mathtt{B}_\pm = \nabla\times\mathtt{A}_\pm$, one obtains from (8.7), (8.8) and the commutation rules (3.33)

$$\begin{aligned}\left[\mathtt{E}^\alpha(\boldsymbol{x},t),\mathtt{B}^\beta(\boldsymbol{y},s)\right] &= \left[\mathtt{E}^\alpha_+(\boldsymbol{x},t),\mathtt{B}^\beta_-(\boldsymbol{y},s)\right] + \left[\mathtt{E}^\alpha_-(\boldsymbol{x},t),\mathtt{B}^\beta_+(\boldsymbol{y},s)\right] \\ &= -\frac{\mathrm{i}\hbar}{\varepsilon_0 L^3}\sum_{\boldsymbol{k}\lambda} e^\alpha_\lambda(\boldsymbol{k})\left[\boldsymbol{k}\times\boldsymbol{e}_\lambda(\boldsymbol{k})\right]^\beta \sin\left[\boldsymbol{k}\cdot(\boldsymbol{x}-\boldsymbol{y})-\omega_{\boldsymbol{k}}(t-s)\right] \quad .\end{aligned} \tag{8.28}$$

The sum over the polarization indices is carried out, noting that the three vectors $\hat{\boldsymbol{k}},\boldsymbol{e}_1(\boldsymbol{k}),\boldsymbol{e}_2(\boldsymbol{k})$ are orthogonal

$$\begin{aligned}\hat{\boldsymbol{k}}\times\boldsymbol{e}_1(\boldsymbol{k}) &= \boldsymbol{e}_2(\boldsymbol{k}) \quad , \\ \hat{\boldsymbol{k}}\times\boldsymbol{e}_2(\boldsymbol{k}) &= -\boldsymbol{e}_1(\boldsymbol{k}) \quad , \\ \boldsymbol{e}_1(\boldsymbol{k})\times\boldsymbol{e}_2(\boldsymbol{k}) &= \hat{\boldsymbol{k}} \quad ,\end{aligned} \tag{8.29}$$

so that

$$\begin{aligned}\sum_{\lambda=1}^{2} e^\alpha_\lambda(\boldsymbol{k})\left[\boldsymbol{k}\times\boldsymbol{e}_\lambda(\boldsymbol{k})\right]^\beta &= |\boldsymbol{k}|\left(e^\alpha_1(\boldsymbol{k})e^\beta_2(\boldsymbol{k}) - e^\beta_1(\boldsymbol{k})e^\alpha_2(\boldsymbol{k})\right) \\ &= |\boldsymbol{k}|\left(\boldsymbol{e}_1(\boldsymbol{k})\times\boldsymbol{e}_2(\boldsymbol{k})\right)^\gamma = \begin{cases} k^\gamma & \alpha\neq\beta \\ 0 & \alpha=\beta \end{cases} \quad ,\end{aligned} \tag{8.30}$$

where α,β,γ is a cyclic permutation of $1,2,3$. One sees that components of the same index commute

$$\left[\mathtt{E}^\alpha(\boldsymbol{x},t),\mathtt{B}^\alpha(\boldsymbol{y},s)\right] = 0 \tag{8.31}$$

but for $\alpha \neq \beta$ one obtains

$$\begin{aligned}\left[\mathrm{E}^{\alpha}(\boldsymbol{x},t),\mathrm{B}^{\beta}(\boldsymbol{y},s)\right] &= \frac{\mathrm{i}\hbar}{\varepsilon_0 L^3}\frac{\partial}{\partial x^{\gamma}}\sum_{\boldsymbol{k}}\cos\left[\boldsymbol{k}\cdot(\boldsymbol{x}-\boldsymbol{y})-\omega_{\boldsymbol{k}}(t-s)\right] \\ &= \frac{\mathrm{i}\hbar}{\varepsilon_0}\frac{\partial}{\partial x^{\gamma}}\frac{\partial}{\partial t}D(\boldsymbol{x}-\boldsymbol{y},t-s) \quad , \end{aligned} \tag{8.32}$$

with

$$D(\boldsymbol{x},t) = -\frac{1}{L^3}\sum_{\boldsymbol{k}}\frac{\sin(\boldsymbol{k}\cdot\boldsymbol{x}-\omega_{\boldsymbol{k}}t)}{\omega_{\boldsymbol{k}}} \quad . \tag{8.33}$$

The reason for writing the commutation relation in the form (8.33) with the introduction of the function $D(\boldsymbol{x},t)$ is the following: *$D(\boldsymbol{x},t)$ is invariant under the Lorentz transformations of special relativity and is non-zero only on the light cone $(ct)^2 - |\boldsymbol{x}|^2 = 0$.* When taking the infinite volume limit to evaluate $D(\boldsymbol{x},t)$, one finds

$$\begin{aligned} D(\boldsymbol{x},t) &= -\frac{1}{(2\pi)^3}\int \mathrm{d}\boldsymbol{k}\,\frac{\sin(\boldsymbol{k}\cdot\boldsymbol{x}-c|\boldsymbol{k}|t)}{c|\boldsymbol{k}|} \\ &= -\frac{1}{(2\pi)^3}\int \mathrm{d}\boldsymbol{k}\,\frac{\mathrm{e}^{\mathrm{i}(\boldsymbol{k}\cdot\boldsymbol{x}-c|\boldsymbol{k}|t)}}{2\mathrm{i}c|\boldsymbol{k}|} + \text{complex conjugate} \\ &= -\frac{1}{(2\pi)^2}\int_0^{\infty}\frac{k\mathrm{d}k}{2\mathrm{i}c}\int_{-1}^{+1}\mathrm{d}u\,\mathrm{e}^{\mathrm{i}(k|\boldsymbol{x}|u-ckt)} + \text{c.c.} \\ &= \frac{1}{(2\pi)^2}\frac{1}{2c|\boldsymbol{x}|}\int_0^{\infty}\mathrm{d}k\,\left(\mathrm{e}^{\mathrm{i}k(|\boldsymbol{x}|-ct)} - \mathrm{e}^{-\mathrm{i}k(|\boldsymbol{x}|+ct)}\right) + \text{c.c.} \\ &= \frac{1}{(2\pi)^2}\frac{1}{2c|\boldsymbol{x}|}\left(\int_{-\infty}^{\infty}\mathrm{d}k\,\mathrm{e}^{\mathrm{i}k(|\boldsymbol{x}|-ct)} - \int_{-\infty}^{\infty}\mathrm{d}k\,\mathrm{e}^{\mathrm{i}k(|\boldsymbol{x}|+ct)}\right) \\ &= \frac{\delta(|\boldsymbol{x}|-ct)-\delta(|\boldsymbol{x}|+ct)}{4\pi c|\boldsymbol{x}|} = \frac{\varepsilon(t)}{2\pi c}\delta\left[(ct)^2-|\boldsymbol{x}|^2\right] \quad , \end{aligned} \tag{8.34}$$

where

$$\varepsilon(t) = \begin{cases} -1 \ , & t<0 \\ 0 \ , & t=0 \\ 1 \ , & t>0 \ . \end{cases}$$

As $(ct)^2 - |\boldsymbol{x}|^2$ and the sign of time are invariant, $D(\boldsymbol{x},t)$ does not change form under Lorentz transformations of its arguments. The function $D(\boldsymbol{x},t)$ is called the *Jordan and Pauli commutation function.* It was introduced and discussed by these authors in 1928. One can also calculate commutators of the various electric and magnetic field components. They can all be written in terms of $D(\boldsymbol{x},t)$ by expressions analogous to (8.32).

Microcausality

This result leads to the following important observation. Since $D(\boldsymbol{x}-\boldsymbol{y}, t-s)$ only depends on arguments $|\boldsymbol{x}-\boldsymbol{y}|-c(t-s)$ and $|\boldsymbol{x}-\boldsymbol{y}|+c(t-s)$, it is only non-zero if the two points $(\boldsymbol{x}, t)$ and $(\boldsymbol{y}, s)$ at which the fields are considered can be linked by a light signal (t can come before or after s). As a consequence, fields taken at two space–time points which cannot be linked by a light signal commute. As $\mathtt{E}(\boldsymbol{x}, t)$ and $\mathtt{B}(\boldsymbol{y}, s)$ are quantum observables of the free photon field, this expresses the fact that *measurements taken in regions of space–time which cannot be joined by light signals cannot perturb each other.* The same property holds for all pairs of observables $\mathtt{O}(\boldsymbol{x}, t)$ and $\mathtt{O}(\boldsymbol{y}, s)$ written in terms of the fields. Free photons propagate at exactly the speed of light c. In the general case of massive fields in interaction, it holds true according to the causality principle of special relativity that no information can propagate at a speed greater than c: observations made at $(\boldsymbol{x}, t)$ and $(\boldsymbol{y}, s)$ do not affect each other if $[c(t-s)]^2 - |\boldsymbol{x}-\boldsymbol{y}|^2 < 0$. For these fundamental reasons, it must be true in all generality that (see the scalar massive field in Sect. 8.3.1)

$$[\mathtt{O}(\boldsymbol{x}, t), \mathtt{O}(\boldsymbol{y}, s)] = 0 \quad , \quad [c(t-s)]^2 - |\boldsymbol{x}-\boldsymbol{y}|^2 < 0 \tag{8.35}$$

for any kind of field. Property (8.35) is called the *microcausality principle.* It brings together the commutativity of the fields (their fundamentally quantum aspect) with the causality of special relativity.

It is important to make a remark regarding the mathematical nature of the quantum fields: because of the improper Dirac function which appears in the commutators (8.25) and (8.34), they cannot be considered as regular operators in Fock space. In precise terms, the fields are operator-valued distributions, just as $\delta(\boldsymbol{x})$ is a distribution or generalized function, with real values. In general, the product of two distributions or two fields taken at the same point gives a divergent result, as shown in the fluctuation calculation (8.19). This fact is the source of delicate mathematical problems which we will not address here. It suffices to say that in what follows, the use of the field (8.7) and singular commutators can be justified in a similar manner to the use of the improper vector $|\boldsymbol{x}\rangle$ and the Dirac function in quantum mechanics.

8.2.4 Emission of Photons by a Classical Source

Two Characteristic Situations

In the following sections, we study the interaction of radiation and non-relativistic matter and the emission (or absorption) of photons in two different characteristic situations. In the first case, we suppose that *the source of photons is described by a classical current distribution $\boldsymbol{j}(\boldsymbol{x}, t)$, prescribed at will.* In this case, it is the emitted radiation which is of interest and one can completely neglect the effect of the photons on the source. Classically, one must solve the Maxwell equations (1.88) and (1.89) with $\boldsymbol{j}(\boldsymbol{x},t)$ prescribed without taking into account the effect of the field on the charges (1.90). In this case, the energy and the momentum of the photons are not conserved since they are produced by an external source.

In the second case, one considers the emission or absorption of a photon by an atom. Here, one cannot neglect the exchange of energy between the photon and the electron and must simultaneously treat the photons and the electrons with the total Hamiltonian (1.124). The energy and the momentum of the total system, atom + photons, is thus conserved. We will see that the state of the photons will be very different in these two cases. The state produced by a prescribed current is a *coherent state* where the quantum-electromagnetic field takes on the most classical appearance possible and the photons are of undetermined number. However, when an atom interacts with the electromagnetic field, *states with a well determined number* of photons play a very important role.

Evolution of the Field in the Presence of a Classical Current

The current distribution $\boldsymbol{j}(\boldsymbol{x},t)$ can, according to (1.113), be chosen to be transverse in the Coulomb gauge: $\nabla\cdot\boldsymbol{j}(\boldsymbol{x},t)=0$. Suppose that the current is switched on at time $t=0$, that is, $\boldsymbol{j}(\boldsymbol{x},t)=0,\ t<0$. The problem consists of finding the operators $\mathtt{A}(\boldsymbol{x},t)$ which are solutions of the sourced wave equation (1.114)

$$\frac{1}{c^2}\frac{\partial^2}{\partial t^2}\mathtt{A}(\boldsymbol{x},t)-\nabla^2\mathtt{A}(\boldsymbol{x},t)=\mu_0\boldsymbol{j}(\boldsymbol{x},t) \quad . \tag{8.36}$$

Moreover, $\mathtt{A}(\boldsymbol{x},t)$ is given by a quantum law of evolution,

$$\mathtt{A}(\boldsymbol{x},t)=\mathtt{U}^{-1}(t)\mathtt{A}(\boldsymbol{x})\mathtt{U}(t) \quad , \tag{8.37}$$

where $\mathtt{A}(\boldsymbol{x})=\mathtt{A}(\boldsymbol{x},0)$ is the field (8.7) at time $t=0$, and $\mathtt{U}(t)$ is the evolution operator to be determined. Note that $\mathtt{U}(t)$ is no longer generated by the free Hamiltonian $\mathtt{H}_0$ as in the case of a free field (8.13), but by a new Hamiltonian $\mathtt{H}(t)$ which includes the coupling of the photons to the current ($\mathtt{H}(t)$ depends explicitly on time, and the system is no longer conservative).

To solve (8.36), one proceeds exactly as in Sect. 1.3.5, introducing the Fourier components of the current

$$\boldsymbol{j}_{\boldsymbol{k}}(t)=\frac{1}{L^{3/2}}\int \mathrm{d}\boldsymbol{x}\,\boldsymbol{j}(\boldsymbol{x},t)\mathrm{e}^{-\mathrm{i}\boldsymbol{k}\cdot\boldsymbol{x}} \quad . \tag{8.38}$$

It follows from transversality and the fact that the current must be real that

$$\boldsymbol{j}_{\boldsymbol{k}}(t)=\sum_{\lambda} j_{\boldsymbol{k}\lambda}(t)\boldsymbol{e}_{\lambda}(\boldsymbol{k}) \quad , \tag{8.39}$$

$$\boldsymbol{j}^*_{\boldsymbol{k}}(t)=\boldsymbol{j}_{-\boldsymbol{k}}(t) \quad . \tag{8.40}$$

We now write the expansion

$$\mathtt{A}(\boldsymbol{x},t) = \left(\frac{\hbar}{\varepsilon_0 L^3}\right)^{1/2} \sum_{\boldsymbol{k}\lambda} \frac{1}{\sqrt{2\omega_{\boldsymbol{k}}}} \mathrm{e}^{\mathrm{i}\boldsymbol{k}\cdot\boldsymbol{x}} \boldsymbol{e}_\lambda(\boldsymbol{k}) a_{\boldsymbol{k}\lambda}(t)$$
$$+ \text{Hermitian conjugate} \quad , \tag{8.41}$$

where the time dependence of the creation and annihilation operators

$$a_{\boldsymbol{k}\lambda}(t) = \mathtt{U}^{-1}(t) a_{\boldsymbol{k}\lambda} \mathtt{U}(t) \quad , \quad a^*_{\boldsymbol{k}\lambda}(t) = \mathtt{U}^{-1}(t) a^*_{\boldsymbol{k}\lambda} \mathtt{U}(t) \quad , \tag{8.42}$$

must be determined by (8.36). As in (1.134) and (1.135), we denote the Fourier transform of the field by $\boldsymbol{c}_{\boldsymbol{k}}(t)$

$$\boldsymbol{c}_{\boldsymbol{k}}(t) = \frac{1}{L^{3/2}} \int \mathrm{d}\boldsymbol{x}\, \mathtt{A}(\boldsymbol{x},t) \mathrm{e}^{-\mathrm{i}\boldsymbol{k}\cdot\boldsymbol{x}}$$
$$= \left(\frac{\hbar}{2\varepsilon_0 \omega_{\boldsymbol{k}}}\right)^{1/2} \left[\boldsymbol{a}_{\boldsymbol{k}}(t) + \boldsymbol{a}^*_{-\boldsymbol{k}}(t)\right] \quad , \tag{8.43}$$

with

$$\boldsymbol{a}_{\boldsymbol{k}}(t) = \sum_\lambda \boldsymbol{e}_\lambda(\boldsymbol{k}) a_{\boldsymbol{k}\lambda}(t) \quad . \tag{8.44}$$

Applying the Fourier transformation, the propagation equation (8.36) is written as

$$\frac{1}{c^2} \frac{\mathrm{d}^2}{\mathrm{d}t^2} \boldsymbol{c}_{\boldsymbol{k}}(t) + |\boldsymbol{k}|^2 \boldsymbol{c}_{\boldsymbol{k}}(t) = \mu_0 \boldsymbol{j}_{\boldsymbol{k}}(t) \quad , \tag{8.45}$$

which is equivalent to (1.136) in the presence of sources. One verifies that (8.45) is satisfied if the creation and annihilation operators obey the differential equation of first order in time

$$\frac{\mathrm{d}}{\mathrm{d}t} a_{\boldsymbol{k}\lambda}(t) = -\mathrm{i}\omega_{\boldsymbol{k}} a_{\boldsymbol{k}\lambda}(t) + \mathrm{i}\left(\frac{1}{2\varepsilon_0 \hbar \omega_{\boldsymbol{k}}}\right)^{1/2} j_{\boldsymbol{k}\lambda}(t) \quad . \tag{8.46}$$

To see this, we differentiate (8.43) with respect to time

$$\frac{\mathrm{d}}{\mathrm{d}t} \boldsymbol{c}_{\boldsymbol{k}}(t) = -\mathrm{i}\left(\frac{\hbar}{2\varepsilon_0 \omega_{\boldsymbol{k}}}\right)^{1/2} \omega_{\boldsymbol{k}} \left[\boldsymbol{a}_{\boldsymbol{k}}(t) - \boldsymbol{a}^*_{-\boldsymbol{k}}(t)\right] \quad , \tag{8.47}$$

where we have taken into account (8.46) and the fact that the current is real. Taking the derivative of (8.47) one more time leads to

$$\frac{\mathrm{d}^2}{\mathrm{d}t^2} \boldsymbol{c}_{\boldsymbol{k}}(t) = -\left(\frac{\hbar}{2\varepsilon_0 \omega_{\boldsymbol{k}}}\right)^{1/2} \omega_{\boldsymbol{k}}^2 \left[\boldsymbol{a}_{\boldsymbol{k}}(t) + \boldsymbol{a}^*_{-\boldsymbol{k}}(t)\right]$$
$$+ \frac{1}{2\varepsilon_0} \left[\boldsymbol{j}_{\boldsymbol{k}}(t) + \boldsymbol{j}^*_{\boldsymbol{k}}(t)\right] \quad , \tag{8.48}$$

which, with (8.43), (8.40) and $\omega_{\boldsymbol{k}} = c|\boldsymbol{k}|$ is indeed equal to (8.45).

Interaction Hamiltonian

The system of equations (8.46) for $a_{\boldsymbol{k}\lambda}(t)$ is equivalent to the sourced wave equation (8.36). However, in (8.46), all the modes $(\boldsymbol{k}, \lambda)$ are decoupled and each one follows exactly the equation of motion of a forced quantum oscillator (1.67), under the condition that, for each mode, one identifies $f(t)$ with

$$f_{\boldsymbol{k}\lambda}(t) = -\left(\frac{\hbar}{2\varepsilon_0\omega_{\boldsymbol{k}}}\right)^{1/2} j_{\boldsymbol{k}\lambda}(t) \quad . \tag{8.49}$$

From this, according to (1.64), the interaction term responsible for the evolution of the mode $\boldsymbol{k}, \lambda$ is $a^*_{\boldsymbol{k}\lambda} f_{\boldsymbol{k}\lambda}(t) + a_{\boldsymbol{k}\lambda} f^*_{\boldsymbol{k}\lambda}(t)$. As the modes evolve independently, and as the operators acting on different modes commute, the interaction Hamiltonian of the field and the current is given by the sum over all modes,

$$\begin{aligned} \mathtt{H}_{\mathrm{I}}(t) &= -\left(\frac{\hbar}{\varepsilon_0}\right)^{1/2} \sum_{\boldsymbol{k}\lambda} \frac{1}{\sqrt{2\omega_{\boldsymbol{k}}}} \left[a^*_{\boldsymbol{k}\lambda} j_{\boldsymbol{k}\lambda}(t) + a_{\boldsymbol{k}\lambda} \boldsymbol{j}^*_{\boldsymbol{k}\lambda}(t)\right] \\ &= -\int \mathrm{d}\boldsymbol{x}\, \boldsymbol{j}(\boldsymbol{x}, t) \cdot \mathtt{A}(\boldsymbol{x}) \quad , \end{aligned} \tag{8.50}$$

where $\mathtt{A}(\boldsymbol{x})$ is the free field (8.7) at time $t = 0$. One finds the form (8.50) by returning to the configuration space with the help of (8.38) and the orthogonality of plane waves. The complete Hamiltonian is thus

$$\mathtt{H}(t) = \mathtt{H}_0 - \int \mathrm{d}\boldsymbol{x}\, \boldsymbol{j}(\boldsymbol{x}, t) \cdot \mathtt{A}(\boldsymbol{x}) \quad . \tag{8.51}$$

Expression (8.50) brings to light several important characteristics of the interaction Hamiltonian. *It is only written in terms of the vector potential* $\mathtt{A}$ and not in terms of $\mathtt{E}$ and $\mathtt{B}$. As with the free Hamiltonian, it can be represented by an integral over an Hamiltonian density. *The current and the field are only coupled at the same point x in space: such an interaction is called local.* Locality is a general property of quantum fields. Since $\mathtt{A}(\boldsymbol{x})$ is linear in the creation and annihilation operators, the interaction Hamiltonian $\mathtt{H}_{\mathrm{I}}$ *does not conserve the number of photons.* This allows for transitions between states with different numbers of photons; it can thus account for the emission or absorption of photons.

8.2.5 Coherent States of Photons

We now look for the photon state produced by the current $\boldsymbol{j}(\boldsymbol{x}, t)$. At $t = 0$, the current is zero and the state is in the vacuum. At time t, the state is thus

$$|\Phi_t\rangle = \mathtt{U}(t)\,|0\rangle \quad . \tag{8.52}$$

Suppose first of all, for the sake of simplicity, that the current has only one non-zero Fourier amplitude $j_{\boldsymbol{k}\lambda}(t)$. The evolution operator is thus identical to the forced oscillator (1.71). The photon state of a mode $\boldsymbol{k}, \lambda$ at time t is the coherent state (1.72)

$$\mathtt{U}_{\boldsymbol{k},\lambda}(t) = \mathrm{e}^{\mathrm{i}\chi_{\boldsymbol{k}\lambda}(t)} \mathtt{D}\left[\alpha_{\boldsymbol{k}\lambda}(t)\right] |0\rangle \quad , \tag{8.53}$$

where the amplitude $\alpha_{\boldsymbol{k}\lambda}(t)$ is determined by the current according to (1.69) and (8.49)

$$\alpha_{\boldsymbol{k}\lambda}(t) = \mathrm{i}\left(\frac{1}{2\varepsilon_0 \hbar\omega_{\boldsymbol{k}}}\right)^{1/2} \int_0^t \mathrm{d}s\, \mathrm{e}^{-\mathrm{i}\omega_{\boldsymbol{k}}(t-s)} j_{\boldsymbol{k}\lambda}(s) \quad . \tag{8.54}$$

If all the modes are excited, the evolution operator corresponding to (8.51) is a product $\prod_{\boldsymbol{k}\lambda} \mathtt{U}_{\boldsymbol{k}\lambda}(t)$ since the modes are independent. Taking into account (8.53) and (1.55), the state of the radiated photons is given by

$$\begin{aligned} \mathtt{U}(t)\,|0\rangle &= \mathrm{e}^{\mathrm{i}\chi(t)} \prod_{\boldsymbol{k}\lambda} \mathtt{D}\left[\alpha_{\boldsymbol{k}\lambda}(t)\right] |0\rangle \\ &= \mathrm{e}^{\mathrm{i}\chi(t)} \exp\left[\sum_{\boldsymbol{k}\lambda} \left(\alpha_{\boldsymbol{k}\lambda}(t) a^*_{\boldsymbol{k}\lambda} - \alpha^*_{\boldsymbol{k}\lambda}(t) a_{\boldsymbol{k}\lambda}\right)\right] |0\rangle \\ &= \mathrm{e}^{\mathrm{i}\chi(t)} \exp\left[a^*(\varphi_t) - a(\varphi_t)\right] |0\rangle \quad , \end{aligned} \tag{8.55}$$

where $|\varphi_t\rangle$ is a state in a one-photon space of amplitude $\langle \boldsymbol{k}\lambda|\varphi_t\rangle = \alpha_{\boldsymbol{k}\lambda}(t)$ and $\chi(t)$ is a phase unimportant for this discussion. Comparison with (3.48) and (3.49) shows that $\mathtt{U}(t)\,|0\rangle = \mathrm{e}^{\mathrm{i}\chi(t)}\,|\varphi_t\rangle_{\mathrm{coh}}$ is a coherent state in the Fock space of the photons: *photons radiated from a classical source are found in a coherent state, with each of them occupying the same normalized individual state* $|\tilde{\varphi}_t\rangle = (\langle\varphi_t|\varphi_t\rangle)^{-1/2}\,|\varphi_t\rangle$.

The Quantum Field and the Classical Field

As $a_{\boldsymbol{k}\lambda}\,|\varphi_t\rangle_{\mathrm{coh}} = \alpha_{\boldsymbol{k}\lambda}(t)\,|\varphi_t\rangle_{\mathrm{coh}}$ (see (3.52)), it is clear that the average value of $\mathtt{A}(\boldsymbol{x})$

$$\begin{aligned} {}_{\mathrm{coh}}\langle\varphi_t|\mathtt{A}(\boldsymbol{x})|\varphi_t\rangle_{\mathrm{coh}} = \left(\frac{\hbar}{\varepsilon_0 L^3}\right)^{1/2} \sum_{\boldsymbol{k}\lambda} \frac{1}{\sqrt{2\omega_{\boldsymbol{k}}}} \boldsymbol{e}_\lambda(\boldsymbol{k}) \alpha^*_{\boldsymbol{k}\lambda} \mathrm{e}^{-\mathrm{i}\boldsymbol{k}\cdot\boldsymbol{x}} \\ + \text{complex conjugate} \end{aligned} \tag{8.56}$$

satisfies the classical wave equation with source $\boldsymbol{j}(\boldsymbol{x}, t)$. The average number of photons $\langle\mathtt{N}\rangle$ and the radiation energy $E^{\mathrm{em}} = \langle\mathtt{H}_0\rangle$ are just as easy to calculate, with the help of (3.52),

$$\langle \mathtt{N} \rangle = {}_{\text{coh}}\langle \varphi_t | \sum_{\boldsymbol{k}\lambda} a^*_{\boldsymbol{k}\lambda} a_{\boldsymbol{k}\lambda} | \varphi_t \rangle_{\text{coh}}$$

$$= \sum_{\boldsymbol{k}\lambda} |\alpha_{\boldsymbol{k}\lambda}(t)|^2 = \langle \varphi_t | \varphi_t \rangle \quad , \tag{8.57}$$

$$E^{\text{em}} = {}_{\text{coh}}\langle \varphi_t | \sum_{\boldsymbol{k}\lambda} \hbar\omega_{\boldsymbol{k}} a^*_{\boldsymbol{k}\lambda} a_{\boldsymbol{k}\lambda} | \varphi_t \rangle_{\text{coh}}$$

$$= \sum_{\boldsymbol{k}\lambda} \hbar\omega_{\boldsymbol{k}} |\alpha_{\boldsymbol{k}\lambda}(t)|^2 = \langle \mathtt{N} \rangle \, \langle \tilde{\varphi}_t | H_\gamma | \tilde{\varphi}_t \rangle \quad , \tag{8.58}$$

where H_γ is the Hamiltonian of the photon (1.160).

These results clarify and confirm the interpretation of the classical field in terms of photons presented in Sect. 1.3.6, specifically that the energy associated to one mode is that of one photon multiplied by the average number of photons. In a general manner, one sees that *there is a one-to-one correspondence between a classical field of amplitudes $\alpha_{\boldsymbol{k}\lambda}$ and a coherent state of photons in an individual state $|\varphi\rangle$ such that $\langle \boldsymbol{k}\lambda | \varphi \rangle = \alpha_{\boldsymbol{k}\lambda}$: the average values of the fields are identical to the classical expressions.* In particular, for each solution of (1.114), one can make a correspondence to the coherent state given by (8.55) and (8.54), in which the average values of the fields obey the classical Maxwell equations. It is in this way that the classical theory of radiation appears as a particular case of quantum-electromagnetic field theory.

8.2.6 Emission and Absorption of Photons by an Atom

Hamiltonian of an Atom Coupled to Photons

The classical Hamiltonian of a system of electrons in the presence of an electromagnetic field is given in Sect. 1.3.4. To simplify matters, we consider the case of a single electron of charge e and mass m under the influence of the potential $V(\boldsymbol{q})$ of an atomic nucleus

$$\mathtt{H} = \frac{1}{2m} |\boldsymbol{p} - e\mathtt{A}(\boldsymbol{q})|^2 + V(\boldsymbol{q}) + \mathtt{H}_0^{\text{ph}}$$

$$= \frac{1}{2m} |\boldsymbol{p}|^2 + V(\boldsymbol{q}) + \mathtt{H}_0^{\text{ph}} - \frac{e}{m} \boldsymbol{p} \cdot \mathtt{A}(\boldsymbol{q}) + \frac{e^2}{2m} |\mathtt{A}(\boldsymbol{q})|^2 \quad . \tag{8.59}$$

One obtains the Hamiltonian of the quantum system by the correspondence principle: $\boldsymbol{p}$ and $\boldsymbol{q}$ become canonical operators of a quantum particle, $\mathtt{A}(\boldsymbol{q})$ is the quantum field (8.7) taken at time $t = 0$ with $\boldsymbol{x}$ replaced by $\boldsymbol{q}$. $\mathtt{H}_0^{\text{ph}}$ is the Hamiltonian (8.6) of free photons. The state space of the coupled system is $\mathcal{L}^2(\mathcal{R}^3) \otimes \mathcal{F}_+$, where $\mathcal{F}_+$ is the Fock space of the photons (we do not take into account the spin of the electron in this section). The total system is conservative and evolves according to the quantum dynamics associated with $\mathtt{H}$. Note that here the argument $\boldsymbol{q}$ of $\mathtt{A}(\boldsymbol{q})$ is the position of the electron, which does not commute with $\boldsymbol{p}$. However, $\boldsymbol{p}$ commutes with $\mathtt{A}(\boldsymbol{q})$ by virtue of the transversality condition

$$\boldsymbol{p} \cdot \boldsymbol{A}(\boldsymbol{q}) - \boldsymbol{A}(\boldsymbol{q}) \cdot \boldsymbol{p} = -\mathrm{i}\hbar\nabla \cdot \boldsymbol{A}(\boldsymbol{q}) = 0 \quad , \tag{8.60}$$

which shows that the order of these operators in (8.59) is of no importance.

The first two terms of (8.59),

$$H^{\mathrm{at}} = \frac{|\boldsymbol{p}|^2}{2m} + V(\boldsymbol{q}) \quad , \tag{8.61}$$

describe the Hamiltonian of an isolated atom in the absence of radiation. One can write (8.59) as

$$\begin{aligned} \mathtt{H} &= \mathtt{H}_0 + \mathtt{H}_1 \quad , \\ \mathtt{H}_1 &= -\frac{e}{m}\boldsymbol{p} \cdot \mathtt{A}(\boldsymbol{q}) + \frac{e^2}{2m}|\mathtt{A}(\boldsymbol{q})|^2 \quad , \end{aligned} \tag{8.62}$$

where $\mathtt{H}_1$ represents the electromagnetic coupling and $\mathtt{H}_0 = H^{\mathrm{at}} + \mathtt{H}_0^{\mathrm{ph}}$ is the Hamiltonian of the uncoupled atom and photons. We will apply the perturbative method of quantum mechanics to (8.62). It should be emphasized immediately that in reality *one can never isolate an atom from the radiation* since the coupling, which is proportional to the charge of the electron e, can never be zero. The perturbative treatment is justified by the presence of a weak coupling constant which, once expressed as a dimensionless quantity (the fine-structure constant α), is equal to

$$\alpha = \frac{e^2}{4\pi\varepsilon_0\hbar c} \simeq \frac{1}{137} \quad . \tag{8.63}$$

First-Order Transitions

The eigenstates of the unperturbed Hamiltonian can be constructed with the help of taking products of eigenstates of the atomic Hamiltonian and the photon Hamiltonian. If we let ν be the quantum number which labels the bound states $|\nu\rangle$ of the electron and let ε_ν be the corresponding eigenvalues, then

$$H^{\mathrm{at}}\,|\nu\rangle = \varepsilon_\nu\,|\nu\rangle \quad . \tag{8.64}$$

A basis of eigenvectors of $\mathtt{H}_0^{\mathrm{ph}}$ is given by the states (3.38) labelled by the occupation numbers $n_{\boldsymbol{k}\lambda}$ of each mode of the electromagnetic field. The eigenstates of the unperturbed Hamiltonian are thus

$$|\nu, \{n_{\boldsymbol{k}\lambda}\}\rangle = |\nu\rangle \otimes |\{n_{\boldsymbol{k}\lambda}\}\rangle \quad , \tag{8.65}$$

corresponding to energies $\varepsilon_\nu + \sum_{\boldsymbol{k}\lambda} n_{\boldsymbol{k}\lambda}\hbar\omega_{\boldsymbol{k}\lambda}$. The fundamental mechanism for the interaction of radiation with atomic matter is the following: In the absence of any coupling, the atomic levels are stationary and their energies remain constant over time. Once we consider the complete Hamiltonian (8.59), however, this is no longer the case, because the states no longer diagonalize $\mathtt{H}$. As in usual quantum mechanics, the interaction induces finite probabilities for transitions to occur between the levels (8.65) of the unperturbed Hamiltonian. *During such a transition, the electron passes from one atomic level to another with a subsequent modification in the number of photons, since the interaction Hamiltonian* $\mathtt{H}_\mathrm{I}$ *does not conserve their number.* We can calculate the transition probability per unit time w_{fi} between an initial state i and a final state f, both of type (8.65), by a formula known as Fermi's golden rule

$$w_{fi} = \frac{2\pi}{\hbar} |\langle f|\mathtt{H}_\mathrm{I}|i\rangle|^2 \delta(E_f - E_i) \quad , \tag{8.66}$$

where E_i and E_f are the total energies of the initial and final states, respectively. Formula (8.66), which results from first-order perturbation theory, can be applied as long as the time for the transition to take place is much longer than the time of atomic oscillations. The validity of this condition is demonstrated at the end of Sect. 8.2.7. The interaction term, to lowest order in the coupling constant e, is $-(e/m)\boldsymbol{p} \cdot \mathtt{A}(\boldsymbol{q})$. This is the term that we will keep from (8.62). The term $(e^2/m)|\mathtt{A}(\boldsymbol{q})|^2$ represents the mutual interaction energy between different modes due to the coupling of the electron to the field, and can be neglected to first approximation in the study of the emission and absorption of photons.

If we take an initial state $|i\rangle$ containing exactly $n_{\boldsymbol{k}\lambda}$ photons of momentum $\hbar\boldsymbol{k}$ and polarization λ, with no photons in the other modes, and suppose that the electron is in the state $|\nu_1\rangle$, then

$$\begin{aligned} |i\rangle &= |\nu_1, n_{\boldsymbol{k}\lambda}\rangle \quad , \\ E_i &= \varepsilon_1 + n_{\boldsymbol{k}\lambda}\hbar\omega_{\boldsymbol{k}} \quad . \end{aligned} \tag{8.67}$$

As $-(e/m)\boldsymbol{p}\cdot\mathtt{A}(\boldsymbol{q})$ is linear in $a^*_{\boldsymbol{k}\lambda}$ and $a_{\boldsymbol{k}\lambda}$, the only non-zero matrix elements are those for which the number of initial and final state photons differ by one. Let us first treat the case of the emission of one photon of momentum $\hbar\boldsymbol{k}$ and polarization λ (Fig. 8.1), for which

$$\begin{aligned} |f\rangle &= |\nu_2, n_{\boldsymbol{k}\lambda} + 1\rangle \quad , \\ E_f &= \varepsilon_2 + (n_{\boldsymbol{k}\lambda} + 1)\hbar\omega_{\boldsymbol{k}} \quad , \quad \varepsilon_1 > \varepsilon_2 \quad . \end{aligned} \tag{8.68}$$

If one introduces the expansion (8.7) of $\mathtt{A}(\boldsymbol{q}) = \mathtt{A}_+(\boldsymbol{q}) + \mathtt{A}_-(\boldsymbol{q})$ and takes into account (3.39) and (3.40), one sees that only the creator $a^*_{\boldsymbol{k}\lambda}$ contributes, with $\langle n_{\boldsymbol{k}\lambda} + 1|a^*_{\boldsymbol{k}\lambda}|n_{\boldsymbol{k}\lambda}\rangle = \sqrt{n_{\boldsymbol{k}\lambda} + 1}$, so

$$\begin{aligned} &\langle \nu_1, n_{\boldsymbol{k}\lambda} + 1| \left(-\frac{e}{m}\boldsymbol{p}\cdot\mathtt{A}(\boldsymbol{q})\right) |\nu_1, n_{\boldsymbol{k}\lambda}\rangle \\ &= -\frac{1}{L^{3/2}}\frac{e}{m}\left(\frac{\hbar}{2\varepsilon_0\omega_{\boldsymbol{k}}}\right)^{1/2} \sqrt{n_{\boldsymbol{k}\lambda}+1}\,\langle\nu_2|\boldsymbol{p}\cdot\boldsymbol{e}_\lambda(\boldsymbol{k})\mathrm{e}^{-\mathrm{i}\boldsymbol{k}\cdot\boldsymbol{q}}|\nu_1\rangle \quad . \end{aligned} \tag{8.69}$$

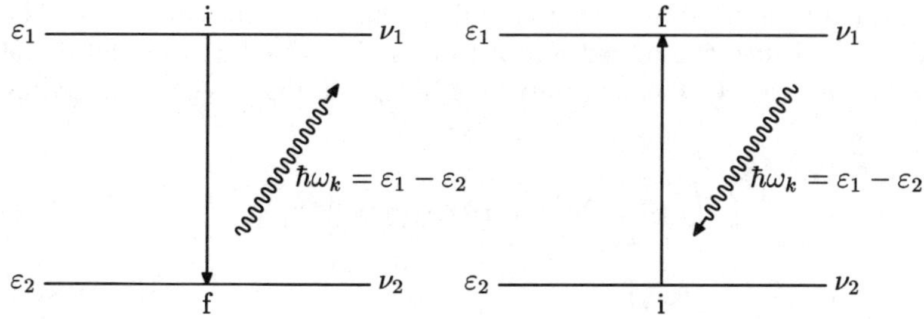

Fig. 8.1. Diagrams of the emission (*left*) and absorption (*right*) of a photon of momentum $\hbar\omega_{\boldsymbol{k}}$

According to (8.66), the probability w_{fi}^{em} of emission per unit time is

$$\begin{aligned} w_{fi}^{\mathrm{em}} &= (n_{\boldsymbol{k}\lambda}+1)\frac{\pi e^2}{\varepsilon_0 L^3 m^2 \omega_{\boldsymbol{k}}} |\langle \nu_2|\boldsymbol{p}\cdot\boldsymbol{e}_\lambda(\boldsymbol{k})\mathrm{e}^{-\mathrm{i}\boldsymbol{k}\cdot\boldsymbol{q}}|\nu_1\rangle|^2 \delta(\varepsilon_1-\varepsilon_2-\hbar\omega_{\boldsymbol{k}}) \\ &= w_{fi}^{\mathrm{ind}} + w_{fi}^{\mathrm{sp}} \quad . \end{aligned} \tag{8.70}$$

The factor $(n_{\boldsymbol{k}\lambda}+1)$ suggests the decomposition of w_{if}^{em} into two terms. *The first term w_{fi}^{ind} is proportional to the number $n_{\boldsymbol{k}\lambda}$ of photons already present in the initial state*; it describes the process of *induced emission* by the electromagnetic field. *The second term w_{fi}^{sp} is independent of the number of photons present and is non-zero even if the initial state contains no photons*; this phenomenon is known as *spontaneous emission.*

Finally, the process of absorption is characterized by $|i\rangle = |\nu_2, n_{\boldsymbol{k}\lambda}\rangle$, $|f\rangle = |\nu_1, n_{\boldsymbol{k}\lambda}-1\rangle$ (Fig. 8.1). In this case, only $a_{\boldsymbol{k}\lambda}$ contributes to the matrix element, with $\langle n_{\boldsymbol{k}\lambda}-1|a_{\boldsymbol{k}\lambda}|n_{\boldsymbol{k}\lambda}\rangle = \sqrt{n_{\boldsymbol{k}\lambda}}$, giving

$$w_{fi}^{\mathrm{abs}} = n_{\boldsymbol{k}\lambda}\frac{\pi e^2}{\varepsilon_0 L^3 m^2 \omega_{\boldsymbol{k}}} |\langle \nu_1|\boldsymbol{p}\cdot\boldsymbol{e}_\lambda(\boldsymbol{k})\mathrm{e}^{\mathrm{i}\boldsymbol{k}\cdot\boldsymbol{q}}|\nu_2\rangle|^2 \delta(\varepsilon_1-\varepsilon_2-\hbar\omega_{\boldsymbol{k}}) \quad . \tag{8.71}$$

As $\langle \nu_2|\boldsymbol{p}\cdot\boldsymbol{e}_\lambda(\boldsymbol{k})\mathrm{e}^{-\mathrm{i}\boldsymbol{k}\cdot\boldsymbol{q}}|\nu_1\rangle = (\langle \nu_1|\mathrm{e}^{\mathrm{i}\boldsymbol{k}\cdot\boldsymbol{q}}\boldsymbol{p}\cdot\boldsymbol{e}_\lambda(\boldsymbol{k})|\nu_2\rangle)^*$ and since the order of the operators $\boldsymbol{p}\cdot\boldsymbol{e}_\lambda(\boldsymbol{k})$ and $\mathrm{e}^{\mathrm{i}\boldsymbol{k}\cdot\boldsymbol{q}}$ is irrelevant due to transversality (see (8.60)), one notes from (8.70) and (8.71) that the probabilities of induced absorption and emission per unit time are equal. To obtain the transition rate in a specific situation, it is necessary to sum w_{fi} over all accessible final states and to take the average over the initial states if the latter are statistically distributed.

8.2.7 Spontaneous Emission

Lifetime

As an example, we will calculate the decay rate τ^{-1} per unit time from level ν_1 to level ν_2 by spontaneous emission. One obtains $\tau^{-1} = \sum_f w_{fi}^{\mathrm{sp}}$, summing over all available final states, that is, all possible one-photon states. The quantity τ, which has the dimension of time, is identified as the lifetime of level ν_1 relative to level ν_2. Going to the continuum limit, this gives according to (8.70)

$$\tau^{-1} = \frac{\pi e^2}{\varepsilon_0 m^2} \frac{1}{(2\pi)^3} \int d\boldsymbol{k} \, \frac{1}{\omega_{\boldsymbol{k}}} \sum_{\lambda} | \langle \nu_2 | \boldsymbol{p} \cdot \boldsymbol{e}_\lambda(\boldsymbol{k}) e^{-i\boldsymbol{k}\cdot\boldsymbol{q}} | \nu_1 \rangle |^2 \times \delta(\varepsilon_1 - \varepsilon_2 - \hbar\omega_{\boldsymbol{k}}) \quad . \tag{8.72}$$

Dipolar Approximation

One can simplify the evaluation of (8.72) by making the dipolar approximation for the atomic matrix element. The atomic wavefunctions $\langle \boldsymbol{x} | \nu_1 \rangle$ and $\langle \boldsymbol{x} | \nu_2 \rangle$ are very small outside the atomic diameter which is of the order of $a_B \simeq 10^{-8}$ cm. If the transition takes place in the optical region, the wave vector $|\boldsymbol{k}|$ is of the order of 10^5 cm^{-1}. In the calculation of the atomic matrix element, the integral is thus performed in a region $\boldsymbol{k} \cdot \boldsymbol{x} \simeq 10^{-3}$, so that one only needs to keep the first term in the expansion $e^{-i\boldsymbol{k}\cdot\boldsymbol{q}} = 1 - i\boldsymbol{k} \cdot \boldsymbol{q} + \dots$. The higher-order terms are multipolar corrections. One thus obtains

$$\langle \nu_2 | \boldsymbol{p} \cdot \boldsymbol{e}_\lambda(\boldsymbol{k}) e^{-i\boldsymbol{k}\cdot\boldsymbol{q}} | \nu_1 \rangle \simeq \langle \nu_2 | \boldsymbol{p} | \nu_1 \rangle \cdot \boldsymbol{e}_\lambda(\boldsymbol{k}) \quad . \tag{8.73}$$

The matrix element $\langle \nu_2 | \boldsymbol{p} | \nu_1 \rangle$ is related to the dipolar electric moment $\boldsymbol{d} = e \langle \nu_2 | \boldsymbol{q} | \nu_1 \rangle$ by the relation

$$\frac{e\hbar}{m} \langle \nu_2 | \boldsymbol{p} | \nu_1 \rangle = i(\varepsilon_2 - \varepsilon_1) \boldsymbol{d} \quad , \tag{8.74}$$

which results from the identity

$$\left[\frac{|\boldsymbol{p}|^2}{2m} + V(\boldsymbol{q}), \boldsymbol{q} \right] = -i\hbar \frac{\boldsymbol{p}}{m} \tag{8.75}$$

evaluated between the states $|\nu_1\rangle$ and $|\nu_2\rangle$.

We choose the polarization vector $\boldsymbol{e}_1(\boldsymbol{k})$ in the $(\boldsymbol{d}, \boldsymbol{k})$ plane. Using (8.74) and the geometry of Fig. 8.2, one has

$$\sum_{\lambda} | \langle \nu_2 | \boldsymbol{p} | \nu_1 \rangle \cdot \boldsymbol{e}_\lambda(\boldsymbol{k}) |^2 = \left(\frac{m}{\hbar e} \right)^2 |\boldsymbol{d}|^2 (\varepsilon_1 - \varepsilon_2)^2 (\sin\theta)^2 \quad . \tag{8.76}$$

One can thus perform the integral (8.72) in spherical coordinates, with the third axis in the direction of $\boldsymbol{d}$

$$\begin{aligned} \tau^{-1} &= \frac{|\boldsymbol{d}|^2 (\varepsilon_2 - \varepsilon_1)^2}{2\pi\varepsilon_0 \hbar^2} \int_0^\infty k^2 dk \int_0^\pi \sin\theta d\theta \, \frac{(\sin\theta)^2}{2ck} \delta(\varepsilon_1 - \varepsilon_2 - \hbar c k) \\ &= \frac{1}{3\pi c^3 \varepsilon_0 \hbar^4} (\varepsilon_1 - \varepsilon_2)^3 |\boldsymbol{d}|^2 \quad . \end{aligned} \tag{8.77}$$

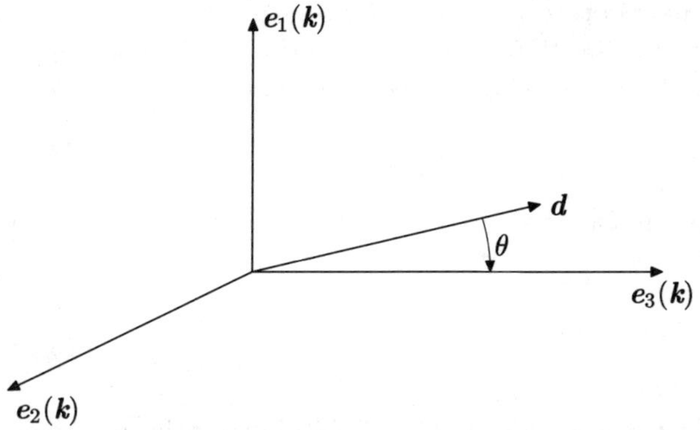

Fig. 8.2. The polarization vector $\boldsymbol{e}_1(\boldsymbol{k})$ is taken in the $(\boldsymbol{d}, \boldsymbol{k})$ plane

In terms of the relevant orders of magnitude we have, $|\boldsymbol{d}| \simeq a_{\mathrm{B}}$, where a_{B} is the Bohr radius. If the transition $\nu_1 \to \nu_2$ is in the optical region $|\boldsymbol{k}| \simeq 10^5\ \mathrm{cm}^{-1}$, the lifetime τ is of the order of 10^{-8} s. The period of atomic frequencies is of the order of $(c|\boldsymbol{k}|)^{-1} \simeq 10^{-15}$ s, which indeed allows us to use the perturbation formula (8.66).

8.2.8 Photons and Matter in Equilibrium

One important case is when the photons are in thermal equilibrium with the atoms. The initial state of the radiation field does not have a fixed number of photons, but corresponds to a statistical mixture given by the thermal density matrix (3.127). To obtain the transition probability w_{21}^{em} of the atom from state ν_1 to state ν_2 by emission in the presence of thermal photons, one must take the average of (8.70) in the distribution (8.87) and sum over all photon states. This amounts to replacing $n_{\boldsymbol{k}\lambda}$ in (8.70) by the average number of photons

$$\tilde{n}(\omega_{\boldsymbol{k}}) = \frac{1}{\mathrm{e}^{\beta\hbar\omega_{\boldsymbol{k}}} - 1} \tag{8.78}$$

then to carry out the summation in a similar manner to (8.72). Because of the energy conservation function $\delta(\varepsilon_1 - \varepsilon_2 - \hbar\omega_{\boldsymbol{k}})$, one of course finds, following the calculation of the preceding section,

$$w_{21}^{\mathrm{em}} = \tau^{-1}\left[\tilde{n}(\varepsilon_1 - \varepsilon_2) + 1\right] = w_{21}^{\mathrm{ind}} + w_{21}^{\mathrm{sp}} \quad . \tag{8.79}$$

Similarly, one can calculate the transition probability for absorption w_{12}^{abs} from state ν_2 to state ν_1 from (8.71)

$$w_{12}^{\mathrm{abs}} = \tau^{-1}\tilde{n}(\varepsilon_1 - \varepsilon_2) \quad . \tag{8.80}$$

For equilibrium, the number of transitions per unit time due to emission must be equal to the number of transitions due to absorption. If N_1 and N_2 are the atomic populations of levels ν_1 and ν_2, one must have

$$w_{21}^{\text{em}} N_1 = w_{12}^{\text{abs}} N_2 \quad . \tag{8.81}$$

From (8.78)–(8.80), one can easily show that the ratio between the populations is given by

$$\frac{N_1}{N_2} = \frac{w_{12}^{\text{abs}}}{w_{21}^{\text{em}}} = \mathrm{e}^{-\beta(\varepsilon_1 - \varepsilon_2)} \quad . \tag{8.82}$$

This relation satisfies the condition for thermal equilibrium for the atomic populations as it must (note that in this treatment the atoms are distinguishable and subject to the classical Maxwell–Boltzmann statistics). These considerations highlight *the fundamental link between Bose statistics of photons and the existence of spontaneous emission* or, in other terms, the compatibility between formulae (8.78), (8.79), (8.80) and (8.82). It is instructive to recall Einstein's reasoning: in postulating spontaneous emission in (8.79) and the equality $w_{21}^{\text{ind}} = w_{12}^{\text{abs}}$, he deduced precisely the Planck distribution by assuming the thermal equilibrium (8.82) of matter. Conversely, if one assumes Planck's law (8.78), the spontaneous emission term is necessary to correctly obtain (8.82). But, as we have seen, this term is due to the commutation rules of the electromagnetic field, and thus due to its essentially quantum nature. *Quantization of the electromagnetic field thus explains spontaneous emission and the black body law.*

8.2.9 Photon Statistics

The number of photons is subject to fluctuations in a coherent state as well as in a thermal state. The coherent state is a pure quantum state, but made up from a superposition of states of different photon number, whereas the thermal state is a statistical mixture. To simplify, we will only consider a single mode $\boldsymbol{k}, \lambda$ of the radiation field and will omit the index $\boldsymbol{k}\lambda$ in the following.

Coherent State

The probability of having n photons in the coherent state $|\alpha\rangle$ is, according to (1.34),

$$p_n^{\text{coh}} = |\langle n|\alpha\rangle|^2 = \frac{|\alpha|^{2n}}{n!} \mathrm{e}^{-|\alpha|^2} \quad . \tag{8.83}$$

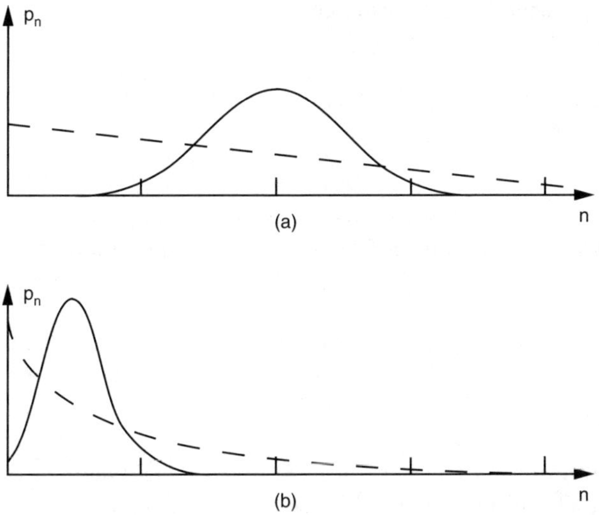

Fig. 8.3. The solid lines represent Poisson distributions $p_n^{\rm coh}$ and the dashed lines represent geometrical distributions $p_n^{\rm th}$ in the thermal state for average values (**a**) $\tilde{n} = 20$ and (**b**) $\tilde{n} = 5$

As the average number of photons $\tilde{n}$ in the state $|\alpha\rangle$ is given by (see (1.35))

$$\tilde{n} = \langle\alpha|a^* a|\alpha\rangle = |\alpha|^2 \quad , \tag{8.84}$$

one notes that $p_n^{\rm coh}$ is a *Poisson distribution* with mean value $\tilde{n}$ (Fig. 8.3)

$$p_n^{\rm coh} = \frac{\tilde{n}^n}{n!}{\rm e}^{-\tilde{n}} \quad . \tag{8.85}$$

It is straightforward to obtain the corresponding mean square deviation

$$(\Delta n)^2 = \sum_{n=0}^{\infty}(n-\tilde{n})^2 p_n^{\rm coh} = \tilde{n} \quad . \tag{8.86}$$

Thermal State

The probability of finding n photons in the thermal state given by $\rho = {\rm e}^{-\beta\hbar\omega a^* a}/Tr\ {\rm e}^{-\beta\hbar\omega a^* a}$ is

$$p_n^{\rm th} = \langle n|\rho|n\rangle = {\rm e}^{-\beta\hbar\omega n}(1-{\rm e}^{-\beta\hbar\omega}) \quad . \tag{8.87}$$

One can also express p_n^{th} in terms of the average number of photons

$$\tilde{n} = \sum_{n=0}^{\infty} n p_n^{\text{th}} = \frac{1}{e^{\beta\hbar\omega} - 1} \quad ,$$

$$p_n^{\text{th}} = \frac{1}{1+\tilde{n}} \left(\frac{\tilde{n}}{1+\tilde{n}} \right)^n \quad . \tag{8.88}$$

This is a *geometrical distribution* (Fig. 8.3) and the mean square deviation in this case is given by

$$(\Delta n)^2 = \tilde{n} + \tilde{n}^2 \quad . \tag{8.89}$$

Origin of Fluctuations

Fluctuations of the photon number in general have two origins. One relates to the quantum aspect of the radiation field, the other is due to the random nature of the emission process itself. Schematically one can distinguish two types of sources. The usual sources (black body, lamp filament, gaseous discharge) create photons incoherently: the atoms emit independently from each other, themselves being submitted to various perturbations, such as random collisions or variations in velocity. Radiation emitted by such sources has sizeable intensity fluctuations, which are produced on a time scale τ_c called the coherence time. The other type of source, represented by a laser operating above threshold, is characterized by a coherent radiation and the absence of intensity fluctuations.

Observing the photon statistics, by photon-number counting, allows one to study the source characteristics. One can show (and confirm by experiment) that photons emitted from a laser follow the Poisson distribution (8.85), whereas those emitted from a random source obey the geometric distribution (8.88) (in practice, one can not make instantaneous measurements, but records the number of photons in an interval of time $\Delta t \ll \tau_c$). As we have seen in Sect. 8.2.5 the coherent state corresponds to a source producing radiation without any fluctuations (a well determined current $\boldsymbol{j}(\boldsymbol{x}, t)$). *In this case, any fluctuations in photon number must be attributed to the intrinsic quantum nature of the electromagnetic field.* In particular, the mean square deviation (8.86), proportional to the intensity, persists in all situations. It can only increase because of the statistical properties of the source, as shown in (8.89), where the term $\tilde{n}^2$ must be attributed to the disorderly nature of the emission process.

8.3 Massive Scalar Field

8.3.1 Neutral Scalar Field

Scalar Particles

In the interactions between elementary particles, one observes the production and absorption of particles without spin, such as the π, K and Λ^0 mesons, for example:

$$p + p \to p + p + \pi^0$$
$$p + p \to p + n + \pi^+ \quad ,$$

where p and n represent a proton and a neutron, respectively. Experimentally, these particles have the following properties:

(i) they are *massive and spinless*;
(ii) they are *nearly stable*: π^+ has a weak decay channel $\pi^+ \to \mu^+ + \nu_\mu$ (μ = muon, ν_μ = muon neutrino). Because of the small magnitude of the weak coupling constant, π^+ has a lifetime of the order of 10^{-8}s. This time is much longer than the natural unit of time $\hbar/mc^2 \simeq 10^{-23}$s defined with the help of $\hbar$, c and the mass of the π. Similarly, the lifetime of the $\pi^0(\pi^0 \to \gamma + \gamma)$ is $10^{-16}\text{s} \gg 10^{-23}\text{s}$;
(iii) these particles are produced at energies which require the use of *relativistic kinetic energy* $c\sqrt{|\boldsymbol{p}|^2 + (mc)^2}$;
(iv) they can be *created and absorbed in varied numbers*;
(v) they obey *Bose statistics*.

We will examine first the case of a neutral particle, such as the π^0. The characteristics (i)–(v) listed above lead to a description of the π^0 by a quantum field analogous to the photons. There are two differences.

First of all, being a spin zero particle, the wavefunction of the π^0 is a pseudoscalar (one single component rather than the three components of the vector potential). The difference between a scalar and a pseudoscalar, which changes sign under the inversion operation, is not important here and we will not make the distinction in what follows. Secondly, as the mass of the π^0 is non-zero, the energy $\hbar c|\boldsymbol{k}|$ of the photon must be replaced by

$$\hbar\omega_{\boldsymbol{k}} = c\sqrt{(\hbar|\boldsymbol{k}|)^2 + (mc)^2} = \hbar c\sqrt{|\boldsymbol{k}|^2 + \mu^2} \quad , \quad \mu = \frac{mc}{\hbar} \quad . \tag{8.90}$$

With these modifications, the space of one-meson states $\mathcal{H}_\pi$ is defined in a similar way to the photon case (see (1.162)). In a cubic box with periodic boundary conditions, the eigenstates $|\boldsymbol{k}\rangle$ of momentum and the Hamiltonian of the meson satisfy

$$\boldsymbol{p}\,|\boldsymbol{k}\rangle = \hbar\boldsymbol{k}\,|\boldsymbol{k}\rangle \quad , \tag{8.91}$$
$$H_\pi\,|\boldsymbol{k}\rangle = \hbar\omega_{\boldsymbol{k}}\,|\boldsymbol{k}\rangle \quad . \tag{8.92}$$

They generate the single-particle space

$$\mathcal{H}_\pi = \left\{ |\varphi\rangle \, ; \sum_{\boldsymbol{k}} | \langle \boldsymbol{k}|\varphi\rangle |^2 < \infty \right\} \quad . \tag{8.93}$$

Definition of the Scalar Field

Word for word one can repeat the construction that was made for the electromagnetic field. One introduces the Fock space $\mathcal{F}_+(\mathcal{H}_\pi)$ of mesons with the total Hamiltonian

$$\mathtt{H}_0 = \sum_{\boldsymbol{k}} \hbar\omega_{\boldsymbol{k}} a_{\boldsymbol{k}}^* a_{\boldsymbol{k}} \quad . \tag{8.94}$$

In analogy with the electromagnetic vector potential (8.7), we define the *free scalar field*

$$\Phi(\boldsymbol{x}, t) = \left(\frac{\hbar}{L^3}\right)^{1/2} \sum_{\boldsymbol{k}} \frac{1}{\sqrt{2\omega_{\boldsymbol{k}}}} \left(a_{\boldsymbol{k}}^* \mathrm{e}^{-\mathrm{i}(\boldsymbol{k}\cdot\boldsymbol{x} - \omega_{\boldsymbol{k}} t)} + a_{\boldsymbol{k}} \mathrm{e}^{\mathrm{i}(\boldsymbol{k}\cdot\boldsymbol{x} - \omega_{\boldsymbol{k}} t)} \right) \quad , \tag{8.95}$$

$$\Phi(\boldsymbol{x}, t) = \Phi_+(\boldsymbol{x}, t) + \Phi_-(\boldsymbol{x}, t) \quad . \tag{8.96}$$

The components $\Phi_+(\boldsymbol{x}, t)$ and $\Phi_-(\boldsymbol{x}, t) = (\Phi_+(\boldsymbol{x}, t))^*$ again denote the contribution from creators and annihilators of the field, which is itself a Hermitian operator. It follows from the relativistic energy law (8.90) that the field satisfies the propagation equation

$$\left(\frac{1}{c^2} \frac{\partial^2}{\partial t^2} - \nabla^2 \right) \Phi(\boldsymbol{x}, t) + \mu^2 \Phi(\boldsymbol{x}, t) = 0 \quad . \tag{8.97}$$

This is the *Klein–Gordon equation*: it reduces to the wave equation (8.11) if $m = 0$. The field at time t is derived from $\Phi(\boldsymbol{x}) = \Phi(\boldsymbol{x}, t = 0)$ by Heisenberg's evolution law

$$\Phi(\boldsymbol{x}, t) = \exp\left(\frac{\mathrm{i}\mathtt{H}_0 t}{\hbar}\right) \Phi(\boldsymbol{x}) \exp\left(-\frac{\mathrm{i}\mathtt{H}_0 t}{\hbar}\right) \quad , \tag{8.98}$$

where $\mathtt{H}_0$ is the Hamiltonian of the mesons (8.94).

Commutation Relations

The field can be described with the help of its canonical variables. In analogy with the electromagnetic field where we had the canonical (transverse) variables $\mathtt{A}(\boldsymbol{x})$ and $\dot{\mathtt{A}}(\boldsymbol{x}) = (\partial/\partial t)\mathtt{A}(\boldsymbol{x}, t)|_{t=0}$, we consider the pair of operators $\Phi(\boldsymbol{x})$ and $\dot{\Phi}(\boldsymbol{x}) = (\partial/\partial t)\Phi(\boldsymbol{x}, t)|_{t=0}$. Following the same calculation as in Sect. 8.2.2, one immediately establishes that $\Phi(\boldsymbol{x})$ and $\dot{\Phi}(\boldsymbol{y})$ satisfy the *canonical commutation relations*

$$[\Phi(\boldsymbol{x}), \dot{\Phi}(\boldsymbol{y})] = \mathrm{i}\hbar\delta(\boldsymbol{x} - \boldsymbol{y}) \quad , \tag{8.99}$$

$$[\Phi(\boldsymbol{x}), \Phi(\boldsymbol{y})] = [\dot{\Phi}(\boldsymbol{x}), \dot{\Phi}(\boldsymbol{y})] = 0 \quad . \tag{8.100}$$

These relations are preserved throughout the temporal evolution. In this way, *the meson field is also described by a family of dynamical variables attached to the points $\boldsymbol{x}$ of space, and quantized with the canonical rules*. Finally, we express the Hamiltonian of the mesons (8.94) with the help of the canonical variables. By a calculation completely analogous to the one which lead to (8.16), $\mathtt{H}_0$ is written as an integral over a Hamiltonian density

$$\mathtt{H}_0 = \frac{1}{2} \int \mathrm{d}\boldsymbol{x} : \left[\left(\frac{\partial \Phi}{\partial t} \right)^2 + c^2 |\nabla \Phi|^2 + (\mu c)^2 \Phi^2 \right] : \quad . \tag{8.101}$$

The only difference with the electromagnetic Hamiltonian (8.16) is the mass term and the fact that the field has only one component.

The commutation function $\Delta(\boldsymbol{x} - \boldsymbol{y}, t - s)$ of two fields taken at different points $\boldsymbol{x}, t$ and $\boldsymbol{y}, s$ of space–time

$$[\Phi(\boldsymbol{x}, t), \Phi(\boldsymbol{y}, s)] = \mathrm{i}\hbar\Delta(\boldsymbol{x} - \boldsymbol{y}, t - s) \tag{8.102}$$

has the same expression as (8.33)

$$\Delta(\boldsymbol{x}, t) = -\frac{1}{L^3} \sum_{\boldsymbol{k}} \frac{\sin(\boldsymbol{k} \cdot \boldsymbol{x} - \omega_{\boldsymbol{k}} t)}{\omega_{\boldsymbol{k}}} \tag{8.103}$$

but here $\omega_{\boldsymbol{k}}$ is given by (8.90). In the continuum limit, one can explicitly calculate the integral, with the result

$$\begin{aligned} \Delta(\boldsymbol{x}, t) &= \frac{1}{2\pi c} \varepsilon(t) \delta(\lambda) - \frac{\mu}{4\pi c \sqrt{\lambda}} \theta(\lambda) \varepsilon(t) J_1(\mu \sqrt{\lambda}) \quad , \\ \lambda &= (ct)^2 - |\boldsymbol{x}|^2 \quad , \\ \theta(\lambda) &= \begin{cases} 0 \;, & \lambda < 0 \\ 1 \;, & \lambda > 0 \end{cases} \quad , \quad \varepsilon(t) = \begin{cases} -1 \;, & t < 0 \\ 0 \;, & t = 0 \\ +1 \;, & t > 0 \end{cases} \end{aligned} \tag{8.104}$$

and J_1 is the Bessel function of order 1.

The form (8.104) shows the following important facts: $\Delta(\boldsymbol{x}, t)$ is invariant under Lorentz transformations since it only depends on the invariant $\lambda = (ct)^2 - |\boldsymbol{x}|^2$ and the sign of time, and $\Delta(\boldsymbol{x}, t)$ reduces to the Jordan and Pauli function (8.34) when $m = 0$. Moreover, $\Delta(\boldsymbol{x}, t)$ satisfies

$$\Delta(\boldsymbol{x}, t) = 0 \quad \text{if} \quad (ct)^2 - |\boldsymbol{x}|^2 < 0 \quad , \tag{8.105}$$

which shows that *the scalar field also satisfies the microcausality principle* (8.35) discussed in Sect. 8.2.3.

8.3.2 The Yukawa Potential

A Classical Interaction Model

The π^0 mesons can be emitted and absorbed by nucleons. To clarify certain fundamental aspects of these interactions, we will first of all consider a model inspired by electrodynamics, in which the (spinless) nucleon and the scalar field $\Phi(\boldsymbol{x}, t)$ are treated classically. The total Hamiltonian $\mathtt{H}$ is thus composed of three terms

$$\mathtt{H} = \mathtt{H}_0 + H^{\text{part}}(\boldsymbol{q}, \boldsymbol{p}) + \mathtt{H}_\text{I} \quad . \tag{8.106}$$

The Hamiltonian $H^{\text{part}}(\boldsymbol{q}, \boldsymbol{p})$ describes a classical particle with canonical variables $\boldsymbol{q}$ and $\boldsymbol{p}$ and $\mathtt{H}_0$ is the energy of the free field. As for the interaction Hamiltonian $\mathtt{H}_\text{I}$, it can, in principle, depend on Φ, $(\partial/\partial t)\Phi$, $\boldsymbol{q}$ and $\boldsymbol{p}$. To determine its form, we proceed in analogy to the case of the coupling (8.50) of the current to photons. We suppose a linear and local coupling between the field and the particle density $\rho(\boldsymbol{x})$

$$\mathtt{H}_\text{I} = g \int \mathrm{d}\boldsymbol{x}\, \rho(\boldsymbol{x}) \Phi(\boldsymbol{x}) \quad . \tag{8.107}$$

Here, as the field is scalar, it is natural to couple it to the density and not to the vector current. The constant g controls the strength of the interaction: this is thus a coupling constant. If the particle is point-like, one has

$$\rho(\boldsymbol{x}) = \delta(\boldsymbol{x} - \boldsymbol{q}) \quad , \quad \mathtt{H}_\text{I} = g\Phi(\boldsymbol{q}) \quad . \tag{8.108}$$

The analogies with electromagnetism are presented in Table 8.1.

One concludes from (8.106) and (8.108) that the Hamiltonian equations are written as

$$\begin{aligned} \frac{\mathrm{d}\boldsymbol{q}}{\mathrm{d}t} &= \frac{\boldsymbol{p}}{m} \quad , \\ \frac{\mathrm{d}\boldsymbol{p}}{\mathrm{d}t} &= -\nabla_q \mathtt{H} = -\nabla_q H^{\text{part}} - g\nabla_q \Phi \quad . \end{aligned} \tag{8.109}$$

Thus, for a classical particle, the scalar field acts as a potential. It is therefore of some interest to determine the static field $\Phi(\boldsymbol{x})$ created by a point particle situated at $\boldsymbol{r}$. This field satisfies the equation

$$-\nabla^2 \Phi(\boldsymbol{x}) + \mu^2 \Phi(\boldsymbol{x}) = -\frac{g}{c^2} \delta(\boldsymbol{x} - \boldsymbol{r}) \quad . \tag{8.110}$$

Up to the mass term, this equation is analogous to the Poisson equation (1.108) with g and $1/c^2$ playing the same roles as $-e$ and $1/\varepsilon_0$. The Fourier transform $\tilde{\Phi}(\boldsymbol{k})$ of $\Phi(\boldsymbol{x})$ is obtained by Fourier transforming equation (8.110)

$$\tilde{\Phi}(\boldsymbol{k}) = \int \mathrm{d}\boldsymbol{x}\, \Phi(\boldsymbol{x}) \mathrm{e}^{-\mathrm{i}\boldsymbol{k}\cdot\boldsymbol{x}} = -\frac{g}{c^2} \frac{\mathrm{e}^{-\mathrm{i}\boldsymbol{k}\cdot\boldsymbol{r}}}{|\boldsymbol{k}|^2 + \mu^2} \quad . \tag{8.111}$$

Table 8.1. Analogies between the electromagnetic field coupled to the charged current and the meson field coupled to the nucleons

	Electromagnetic field	Scalar field
Hamiltonian of the free field $\mathtt{H}_0$	$\frac{1}{2}\sum_{\alpha=1}^{3}\int \mathrm{d}\boldsymbol{x} : \left[\varepsilon_0 \left(\frac{\partial}{\partial t}\mathtt{A}^\alpha\right)^2 + \mu_0^{-1}\|\nabla \mathtt{A}^\alpha\|^2 \right] :$	$\frac{1}{2}\int \mathrm{d}\boldsymbol{x} : \left[\left(\frac{\partial}{\partial t}\Phi\right)^2 + c^2\|\nabla\Phi\|^2 + (\mu c)^2\Phi^2 \right] :$
Equation of free motion	$\frac{1}{c^2}\frac{\partial^2}{\partial t^2}\mathtt{A} - \nabla^2\mathtt{A} = 0$	$\frac{1}{c^2}\frac{\partial^2}{\partial t^2}\Phi - \nabla^2\Phi + \mu^2\Phi = 0$
Interaction Hamiltonian $\mathtt{H}_\mathrm{I}$	$-\int \mathrm{d}\boldsymbol{x}\, \boldsymbol{j}(\boldsymbol{x}) \cdot \mathtt{A}(\boldsymbol{x})$	$g\int \mathrm{d}\boldsymbol{x}\, \rho(\boldsymbol{x})\Phi(\boldsymbol{x})$
Equation of motion with sources	$\frac{1}{c^2}\frac{\partial^2}{\partial t^2}\mathtt{A} - \nabla^2\mathtt{A} = \mu_0\boldsymbol{j}$	$\frac{1}{c^2}\frac{\partial^2}{\partial t^2}\Phi - \nabla^2\Phi + \mu^2\Phi = -\frac{g}{c^2}\rho$

The inverse transform of (8.111) gives rise to the *Yukawa potential*

$$\Phi(\boldsymbol{x}) = -\frac{g}{4\pi c^2}\frac{\mathrm{e}^{-\mu|\boldsymbol{x}-\boldsymbol{r}|}}{|\boldsymbol{x}-\boldsymbol{r}|} \quad . \tag{8.112}$$

This potential is a Coulomb potential at short distance and decreases exponentially with a range $\mu^{-1} = \hbar/mc$. As the order of magnitude of the range of nuclear forces is 10^{-13} cm, this corresponds to a mass m approximately 200 times that of an electron, which is comparable to the pion mass ($m_\pi \simeq 273 m_\mathrm{e}$). Concerning the interaction energy between two particles situated at $\boldsymbol{r}_1$ and $\boldsymbol{r}_2$, it can be written in analogy with the Coulomb case as

$$E^{\mathrm{pot}} = -\frac{g^2}{4\pi c^2}\frac{\mathrm{e}^{-\mu|\boldsymbol{r}_1-\boldsymbol{r}_2|}}{|\boldsymbol{r}_1-\boldsymbol{r}_2|} \quad . \tag{8.113}$$

This potential energy is negative: it describes an attraction at short distance between the two particles.

A Quantum Interaction Model

The actual interaction between nucleons is much more complicated than the simplified model which leads to the potential (8.112). In reality, the nucleons and the pions are composite particles made up of quarks. *The strong interaction between the nucleons is the result of more fundamental fields which are described by the theory of quantum chromodynamics.* It is nonetheless instructive as a pedagogical example to consider a quantum version of the simple model which was introduced in the previous section. In this case, the nucleon is still treated as a classical point particle, but the quantum nature of the scalar field is taken into account.

Suppose we have two static nucleons fixed at points $\boldsymbol{r}_1$ and $\boldsymbol{r}_2$, that is

$$\rho(\boldsymbol{x}) = \delta(\boldsymbol{x} - \boldsymbol{r}_1) + \delta(\boldsymbol{x} - \boldsymbol{r}_2) \quad . \tag{8.114}$$

Taking into account (8.107) and (8.95), the interaction Hamiltonian becomes

$$\mathtt{H}_\mathrm{I} = g \left(\frac{\hbar}{L^3} \right)^{1/2} \sum_{\boldsymbol{k}} \frac{1}{\sqrt{2\omega_{\boldsymbol{k}}}} \sum_{j=1}^{2} (a^*_{\boldsymbol{k}} \mathrm{e}^{-\mathrm{i}\boldsymbol{k}\cdot\boldsymbol{r}_j} + a_{\boldsymbol{k}} \mathrm{e}^{\mathrm{i}\boldsymbol{k}\cdot\boldsymbol{r}_j}) \quad . \tag{8.115}$$

The complete Hamiltonian is given by

$$\mathtt{H} = E_0 + \mathtt{H}_0 + \mathtt{H}_\mathrm{I} \quad , \tag{8.116}$$

where E_0 is the energy of an independent pair of nucleons in the absence of coupling and $\mathtt{H}_0$ is the free Hamiltonian (8.94) of the pions.

We would like to perturbatively calculate the energy of this pair in the presence of the pion field. The unperturbed state of energy E_0 is simply the vacuum $|0\rangle$ of the pions (the nucleons do not have a dynamical state). According to the usual perturbation formulae of a non-degenerate quantum state, the first-order correction is zero

$$E_0^{(1)} = \langle 0|\mathtt{H}_\mathrm{I}|0\rangle = 0 \quad , \tag{8.117}$$

since $\langle 0|a_{\boldsymbol{k}}|0\rangle = \langle 0|a^*_{\boldsymbol{k}}|0\rangle = 0$. The second-order term is equal to

$$E_0^{(2)} = \sum_{r\neq 0} \frac{|\langle 0|\mathtt{H}_\mathrm{I}|\Phi_r\rangle|^2}{E_0 - E_r} \quad , \tag{8.118}$$

where the sum is carried out over the basis of eigenstates of $\mathtt{H}_0$ in Fock space, with the exception of $|\Phi_0\rangle = |0\rangle$. As only those matrix elements of $\mathtt{H}_\mathrm{I}$ between the vacuum and the one-particle states are non-zero, one needs only make $|\Phi_r\rangle$ run through the one-particle states $|\boldsymbol{k}\rangle$. With $\langle 0|a_{\boldsymbol{k}'}|\boldsymbol{k}\rangle = \delta_{\boldsymbol{k},\boldsymbol{k}'}$, one obtains

$$\langle 0|\mathtt{H}_\mathrm{I}|\boldsymbol{k}\rangle = g \left(\frac{\hbar}{2L^3\omega_{\boldsymbol{k}}} \right)^{1/2} \sum_{j=1}^{2} \mathrm{e}^{\mathrm{i}\boldsymbol{k}\cdot\boldsymbol{r}_j} \quad , \tag{8.119}$$

from which, since $E_0 - E_{\boldsymbol{k}} = -\hbar\omega_{\boldsymbol{k}}$, one obtains

$$E_0^{(2)} = -\frac{g^2}{2}\frac{1}{L^3}\sum_{\boldsymbol{k}}\frac{1}{\omega_{\boldsymbol{k}}^2}\left|\sum_{j=1}^{2}\mathrm{e}^{\mathrm{i}\boldsymbol{k}\cdot\boldsymbol{r}_j}\right|^2$$
$$= -g^2\frac{1}{L^3}\sum_{\boldsymbol{k}}\frac{1}{\omega_{\boldsymbol{k}}^2} - g^2\frac{1}{L^3}\sum_{\boldsymbol{k}}\frac{\cos\boldsymbol{k}\cdot(\boldsymbol{r}_1-\boldsymbol{r}_2)}{\omega_{\boldsymbol{k}}^2} \quad . \qquad (8.120)$$

The first term of (8.120), which is infinite, does not depend on the position of the nucleons and would be finite if the nucleons were extended particles. Here we have a typical example of a divergence occurring in field theory when one treats the interaction of point particles. This infinite constant can be combined with E_0 by a redefinition of the energy of independent nucleons (see also the problem of renormalization in Sect. 9.4.3). The term which is physically interesting is the second term of (8.120): *it represents the mutual interaction energy of the two nucleons* and in the continuum limit is equal to

$$-\frac{g^2}{(2\pi)^3}\int \mathrm{d}\boldsymbol{k}\,\frac{\cos\boldsymbol{k}\cdot(\boldsymbol{r}_1-\boldsymbol{r}_2)}{c^2(|\boldsymbol{k}|^2+\mu^2)} = -\frac{g^2}{4\pi c^2}\frac{\mathrm{e}^{-\mu|\boldsymbol{r}_1-\boldsymbol{r}_2|}}{|\boldsymbol{r}_1-\boldsymbol{r}_2|} \quad . \qquad (8.121)$$

This is the same as the result (8.113). The similarity is striking, but one must not forget that, in the quantum case, we have only calculated the first non-vanishing term of the perturbation series. It is thus necessary to determine the order of magnitude of the coupling constant g.

The Strong Interaction Coupling Constant

Comparing the interaction Hamiltonians in Table 8.1, one notes that the coupling constant g is made explicit in the case of the scalar field while, in the case of the electromagnetic field, the same role is played by the charge e, which is included in the definition of j, linear in e. In Sect. 8.2.6, we mentioned that the fine structure constant $\alpha = e^2/(4\pi\varepsilon_0\hbar c) \simeq 1/137$ gives a measure of the coupling force, since α is dimensionless: in quantum electrodynamics the physical quantities are given in the form of a powers series in α. As $\alpha \ll 1$, the perturbation calculation gives rise to predictions which have excellent agreement with experimental measurements.

To determine the dimensionless coupling constant which plays the same role for the strong interactions, note that α can be defined as the ratio of the potential energy of the electron in the hydrogen atom to its mass energy

$$\frac{e^2}{4\pi\varepsilon_0}\frac{1}{a_{\mathrm{B}}}\frac{1}{mc^2} = \alpha^2 \quad , \qquad (8.122)$$

where $a_{\mathrm{B}} = 4\pi\varepsilon_0\hbar^2/(me^2)$ is the Bohr radius. Note that α is independent of the mass. If one follows the same line of reasoning for the scalar field, it is necessary to evaluate the potential energy of a particle of mass m under the influence of the strong interaction of a nucleon situated at a distance $d \simeq \mu^{-1} = \hbar/(mc)$. If one makes use of the Yukawa potential (8.112), this becomes

$$\frac{g^2}{4\pi c^2}\frac{\mathrm{e}^{-\mu d}}{d}\frac{1}{m_0 c^2} \simeq \frac{g^2}{4\pi\hbar c^3}\frac{m}{m_0} \quad . \tag{8.123}$$

In the case of electromagnetism, the Coulomb force is long range: the only natural length is the Bohr radius a_B, so that the energy ratio (8.122) is proportional to e^4. In the case of the scalar field, the natural distance d is the range μ^{-1} of the potential, independent of g. As a result, the ratio (8.123) is quadratic in g. The dimensionless constant, independent of the mass of the particles which measures the coupling force, is thus

$$\frac{g^2}{4\pi\hbar c^3} \simeq 14.8 \quad . \tag{8.124}$$

The high value of this constant justifies the name of the strong interaction, but renders all perturbation calculations illusory.

Particles Mediating the Force

In spite of our reservations regarding the application of perturbation methods to the strong interactions, the quantum model defined by (8.114)–(8.116), as well as the comparison between (8.121) and (8.113), reveals a fundamental property of quantum fields. We see that the ability of nucleons to emit or absorb pions is the source of a force between the particles: *one says that the force is due to the exchange of virtual particles.* The pions participating in this exchange correspond to transitions that are possible between states where the energy differs by $\Delta E \geq mc^2$ (see (8.118)). They can only exist for a time $\Delta t \simeq \hbar/(mc^2)$ and, in the figurative sense, cover a distance after their emission equal to $\Delta x \simeq c\Delta t = \hbar/(mc) = \mu^{-1}$. These pions thus cannot be detected in a free state, hence their name *virtual particles.* One can equally imagine these pions as crossing the distance between nucleons by the tunnel effect, as in the case of two electrons which secure the binding between the protons of a hydrogen molecule. This idea of virtual particles is fundamental to particle physics: *any force law between particles can be attributed to another field of particles which mediate that force.* According to (8.90), *the range μ^{-1} of the force is inversely proportional to the mass of the exchanged particle.*

For example, when $m = 0$, $\mu^{-1} = \infty$ and (8.113) reduces to the Coulomb potential: one can thus interpret the Coulomb potential as an exchange of particles of zero mass. This interpretation becomes precise when one quantizes the electromagnetic field in the covariant Lorenz gauge. In this gauge, $\boldsymbol{A}$ and V form components of a four-vector. In addition to the transverse photons, longitudinal photons appear which are responsible for the Coulomb force. The exchange of scalar particles gives rise to the attractive Yukawa potential. If the exchange particles carry internal quantum numbers (spin, isospin, etc.), the force can take on an attractive or repulsive nature, as we have seen for the Coulomb force between electric charges.

Another example is the weak interactions of very short range ($\mu^{-1} \simeq 10^{-15}$ cm). The corresponding exchanged particles, the charged $W^{\pm}$ and neutral Z^0 vector bosons which were observed in 1983, have a mass of the order of 80 GeV ($W^{\pm}$) and 90 GeV (Z^0). This concept also applies to condensed matter physics: the effective interaction between the electrons by the exchange of phonons is responsible for superconductivity (Sect. 9.4.4).

8.3.3 Charged Scalar Field

Now, consider the case when we have two scalar particles of opposite charge $\pm e$ and of the same mass m, as with the pions π^+ and π^-. The state of each of these two particles will be a relativistic scalar meson with momentum and energy given by (8.91) and (8.92). As particles with opposite charge are distinguishable, it is only necessary to impose Bose statistics for each of the particles separately. Naturally, one still imposes that the creation and annihilation operators relative to particles of opposite charge commute. This gives rise to the following relations

$$\begin{aligned} &[a_{\boldsymbol{k}}, a^*_{\boldsymbol{k}'}] = \delta_{\boldsymbol{k},\boldsymbol{k}'} \quad , \quad [b_{\boldsymbol{k}}, b^*_{\boldsymbol{k}'}] = \delta_{\boldsymbol{k},\boldsymbol{k}'} \quad , \\ &[a_{\boldsymbol{k}}, a_{\boldsymbol{k}'}] = [b_{\boldsymbol{k}}, b_{\boldsymbol{k}'}] = [a_{\boldsymbol{k}}, b_{\boldsymbol{k}'}] = [a_{\boldsymbol{k}}, b^*_{\boldsymbol{k}'}] = 0 \quad , \end{aligned} \tag{8.125}$$

where the $a_{\boldsymbol{k}}$, $a^*_{\boldsymbol{k}}$ ($b_{\boldsymbol{k}}$, $b^*_{\boldsymbol{k}}$) correspond to π^+ (π^-). Systems of π^+ and π^- are described by states in the product $\mathcal{F}_+(\mathcal{H}) \otimes \mathcal{F}_+(\mathcal{H})$ of two copies of the Fock space $\mathcal{F}_+(\mathcal{H})$, where $\mathcal{H}$ is the space of one-meson states (8.93).

The number of π^+ and π^- are given by the operators

$$\mathtt{N}_+ = \sum_{\boldsymbol{k}} a^*_{\boldsymbol{k}} a_{\boldsymbol{k}} \quad , \quad \mathtt{N}_- = \sum_{\boldsymbol{k}} b^*_{\boldsymbol{k}} b_{\boldsymbol{k}} \quad , \tag{8.126}$$

and the total kinetic energy is given by

$$\mathtt{H}_0 = \sum_{\boldsymbol{k}} \hbar\omega_{\boldsymbol{k}} (a^*_{\boldsymbol{k}} a_{\boldsymbol{k}} + b^*_{\boldsymbol{k}} b_{\boldsymbol{k}}) \quad . \tag{8.127}$$

Now, we add to these observables the total electric charge, which is obviously

$$\mathtt{Q} = e(\mathtt{N}_+ - \mathtt{N}_-) \quad . \tag{8.128}$$

The non-Hermitian *free charged field* at time t is defined by

$$\Phi^*(\boldsymbol{x}, t) = \left(\frac{\hbar}{L^3}\right)^{1/2} \sum_{\boldsymbol{k}} \frac{1}{\sqrt{2\omega_{\boldsymbol{k}}}} \left(a^*_{\boldsymbol{k}} \mathrm{e}^{-\mathrm{i}(\boldsymbol{k}\cdot\boldsymbol{x} - \omega_{\boldsymbol{k}} t)} + b_{\boldsymbol{k}} \mathrm{e}^{\mathrm{i}(\boldsymbol{k}\cdot\boldsymbol{x} - \omega_{\boldsymbol{k}} t)}\right) \quad . \tag{8.129}$$

The positive and negative frequency components now correspond to the π^+ creators and the π^- annihilators, respectively. It is clear that $\Phi^*(\boldsymbol{x},t)$ and its adjoint $\Phi(\boldsymbol{x},t)$ still obey the Klein Gordon equation (8.97).

There are two fundamental reasons which lead us to introduce the non-Hermitian field (8.129) and not a pair of Hermitian fields associated separately with π^+ and π^-. The first comes from the charge conservation law in the reactions between elementary particles. The field $\Phi^*(\boldsymbol{x})$ is subject to symmetry transformations generated by the charge operator $\mathtt{Q}$, the pure phase transformations (or gauge transformations of the first type). With (3.95), (3.96) and (8.129), one has

$$\mathrm{e}^{\mathrm{i}\alpha\mathtt{Q}}\Phi^*(\boldsymbol{x})\mathrm{e}^{-\mathrm{i}\alpha\mathtt{Q}} = \exp(\mathrm{i}\alpha e)\Phi^*(\boldsymbol{x}) \tag{8.130}$$

and in infinitesimal form

$$[\mathtt{Q},\Phi^*(\boldsymbol{x})] = e\Phi^*(\boldsymbol{x}) \quad . \tag{8.131}$$

Applying (8.131) to an eigenstate of total charge, one sees that $\Phi^*(\boldsymbol{x})$ *increases the charge by one unit*, and similarly that $\Phi(\boldsymbol{x})$ *decreases the charge by one unit* (which is also obvious from the definition (8.129)).

Transformation (8.130) implies that bilinear expressions in Φ and Φ^* commute with $\mathtt{Q}$, and are thus natural candidates to form the interaction Hamiltonians which conserve the total charge. Indeed, the free Hamiltonian (8.127) and the charge (8.128) are the bilinear expressions

$$\mathtt{H}_0 = \int \mathrm{d}\boldsymbol{x} : \left[\frac{\partial\Phi^*}{\partial t}\frac{\partial\Phi}{\partial t} + c^2\nabla\Phi^*\cdot\nabla\Phi + (\mu c)^2\Phi^*\Phi\right] : \tag{8.132}$$

$$\mathtt{Q} = \frac{\mathrm{i}e}{\hbar}\int \mathrm{d}\boldsymbol{x} : \left[\Phi^*\left(\frac{\partial\Phi}{\partial t}\right) - \left(\frac{\partial\Phi^*}{\partial t}\right)\Phi\right] : \quad . \tag{8.133}$$

These are obtained like (8.16) and (8.101) with help of the expansion (8.129) and the orthogonality of the plane waves. One can see that $\mathtt{Q}$ is given by the integral of the density

$$\rho_e(\boldsymbol{x},t) = \frac{\mathrm{i}e}{\hbar} : \left[\Phi^*(\boldsymbol{x},t)\left(\frac{\partial}{\partial t}\Phi(\boldsymbol{x},t)\right) - \left(\frac{\partial}{\partial t}\Phi^*(\boldsymbol{x},t)\right)\Phi(\boldsymbol{x},t)\right] : \tag{8.134}$$

interpreted as the charge density operator. Finally, one defines the current density operator

$$\boldsymbol{j}(\boldsymbol{x},t) = -\mathrm{i}\frac{c^2 e}{\hbar} : \{\Phi^*(\boldsymbol{x},t)\,[\nabla\Phi(\boldsymbol{x},t)] - [\nabla\Phi^*(\boldsymbol{x},t)]\,\Phi(\boldsymbol{x},t)\} : \tag{8.135}$$

and verifies, from the equation of motion of the free field, the validity of the continuity equation

$$\frac{\partial}{\partial t}\rho_e(\boldsymbol{x},t) + \nabla\cdot\boldsymbol{j}(\boldsymbol{x},t) = 0 \quad . \tag{8.136}$$

Antiparticles

A second symmetry, which is possessed by definition (8.129) of the non-Hermitian field $\Phi(\boldsymbol{x})$, is the *particle–antiparticle symmetry*. One can define a canonical transformation by means of the unitarity operator $\mathtt{U}_c$ which exchanges the $a^*_{\boldsymbol{k}}$ for the $b^*_{\boldsymbol{k}}$, and leaves the commutation rules (8.125) invariant

$$\mathtt{U}_c a^*_{\boldsymbol{k}} \mathtt{U}^*_c = \eta b^*_{\boldsymbol{k}} \quad , \\ \mathtt{U}_c b^*_{\boldsymbol{k}} \mathtt{U}^*_c = \eta^* a^*_{\boldsymbol{k}} \quad , \tag{8.137}$$

where η is an arbitrary phase factor $|\eta| = 1$. Note that the vacuum is invariant under the action of $\mathtt{U}_c$

$$\mathtt{U}_c |0\rangle = \mathtt{U}^*_c |0\rangle = |0\rangle \quad . \tag{8.138}$$

The effect of $\mathtt{U}_c$ is to interchange the roles of π^+ and π^- by transforming any state of $n\pi^+$ into the identical quantum state formed of $n\pi^-$ up to a phase factor. For example, for a state of two π^+ of momenta $\hbar\boldsymbol{k}_1$ and $\hbar\boldsymbol{k}_2$, relations (8.137) and (8.138) give

$$\mathtt{U}_c a^*_{\boldsymbol{k}_1} a^*_{\boldsymbol{k}_2} |0\rangle = \eta^2 b^*_{\boldsymbol{k}_1} b^*_{\boldsymbol{k}_2} |0\rangle \quad . \tag{8.139}$$

Furthermore, it follows from (8.137) that $\mathtt{U}_c a^*_{\boldsymbol{k}} a_{\boldsymbol{k}} \mathtt{U}^*_c = b^*_{\boldsymbol{k}} b_{\boldsymbol{k}}$ and the converse relation. It thus follows from (8.127), (8.126) and (8.128) that the transformation $\mathtt{U}_c$ leaves the Hamiltonian invariant, but changes the sign of the charge

$$\mathtt{U}_c \mathtt{H}_0 \mathtt{U}^*_c = \mathtt{H}_0 \quad , \\ \mathtt{U}_c \mathtt{Q} \mathtt{U}^*_c = -\mathtt{Q} \quad . \tag{8.140}$$

The symmetry operation $\mathtt{U}_c$ is called *charge conjugation*. The particle of opposite charge obtained by charge conjugation is the corresponding antiparticle.

One sees from (8.129) that the field and its adjoint are related by charge conjugation

$$\mathtt{U}_c \Phi^*(\boldsymbol{x}, t) \mathtt{U}^*_c = \eta \Phi(\boldsymbol{x}, t) \quad . \tag{8.141}$$

In general, the non-Hermitian field is always formed from the combination of the particle creator and the annihilator of its associated antiparticle. This notation assures that, as far as interactions invariant under charge conjugation are concerned, *antiparticles behave identically to particles, except for effects depending on the sign of the charge*. The Hermitian field corresponds to the special case where particles are identical to their antiparticles, such as for the photon or the π^0 mesons. *The possibility of treating particles and antiparticles in the same manner comes from the relativistic nature of the field*: the Klein–Gordon equation, which is a second-order differential equation in time, has the general solution which is a linear combination of terms of positive and negative frequencies with complex coefficients. We have made use of this freedom in (8.129) to abandon real fields and to naturally extend the formalism to the description of antiparticles, by complex (or non-Hermitian) fields.

Antimatter

As was emphasized at the beginning of Sect. 8.3.1 for the case of spinless, massive particles (the pions), the existence of particles distinct from their antiparticles is firmly based on experimental grounds. Moreover, this situation is known to be valid for other cases, namely for particles bearing non-vanishing spin, either charged or uncharged. In the latter case, the antiparticle may or may not differ from its particle.

We have also just seen that the description of antiparticles occurs in a very natural way in the form (8.129) of the relativistic charged scalar field. Let us make a short historical digression on the emergence of the antiparticle concept (namely the positron) in the context of the single-electron relativistic wave equation.

At the beginning of 1928, P.A.M. Dirac published what would perhaps become his most important work, namely "The quantum theory of the electron".[1] In this paper, Dirac wrote down for the first time what is now known as the "Dirac equation". This equation describes the electron as a massive point charge. It was the first successful attempt to generalize Schrödinger's equation in order to make it Lorentz invariant. By design, Dirac's equation describes a single particle.

The first success of Dirac's equation was unexpected. Dirac soon noticed that the existence of spin would follow in a quite straightforward manner from the very structure of the equation. The spin arose as soon as the equation was generalized in order to include the effect of an external magnetic field upon the electron. Previously, spin had to be introduced separately and, say, in an artificial way into the description of the electron or the proton. Now, it appeared as a necessary consequence of the coupling of quantum physics and relativity.

Attaching spin to the electron, however, was not the only beautiful consequence of Dirac's achievement. Dirac himself soon discovered that his equation allowed the existence of new solutions which seemed to describe electrons with negative energy. A reinterpretation of these solutions was possible, but Dirac did not achieve it in one step. He first showed that these solutions could be associated with positively charged particles whose energy was positive. Dirac then related these solutions to the proton. At about the same time, however, Dirac seriously considered the possible existence of a yet unknown particle having the same mass as the electron but opposite charge. He

[1] P.A.M. Dirac, Proceedings of the Royal Society *A117*, 351 (1928)

proposed the name "anti-electron" for this elusive particle. Soon afterwards, the positron was discovered by Anderson (1932).

The Dirac equation indeed implies that the charge of the positron is equal but opposite in sign to the electron and that the masses of both particles is the same. Creation of electron–positron pairs in an electromagnetic field vindicated this prediction. It is now well understood that any quantum-field theory of charged particles compatible with relativistic invariance, as is quantum electrodynamics (QED), cannot survive without the simultaneous existence of particles and their antiparticles. The type of arguments leading to this result are not very different from those which are used in order to show the connection between spin and statistics (Sect. 8.3.4).

Since the beginning of the 1930's, a lot of particles have been discovered. To each particle, there corresponds an antiparticle with the same mass, the same spin but opposite charge. As emphasized above, neutral particles may or may not be identical to their antiparticles. This fact has far reaching consequences. A telescope making use of electromagnetic waves of any wavelength cannot tell the difference between galaxies and anti-galaxies, that is galaxies made of antimatter, more precisely of anti-stars turning anti-hydrogen into anti-helium and of anti-earths bearing anti-life. These (possible?) anti-galaxies would emit anti-photons which, as mentioned above, do not differ from photons. As a consequence, the image of galaxies and of anti-galaxies would not be distinct.

Generalizing the case of the photon to any neutral particle would lead to flatly wrong conclusions. The anti-neutron, being built up from three anti-quarks, is not identical to the neutron and its three quarks. As a consequence, helium and anti-helium can annihilate in the same way as electrons and positrons do. Now the neutron and the helium nucleus are composite, made up from charged units. This is very different from the case of the photon, which is truly elementary, and one could expect that they behave differently. The case of the neutrino, another elementary neutral particle, is therefore rather surprising.

In the macroscopic world, physical laws are invariant with respect to mirror reflection (MR): for any physical process not invariant with respect to a mirror reflection, there exists an equivalent physical process obtained from the first one through application of MR. Both processes are equally valid in the sense that both obey known classical physical laws. For any boomerang following an helical trajectory, it is possible to carve a "mirror-reflected" boomerang following another helix, obtained from the first by a mirror reflection.

From the point of view of quantum physics, this means that gravity and the electromagnetic interaction are invariant under the application of P, the parity operator, defined as the unitary operator changing the sign of all spatial coordinates: P is equivalent to the product of the operator corresponding to a mirror reflection by a 180°-rotation about the axis perpendicular to the mirror. As a consequence for the present discussion, quantum physics does not require any distinction between the parity operator and the operator corresponding to a mirror reflection. Particle physicists often say that parity is conserved by the electromagnetic interaction (gravity does not play any role at the particle level).

The weak interaction, however, does not conserve parity. This unexpected result was first shown in 1957 in an experiment run by C.S. Wu (Mrs. Yuan) and her colleagues. The most spectacular consequence of this lack of invariance becomes clear if one considers the properties of neutrinos. Soon after the experiment was performed, it was realized that the results were equivalent to the fact that the helicity of (electric) neutrinos is always equal to $-1/2$ while their anti-particles have opposite helicity.[2]

Neutrinos are neutral leptons with spin 1/2 and masses very small, even compared with the electron. Like the charged leptons, they belong to three different families, characterized by a different flavour (electronic, muonic and tauic neutrinos). The possible existence of a non-vanishing mass had been first suggested in 1998 by an experiment in Kamioka (Japan). In the following, we shall disregard the possible consequences of non-vanishing neutrino masses.

Under the assumption of vanishing mass, a fixed value of the helicity is quite natural: the neutrino cannot change its velocity and therefore its helicity. In this case, opposite signs for the helicities of the neutrino and the anti-neutrino are significant. This fact implies that transforming a neutrino state into an anti-neutrino state involves the product CP of the charge-conjugation operator C by the parity P. C has been defined through (8.137) where it appears as $\mathtt{U}_c$. It transforms a particle into its anti-particle with opposite charge.

From a general point of view, the fact that the neutrino and its anti-particle are distinct has interesting consequences. It could in principle allow us to distinguish between signals from galaxies and anti-galaxies, at least if these signals would consist of fluxes of neutrinos or antineutrinos. The field of "neutrino-astronomy" already exists. Unfortunately, it is still in an embryonic state, and no conclusions can be drawn concerning the distribution of matter and antimatter in the universe from (anti-)neutrino captures. Nevertheless, other strong astrophysical arguments lead to the general belief that antimatter is very scarce in the universe, i.e. that any galaxy and any stellar system is made of matter and not of antimatter.

This latter fact, however, raises an important question. If all antimatter is the exact image of matter through CP, the universal presence of matter rather than antimatter in the universe would be a riddle for cosmologists. However, the non-conservation of the product CP itself was discovered in

[2] Helicity has been defined in Sect. 1.3.6: it is the projection of the total angular momentum along the direction of motion.

some decays due to the weak interaction.[3] Soon after this result was known, A. Sakharov called on this new symmetry breaking in order to explain the apparent lack of symmetry between the distribution of matter and antimatter in the universe[4] (actually Sakharov had to make further assumptions to reach a tentative explanation of this absence of symmetry).[5]

At the dawn of the third millennium, a better understanding of the relation between matter and antimatter and of the range of validity of the Standard Model of particle physics demands new results on the violation of CP. Investigations of the decay of B-mesons are currently carried out.

Violation of CP implies some lack of symmetry between matter and antimatter, which may also appear somewhere else than in their decay products. We know that a particle has the same mass as its antiparticle. We also know that an antiparticle behaves in the same way as its corresponding particle in the gravitational field of the Earth (it "falls down" as the particle does). But the experimental proofs of both claims are scarce and not very precise. The best way to get new insights into the relation between the masses of particles and their antiparticles would involve the determination of the spectrum of antihydrogen (the bound state of an anti-proton and a positron). This experiment is difficult because it is far from easy to hold onto a large number of anti-hydrogen atoms for a long enough time. There is, however, a project at CERN called ATHENA which led to the production of a sufficient amount of anti-atoms during the year 2002. One then expects that precise measurements of the electromagnetic spectrum of anti-hydrogen could be compared with the well-known spectrum of hydrogen.[6] Will they coincide within the limits of the precision attained?

8.3.4 Spin and Statistics

In relativistic physics, there exists a profound connection between the spin that a particle carries and whether that particle is a boson or a fermion: the microcausality principle, combined with the Lorentz invariance, implies that *particles of integer spin are bosons and particles of half-integer spin are fermions.* Without proving this theorem in complete generality here, we can give an argument that is valid for the charged scalar field. The argument consists of showing that microcausality (8.35) would be violated if the particles associated with the scalar field, which have zero spin, were fermions rather than bosons.

The typical local observables appearing in the Hamiltonian density (8.101) are the quadratic quantities

---

[3] J.H. Christenson, J.W. Cronin, V.L. Fitch, R. Turlay, Phys. Rev. Lett. *13*, 138 (1964).

[4] A.D. Sakharov, JETP Lett. *6*, 24 (1967).

[5] M.E. Shaposhnikov, Contemporary Physics *39*, 177 (1998).

[6] P.T. Greenland, Contemporary Physics 38, 181 (1997).

$$\mathsf{O}(\boldsymbol{x},t) = {:}\Phi^*(\boldsymbol{x},t)\Phi(\boldsymbol{x},t){:} \quad . \tag{8.142}$$

A little algebra allows one to establish the following identity (one notes that the normal product of two fields only differs from an ordinary product by a scalar which disappears from the commutator)

$$\begin{aligned}[\mathsf{O}(\boldsymbol{x},t),\mathsf{O}(\boldsymbol{y},s)]_- = {}& \Phi^*(\boldsymbol{x},t)[\Phi(\boldsymbol{x},t),\Phi^*(\boldsymbol{y},s)]_+\Phi(\boldsymbol{y},s) \\ & -\Phi^*(\boldsymbol{x},t)\Phi^*(\boldsymbol{y},s)[\Phi(\boldsymbol{x},t),\Phi(\boldsymbol{y},s)]_+ \\ & +[\Phi^*(\boldsymbol{x},t),\Phi^*(\boldsymbol{y},s)]_+\Phi(\boldsymbol{y},s)\Phi(\boldsymbol{x},t) \\ & -\Phi^*(\boldsymbol{y},s)[\Phi^*(\boldsymbol{x},t),\Phi(\boldsymbol{y},s)]_+\Phi(\boldsymbol{x},t) \quad , \end{aligned} \tag{8.143}$$

with $[A,B]_\mp = AB \mp BA$. If the particles had been fermions (that is, if they obeyed rules (8.100) with the commutators replaced by the anticommutators), one would have

$$[\Phi(\boldsymbol{x},t),\Phi(\boldsymbol{y},s)]_+ = [\Phi^*(\boldsymbol{x},t),\Phi^*(\boldsymbol{y},s)]_+ = 0 \tag{8.144}$$

and by a similar calculation to the one leading to (8.102)

$$[\Phi^*(\boldsymbol{x},t),\Phi(\boldsymbol{y},s)]_+ = [\Phi(\boldsymbol{x},t),\Phi^*(\boldsymbol{y},s)]_+ = \hbar\Delta_1(\boldsymbol{x}-\boldsymbol{y},t-s) \quad , \tag{8.145}$$

with

$$\Delta_1(\boldsymbol{x},t) = \frac{1}{(2\pi)^3}\int \mathrm{d}\boldsymbol{k}\,\frac{\cos(\boldsymbol{k}\cdot\boldsymbol{x}-\omega_{\boldsymbol{k}}t)}{\omega_{\boldsymbol{k}}} \quad . \tag{8.146}$$

Thus (8.143) would reduce to

$$\begin{aligned}&[\mathsf{O}(\boldsymbol{x},t),\mathsf{O}(\boldsymbol{y},s)]_- \\ &\quad = \hbar\Delta_1(\boldsymbol{x}-\boldsymbol{y},t-s)\,[\Phi^*(\boldsymbol{x},t)\Phi(\boldsymbol{y},s)-\Phi^*(\boldsymbol{y},s)\Phi(\boldsymbol{x},t)] \quad . \end{aligned} \tag{8.147}$$

For fermion fields, the combination occurring in (8.147) does not vanish in general. Moreover, unlike $\Delta(\boldsymbol{x},t)$, the function $\Delta_1(\boldsymbol{x},t)$ is non-zero at $t=0$

$$\Delta_1(\boldsymbol{x},0) = \frac{1}{(2\pi)^3}\int \mathrm{d}\boldsymbol{k}\,\frac{\cos\boldsymbol{k}\cdot\boldsymbol{x}}{c\sqrt{|\boldsymbol{k}|^2+\mu^2}} = \frac{\mu}{2\pi^2 c|\boldsymbol{x}|}K_1(\mu^{-1}|\boldsymbol{x}|) \quad , \tag{8.148}$$

where K_1 is the modified first-order Bessel function. Therefore the commutator (8.147) cannot vanish in the entire region $[c(t-s)]^2 - |\boldsymbol{x}-\boldsymbol{y}|^2 < 0$, thus violating the microcausality principle.

This argument extends to fields of particles of spin 1/2 or above. In the case of a spin-1/2 particle (Dirac's electron field), the observables are also bilinear expression of the fields. The calculation of the commutators and anticommutators requires spinor algebra. One shows this time that the commutator $[\mathsf{O}(\boldsymbol{x},t),\mathsf{O}(\boldsymbol{y},s)]_-$ of local observables could not vanish for $[c(t-s)]^2 - |\boldsymbol{x}-\boldsymbol{y}|^2 < 0$ if particles of spin 1/2 were bosons.

8.3.5 The Lagrangian Formalism

The Lagrangian formalism offers several advantages when formulating field theory. *It provides a method for quantizing the classical field and allows the symmetry properties of the theory to be easily expressed.* We limit ourselves here to recalling the principal aspects of the formalism.

Classical Mechanics

In classical mechanics, we know that the equations of motion can be derived from a Lagrangian function $L(\{\boldsymbol{q}_i(t), \boldsymbol{v}_i(t)\})$ which is dependent on the positions $\boldsymbol{q}_i(t)$ and velocities $\boldsymbol{v}_i(t)$ of n particles, with the help of the *Euler–Lagrange equations*

$$\frac{\partial L}{\partial q_i^\alpha} = \frac{\partial}{\partial t}\left(\frac{\partial L}{\partial v_i^\alpha}\right), \quad i = 1, \dots, n, \quad \alpha = 1, 2, 3 \quad . \tag{8.149}$$

These equations are obtained by requiring the action $\int_{t_1}^{t_2} \mathrm{d}t L(\{\boldsymbol{q}_i(t), \boldsymbol{v}_i(t)\})$ to be stationary under variations of trajectories with fixed endpoints at t_1 and t_2. The canonically conjugate variable to q_i^α, the momentum p_i^α, is defined by

$$p_i^\alpha = \frac{\partial L}{\partial v_i^\alpha} \quad , \tag{8.150}$$

and the Hamiltonian is expressed as a function of the canonical variables $\boldsymbol{q}_i(t)$ and $\boldsymbol{p}_i(t)$ by

$$H = \sum_{i=1}^{n} \boldsymbol{p}_i(t) \cdot \frac{\mathrm{d}\boldsymbol{q}_i(t)}{\mathrm{d}t} - L \quad . \tag{8.151}$$

For classical particles under the influence of a potential $V(\boldsymbol{q}_1, \dots \boldsymbol{q}_n)$, one simply has $L = \sum_{i=1}^{n} m|\boldsymbol{v}_i|^2/2 - V(\boldsymbol{q}_1, \dots, \boldsymbol{q}_n)$.

Classical Field

The extension of the Lagrangian formalism to a field $\Phi(\boldsymbol{x}, t)$ can be made by considering an analogy with a continuous medium in which the Lagrangian function can be represented by the integral of a Lagrangian density $\mathcal{L}(\boldsymbol{x}, t)$

$$L(t) = \int \mathrm{d}\boldsymbol{x}\, \mathcal{L}(\boldsymbol{x}, t) \quad . \tag{8.152}$$

One supposes that

$$\mathcal{L}(\boldsymbol{x}, t) = \mathcal{L}\left(\{\Phi(\boldsymbol{x}, t), \partial_{x^\alpha}\Phi(\boldsymbol{x}, t), \partial_t\Phi(\boldsymbol{x}, t)\}\right) \tag{8.153}$$

only depends on $\boldsymbol{x}$ or t through the field and its first derivatives $\partial_{x^\alpha}\Phi = \partial\Phi/\partial x^\alpha$ and $\partial_t\Phi = \partial\Phi/\partial t$. By requiring that the action $\int_{t_1}^{t_2} \mathrm{d}t\, L(t) = \int_{t_1}^{t_2} \mathrm{d}t \int \mathrm{d}\boldsymbol{x}\, \mathcal{L}(\boldsymbol{x}, t)$ be stationary under variations of the field holding the endpoints fixed, one obtains the Euler–Lagrange equations

$$\frac{\partial \mathcal{L}}{\partial \Phi(\boldsymbol{x},t)} - \sum_{\alpha=1}^{3} \frac{\partial}{\partial x^\alpha} \left(\frac{\partial \mathcal{L}}{\partial[\partial_{x^\alpha}\Phi(\boldsymbol{x},t)]} \right) - \frac{\partial}{\partial t} \left(\frac{\partial \mathcal{L}}{\partial[\partial_t\Phi(\boldsymbol{x},t)]} \right) = 0 \quad , \tag{8.154}$$

which gives the equation of motion of the field. The canonically conjugate variable to the field is defined by

$$\Pi(\boldsymbol{x},t) = \frac{\partial \mathcal{L}}{\partial[\partial_t\Phi(\boldsymbol{x},t)]} \quad . \tag{8.155}$$

From this, one obtains the Hamiltonian density, expressed as a function of $\Phi(\boldsymbol{x},t)$ and $\Pi(\boldsymbol{x},t)$, by the formula

$$\mathcal{H}(\boldsymbol{x},t) = \Pi(\boldsymbol{x},t)\partial_t\Phi(\boldsymbol{x},t) - \mathcal{L}(\boldsymbol{x},t) \quad . \tag{8.156}$$

Neutral Scalar Field

The Lagrangian density for which the Euler–Lagrange equations produce the Klein–Gordon equations is

$$\mathcal{L}(\boldsymbol{x},t) = \frac{1}{2}\left\{[\partial_t\Phi(\boldsymbol{x},t)]^2 - c^2|\nabla\Phi(\boldsymbol{x},t)|^2 - (\mu c)^2\,[\Phi(\boldsymbol{x},t)]^2\right\} \quad , \tag{8.157}$$

which can be immediately verified starting from (8.154). According to (8.155), the canonical variable associated with $\Phi(\boldsymbol{x},t)$ is $\Pi(\boldsymbol{x},t) = \partial_t\Phi(\boldsymbol{x},t)$ and the canonical commutation rules indeed correspond to (8.99). Finally, (8.156) gives the Hamiltonian density which appears in (8.101), up to Wick ordering.

Before giving other examples, it is useful to introduce the abbreviated notation $x = \{x^\mu\}, \mu = 0,1,2,3$, where x is the four-vector

$$\{x^\mu\} = \{ct, \boldsymbol{x}\} \tag{8.158}$$

having time $x^0 = ct$ as its first component. The scalar product of any two four-vectors x^μ and y^μ is given by a form which is invariant under Lorentz transformations

$$x_\mu y^\mu = x^0 y^0 - \boldsymbol{x}\cdot\boldsymbol{y} \quad , \tag{8.159}$$

where one sums over repeated indices and, by definition,

$$\{x_\mu\} = \{ct, -\boldsymbol{x}\} \quad . \tag{8.160}$$

The gradient operators have the components

$$\{\partial_\mu\} = \left\{\frac{\partial}{\partial x^\mu}\right\} = \left\{\frac{1}{c}\frac{\partial}{\partial t}, \nabla\right\} \quad ,$$
$$\{\partial^\mu\} = \left\{\frac{\partial}{\partial x_\mu}\right\} = \left\{\frac{1}{c}\frac{\partial}{\partial t}, -\nabla\right\} \quad . \tag{8.161}$$

To further abbreviate the notation in the following sections, we have chosen a system of units for which $\hbar = c = 1$. In this condensed notation, the Lagrangian density (8.157) is written simply as

$$\mathcal{L}(x) = \frac{1}{2}\left[\partial_\mu \Phi(x)\partial^\mu \Phi(x) - m^2 \left(\Phi(x)\right)^2\right] \quad . \tag{8.162}$$

Charged Scalar Field

For the case of a charged scalar field, it is convenient to consider the complex field $\Phi(x)$ and its conjugate $\Phi^*(x)$ as independent dynamical variables. The appropriate Lagrangian density is

$$\mathcal{L}(x) = \partial_\mu \Phi^*(x)\partial^\mu \Phi(x) - m^2 \Phi^*(x)\Phi(x) \quad . \tag{8.163}$$

The non-Hermitian conjugate variables $\Pi(x) = \partial_0 \Phi^*(x)$ and $\Pi^*(x) = \partial_0 \Phi(x)$ give rise to the Hamiltonian density

$$\mathcal{H}(x) = \Pi(x)\partial_0\Phi(x) + \Pi^*(x)\partial_0\Phi^*(x) - \mathcal{L}(x) \quad , \tag{8.164}$$

which is the same as that in (8.132). One must remember that $\Phi = (\Phi_1 + \mathrm{i}\Phi_2)/\sqrt{2}$ is equivalent to a pair of Hermitian fields, Φ_1 and Φ_2, and that it is for these fields that one must write the canonical commutation rules.

Electromagnetic Field

If one introduces the four-vector potential with first component $A^0 = V$,

$$\{A^\mu\} = \{V, \boldsymbol{A}\} \quad , \tag{8.165}$$

the Lagrangian density of the free field is written as

$$\mathcal{L}(x) = -\frac{1}{4}\left[\partial_\nu A_\mu(x) - \partial_\mu A_\nu(x)\right]\left[\partial^\nu A^\mu(x) - \partial^\mu A^\nu(x)\right]$$
$$= \frac{1}{2}\left(|\boldsymbol{E}(x)|^2 - |\boldsymbol{B}(x)|^2\right) \tag{8.166}$$

and only depends on derivatives of the field A^μ. The Euler–Lagrange equations thus lead to

$$\frac{\partial \mathcal{L}}{\partial A_\mu} = \partial_\nu\left(\frac{\partial \mathcal{L}}{\partial(\partial_\nu A_\mu)}\right) = -\partial_\nu(\partial^\nu A^\mu - \partial^\mu A^\nu) = 0 \quad , \tag{8.167}$$

which is identical to the sourceless Maxwell equations.

Usefulness of the Lagrangian Formalism

Using the Lagrangian formalism allows for the quantization of a classical field according to the following procedure:

(i) write down a Lagrangian density which is dependent on the classical field and its derivatives;
(ii) determine the canonical variable associated with the field;
(iii) the quantized field theory is obtained by postulating the canonical commutation rules at fixed time;
(iv) the quantum Hamiltonian is obtained by replacing the fields in the classical expression by their quantum equivalents, Wick ordered.

For a free field, the state space on which the field operators act is identical to a Fock space (if one postulates the existence of a vacuum state). One thus interprets it as the state space associated with the field. The fact that $\mathcal{L}$ only depends on the field and its first derivatives follows from the fact that the equation of motion is a second-order differential equation. In general, if the Lagrangian only depends on the field and on a finite number of its derivatives (that is, if $\mathcal{L}$ only depends on the field and an infinitesimal neighbourhood of x), the theory is said to be local.

The main interest of the Lagrangian formalism lies in its ability to express the symmetric properties of the theory. *A transformation $\Phi(x) \to \Phi'(x)$ of the field is a symmetry of the theory if the equations of motion for $\Phi(x)$ and $\Phi'(x)$ have the same form.* By virtue of the Euler–Lagrange equations, this is obviously the case if *the Lagrangian density is itself invariant:*

$$\mathcal{L}\left(\{\Phi(x), \partial_\mu \Phi(x)\}\right) = \mathcal{L}\left(\{\Phi'(x), \partial_\mu \Phi'(x)\}\right) \quad . \tag{8.168}$$

Some examples include phase transformations of the charged field $\Phi'(x) = \exp(-\mathrm{i}\alpha e)\Phi(x)$ and gauge transformations of the potentials $A'^\mu(x) = A^\mu(x) - \partial^\mu \chi(x)$. It is clear that Lagrangians (8.162) and (8.166) are invariant under such transformations. The same goes for changes of reference frame $x \to x'$, for example, translations $x' = x + a$ or Lorentz transformations. If one defines the transformed field by $\Phi'(x') = \Phi(x)$, the equality $\mathcal{L}(\{\Phi'(x'), \partial'_\mu \Phi'(x')\}) = \mathcal{L}(\{\Phi(x), \partial_\mu \Phi(x)\})$ ensures that the equations of motion have the same form in all reference frames. In relativistic physics, a property such as this would be difficult to express using the Hamiltonian which is not a scalar, but only a component of the energy–momentum four-vector.

8.3.6 The Gauge Invariance Principle and Field Interactions

Elementary examples of interactions involving quantum fields have been given in Sects. 8.2.4 and 8.3.2: the couplings of the electric current to photons (8.50) and static nucleons to with pions (8.115). We propose here to return to the study of the principles which govern quantum-field interactions on a more fundamental basis. The coupling of the electromagnetic field to matter will serve as our guide.

Gauge Invariance in Non-Relativistic Quantum Mechanics

We know that the Hamiltonian of a massive non-relativistic particle under the influence of the electromagnetic potentials $\boldsymbol{A}(\boldsymbol{q},t)$ and $V(\boldsymbol{q},t)$ can be written (see Sect. 1.3.4) as

$$H = \frac{|\boldsymbol{p} - e\boldsymbol{A}(\boldsymbol{q},t)|^2}{2m} + eV(\boldsymbol{q},t) \quad . \tag{8.169}$$

For a quantum particle, this Hamiltonian gives rise to the Schrödinger equation

$$\mathrm{i}\hbar\frac{\partial}{\partial t}\varphi(\boldsymbol{x},t) = \frac{1}{2m}\left|\mathrm{i}\hbar\nabla + e\boldsymbol{A}(\boldsymbol{x},t)\right|^2\varphi(\boldsymbol{x},t) + eV(\boldsymbol{x},t)\varphi(\boldsymbol{x},t) \tag{8.170}$$

or

$$\mathrm{i}\left(\hbar\frac{\partial}{\partial t} + \mathrm{i}eV(\boldsymbol{x},t)\right)\varphi(\boldsymbol{x},t) + \frac{1}{2m}\left|\hbar\nabla - \mathrm{i}e\boldsymbol{A}(\boldsymbol{x},t)\right|^2\varphi(\boldsymbol{x},t) = 0 \quad . \tag{8.171}$$

This is obtained by starting from a free particle equation $\mathrm{i}\hbar(\partial/\partial t)\varphi(\boldsymbol{x},t) + (\hbar^2/2m)\nabla^2\varphi(\boldsymbol{x},t)$ and then making the following simple substitutions

$$\begin{aligned}
&\hbar\frac{\partial}{\partial t} \to \hbar\frac{\partial}{\partial t} + \mathrm{i}eV(\boldsymbol{x},t) \quad , \\
&\hbar\nabla \to \hbar\nabla - \mathrm{i}e\boldsymbol{A}(\boldsymbol{x},t) \quad .
\end{aligned} \tag{8.172}$$

Now, let us examine the behaviour of the Schrödinger equation (8.171) under a gauge transformation

$$\begin{aligned}
V'(\boldsymbol{x},t) &= V(\boldsymbol{x},t) - \hbar\frac{\partial}{\partial t}\chi(\boldsymbol{x},t) \quad , \\
\boldsymbol{A}'(\boldsymbol{x},t) &= \boldsymbol{A}(\boldsymbol{x},t) + \hbar\nabla\chi(\boldsymbol{x},t) \quad .
\end{aligned} \tag{8.173}$$

The form of equation (8.171) must be the same for any choice of gauge: there must therefore exist a new wavefunction φ' which satisfies (8.171) written in terms of the transformed potentials V' and $\boldsymbol{A}'$. To see this, we note the identity

$$\begin{aligned}
&(\hbar\nabla - \mathrm{i}e\boldsymbol{A}')\varphi\exp(\mathrm{i}e\chi) \\
&\quad = \exp(\mathrm{i}e\chi)(\mathrm{i}e\hbar\varphi\nabla\chi + \hbar\nabla\varphi) - \mathrm{i}e\varphi\exp(\mathrm{i}e\chi)(\boldsymbol{A} + \hbar\nabla\chi) \\
&\quad = \exp(\mathrm{i}e\chi)\,(\hbar\nabla - \mathrm{i}e\boldsymbol{A})\varphi \quad .
\end{aligned} \tag{8.174}$$

This relation can be written in the condensed form

$$(\hbar\nabla - \mathrm{i}e\boldsymbol{A}')\exp(\mathrm{i}e\chi) = \exp(\mathrm{i}e\chi)(\hbar\nabla - \mathrm{i}e\boldsymbol{A}) \quad , \tag{8.175}$$

where the equality signifies that application of either of the two sides of (8.175) on a function of space and time give the same result. It follows from (8.175) that the same identity is valid for all powers

$$(\hbar\nabla^\alpha - \mathrm{i}eA'^\alpha)^n \exp(\mathrm{i}e\chi) \\ = \exp(\mathrm{i}e\chi)(\hbar\nabla^\alpha - \mathrm{i}eA^\alpha)^n \quad , \quad n = 1, 2, \dots \quad . \tag{8.176}$$

Similarly, we see that

$$\left(\hbar\frac{\partial}{\partial t} + \mathrm{i}eV'\right)^n \exp(\mathrm{i}e\chi) \\ = \exp(\mathrm{i}e\chi)\left(\hbar\frac{\partial}{\partial t} + \mathrm{i}eV\right)^n \quad , \quad n = 1, 2, \dots \quad . \tag{8.177}$$

If we define the transformed wavefunction by

$$\varphi'(\boldsymbol{x}, t) = \exp\left[\mathrm{i}e\chi(\boldsymbol{x}, t)\right]\varphi(\boldsymbol{x}, t) \quad , \tag{8.178}$$

it is evident from (8.176) and (8.177) that φ' *obeys the same equation (8.171) written in terms of the transformed quantities*

$$\mathrm{i}\left(\hbar\frac{\partial}{\partial t} + \mathrm{i}eV'\right)\varphi' + \frac{1}{2m}\left|\hbar\nabla - \mathrm{i}e\boldsymbol{A}'\right|^2\varphi' = 0 \quad . \tag{8.179}$$

Invariance of the form of the Schrödinger equation under the simultaneous transformation of the potentials (8.173) and the wavefunction (8.178) ensures that all predictions of the theory are independent of the choice of gauge. Indeed, since $(\boldsymbol{q}, \boldsymbol{v} = (\boldsymbol{p} - e\boldsymbol{A})/m)$ and $(\boldsymbol{q}' = \boldsymbol{q}, \boldsymbol{v}' = (\boldsymbol{p} - e\boldsymbol{A}')/m)$ are the position and velocity observables in the two gauges, one verifies with the help of (8.175) and (8.178) that their average values are the same

$$\langle\varphi|\boldsymbol{q}|\varphi\rangle = \langle\varphi'|\boldsymbol{q}'|\varphi'\rangle \quad , \quad \langle\varphi|\boldsymbol{v}|\varphi\rangle = \langle\varphi'|\boldsymbol{v}'|\varphi'\rangle \quad . \tag{8.180}$$

The same goes for any other physical observable which can be expressed as a function of position and velocity.

Interaction of Charged Scalar and Electromagnetic Fields

These considerations extend naturally to the coupling of relativistic massive scalar particles to the electromagnetic field. One applies the substitution rule (8.172) to the Klein–Gordon equation (8.97). The former can be written in the more condensed notation (8.161), (8.165) (with $\hbar = c = 1$)

$$\partial_\mu \to \partial_\mu + \mathrm{i}eA_\mu(x) \tag{8.181}$$

and the equation of motion for the field in the presence of an electromagnetic potential becomes

$$[\partial_\mu + \mathrm{i}eA_\mu(x)]\,[\partial^\mu + \mathrm{i}eA^\mu(x)]\,\Phi(x) + m^2\Phi(x) = 0 \quad . \tag{8.182}$$

It clearly follows from relations (8.176) and (8.177) that the form of this equation is preserved under simultaneous transformations of the potentials A^μ and the field Φ

$$\begin{aligned} A'^\mu(x) &= A^\mu - \partial^\mu\chi(x) \quad , \\ \Phi'(x) &= \exp\left(\mathrm{i}e\chi(x)\right)\Phi(x) \quad . \end{aligned} \tag{8.183}$$

As indicated in the preceding section, the gauge symmetry of the theory is expressed by a Lagrangian which is invariant under the transformations (8.183). One such Lagrangian, leading to equation (8.182), is

$$\begin{aligned} \mathcal{L} &= [(\partial_\mu + \mathrm{i}eA_\mu)\Phi]^*\,[(\partial^\mu + \mathrm{i}eA^\mu)\Phi] - m^2\Phi^*\Phi + \mathcal{L}^{\mathrm{em}} \\ &= \mathcal{L}^\Phi + \mathcal{L}^{\mathrm{em}} \quad , \end{aligned} \tag{8.184}$$

where the first term is obtained from (8.163) by rule (8.181) and $\mathcal{L}^{\mathrm{em}}$ is the electromagnetic Lagrangian (8.166). Since

$$-\frac{\partial\mathcal{L}^\Phi}{\partial A_\mu} = \mathrm{i}e\left\{\Phi^*(\partial^\mu + \mathrm{i}eA^\mu)\Phi - [(\partial^\mu + \mathrm{i}eA^\mu)\Phi]^*\,\Phi\right\} = j^\mu \quad , \tag{8.185}$$

one sees with (8.167) that the Euler–Lagrange equations for the electromagnetic field are identical to the Maxwell equations with the source term defined by (8.185)

$$\partial_\nu(\partial^\nu A^\mu - \partial^\mu A^\nu) = j^\mu \quad . \tag{8.186}$$

It follows from (8.186) that the current density j^μ obeys the conservation equation

$$\partial_\mu j^\mu = 0 \quad . \tag{8.187}$$

The non-linear coupled equations for Φ and A^μ (8.182) and (8.186) are the fundamental equations of the electromagnetic interactions of a relativistic charged scalar field.

The Gauge Principle

Note that, for both the Schrödinger particle and for the scalar field coupled to an electromagnetic potential, the theory is invariant under phase transformations (8.178) and (8.183) which depend on space and time. Such transformations of the wavefunction or the field only modify the gauge of the potentials, not the electromagnetic fields themselves. This invariance is not obeyed by the free particle or the free field. We know that the theories for the free particle and the free field are invariant under phase transformations

$$\varphi'(x) = \mathrm{e}^{\mathrm{i}\alpha}\varphi(x) \quad , \tag{8.188}$$

$$\Phi'(x) = \mathrm{e}^{\mathrm{i}\alpha}\Phi(x) \quad , \tag{8.189}$$

where α is a constant with the same value at all space–time points. Indeed, (8.188) expresses the fact that, in quantum mechanics, the wavefunction is defined up to a phase factor, and the invariance (8.189) is related to charge conservation (see (8.130)). The constant phase transformations (8.188) and (8.189) are called global, while those which depend on space and time are called local. Within the framework of a free theory, it is not possible to extend global invariance to local transformations (the forms of the free Schrödinger and Klein–Gordon equations are not preserved under local transformations). One thus arrives at the following remarkable conclusion: *the theory of particles interacting with the electromagnetic field possesses a larger symmetry than that of free particles: it is invariant under all local phase transformations.* In what has just preceded, the conclusion was drawn from the form of the electromagnetic interaction (8.169), which is assumed to be well known. At this point, one can reverse the argument: one demands invariance of the theory under the local transformations of the wavefunction φ or of the scalar field Φ

$$\begin{aligned} \varphi'(x) &= \mathrm{e}^{\mathrm{i}\alpha(x)}\varphi(x) \quad , \\ \Phi'(x) &= \mathrm{e}^{\mathrm{i}\alpha(x)}\Phi(x) \quad . \end{aligned} \tag{8.190}$$

This constraint, elevated to the rank of principle, the *gauge principle*, allows one to deduce the form of the interactions.

Applying the gauge principle enables one to regain the electromagnetic coupling (8.171) and (8.182). For instance, to render the Schrödinger theory invariant under local phase transformations, it is necessary to modify the equation of the free particle by the introduction of the fields V and $\boldsymbol{A}$ as in (8.171) and, according to (8.176) and (8.177), the fields must transform as

$$\begin{aligned} V'(x) &= V(x) - \frac{\hbar}{e}\frac{\partial}{\partial t}\alpha(x) \quad , \\ \boldsymbol{A}'(x) &= \boldsymbol{A}(x) + \frac{\hbar}{e}\nabla\alpha(x) \quad . \end{aligned} \tag{8.191}$$

The new fields, introduced to assure local invariance, are called *gauge fields*, and it is necessary to give them a physical interpretation. In the present case, V and $\boldsymbol{A}$ must naturally be identified to the electromagnetic potentials.

The gauge principle extends to the case in which the field $\Phi_a(x)$ has m internal degrees of freedom $a = 1, 2, \ldots, m$ (isospin or quark "colour", for example). Invariance under global phase transformations $\mathrm{e}^{\mathrm{i}\alpha}$ is replaced by invariance under certain group transformations of internal degrees of freedom, represented by unitary matrices U_{ab} given by

$$\Phi'_a(x) = \sum_{b=1}^{m} U_{ab}\Phi_b(x) \quad . \tag{8.192}$$

For isospin, the U_{ab} matrices are given by the $SU(2)$ group; for colour, by the $SU(3)$ group. The gauge principle consists of demanding that the theory also be invariant under all local transformations

$$\Phi'_a(x) = \sum_{b=1}^{m} U_{ab}(x)\Phi_b(x) \quad , \tag{8.193}$$

where the elements of the group $U_{ab}(x)$ can be chosen at each space–time point x. In general, the U_{ab} matrices do not commute with each other, and one speaks of a non-Abelian gauge theory. In the current state of knowledge, it appears that the gauge principle governs the interactions between elementary particles (electroweak theory, quantum chromodynamics). For this reason, and in virtue of its simplicity and its beauty, the gauge principle plays a major role in quantum-field theory.

8.3.7 Mass Generation

Massive Vector Field

The gauge invariance of Maxwell's equations is intimately related to the fact that the photon has zero mass. Suppose instead that the photon had a mass $m > 0$. The corresponding equation of motion of the free field would be obtained by adding a mass term $m^2 A^\mu$ to Maxwell's equations (8.167), similar to that which appears in the Klein–Gordon equation (8.97)

$$(\partial_\nu \partial^\nu + m^2)A^\mu - \partial^\mu(\partial_\nu A^\nu) = 0 \quad . \tag{8.194}$$

If m is non-zero, the Lorentz relation $\partial_\nu A^\nu = 0$ is no longer an independent condition, but is determined by (8.194): one sees this immediately by taking the divergence of (8.194), which then reduces to $m^2 \partial_\mu A^\mu = 0$. When m is non-zero, (8.194) is thus equivalent to the pair of equations

$$\begin{aligned} &(\partial_\nu \partial^\nu + m^2)A^\mu = 0 \quad , \\ &\partial_\nu A^\nu = 0 \quad . \end{aligned} \tag{8.195}$$

These equations describe the *free massive vector field*.

If the theory is gauge invariant, (8.195) must still preserve their form under restricted gauge transformations $A'^\mu = A^\mu - \partial^\mu \chi$ with $\partial_\mu \partial^\mu \chi = 0$. It is clear from (8.195) that this is the case only if $m = 0$. One thus concludes that *the particles associated with a gauge field must have zero mass.* This conclusion, at first glance, renders the gauge principle unsuitable to describe the weak interactions. We know that the weak force has a very short range. The particles mediating the force must therefore be massive, according to the principle that the range of the force is inversely proportional to the mass (Sect. 8.3.2); these particles thus could not be associated with a gauge field.

This line of reasoning is based on the analysis of the free vector field (8.195); it does not necessarily apply when the gauge field is coupled with other fields in an appropriate manner. We are going to see that under specific conditions, the particles of a gauge field in interaction can acquire a mass: this is the phenomenon of *mass generation*, or the *Higgs mechanism*. For reasons of simplicity, we will limit ourselves here to illustrating this mechanism in the theoretical framework of an Abelian gauge field W^μ coupled to a complex scalar field Φ. The generalization of this mechanism to the non-Abelian case is an essential element to the electroweak gauge theory.

According to (8.185) and (8.186), the equation of motion of a vector field W^μ can be written under the form

$$\partial_\nu \partial^\nu W^\mu - \partial^\mu(\partial_\nu W^\nu) = j^\mu \quad , \tag{8.196}$$

$$j^\mu = \mathrm{i}g\,(\Phi^*(\partial^\mu \Phi) - (\partial^\mu \Phi^*)\Phi) - 2g^2 W^\mu |\Phi|^2 \quad , \tag{8.197}$$

where g is the coupling constant. Suppose that, for some reason, the field Φ is maintained in a configuration in which its modulus remains constant over space and time

$$|\Phi(x)| = \Phi_0 \quad , \tag{8.198}$$

where Φ_0 is a real positive constant independent of $\boldsymbol{x}$ and t. First of all, we examine the consequences of condition (8.198) when it is satisfied by the particular choice

$$\Phi(x) = \Phi_0 \quad , \tag{8.199}$$

where $\Phi(x)$ is real. In this situation, if one sets $M^2 = 2g^2\Phi_0^2$, the current becomes proportional to the field W^μ

$$j^\mu = -M^2 W^\mu \quad . \tag{8.200}$$

This relation, together with the conservation equation (8.187) implies the Lorenz condition $\partial_\mu W^\mu = 0$. *Equation (8.196) thus describes a vector field of mass M*

$$\begin{aligned} &(\partial_\nu \partial^\nu + M^2) W^\mu = 0 \quad , \\ &\partial_\nu W^\nu = 0 \quad . \end{aligned} \tag{8.201}$$

In this way, the gauge field W^μ acquires mass through its coupling to the field Φ, provided the latter remains "frozen" in the particular configuration $\Phi(x) = \Phi_0$.

In the general case, condition (8.198) is satisfied by

$$\Phi(x) = \Phi_0 \mathrm{e}^{\mathrm{i}\alpha(x)} \quad , \quad \Phi_0 > 0 \quad , \tag{8.202}$$

with some choice of phase $\alpha(x)$. Of course, two different choices of phase must lead to the same theory, albeit written in two different gauges. To see this, we calculate the current (8.197) with the choice of phase given by (8.202)

$$j^\mu = -2g^2\Phi_0^2 \left(W^\mu + \frac{1}{g}\partial^\mu \alpha \right) \quad . \tag{8.203}$$

After the gauge transformation $W'^\mu = W^\mu + \partial^\mu \alpha / g$ and $\Phi' = \mathrm{e}^{-\mathrm{i}\alpha}\Phi$, (8.202) and (8.203) become

$$\begin{aligned} \Phi'(x) &= \Phi_0 \quad , \\ j'^\mu(x) &= -M^2 W'^\mu(x) \quad , \end{aligned} \tag{8.204}$$

which, in the new gauge, are identical to (8.199) and (8.200). As a consequence, the transformed field $\Phi'(x)$ again obeys the equation of a massive vector field. Thus, all choices of phase (8.202) give rise to equivalent physical theories since they only differ by gauge transformations. In the following, there will thus be no loss of generality in making the choice (8.199) for which $\Phi(x)$ is real.

The central question which remains to be answered is: how can the field $\Phi(x)$ take on a constant non-zero value?

The Nature of the Ground State

The preceding discussion was in terms of classical fields. As quantum fields are operators, (8.199) can only make sense as an expectation value, and it must be replaced by

$$\langle \Phi(x) \rangle = \Phi_0 \quad . \tag{8.205}$$

The expectation value is taken in the ground state, that is, the "vacuum" of the field Φ. First of all we are interested in the simplest case for the dynamics of a gauge field where the quanta of the field Φ are unexcited. We have seen that the expectation value of a free field in the vacuum $|0\rangle$ of Fock space is zero (see (8.18)). If $\langle \Phi(x) \rangle$ is non-zero, we must be ready to allow for the fact that this new "vacuum" possesses properties which are quite different from a Fock space.

At this point, an instructive analogy can be made with the theory of superconductivity. Note that the linear relation between the current and the field (8.203) resembles the phenomenological equation of superconductivity (5.15). The reasoning which has led from (8.196) to (8.201) is in fact identical to how the London equation is derived, beginning by hypothesizing of the existence of a macroscopic wavefunction for the superconductor $\varphi(\boldsymbol{x}, t)$ corresponding to a uniform density ρ (see (5.23)); the field Φ and the wave function φ play corresponding roles.

In both cases, we highlight the existence of a screening phenomenon. In a superconducting medium, the magnetic field creates a current which exactly compensates for the effect of the field inside the sample: this is manifested in the "mass" δ^{-2} which the magnetic field acquires in the superconductor (see (5.20)). In the same way, one can say that *the force transmitted by the gauge particles create a "screening current" (8.200) in the "vacuum" which, by generating a mass, causes a reduction of the range of the force. In this sense, mass generation is the relativistic analogue of the Meissner effect.*

In the superconductor, the diamagnetic current is carried by a polarizable medium, the superconducting electrons of the BCS state. Although the "vacuum" associated with the field Φ does not represent a material medium, one tends to consider that the "vacuum" is not an inert substrate without structure: it appears more like the ground state of an interacting system with remarkable properties. Today, there does not yet exist a complete theory which establishes the properties of this "vacuum" starting from fundamental interactions of fields, as in the case of BCS theory which starts from microscopic electronic interactions. We thus adopt $\langle\Phi(x)\rangle = \Phi_0 \neq 0$ as a working hypothesis.

A Model of Broken Symmetry

One can, however, construct a simple phenomenological model which possesses minimal-energy configurations for which the field Φ is non-zero. For this, we return to a classical description of the field and neglect the quantum fluctuations. As a first example, consider a classical particle in two-dimensional space under the influence of the potential displayed in Fig. 8.4

$$V(r) = -ar^2 + b^2r^4 \quad , \quad r^2 = x^2 + y^2 \quad , \quad a > 0 \quad . \tag{8.206}$$

When a is positive, the particle is in equilibrium, equivalent at all points on the circle of radius $r_0 = a/2b^2$, such that $(\mathrm{d}/\mathrm{d}r)V(r_0) = 0$. These equilibrium positions are degenerate with respect to the phase $tg^{-1}(y/x)$.

Analogously, suppose that the field Φ is under the influence of an effective interaction given by

$$V(\Phi) = -a|\Phi|^2 + b^2|\Phi|^4 \tag{8.207}$$

and the Lagrangian (8.184) is replaced by

$$\mathcal{L} = [(\partial_\mu + \mathrm{i}gW^\mu)\Phi]^* \, [(\partial^\mu + \mathrm{i}gW^\mu)\Phi] - V(\Phi) + \mathcal{L}^W \quad , \tag{8.208}$$

where $\mathcal{L}^W$ is the Lagrangian of the gauge field W^μ, of the same form as (8.166). All configurations of the field

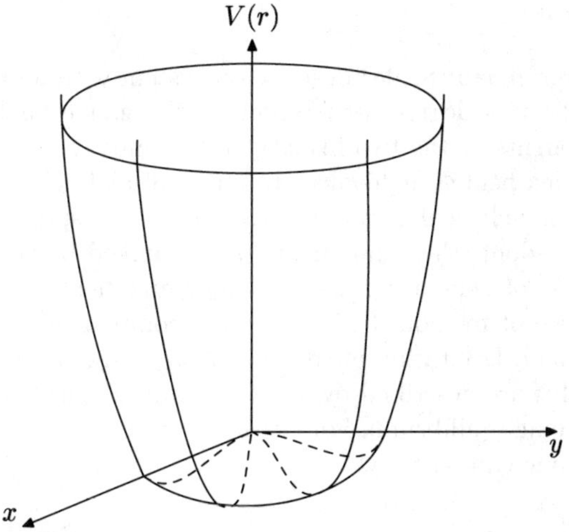

Fig. 8.4. Example of a potential with a minimum on a circle of radius $r_0 = a/2b^2$ and a local maximum at $r = 0$

$$\Phi(x) = \Phi_0 e^{i\alpha(x)} \quad , \quad \Phi_0 = \frac{a}{2b^2} \tag{8.209}$$

with a constant modulus and any phase are equivalent minima of the potential energy $V(\Phi)$. Each of them is an acceptable candidate for the "vacuum", and no principle dictates the choice of phase. In fact, two choices of "vacuum" which differ by a phase factor give rise to the same physical theory, although written in two different gauges, as we have already shown (see (8.204)). Thus, in the theory defined by the Lagrangian (8.208), all of the "vacuums" are physically equivalent, as must be the case.

When a phase is chosen in (8.209), *one is said to spontaneously break the symmetry*: this particular ground state no longer has the gauge invariance which is manifest in the Lagrangian (8.208). The situation can be compared to a ferromagnet at low temperature, where the phase plays the role of the orientation of the spontaneous magnetization. A sample of the ferromagnet can be prepared by applying a magnetic field which induces the alignment of the spins in the same direction. However, this is not the case here: the choice of phase has no physical realization. The gauge symmetry, if it is no longer manifest due to the fixing of the phase, remains implicitly present by the fact that any choice of phase is possible. The normal vacuum (for which the fields have an average value equal to zero) corresponds to the local maximum $r = 0$ of Fig. 8.4. As this point in unstable, it seems that just as in superconductivity, the real ground state cannot be obtained by a perturbation calculation.

The Higgs Field and Particle

There is one final step. The Lagrangian (8.208) describes a system with four internal degrees of freedom: the two degrees of freedom of the gauge field W^ν (the two transverse components or the two helicity states) plus the two degrees of freedom of the complex field Φ (equivalent to a pair of real fields). Furthermore, the massive vector field, corresponding to particles of spin 1, has three internal degrees of freedom (the components of W^μ linked by the Lorentz relation). In the process of generating a mass, the gauge field thus acquires a supplementary degree of freedom (a longitudinal component) to the detriment of the complex field. Excitation of the field $\Phi(x)$ in the neighbourhood of its equilibrium value are described by a new dynamical variable $h(x)$, which is defined to be zero at equilibrium. For example, with the choice (8.199) for which $\Phi(x)$ is real, one can write

$$\Phi(x) = \Phi_0 + h(x) \tag{8.210}$$

and $h(x)$ is a real scalar field equivalent to a single internal degree of freedom. In general, when a symmetry is broken and the phase is fixed, $h(x)$ is the only degree of freedom of the field $\Phi(x)$ which persists, while the second degree of freedom is transferred to the longitudinal part of the vector field $W^\mu(x)$. The number of degrees of freedom naturally remains as four. The existence of the scalar field $h(x)$ and of its associated particles (the Higgs field and bosons) is a necessary consequence of the process of mass generation by a broken symmetry.

The process of mass generation plays a central role in the electroweak theory which unifies quantum electrodynamics and the weak interaction. In this way three new massive particles have been added to the lonely photon, namely the charged $W^\pm$ and the neutral Z^0 whose respective masses (80.2 and 91.2 GeV) are huge when compared with the proton mass (0, 938 GeV). These particles were observed in 1983 (Sect. 8.3.2).

There is, however, a price to pay for breaking the symmetry between a massless photon and the heavy bosons $W^\pm$ and Z^0, namely the introduction of a new quantum scalar field, the Higgs field Φ, which must be given a physical interpretation. This undertaking is not easy. From a very general point of view, one can associate a definite particle to a given quantum field. At present, many groups of physicists, with the help of the current accelerator facilities, are searching for evidence of the decay of the Higgs particle H, predicted to be spinless (Φ is a scalar field) and neutral. Its own mass M_H is unknown. Lower bounds on M_H have been more or less determined by the lack of evidence of Higgs particle decays in the data of experiments running in several different mass ranges. As a consequence, around the beginning of the year 2000, the lower bound on the Higgs mass has been determined to be $M_H \geq 95$ GeV. On the other hand, particle physicists predict some interactions H may have with $W^\pm$ and Z^0, whose own properties are now well known. They conclude that M_H should not be much larger than 100 GeV. Eventually, experiments should settle the question.

At this point, it is worth returning to (8.198) because this relation is new from the elementary point of view assumed in this book. Relation (8.198) relates to the properties of the vacuum just as much as it concerns the properties of the still hypothetical Higgs field. At this stage, an analogy with the gravitational field could be useful.

The gravitational field is a concept well understood in classical physics. Its presence changes the properties of the classical vacuum which then loses isotropy and homogeneity. The quantum counterpart of the gravitational field is in principle associated with a hypothetical particle, the graviton, namely the quantum constituent of the gravitational waves, whose own existence was settled through observations of the so-called binary pulsar.[7] As with the photon, the graviton would be massless (the gravitational field has a long range as the Coulomb potential). Individual gravitons have never been observed: in the present universe, it is almost impossible to imagine any physical process generating gravitational waves with a frequency ν high enough for the product $\hbar\nu$ to be observable. Moreover, from the point of view of quantum field theory, the properties of the gravitational field lead to difficulties which have not yet been solved.

Returning to the Higgs field, one faces an analogous, though different situation. Unlike the graviton, the Higgs particle is likely to be observed, because its mass may lie within a range currently accessible to experiments. Moreover, the Higgs field is a purely quantum concept, having no classical counterpart. On the other hand, the Higgs field and the gravitational field share the common property that both of them do not just propagate through space–time, but also deeply modify its structure. Even if other fields lead to non-trivial properties of the vacuum (e.g. vacuum polarization by virtual electron–positron pairs in quantum electrodynamics, Sect. 9.4.1) the interrelation between the Higgs field or the gravitational field and a highly structured vacuum plays a predominant role. This interrelation manifests itself at a more phenomenological level by the model of broken symmetry described above using the effective interaction $V(\Phi)$ (8.207).

---

[7] We have an indirect confirmation that gravity waves exist through the energy they carry away from a whirling pair of pulsars discovered by 1974 by Hulse and Taylor. The rotation period of a spinning neutron star constitutes a highly precise clock. In the case of the binary pulsar discovered by Hulse and Taylor, this clock reveals the slowing down of the spin of one member of the pair (the second one is not observable). This change of angular velocity corresponds to the energy loss one would expect in the case of an emission of gravity waves from the system (see J.H. Taylor et al., Nature *355*, 132 (1992); for a tutorial approach, see C. Will, *The renaissance of general relativity*, in The New Physics, P. Davies ed., Cambridge University Press (1989)).

8.4 Electrons and Phonons

8.4.1 Non-Relativistic Fermi Field

Problems involving N massive non-relativistic particles whose number is conserved do not in principle require the notion of the quantum field. Nevertheless, we can benefit from introducing the concept of the quantum field in the non-relativistic context: it provides a useful comparison with elementary particle fields, as well as allowing us to formulate similar calculation techniques.

The Free Field

Consider a collection of electrons described by the states in a Fock space $\mathcal{F}_-(\mathcal{H})$, where $\mathcal{H}$ is the space of non-relativistic one-electron states (Sect. 2.1.2). We define the non-relativistic *Fermi free field* $a^*(\boldsymbol{x},\sigma,t)$ in analogy to the charged scalar field (8.129)

$$a^*(\boldsymbol{x},\sigma,t) = \frac{1}{L^{3/2}}\sum_{\boldsymbol{k}} a^*_{\boldsymbol{k}\sigma} \exp\left[-\mathrm{i}\left(\boldsymbol{k}\cdot\boldsymbol{x} - \frac{\varepsilon_{\boldsymbol{k}} t}{\hbar}\right)\right] \quad , \tag{8.211}$$

where $\varepsilon_{\boldsymbol{k}} = \hbar^2|\boldsymbol{k}|^2/2m$ and $a^*_{\boldsymbol{k}\sigma}$ is the creator of an electron with momentum $\hbar\boldsymbol{k}$ and spin σ. The field $a^*(\boldsymbol{x},\sigma,t)$ is the creator of an electron with spin σ at the point $\boldsymbol{x}$ and at time t

$$a^*(\boldsymbol{x},\sigma,t) = \exp\left(\frac{\mathrm{i}\mathtt{H}_0 t}{\hbar}\right) a^*(\boldsymbol{x},\sigma) \exp\left(\frac{-\mathrm{i}\mathtt{H}_0 t}{\hbar}\right) \quad , \tag{8.212}$$

where $a^*(\boldsymbol{x},\sigma)$ is defined by (3.34) and $\mathtt{H}_0$ is the Hamiltonian (3.72) of the kinetic energy of the electrons. The operators $a(\boldsymbol{x},\sigma,t)$ and $a^*(\boldsymbol{x}',\sigma',t)$ satisfy the anticommutation rules (3.35). The Fermi field is non-Hermitian since it is charged, but the negative frequency components corresponding to the antiparticles do not appear, since positrons are not considered in a non-relativistic theory.

It can be verified from (8.211) that the differential equation of the field is again the Schrödinger equation

$$\mathrm{i}\hbar\frac{\partial}{\partial t}a^*(\boldsymbol{x},\sigma,t) - \frac{\hbar^2}{2m}\nabla^2 a^*(\boldsymbol{x},\sigma,t) = 0 \quad , \tag{8.213}$$

$$\mathrm{i}\hbar\frac{\partial}{\partial t}a(\boldsymbol{x},\sigma,t) + \frac{\hbar^2}{2m}\nabla^2 a(\boldsymbol{x},\sigma,t) = 0 \quad . \tag{8.214}$$

Thus the field formally follows the same law of free evolution as the wavefunction of a free electron. The difference is that the Fourier coefficients, which are scalars for the wavefunction, are now the creation and annihilation operators $a^*_{\boldsymbol{k}\sigma}$ and $a_{\boldsymbol{k}\sigma}$. The wavefunction is thus replaced by an operator in the same way as the classical observables become operators in the transition to quantum mechanics. This is the origin of the term "second quantization". The interpretation is of course given in terms of the Fock space formalism. We now have a charge density (see (3.70))

$$\rho_e(\boldsymbol{x},t) = e\sum_\sigma a^*(\boldsymbol{x},\sigma,t)a(\boldsymbol{x},\sigma,t) = e\,n(\boldsymbol{x},t) \tag{8.215}$$

and a current

$$\boldsymbol{j}(\boldsymbol{x},t) = -\frac{\mathrm{i}e\hbar}{2m}\sum_\sigma \{a^*(\boldsymbol{x},\sigma,t)\nabla a(\boldsymbol{x},\sigma,t) - [\nabla a^*(\boldsymbol{x},\sigma,t)]\,a(\boldsymbol{x},\sigma,t)\} \tag{8.216}$$

which satisfy the continuity equation (8.136). This equation, which expresses the conservation of probability for a wavefunction, is here *the expression of electric charge conservation.*

Interacting Field

It is instructive to derive the equations of motion of the field when the particles interact with a two-body potential

$$\mathtt{H} = \mathtt{H}_0 + \mathtt{V} \quad , \tag{8.217}$$

where $\mathtt{V}$ is given by (3.109). The field at time t is again denoted $a(\boldsymbol{x},\sigma,t)$, and $a^*(\boldsymbol{x},\sigma,t=0) = a^*(\boldsymbol{x},\sigma)$ coincides with expression (8.211) at time $t=0$. As usual in quantum mechanics, the field evolves according to the Heisenberg law

$$a^*(\boldsymbol{x},\sigma,t) = \exp\left(\frac{\mathrm{i}\mathtt{H}t}{\hbar}\right) a^*(\boldsymbol{x},\sigma)\exp\left(-\frac{\mathrm{i}\mathtt{H}t}{\hbar}\right) \quad . \tag{8.218}$$

This has an important consequence: since the evolution $\exp(-\mathrm{i}\mathtt{H}t/\hbar)$ is unitary, the anticommutation relations are also preserved over time for the interacting field

$$a(\boldsymbol{x},\sigma,t)a^*(\boldsymbol{x}',\sigma',t) + a^*(\boldsymbol{x}',\sigma',t)a(\boldsymbol{x},\sigma,t) = \delta_{\sigma,\sigma'}\delta(\boldsymbol{x}-\boldsymbol{x}') \quad . \tag{8.219}$$

To calculate the commutator $[\mathtt{V}, a^*(\boldsymbol{x},\sigma,t)]$ appearing in the Heisenberg equations of motion

$$\begin{aligned}\hbar\frac{\partial}{\partial t}a^*(\boldsymbol{x},\sigma,t) &= \mathrm{i}\,[\mathtt{H}, a^*(\boldsymbol{x},\sigma,t)] \\ &= \mathrm{i}\,[\mathtt{H}_0, a^*(\boldsymbol{x},\sigma,t)] + \mathrm{i}\,[\mathtt{V}, a^*(\boldsymbol{x},\sigma,t)] \quad ,\end{aligned} \tag{8.220}$$

one uses the form (3.109) of the interaction and the anticommutation rules (8.219)

$$\begin{aligned} & \mathrm{i}\hbar \frac{\partial}{\partial t} a^*(\boldsymbol{x},\sigma,t) - \frac{\hbar^2}{2m} \nabla^2 a^*(\boldsymbol{x},\sigma,t) \\ & \quad = -\sum_{\sigma'} \int_\Lambda \mathrm{d}\boldsymbol{x}'\, a^*(\boldsymbol{x},\sigma,t) V(\boldsymbol{x}-\boldsymbol{x}') a^*(\boldsymbol{x}',\sigma',t) a(\boldsymbol{x}',\sigma',t) \quad . \end{aligned} \tag{8.221}$$

Comparing this with (8.213), one sees that the effect of the interaction is to add a term non-linear in the field. This is an important difference to the typical Schrödinger equation. The latter is always a linear equation for the wavefunction, while *the interacting field obeys a non-linear partial differential equation*. We have thus replaced the ensemble of Schrödinger equations for n-particle wavefunctions by a single non-linear equation of motion for the field operators. Needless to say, under this form, the problem remains just as difficult to solve. However, this form can be a source of useful approximation schemes.

Electrons and Holes

In the theory of the electron gas, one frequently evaluates the average values $\langle \Phi_0 | \mathsf{O} | \Phi_0 \rangle$ of observables O in the ground state of the free gas (the filled Fermi sphere)

$$|\Phi_0\rangle = \prod_{\substack{\boldsymbol{k}\sigma \\ |\boldsymbol{k}| \le k_{\mathrm{F}}}} a^*_{\boldsymbol{k}\sigma} |0\rangle \quad . \tag{8.222}$$

In particular, this would be the case for the perturbation calculation at non-zero density for which $|\Phi_0\rangle$ is the non-perturbed reference state (rather than the vacuum of matter $|0\rangle$). We know that such calculations are made much easier by using the Wick theorem in relation to $|\Phi_0\rangle$. The latter requires one to write down a family of creation and annihilation operators for which $|\Phi_0\rangle$ plays the role of the vacuum.

This can be done by considering a new ensemble of operators

$$\begin{aligned} a^{\mathrm{e}}_{\boldsymbol{k}\sigma} &= a_{\boldsymbol{k}\sigma} \quad , \quad |\boldsymbol{k}| > k_{\mathrm{F}} \\ a^{\mathrm{h}}_{\boldsymbol{k}\sigma} &= a^*_{-\boldsymbol{k}\sigma} \quad , \quad |\boldsymbol{k}| < k_{\mathrm{F}} \quad . \end{aligned} \tag{8.223}$$

The operator $(a^{\mathrm{h}}_{\boldsymbol{k}\sigma})^*$ annihilates an electron of wavenumber $|\boldsymbol{k}| < k_{\mathrm{F}}$ in the Fermi sphere. One says that $(a^{\mathrm{h}}_{\boldsymbol{k}\sigma})^*$ *creates a hole*, while $(a^{\mathrm{e}}_{\boldsymbol{k}\sigma})^*$ creates an electron of energy greater than the Fermi energy. Conversely, $a^{\mathrm{h}}_{\boldsymbol{k}\sigma}$ creates an electron in the Fermi sphere, thus *annihilating a hole*. It follows from the definition (8.223) that the anticommutation rules are preserved

$$\begin{aligned} a^{\mathrm{e}}_{\boldsymbol{k}\sigma} (a^{\mathrm{e}}_{\boldsymbol{k}'\sigma'})^* + (a^{\mathrm{e}}_{\boldsymbol{k}'\sigma'})^* a^{\mathrm{e}}_{\boldsymbol{k}\sigma} &= \delta_{\sigma,\sigma'} \delta_{\boldsymbol{k},\boldsymbol{k}'} \quad , \\ a^{\mathrm{h}}_{\boldsymbol{k}\sigma} (a^{\mathrm{h}}_{\boldsymbol{k}'\sigma'})^* + (a^{\mathrm{h}}_{\boldsymbol{k}'\sigma'})^* a^{\mathrm{h}}_{\boldsymbol{k}\sigma} &= \delta_{\sigma,\sigma'} \delta_{\boldsymbol{k},\boldsymbol{k}'} \end{aligned} \tag{8.224}$$

and that

$$a^{\mathrm{e}}_{\boldsymbol{k}\sigma}\,|\Phi_0\rangle = 0 \quad , \quad a^{\mathrm{h}}_{\boldsymbol{k}\sigma}\,|\Phi_0\rangle = 0 \quad . \tag{8.225}$$

One can introduce the following language: we are faced with two types of "particles", electrons and holes. As the creation of a hole is equivalent to the suppression of a charge e, one attributes the charge $-e$ to a hole; $|\Phi_0\rangle$ is the simultaneous "vacuum" of two types of particles. Thus $\langle\Phi_0|\mathsf{O}|\Phi_0\rangle$ can be calculated as the average value of the observable O in the electron–hole "vacuum". In terms of these operators, the kinetic energy takes on the form

$$\begin{aligned}\mathrm{H}_0 &= \sum_{\boldsymbol{k}\sigma} \varepsilon_{\boldsymbol{k}} a^*_{\boldsymbol{k}\sigma} a_{\boldsymbol{k}\sigma}\\ &= \sum_{\boldsymbol{k}\sigma} \theta(|\boldsymbol{k}| - k_{\mathrm{F}})\varepsilon_{\boldsymbol{k}} (a^{\mathrm{e}}_{\boldsymbol{k}\sigma})^* a^{\mathrm{e}}_{\boldsymbol{k}\sigma} - \sum_{\boldsymbol{k}\sigma} \theta(k_{\mathrm{F}} - |\boldsymbol{k}|)\varepsilon_{\boldsymbol{k}} (a^{\mathrm{h}}_{\boldsymbol{k}\sigma})^* a^{\mathrm{h}}_{\boldsymbol{k}\sigma} + E_0 \quad ,\end{aligned} \tag{8.226}$$

where E_0 is the free gas energy (2.94) and $\theta(x) = \begin{cases}1\,, & x > 0\\ 0\,, & x < 0\end{cases}$. The second term represents the kinetic energy of the holes; it is negative since it acts as a subtraction from the Fermi energy. As for the momentum, it is the sum of the constituent electron and hole momenta

$$\mathsf{P} = \sum_{\boldsymbol{k}\sigma} \theta(|\boldsymbol{k}| - k_{\mathrm{F}})\hbar\boldsymbol{k} (a^{\mathrm{e}}_{\boldsymbol{k}\sigma})^* a^{\mathrm{e}}_{\boldsymbol{k}\sigma} + \sum_{\boldsymbol{k}\sigma} \theta(k_{\mathrm{F}} - |\boldsymbol{k}|)\hbar\boldsymbol{k} (a^{\mathrm{h}}_{\boldsymbol{k}\sigma})^* a^{\mathrm{h}}_{\boldsymbol{k}\sigma} \quad . \tag{8.227}$$

The electron-hole formalism allows one to make a useful analogy with the charged field (8.129). One sees this by rewriting the free Fermi field (8.211) in terms of the electron and hole operators (8.223)

$$\begin{aligned}a^*(\boldsymbol{x},\sigma,t) &= \frac{1}{L^{3/2}} \sum_{\boldsymbol{k}} \Bigg\{ \theta(|\boldsymbol{k}| - k_{\mathrm{F}}) (a^{\mathrm{e}}_{\boldsymbol{k}\sigma})^* \exp\left[-\mathrm{i}\left(\boldsymbol{k}\cdot\boldsymbol{x} - \frac{\varepsilon_{\boldsymbol{k}} t}{\hbar}\right)\right]\\ &\quad + \theta(k_{\mathrm{F}} - |\boldsymbol{k}|) a^{\mathrm{h}}_{\boldsymbol{k}\sigma} \exp\left[\mathrm{i}\left(\boldsymbol{k}\cdot\boldsymbol{x} - \frac{\varepsilon_{\boldsymbol{k}} t}{\hbar}\right)\right]\Bigg\}\\ &= (a^{\mathrm{e}}(\boldsymbol{x},\sigma,t))^* + a^{\mathrm{h}}(\boldsymbol{x},\sigma,t) \quad .\end{aligned} \tag{8.228}$$

The Fermi field $a^*(\boldsymbol{x},\sigma,t)$ increases the charge by one unit e either by the creation of an electron or by the annihilation of a hole. By comparing this to the expression (8.129) of the charged scalar field, one sees that *the hole formally plays the same role as an antiparticle*, up to the sign of the energy.

8.4.2 The Quantum-Elastic Field

We have seen in Sect. 1.4 how to define a displacement field $\boldsymbol{u}(\boldsymbol{x},t)$ in the theory of elasticity. In complete analogy with the vector potential $\boldsymbol{A}(\boldsymbol{x},t)$

of electromagnetism in the Coulomb gauge, the classical field $\boldsymbol{u}_{\mathrm{cl}}(\boldsymbol{x},t)$ of an isotropic elastic body is written as

$$\begin{aligned}\boldsymbol{u}_{\mathrm{cl}}(\boldsymbol{x},t) &= \left(\frac{\hbar}{\rho_{\mathrm{M}}L^3}\right)^{1/2}\sum_{\boldsymbol{k}\lambda}\frac{\boldsymbol{e}_\lambda(\boldsymbol{k})}{\sqrt{2\omega_{\boldsymbol{k}\lambda}}}\{\beta_{\boldsymbol{k}\lambda}\exp\left[\mathrm{i}(\boldsymbol{k}\cdot\boldsymbol{x}-\omega_{\boldsymbol{k}\lambda}t)\right] \\ &\quad + \beta^*_{\boldsymbol{k}\lambda}\exp\left[-\mathrm{i}(\boldsymbol{k}\cdot\boldsymbol{x}-\omega_{\boldsymbol{k}\lambda}t)\right]\} \\ &= \boldsymbol{u}_{\mathrm{cl},T}(\boldsymbol{x},t)+\boldsymbol{u}_{\mathrm{cl},L}(\boldsymbol{x},t) \quad ,\end{aligned} \tag{8.229}$$

as a consequence of (1.210), (1.214), (1.218), (1.213) and (1.130). The index λ takes on values $1,2,3$ corresponding to the three orthogonal directions, $\boldsymbol{e}_3(\boldsymbol{k}) = \hat{\boldsymbol{k}}$ being the unit vector in the same direction and orientation as $\boldsymbol{k}$. The longitudinal displacement $\boldsymbol{u}_{\mathrm{cl},L}(\boldsymbol{x},t)$ thus corresponds to the value $\lambda = 3$ in the sum (8.229) and $\boldsymbol{u}_{\mathrm{cl},T}(\boldsymbol{x},t)$ is given by the sum over $\lambda = 1,2$.

As for electromagnetism (8.7), the classical field $\boldsymbol{u}_{\mathrm{cl}}(\boldsymbol{x},t)$ is associated with a family of quantum operators $\boldsymbol{u}(\boldsymbol{x},t)$ defined at time t for each point $\boldsymbol{x}$ (see also (8.1) for the linear harmonic chain)

$$\begin{aligned}\boldsymbol{u}(\boldsymbol{x},t) &= \left(\frac{\hbar}{\rho_{\mathrm{M}}L^3}\right)^{1/2}\sum_{\boldsymbol{k}\lambda}\frac{\boldsymbol{e}_\lambda(\boldsymbol{k})}{\sqrt{2\omega_{\boldsymbol{k}\lambda}}}\{b^*_{\boldsymbol{k}\lambda}\exp\left[-\mathrm{i}(\boldsymbol{k}\cdot\boldsymbol{x}-\omega_{\boldsymbol{k}\lambda}t)\right] \\ &\qquad\qquad + b_{\boldsymbol{k}\lambda}\exp\left[\mathrm{i}(\boldsymbol{k}\cdot\boldsymbol{x}-\omega_{\boldsymbol{k}\lambda}t)\right]\} \\ &= \boldsymbol{u}_{\mathrm{T}}(\boldsymbol{x},t)+\boldsymbol{u}_{\mathrm{L}}(\boldsymbol{x},t) \quad .\end{aligned} \tag{8.230}$$

The operators $b^*_{\boldsymbol{k}\lambda}$ and $b_{\boldsymbol{k}\lambda}$ create and annihilate a phonon in the state $|\boldsymbol{k}\lambda\rangle$

$$\left[b_{\boldsymbol{k}\lambda}, b^*_{\boldsymbol{k}',\lambda'}\right] = \delta_{\lambda,\lambda'}\delta_{\boldsymbol{k},\boldsymbol{k}'} \quad . \tag{8.231}$$

The Hamiltonian $\mathtt{H}_0$ of harmonic vibrations of the isotropic solid is thus given by

$$\mathtt{H}_0 = \sum_{\boldsymbol{k}\lambda}\hbar\omega_{\boldsymbol{k}\lambda}b^*_{\boldsymbol{k}\lambda}b_{\boldsymbol{k}\lambda} \quad . \tag{8.232}$$

This Hamiltonian is obtained by replacing the amplitudes $\beta_{\boldsymbol{k}\lambda}$ in (1.219) by the operators $b_{\boldsymbol{k}\lambda}$ in the same way that (8.230) is obtained from (8.229). The time dependence of $\boldsymbol{u}(\boldsymbol{x},t)$ can be expressed by an evolution operator of the usual form

$$\boldsymbol{u}(\boldsymbol{x},t) = \exp\left(\frac{\mathrm{i}\mathtt{H}_0 t}{\hbar}\right)\boldsymbol{u}(\boldsymbol{x})\exp\left(-\frac{\mathrm{i}\mathtt{H}_0 t}{\hbar}\right) \quad , \tag{8.233}$$

with

$$\boldsymbol{u}(\boldsymbol{x}) = \boldsymbol{u}(\boldsymbol{x},t=0) \quad . \tag{8.234}$$

Recall that, as we have shown in the introduction of this chapter, the continuous field (8.230) obtained from quantizing the classical theory of elasticity must be understood as the limit of the atomic displacement field of a harmonic crystal when the lattice spacing a tends to zero. The creation and annihilation operators are linked to the normal coordinates of the particles, from which the commutation relation (8.231) follows (see Sect. 1.4.6). The continuous description loses its meaning when the wavelength $\lambda = 2\pi/|\boldsymbol{k}|$ associated with the wavenumber $\boldsymbol{k}$ becomes comparable to the distance a between two neighbouring ions. As a result, all of the sums (8.229), (8.230) and (8.232) are in principle limited to values of $|\boldsymbol{k}| \leq a^{-1}$. This is why the constant $1/2(\sum_{\boldsymbol{k}\lambda} \hbar\omega_{\boldsymbol{k}})$, which was omitted in (8.232), is necessarily finite and there is no problem of a divergence.

With the help of $\boldsymbol{u}(\boldsymbol{x}, t)$, one can form the momentum density

$$\begin{aligned} \boldsymbol{p}(\boldsymbol{x}, t) &= \rho_{\mathrm{M}} \frac{\partial}{\partial t} \boldsymbol{u}(\boldsymbol{x}, t) \\ &= \mathrm{i} \left(\frac{\hbar \rho_{\mathrm{M}}}{2L^3} \right)^{1/2} \sum_{\boldsymbol{k}\lambda} \boldsymbol{e}_\lambda(\boldsymbol{k}) \sqrt{\omega_{\boldsymbol{k}\lambda}} \big\{ b^*_{\boldsymbol{k}\lambda} \exp\left[-\mathrm{i}(\boldsymbol{k}\cdot\boldsymbol{x} - \omega_{\boldsymbol{k}\lambda} t)\right] \\ &\quad - b_{\boldsymbol{k}\lambda} \exp\left[\mathrm{i}(\boldsymbol{k}\cdot\boldsymbol{x} - \omega_{\boldsymbol{k}\lambda} t)\right] \big\} \end{aligned} \tag{8.235}$$

and one can easily establish the canonical commutation rules

$$\left[u^\alpha(\boldsymbol{x}), p^\beta(\boldsymbol{y})\right] = \mathrm{i}\hbar \delta_{\alpha,\beta} \delta(\boldsymbol{x} - \boldsymbol{y}) \quad , \quad \alpha, \beta = 1, 2, 3 \quad . \tag{8.236}$$

The Hamiltonian is given by the same formulae (1.206), (1.207), (1.208) as in the classical case after Wick ordering. These formulae are only exact if one sums over all $\boldsymbol{k}$ without taking into account the limitation $|\boldsymbol{k}| \leq a^{-1}$. Up to this limitation, these equations show that *the phonon field has all the same characteristics of a free quantum field.*

8.4.3 Electron–Phonon Interactions

In Chap. 4, we have written the energy of interaction between n charged point particles of the electron-gas and the continuous background (jellium) of positive constant charge density $e\rho$ under the form

$$\mathrm{H}_{\mathrm{I}}^{\mathrm{el\ jel}} = -e^2 \rho \sum_{j=1}^{n} \int_\Lambda \mathrm{d}\boldsymbol{y}\, V(\boldsymbol{x}_j - \boldsymbol{y}) = -\frac{e^2 \rho}{4\pi\varepsilon_0} \sum_{j=1}^{n} \int_\Lambda \mathrm{d}\boldsymbol{y}\, \frac{1}{|\boldsymbol{x}_j - \boldsymbol{y}|} \quad . \tag{8.237}$$

The integration is performed over the volume Λ of the gas. The contribution (8.237) of the electron-gas energy was taken into account in Chap. 4. Imagine now that the positive charge is distributed with a density $\rho(\boldsymbol{x})$ which, rather than being uniform and constant, coincides with that of the ions of a solid, which carry the elastic waves in the continuous description. The interaction energy

$$
\begin{aligned}
\mathtt{H}_{\mathrm{I}}^{\mathrm{el\ ions}} &= -\frac{e^2}{4\pi\varepsilon_0}\sum_{j=1}^{n}\int_{\Lambda}\mathrm{d}\boldsymbol{y}\,\frac{\rho(\boldsymbol{y})}{|\boldsymbol{x}_j-\boldsymbol{y}|} \\
&= -\frac{e^2\rho}{4\pi\varepsilon_0}\sum_{j=1}^{n}\int_{\Lambda}\mathrm{d}\boldsymbol{y}\,\frac{1}{|\boldsymbol{x}_j-\boldsymbol{y}|} - \frac{e^2}{4\pi\varepsilon_0}\sum_{j=1}^{n}\int_{\Lambda}\mathrm{d}\boldsymbol{y}\,\frac{\delta\rho(\boldsymbol{y})}{|\boldsymbol{x}_j-\boldsymbol{y}|} \\
&= \mathtt{H}_{\mathrm{I}}^{\mathrm{el\ jel}} + \mathtt{H}_{\mathrm{I}}^{\mathrm{el\ ph}}
\end{aligned}
\tag{8.238}
$$

is split into two terms. The first term can be identified with (8.237), while the second term describes the interaction between point-like electrons and the deformation $\delta\rho(\boldsymbol{x})$ of the continuous background due to the existence of the lattice vibrations (we are not interested here in other possible types of elastic deformations). If dn is the number of ions which at equilibrium occupy the volume element $dV = dn/\rho$, then

$$
\rho(\boldsymbol{x}) = \frac{\mathrm{d}n}{\mathrm{d}V'(\boldsymbol{x})} \quad , \tag{8.239}
$$

where $\mathrm{d}V'(\boldsymbol{x})$ is the same element whose volume was modified after the deformation. We have seen that, in linear elasticity, $dV'(\boldsymbol{x})$ is expressed with the help of the divergence of the displacement (relation (1.200))

$$
dV'(\boldsymbol{x}) = dV\,[1+\nabla\cdot\boldsymbol{u}_{\mathrm{cl}}(\boldsymbol{x})] = \frac{dn}{\rho}\,[1+\nabla\cdot\boldsymbol{u}_{\mathrm{cl}L}(\boldsymbol{x})] \quad . \tag{8.240}
$$

From the decomposition (1.112) of a vector field, only the longitudinal part can appear in (8.240). Using the linear approximation as usual one can write

$$
\rho(\boldsymbol{x}) = \rho\left(\frac{1}{1+\nabla\cdot\boldsymbol{u}_{\mathrm{cl},L}(\boldsymbol{x})}\right) \simeq \rho - \rho\nabla\cdot\boldsymbol{u}_{\mathrm{cl},L}(\boldsymbol{x}) = \rho + \delta\rho(\boldsymbol{x}) \quad . \tag{8.241}
$$

The interaction of the electrons with the classical elastic waves can thus be written as

$$
\begin{aligned}
\mathtt{H}_{\mathrm{I,cl}}^{\mathrm{el\ ph}} &= \frac{e^2\rho}{4\pi\varepsilon_0}\sum_{j=1}^{n}\int_{\Lambda}\mathrm{d}\boldsymbol{y}\,\frac{\nabla\cdot\boldsymbol{u}_{\mathrm{cl},L}(\boldsymbol{y})}{|\boldsymbol{x}_j-\boldsymbol{y}|} \\
&= \frac{e\rho}{4\pi\varepsilon_0}\int_{\Lambda}\mathrm{d}\boldsymbol{x}\int_{\Lambda}\mathrm{d}\boldsymbol{y}\,\frac{\rho_e(\boldsymbol{x})\nabla\cdot\boldsymbol{u}_{\mathrm{cl},L}(\boldsymbol{y})}{|\boldsymbol{x}-\boldsymbol{y}|} \quad ,
\end{aligned}
\tag{8.242}
$$

where we have introduced the electron charge density

$$
\rho_e(\boldsymbol{x}) = e\sum_{j=1}^{n}\delta(\boldsymbol{x}-\boldsymbol{x}_j) \quad . \tag{8.243}
$$

The quantum electron–phonon interaction Hamiltonian results from replacing $\rho_e(\boldsymbol{x})$ in (8.242) by the quantum density (8.215), and the longitudinal part of the field (8.230) by $\boldsymbol{u}_{\mathrm{cl},L}(\boldsymbol{x})$ (at time $t=0$)

$$\mathtt{H}_{\mathrm{I}}^{\mathrm{el\ ph}} = \frac{e^2\rho}{4\pi\varepsilon_0} \int_\Lambda \mathrm{d}\boldsymbol{x} \int_\Lambda \mathrm{d}\boldsymbol{y}\, \frac{n(\boldsymbol{x})\nabla\cdot\boldsymbol{u}_{\mathrm{L}}(\boldsymbol{y})}{|\boldsymbol{x}-\boldsymbol{y}|} \quad . \tag{8.244}$$

The Fourier representation of $\mathtt{H}_{\mathrm{I}}^{\mathrm{el\ ph}}$ is written as in Sect. 4.2.1 by replacing the Coulomb potential $V(\boldsymbol{x}) = 1/(4\pi\varepsilon_0|\boldsymbol{x}|)$ in (8.244) by the corresponding periodic potential $V_{\mathrm{L}}(\boldsymbol{x})$ (3.112). Taking into account $\boldsymbol{k}\cdot\boldsymbol{e}_\lambda(\boldsymbol{k}) = |\boldsymbol{k}|\delta_{\lambda,3}$, the Fourier transform of the divergence of the phonon field (8.230) is

$$\frac{1}{L^{3/2}} \int_\Lambda \mathrm{d}\boldsymbol{x}\, \mathrm{e}^{-\mathrm{i}\boldsymbol{k}\cdot\boldsymbol{x}} \nabla\cdot\boldsymbol{u}_{\mathrm{L}}(\boldsymbol{x}) = \mathrm{i}\left(\frac{\hbar}{\rho_{\mathrm{M}}}\right)^{1/2} \frac{|\boldsymbol{k}|}{\sqrt{2\omega_{\boldsymbol{k}3}}}(b_{\boldsymbol{k}3} - b^*_{-\boldsymbol{k}3}) \quad , \tag{8.245}$$

so that

$$\mathtt{H}_{\mathrm{I}}^{\mathrm{el\ ph}} = \mathrm{i}e^2\rho \left(\frac{\hbar}{\rho_{\mathrm{M}}L^3}\right)^{1/2} \sum_{\boldsymbol{k}\neq 0} \frac{|\boldsymbol{k}|}{\sqrt{2\omega_{\boldsymbol{k}}}} \tilde{V}_{\mathrm{L}}(\boldsymbol{k})\tilde{n}^*(\boldsymbol{k})(b_{\boldsymbol{k}} - b^*_{-\boldsymbol{k}}) \quad . \tag{8.246}$$

We have omitted the index 3 in $b^*_{\boldsymbol{k}}$, $b_{\boldsymbol{k}}$ and $\omega_{\boldsymbol{k}}$, recalling that from now on that these quantities correspond to longitudinal phonons. The term $\boldsymbol{k} = 0$ can also be omitted from the sum, since $\omega_{\boldsymbol{k}} \simeq c_{\mathrm{L}}|\boldsymbol{k}|$ and, as a consequence, $|\boldsymbol{k}|/\sqrt{2\omega_{\boldsymbol{k}}}$ vanishes for $\boldsymbol{k} = 0$. In (8.246), $\tilde{n}(\boldsymbol{k})$ is the Fourier transform (3.79) of the density.

To obtain the total energy, it is necessary to add to (8.244) the energy of the free phonons (8.232), the kinetic and potential energy of the electrons (the first two terms of (4.28)), as well as the Coulomb energy of the background which is

$$\begin{aligned} e^2 &\int_\Lambda \mathrm{d}\boldsymbol{x} \int_\Lambda \mathrm{d}\boldsymbol{y}\, \frac{\rho(\boldsymbol{x})\rho(\boldsymbol{y})}{|\boldsymbol{x}-\boldsymbol{y}|} \\ &= e^2\rho^2 \int_\Lambda \mathrm{d}\boldsymbol{x} \int_\Lambda \mathrm{d}\boldsymbol{y}\, \frac{1}{|\boldsymbol{x}-\boldsymbol{y}|} + e^2\rho^2 \int_\Lambda \mathrm{d}\boldsymbol{x} \int_\Lambda \mathrm{d}\boldsymbol{y}\, \frac{\nabla\cdot\boldsymbol{u}_{\mathrm{L}}(\boldsymbol{x})\nabla\cdot\boldsymbol{u}_{\mathrm{L}}(\boldsymbol{y})}{|\boldsymbol{x}-\boldsymbol{y}|} \\ &= \mathtt{H}_{\mathrm{I}}^{\mathrm{jel\ jel}} + \mathtt{H}_{\mathrm{I}}^{\mathrm{ph\ ph}} \quad . \end{aligned} \tag{8.247}$$

The second line of (8.247) results from using (8.241) and from the fact that $\int_\Lambda \mathrm{d}\boldsymbol{x}\nabla\cdot\boldsymbol{u}_{\mathrm{L}}(\boldsymbol{x}) = 0$. The term $\mathtt{H}_{\mathrm{I}}^{\mathrm{ph\ ph}}$ represents the Coulomb energy due to the inhomogeneous distribution of the ion charge.

A Simplified Model

The complete Hamiltonian is comprised of all of the Coulomb interactions plus the electron–phonon terms (8.244). To simplify the problem, we would like to separate the study of effects which arise specifically due to the phonons from effects which are related to the Coulomb interaction. With this in mind, we note that the sum of the kinetic and Coulomb energies of the electrons with the first terms of (8.238) and (8.247) form the Hamiltonian (4.28) of jellium. The study of this system shows that the interaction between two charges in the medium is modified by the screening effect which gives rise to an effective short-range potential $V^{\text{eff}}(\boldsymbol{x})$ (Sects. 4.3.4 and 10.2.2). We can take this effect into account by replacing the Coulomb potential in (8.244) by $V^{\text{eff}}(\boldsymbol{x}-\boldsymbol{y})$. Since the quantities a, k_{F}^{-1} and the screening length λ (4.128) are all the same order of magnitude and since only phonons of wavenumber $|\boldsymbol{k}| \leq a^{-1}$ matter, we adopt the form (4.130) of $\tilde{V}^{\text{eff}}(\boldsymbol{k})$

$$\tilde{V}^{\text{eff}}(\boldsymbol{k}) = \frac{1}{\varepsilon_0(|\boldsymbol{k}|^2 + \lambda^{-2})} \simeq \frac{\lambda^2}{\varepsilon_0} \quad . \tag{8.248}$$

For simplicity, we have neglected $|\boldsymbol{k}|$ compared to λ^{-1} in (8.248). After replacing $\tilde{V}_{\text{L}}(\boldsymbol{k})$ in (8.246) by the constant λ^2/ε_0, we find

$$\text{H}_{\text{I}}^{\text{el ph}} = \frac{\mathrm{i}\lambda^2 \rho e^2}{\varepsilon_0} \left(\frac{\hbar}{\rho_{\text{M}} L^3} \right)^{1/2} \sum_{\boldsymbol{k} \neq 0} \frac{|\boldsymbol{k}|}{\sqrt{2\omega_{\boldsymbol{k}}}} \tilde{n}^*(\boldsymbol{k})(b_{\boldsymbol{k}} - b^*_{-\boldsymbol{k}}) \tag{8.249}$$

and if we substitute in the configuration representation of formula (8.245), it becomes

$$\text{H}_{\text{I}}^{\text{el ph}} = \frac{\lambda^2 \rho e^2}{\varepsilon_0} \int_{\Lambda} \mathrm{d}\boldsymbol{x}\, n(\boldsymbol{x}) \nabla \cdot \boldsymbol{u}_{\text{L}}(\boldsymbol{x}) \quad . \tag{8.250}$$

Note that, in the model defined by (8.250), the electron density $n(\boldsymbol{x})$ interacts at a point $\boldsymbol{x}$, not with the displacement $\boldsymbol{u}(\boldsymbol{x})$, but with its divergence $\nabla \cdot \boldsymbol{u}(\boldsymbol{x})$. *The theory is analogous to a local field theory, in which the electron density is coupled at a point $\boldsymbol{x}$ to a scalar field* defined by

$$\begin{aligned} \Phi(\boldsymbol{x},t) &= (\rho_{\text{M}})^{1/2} c_{\text{L}} \nabla \cdot \boldsymbol{u}(\boldsymbol{x},t) \\ &= -\mathrm{i} \left(\frac{\hbar}{L^3} \right)^{1/2} \sum_{\boldsymbol{k}} \sqrt{\frac{\omega_{\boldsymbol{k}}}{2}} \{ b^*_{\boldsymbol{k}} \exp\left[-\mathrm{i}(\boldsymbol{k}\cdot\boldsymbol{x} - \omega_{\boldsymbol{k}} t)\right] \\ &\quad - b_{\boldsymbol{k}} \exp\left[\mathrm{i}(\boldsymbol{k}\cdot\boldsymbol{x} - \omega_{\boldsymbol{k}} t)\right] \} \quad . \end{aligned} \tag{8.251}$$

The latter expression follows from (8.230): here $b^*_{\boldsymbol{k}}$ and $b_{\boldsymbol{k}}$ create and annihilate a longitudinal phonon of frequency $\omega_{\boldsymbol{k}} = c_{\text{L}}|\boldsymbol{k}|$. The factor $(\rho_{\text{M}})^{1/2} c_{\text{L}}$ was introduced for convenience. With definition (8.251), the electron–phonon interaction takes on the form

$$\text{H}_{\text{I}}^{\text{el ph}} = \gamma \int_{\Lambda} \mathrm{d}\boldsymbol{x}\, n(\boldsymbol{x}) \Phi(\boldsymbol{x}) \quad , \tag{8.252}$$

with the coupling constant

$$\gamma = \frac{\lambda^2 \rho e^2}{\varepsilon_0 (\rho_{\text{M}})^{1/2} c_{\text{L}}} \quad . \tag{8.253}$$

If one recalls the relation $\tilde{n}(\boldsymbol{k}) = \sum_{\boldsymbol{k}'\sigma} a^*_{\boldsymbol{k}'\sigma} a_{(\boldsymbol{k}'+\boldsymbol{k})\sigma}$ (see (3.79)), it can be seen that the interaction (8.249) corresponds to the processes displayed in Fig. 8.5,

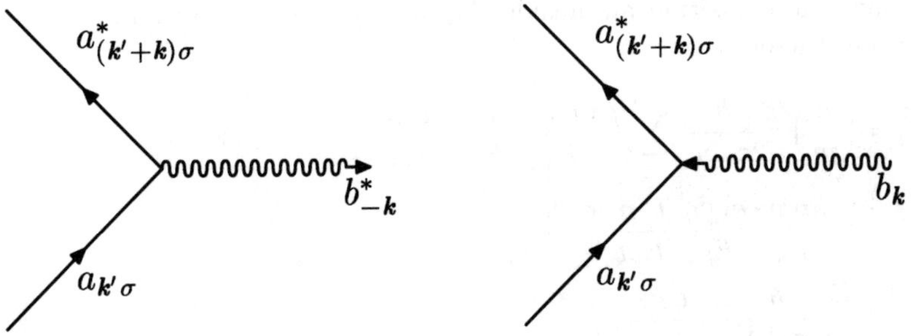

Fig. 8.5. Diagrams of the emission (*left*) or absorption (*right*) of a phonon by an electron. The wavenumbers are conserved in both processes

which represents the emission or absorption of a phonon by an electron. The wavenumbers are conserved during such processes. In the calculations, one includes the limitation $|\boldsymbol{k}| \leq a^{-1}$ by introducing to (8.249) and (8.251) the cut-off function $\theta(\omega_D - \omega_{\boldsymbol{k}})$, where $\omega_D \simeq c_L k_F \simeq c_L a^{-1}$ is the Debye frequency. We note finally that, because of global neutrality, the electron and ion charge densities must be equal. As a consequence, for a system of ions of charge e and mass M, one has $\rho_M = \rho M$, which shows that the coupling constant in (8.253) is inversely proportional to $\sqrt{M}$.

Exercises

1. Scattering of an Electron by a Bound Electron and the Thomson Cross-Section

A photon is scattered elastically by an electron occupying an atomic state $|a\rangle$ with energy E_a. The initial and final states of this process are

$$|i\rangle = |a\rangle \otimes |\boldsymbol{k}_i, \lambda_i\rangle \quad , \quad |f\rangle = |a\rangle \otimes |\boldsymbol{k}_f, \lambda_f\rangle \quad , \quad \boldsymbol{k}_f \neq \boldsymbol{k}_i \quad ,$$

where $|\boldsymbol{k}, \lambda\rangle = a^*_{\boldsymbol{k},\lambda} |0\rangle$ is the state of a photon with wavenumber $\boldsymbol{k}$ and polarization vector $\boldsymbol{e}_\lambda$. In the dipolar approximation, the interaction is

$$\mathtt{H}_I = \mathtt{H}_I^{(1)} + \mathtt{H}_I^{(2)} \quad , \quad \mathtt{H}_I^{(1)} = -\frac{e}{m}\boldsymbol{p} \cdot \mathtt{A}(\boldsymbol{0}) \quad , \quad \mathtt{H}_I^{(2)} = \frac{e^2}{2m}|\mathtt{A}(\boldsymbol{0})|^2 \quad .$$

(i) Calculate the transition amplitude $A_{fi} = A_{fi}^{(1)} + A_{fi}^{(2)}$ to second order in the charge of the electron:

$$A_{fi}^{(1)} = \left(\frac{e}{m}\right)^2 \frac{\hbar}{2\varepsilon_0 \omega_{\boldsymbol{k}_i}} \sum_b \left(\frac{\langle a| \boldsymbol{p}\cdot\boldsymbol{e}_f |b\rangle \langle b| \boldsymbol{p}\cdot\boldsymbol{e}_i |a\rangle}{E_a - E_b + \hbar\omega_{\boldsymbol{k}_i} + \mathrm{i}\varepsilon} + \frac{\langle a| \boldsymbol{p}\cdot\boldsymbol{e}_i |b\rangle \langle b| \boldsymbol{p}\cdot\boldsymbol{e}_f |a\rangle}{E_a - E_b - \hbar\omega_{\boldsymbol{k}_i} + \mathrm{i}\varepsilon} \right) ,$$

$$A_{fi}^{(2)} = \frac{e^2}{2m\,\varepsilon_0 L^3} \frac{\hbar\ \boldsymbol{e}_i\cdot\boldsymbol{e}_f}{\omega_{\boldsymbol{k}_i}} ,$$

where b runs over the intermediate atomic states.

Hint: The transition amplitude, up to second order in $\mathtt{H}_I$, is given by the formula

$$A_{fi} = \langle f|\mathtt{H}_I|i\rangle + \sum_r \frac{\langle f|\mathtt{H}_I|r\rangle \langle r|\mathtt{H}_I|i\rangle}{E_i - E_r + \mathrm{i}\varepsilon} ,$$

where r runs over all intermediate states $r \neq i$ and $E_i = E_f$.

(ii) Show that the contribution of $\mathtt{H}_I^{(2)}$ to the differential cross-section is

$$\frac{d\sigma}{d\Omega_f} = r_e^2 (\boldsymbol{e}_i\cdot\boldsymbol{e}_f)^2 ,$$

where $r_e = e^2/(4\pi\varepsilon_0 mc^2)$, the classical electromagnetic radius of the electron. Moreover, show that the corresponding total cross-section coincides with the classical Thomson total cross-section

$$\sigma = \frac{8\pi}{3} r_e^2 .$$

Hint: The differential cross-section is the probability transition per unit time w_{fi} (8.66) in a solid angle element $d\Omega_f$ around $\boldsymbol{k}_f$, normalized to the incident photon flux.

(iii) Show that the contribution of $H_I^{(1)}$ can be neglected when the energy $\hbar\omega_{\boldsymbol{k}_i}$ of the incident photon is much larger than the ionization energy E_I of the electron.

Hint: Show that the ratio $A_{fi}^{(2)}/A_{fi}^{(1)}$ is of the order $(E_I/\hbar\omega_{\boldsymbol{k}_i})^2$ when $(E_a - E_b)/\hbar \ll \omega_{\boldsymbol{k}_i}$. The kinetic energy of the electron $\langle a| p^2/2m |a\rangle$ does not exceed E_I.

Comment: When the photon frequency is much higher than the atomic frequencies $(E_a - E_b)/\hbar$, the electronic motion around the nucleus is not felt anymore. Then the quantum-mechanical calculation yields the same result as the scattering of a monochromatic wave by a free classical electron (Thomson's scattering).

2. Atom in a Perfect Cavity

Consider a lossless cavity that has a fundamental electromagnetic eigenfrequency ω_0 and real eigenmode $f(\boldsymbol{x})$ with polarization vector $\boldsymbol{e}$, normalized to one. When the other cavity modes are disregarded, the corresponding quantized vector potential is

$$\mathtt{A}(\boldsymbol{x}) = \sqrt{\frac{\hbar}{2\varepsilon_0\omega_0}}\,\boldsymbol{e}\,[f(\boldsymbol{x})a^* + f^*(\boldsymbol{x})a] \quad .$$

An atom is placed in this cavity and undergoes transitions between its ground state $|\alpha\rangle$ and an excited state $|\beta\rangle$ having energies E_α and E_β. The other transitions are neglected and the atomic state space reduces to the two-dimensional space $\mathcal{H}^{\mathrm{at}}$ generated by $|\alpha\rangle$ and $|\beta\rangle$.

(i) Show that the Hamiltonian of this system, to first order in the coupling constant and in the dipolar approximation, is

$$\mathtt{H} = \mathtt{H}_0 + \mathtt{H}_I \quad ,$$

where

$$\mathtt{H}_0 = \hbar\omega_0 a^* a + E_\alpha\,|\alpha\rangle\,\langle\alpha| + E_\beta\,|\beta\rangle\,\langle\beta| \quad ,$$
$$\mathtt{H}_I = (a^* + a)(g_0\,|\alpha\rangle\,\langle\beta| + g_0^*\,|\beta\rangle\,\langle\alpha|) \quad ,$$

with

$$g_0 = -\frac{e}{m}\sqrt{\frac{\hbar}{2\varepsilon_0\omega_0}}\,\langle 1|\,f(\mathbf{0})\boldsymbol{e}\cdot\boldsymbol{p}\,|2\rangle \quad .$$

One splits $\mathtt{H}_I = \mathtt{H}_I^{\mathrm{res}} + \mathtt{H}_I^{\mathrm{nres}}$, where $H_I^{\mathrm{res}} = g_0 a^*\,|\alpha\rangle\,\langle\beta| + g_0^* a\,|\beta\rangle\,\langle\alpha|)$ is the resonant coupling: H_I^{res} describes the process of excitation (de-excitation) of the atom by absorption (emission) of a photon.

(ii) Diagonalize exactly the Hamiltonian $\mathtt{H}^{\mathrm{res}} = \mathtt{H}_0 + \mathtt{H}_I^{\mathrm{res}}$: the eigenenergies and normalized eigenvectors are

$$E_\pm^{(n)} = \frac{E_\alpha + E_\beta}{2} + \left(n - \frac{1}{2}\right)\hbar\omega_0 \pm \sqrt{\frac{\Delta^2}{4} + n|g_0|^2} \quad ,$$

with

$$\Delta = E_2 - E_1 - \hbar\omega_0 \quad ,$$

$$
\begin{aligned}
|+,n\rangle &= c_n\,|\alpha,n\rangle + d_n^*\,|\beta,n-1\rangle \quad , \quad n = 1,2,\ldots \quad , \\
|-,n\rangle &= d_n\,|\alpha,n\rangle - c_n^*\,|\beta,n-1\rangle \quad , \quad n = 1,2,\ldots \quad , \\
c_n &= \frac{2g_0\sqrt{n}}{\sqrt{(R_n+\Delta)^2 + 4n|g_0|^2}} \quad , \quad R_n = \sqrt{\Delta^2 + 4n|g_0|^2} \quad ,
\end{aligned}
$$

where $|\alpha,n\rangle\,,|\beta,n\rangle$ are the eigenstates of $\mathtt{H}_0$.

(iii) Show that the corrections to $E_\pm^{(n)}$ due to $\mathtt{H}_I^{\mathrm{nres}}$ are of the order $|g_0|^2/\hbar\omega_0$.

(iv) Show that the transition probability $p_{1,2}(t)$ between the states $|\alpha,1\rangle$ and $|\beta,0\rangle$ at resonance $\Delta = 0$ is

$$
p_{1,2}(t) = \sin^2\left(\frac{|g_0|t}{\hbar}\right) \quad .
$$

Comments: The states $|\pm,n\rangle$ are “dressed” atomic states, i.e. eigenstates of the coupled atom and the quantized electromagnetic mode. The oscillations between the two eigenstates $|\alpha,1\rangle$ and $|\beta,0\rangle$ of the unperturbed Hamiltonian H_0 generated by the electromagnetic coupling are called vacuum Rabi oscillations. These are due to the process of excitation of the atom into the state $|\beta\rangle$ and the subsequent spontaneous emission of a photon, and are thus a consequence of the vacuum fluctuations of the electromagnetic field. They were observed for a single atom in a cavity in 1992.[8]

[8] R. J. Thompson et al. “Observation of normal mode splitting in an optical cavity”, Phys. Rev. Lett. *68*, 1132 (1992).

9. Perturbative Methods in Field Theory

9.1 Introduction

In the preceding chapters, we have described several physical systems, photons, phonons, electron gas, nuclei, which we treated using semi-quantitative methods: orders of magnitude, models, variational methods. The object of Chaps. 9 and 10 is to develop a more systematic method for calculating the effects of the interactions: the perturbation method. One finds that all *the perturbative calculations used in field theory and in many-body problems are essentially derived from the same common core, the time-dependent perturbative methods of quantum mechanics.* At a strictly formal level, the method offers us a large degree of generality. One must nevertheless take care not to abuse this apparent generality. While the basic formalism is unique, the applications are as diverse as the physical situations being considered. Thus, *before beginning any significant perturbative calculation, it is necessary to first consider carefully the physical effects and their orders of magnitude.* The art of a perturbative calculation lies in the judicious handling of a well-posed physical question. Before performing any calculations, it is important to first determine the physical relevance of the expected results.

Here, one must distinguish particle theory from the study of many-body problems. During a collision between elementary particles, one only observes the initial and final states of the reaction. The moment of the collision and what goes on at that moment are, in general, inaccessible to the experiment; a detailed description of this part of the process is useless. The only important aspect, for a given interaction, is to be able to link the state of the incident particles, before the collision, with that of the outgoing particles, after the collision. This is the information contained in the scattering operator S which, because of this fact, will become the central object that needs to be calculated perturbatively (Sect. 9.3.2). Concerning the many-body problems, we have seen in Sect. 3.2.8 that, in order to calculate the average values of one-body and two-body observables, it is sufficient to know the reduced density matrices. More generally, all of the information accessible from the comparison with experiment can be taken from what are called the Green functions, which generalize the reduced density matrices by including the time dependence. We will thus be performing perturbative calculations on the Green functions.

Table 9.1. Schematic summary of the major connections of the perturbative calculation method

	Particle physics	Many-body problems
Experimentally accessible quantities	Cross-section, lifetime, branching ratio, ...	Response functions, susceptibilities, excitation spectra, ...
Intermediate objects submitted to the perturbation calculation	↑ Scattering operator	↑ Green functions
Theoretical description of the interactions between fields and particles	Evolution operator	

One can see that perturbation theory is applied to those quantities (scattering operators, Green functions) which are both linked to the fundamental interactions as well as being in contact with the measurable quantities. Table 9.1 gives a schematic summary of the major aspects of the theory.

Perturbation theory frequently takes on a forbidding appearance at first because of the long development which is necessary for its elaboration. The use of Feynman diagrams, which provide a suggestive classification of the perturbative series in relation to the physical processes, greatly facilitates the analysis. In Sect. 9.2, we illustrate the main ideas for the case of a particle under the influence of an external potential. Following that, we present the general formalism, as well as some applications to the interaction of two quantum fields. Chapter 10 is dedicated to the Green function in condensed-matter physics.

In the many-body case, the above considerations apply when the system is in its ground state, specifically at zero temperature. It is also important to develop perturbative methods more adapted to the thermal statistical ensemble $Z^{-1}\exp(-\beta H)$ at positive temperature $T = (k_{\mathrm{B}}\beta)^{-1}$, where Z is the partition function. This will not be presented here, but we do note the formal analogy between the quantum devolution $\exp(-\mathrm{i}Ht/\hbar)$ and the operator $\exp(-\beta H)$, the latter being obtained by analytic continuation of the evolution operator to imaginary time $t = -\mathrm{i}\beta\hbar$. Because of this analogy, the corresponding algorithms are similar. Moreover, in the positive temperature case, a useful way to generate perturbation series is through the functional integral representation. Essentially the relevant observation is that for a free field Hamiltonian H_0 as in (8.101), $\exp(-\beta H_0)$ is the exponential of a quadratic form of the field Φ, like a Gaussian weight. The rules for composing the terms in the perturbation series, which will be encoded here in Wick's theorem, will also appear in the functional integral representation as the rules for moments of a Gaussian distribution. Then one can return from the non-zero temperature formalism to propagation in real time using analytic continuation $\beta = \mathrm{i}t/\hbar$.[1]

9.2 The Green Functions

9.2.1 Definition

The Retarded Green Function

In this section, we introduce the principal concepts used in perturbation theory for the case of a single non-relativistic spinless particle. The same concepts extend to much more general situations.

The object which provides the usual quantum description of a particle is the wavefunction $\varphi(\boldsymbol{x}, t)$, governed by the Schrödinger equation

$$\left(\mathrm{i}\hbar\frac{\partial}{\partial t} - H(t)\right)\varphi(\boldsymbol{x}, t) = 0 \quad . \tag{9.1}$$

The Hamiltonian $H(t)$, which is not explicitly specified here, can include the effect of time-dependent external fields. All the properties of the particle at time t can be deduced from knowledge of $\varphi(\boldsymbol{x}, t)$ when one has fixed an initial condition $\varphi(\boldsymbol{x}, t')$ at time t' earlier that t.

We now introduce a new quantity $G(\boldsymbol{x}, t|\boldsymbol{x}', t')$ called the *retarded Green function* or *causal propagator* defined by the relations

$$\varphi(\boldsymbol{x}, t) = \mathrm{i}\int \mathrm{d}\boldsymbol{x}'\, G(\boldsymbol{x}, t|\boldsymbol{x}', t')\varphi(\boldsymbol{x}', t') \quad , \tag{9.2}$$

$$G(\boldsymbol{x}, t|\boldsymbol{x}', t') = 0 \quad , \quad t < t' \quad . \tag{9.3}$$

These two relations allow for an interesting interpretation of the Green function. Suppose that one knows with certainty that a particle is located at $\boldsymbol{x}_0$ at time t'. It is thus characterized by the (improper) function, localized at $\boldsymbol{x}_0$,

$$\varphi(\boldsymbol{x}', t') = \delta(\boldsymbol{x}' - \boldsymbol{x}_0) \quad , \tag{9.4}$$

which, when substituted into (9.2), gives

$$\varphi(\boldsymbol{x}, t) = \mathrm{i}\int \mathrm{d}\boldsymbol{x}'\, G(\boldsymbol{x}, t|\boldsymbol{x}', t')\delta(\boldsymbol{x}' - \boldsymbol{x}_0) = \mathrm{i}G(\boldsymbol{x}, t|\boldsymbol{x}_0, t') \quad . \tag{9.5}$$

---

[1] Imaginary-time quantum-field theory is called Euclidean. For these aspects and the functional integral representation, see recent books cited in the bibliography.

Thus, $G(\boldsymbol{x},t|\boldsymbol{x}',t')$ *is the probability amplitude that the particle be at* $\boldsymbol{x}$ *at time* t*, given that it is located at* $\boldsymbol{x}'$ *at time* $t' < t$. Moreover, (9.2) expresses the linear nature of the evolution law of a quantum wavefunction.

With the causality condition (9.3), $G(\boldsymbol{x},t|\boldsymbol{x}',t')$ governs the *time development of the wavefunction into the future.* By using the step function

$$\theta(t) = \begin{cases} 0 \quad , \quad t<0 \\ 1 \quad , \quad t>0 \end{cases} \quad , \tag{9.6}$$

one can rewrite (9.2) and (9.3) in one single equation

$$\theta(t-t')\varphi(\boldsymbol{x},t) = \mathrm{i}\int \mathrm{d}\boldsymbol{x}'\, G(\boldsymbol{x},t|\boldsymbol{x}',t')\varphi(\boldsymbol{x}',t') \quad . \tag{9.7}$$

To determine the differential equation obeyed by $G(\boldsymbol{x},t|\boldsymbol{x}',t')$, we apply $\mathrm{i}\hbar\partial/\partial t - H(t)$ to both sides of (9.7). As $(d/dt)\theta(t) = \delta(t)$ and as $\varphi(\boldsymbol{x},t)$ is the solution of the Schrödinger equation (9.1), one obtains

$$\mathrm{i}\hbar\delta(t-t')\varphi(\boldsymbol{x},t') = \mathrm{i}\int \mathrm{d}\boldsymbol{x}' \left[\left(\mathrm{i}\hbar\frac{\partial}{\partial t} - H(t)\right) G(\boldsymbol{x},t|\boldsymbol{x}',t')\right]\varphi(\boldsymbol{x}',t') \quad . \tag{9.8}$$

Since (9.8) must be valid for all choices of φ, one deduces that $G(\boldsymbol{x},t|\boldsymbol{x}',t')$ satisfies *the inhomogeneous equation*

$$\left(\mathrm{i}\hbar\frac{\partial}{\partial t} - H(t)\right) G(\boldsymbol{x},t|\boldsymbol{x}',t') = \hbar\delta(t-t')\delta(\boldsymbol{x}-\boldsymbol{x}') \quad . \tag{9.9}$$

It is from (9.9) that $G(\boldsymbol{x},t|\boldsymbol{x}',t')$ gets the name *Green function*: in classical physics, the Green function appears in the solution of the partial differential equations of the potential or of waves emitted by a point source. The qualification "retarded" comes from the condition of causality (9.3).

It is clear that the formulation of the quantum mechanics of a particle either in terms of its wavefunction or in terms of the Green function submitted to (9.9) are equivalent. Knowledge of $\varphi(\boldsymbol{x},t)$ for all $t > t'$ allows one to determine $G(\boldsymbol{x},t|\boldsymbol{x}',t')$ by (9.2). Conversely, the retarded solution of (9.9) determines, according to (9.2), a wavefunction which obeys the Schrödinger equation. The advantage of the second formulation is that *the notion of the Green function can be generalized in many-body problems to situations for which the concept of an individual wavefunction no longer has a well-defined meaning.*

Relation with the Evolution Operator

The Green function is directly related to the operator $U(t,t')$ which governs the evolution of the states

$$|\varphi_t\rangle = U(t,t')\,|\varphi_{t'}\rangle \quad . \tag{9.10}$$

Changing to the configuration representation $\varphi(\boldsymbol{x},t) = \langle \boldsymbol{x}|\varphi_t\rangle$, (9.10) is written as

$$\langle \boldsymbol{x}|\varphi_t\rangle = \int \mathrm{d}\boldsymbol{x}'\, \langle \boldsymbol{x}|U(t,t')|\boldsymbol{x}'\rangle\,\langle \boldsymbol{x}'|\varphi_{t'}\rangle \quad . \tag{9.11}$$

If we multiply this expression by $\theta(t-t')$ we find, by comparison with (9.7),

$$G(\boldsymbol{x},t|\boldsymbol{x}',t') = -\mathrm{i}\theta(t-t')\,\langle \boldsymbol{x}|U(t,t')|\boldsymbol{x}'\rangle = \langle \boldsymbol{x}|G(t,t')|\boldsymbol{x}'\rangle \quad , \tag{9.12}$$

where we have set

$$G(t,t') = -\mathrm{i}\theta(t-t')U(t,t') \quad . \tag{9.13}$$

The operator $G(t,t')$, defined in this way, whose matrix elements (9.12) give the Green function, obeys the operator form of equation (9.9)

$$\left(\mathrm{i}\hbar\frac{\partial}{\partial t} - H(t)\right) G(t,t') = \hbar\delta(t-t') \quad . \tag{9.14}$$

Its use allows one to write the Green function without difficulty in representations other than the configuration representation.

Spectral Representation of the Retarded Green Function

When the Hamiltonian is independent of time, the system is conservative, and it is only necessary to consider the stationary Schrödinger equation

$$H\,|\varphi_r\rangle = \varepsilon_r\,|\varphi_r\rangle \quad . \tag{9.15}$$

It is easy to express the Green function with the help of the energies ε_r and the eigenstates $|\varphi_r\rangle$. First of all, one notes that the evolution operator $U(t,t') = \exp[-\mathrm{i}H(t-t')/\hbar]$ only depends on the time difference $t-t'$. According to (9.12), this also applies for the Green function, and one simply sets

$$G(\boldsymbol{x},t|\boldsymbol{x}',t') = G(\boldsymbol{x},t-t'|\boldsymbol{x}') \quad . \tag{9.16}$$

It thus follows from (9.12) and the fact that the states $|\varphi_r\rangle$ form a complete orthonormal system that

$$G(\boldsymbol{x},t|\boldsymbol{x}') = -\mathrm{i}\theta(t)\sum_r \langle\boldsymbol{x}|\varphi_r\rangle\,\langle\varphi_r|\mathrm{e}^{-\mathrm{i}Ht/\hbar}|\boldsymbol{x}'\rangle$$
$$= -\mathrm{i}\theta(t)\sum_r \langle\boldsymbol{x}|\varphi_r\rangle\,\langle\varphi_r|\boldsymbol{x}'\rangle\,\mathrm{e}^{-\mathrm{i}\varepsilon_r t/\hbar} \quad . \tag{9.17}$$

Conversely, knowledge of the Green function allows one to calculate the energies and the projection operators, ε_r and $|\varphi_r\rangle\,\langle\varphi_r|$. To see this, we introduce the temporal Fourier transform of the function $\mathrm{e}^{-\eta t}G(\boldsymbol{x},t|\boldsymbol{x}')$

$$G(\boldsymbol{x},\omega+\mathrm{i}\eta|\boldsymbol{x}') = \int_{-\infty}^{\infty} \mathrm{d}t\,\mathrm{e}^{\mathrm{i}(\omega+\mathrm{i}\eta)t}G(\boldsymbol{x},t|\boldsymbol{x}') \quad , \quad \eta>0 \quad . \tag{9.18}$$

The integral (9.18) is convergent since, because of the causality enforced by (9.3), the integrand vanishes if $t<0$ and it includes the factor $\mathrm{e}^{-\eta t}$, integrable for $t>0$. In fact, the integral (9.18) defines an analytic function in the complex half-plane $\omega+\mathrm{i}\eta, \eta>0$. Taking into account (9.17) and

$$-\mathrm{i}\int_{-\infty}^{\infty} \mathrm{d}t\,\theta(t)\mathrm{e}^{-\mathrm{i}\lambda t}\mathrm{e}^{\mathrm{i}(\omega+\mathrm{i}\eta)t} = \frac{1}{\omega+\mathrm{i}\eta-\lambda} \quad , \quad \eta>0 \quad , \tag{9.19}$$

one obtains

$$G(\boldsymbol{x},\omega+\mathrm{i}\eta|\boldsymbol{x}') = \hbar\sum_r \frac{\langle\boldsymbol{x}|\varphi_r\rangle\,\langle\varphi_r|\boldsymbol{x}'\rangle}{\hbar(\omega+\mathrm{i}\eta)-\varepsilon_r} \quad . \tag{9.20}$$

Considered as a function of the complex variable $\hbar(\omega+\mathrm{i}\eta)$, $G(\boldsymbol{x},\omega+\mathrm{i}\eta|\boldsymbol{x}')$ has simple poles at the eigenenergies ε_r of the system. The residues at these poles are the projectors on the corresponding eigensubspaces. Thus the singularities of $G(\boldsymbol{x},\omega+\mathrm{i}\eta|\boldsymbol{x}')$ on the real axis of the $\hbar(\omega+\mathrm{i}\eta)$ plane coincide with the energy spectrum of the particle.

The transformation (9.18) can be inverted with the help of the inverse of formula (9.19), which is derived at the end of this section, specifically that

$$\lim_{\eta\to 0}\frac{1}{2\pi}\int_{-\infty}^{\infty} \mathrm{d}\omega\,\mathrm{e}^{-\mathrm{i}\omega t}\frac{1}{\omega+\mathrm{i}\eta-\lambda} = -\mathrm{i}\theta(t)\mathrm{e}^{-\mathrm{i}\lambda t} \quad , \quad \eta>0 \quad . \tag{9.21}$$

Applying (9.21) to (9.20), one finds

$$\lim_{\eta\to 0}\frac{1}{2\pi}\int_{-\infty}^{\infty} \mathrm{d}\omega\,\mathrm{e}^{-\mathrm{i}\omega t}G(\boldsymbol{x},\omega+\mathrm{i}\eta|\boldsymbol{x}') = G(\boldsymbol{x},t|\boldsymbol{x}') \quad , \quad \eta>0 \quad . \tag{9.22}$$

To simplify the notation, we occasionally omit writing the positive imaginary part η, which appears in (9.18) to (9.22), and we refer to (9.18) and its inverse (9.22) as Fourier transforms. It is to be understood that *the Fourier transform (9.18) of the retarded Green function is always defined with the convergence factor $\mathrm{e}^{-\eta t}$ and that its inverse is always obtained by taking the limit (9.22) where the real axis of the plane $\omega+\mathrm{i}\eta$ is approached from positive imaginary values.*

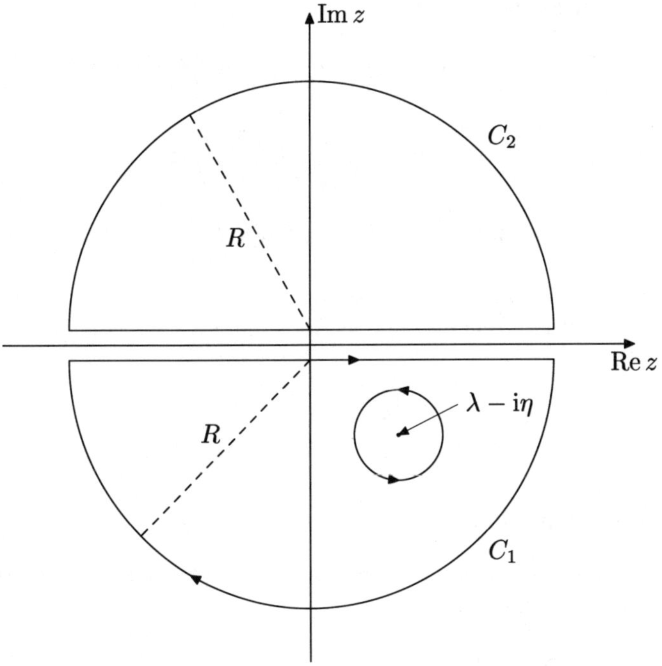

Fig. 9.1. The contours C_1 and C_2 described in the text

Integral Representation of $\theta(t)$

To establish (9.21), consider the integral (9.23) taken over the closed contour C_1 in the complex plane z of Fig. 9.1. If t is positive, the integrand is analytic in the half-plane $\mathrm{Im}z < 0$ except at the simple pole $z = \lambda - \mathrm{i}\eta$

$$\frac{1}{2\pi}\int_{C_1} \mathrm{d}z\, \mathrm{e}^{-\mathrm{i}zt} \frac{1}{z + \mathrm{i}\eta - \lambda} = -\mathrm{i}\mathrm{e}^{-\mathrm{i}\lambda t}\mathrm{e}^{-\eta t} \quad , \quad \eta > 0 \quad , \quad t > 0 \quad . \tag{9.23}$$

In the limit $R \to \infty$, the integral over the semi-circle of radius R tends to zero, giving rise to (9.21) for $t > 0$. If t is negative, the integrand is analytic in the entire half-plane $\mathrm{Im}z > 0$. The same reasoning applies to the contour C_2, showing that the integral (9.21) vanishes in the latter case.

9.2.2 The Free-Particle Green Function

To gain familiarity with the Green function, it is useful to study its structure and its interpretation for the case of a free particle. The Hamiltonian H_0 has the plane waves $|\boldsymbol{k}\rangle$ as eigenstates with energies $\varepsilon_{\boldsymbol{k}} = \hbar^2|\boldsymbol{k}|^2/2m$.

Momentum Representation

The Green function takes on the simplest form in the momentum representation. Taking into account (9.13), one finds

$$\begin{aligned} G_0(\boldsymbol{k},t|\boldsymbol{k}') &= \langle \boldsymbol{k}|G_0(t,0)|\boldsymbol{k}'\rangle = -\mathrm{i}\theta(t)\,\langle \boldsymbol{k}|\exp\left(-\frac{\mathrm{i}H_0 t}{\hbar}\right)|\boldsymbol{k}'\rangle \\ &= -\mathrm{i}\delta_{\boldsymbol{k},\boldsymbol{k}'}\theta(t)\exp\left(-\frac{\mathrm{i}\varepsilon_{\boldsymbol{k}}t}{\hbar}\right) = \delta_{\boldsymbol{k},\boldsymbol{k}'}G_0(\boldsymbol{k},t) \quad . \end{aligned} \tag{9.24}$$

The Green function is diagonal and depends only on a single value of the energy spectrum. From (9.19), one can deduce its time Fourier transform

$$\begin{aligned} G_0(\boldsymbol{k},\omega+\mathrm{i}\eta|\boldsymbol{k}') &= \delta_{\boldsymbol{k},\boldsymbol{k}'}\frac{\hbar}{\hbar(\omega+\mathrm{i}\eta)-\varepsilon_{\boldsymbol{k}}} \quad , \quad \eta>0 \\ &= \delta_{\boldsymbol{k},\boldsymbol{k}'}G_0(\boldsymbol{k},\omega+\mathrm{i}\eta) \quad . \end{aligned} \tag{9.25}$$

Configuration Representation

The Green function in the configuration representation is deduced from (9.20) by introducing the plane waves (2.26)

$$\begin{aligned} G_0(\boldsymbol{x},\omega+\mathrm{i}\eta|\boldsymbol{x}') &= \frac{\hbar}{L^3}\sum_{\boldsymbol{k}}\frac{\mathrm{e}^{\mathrm{i}\boldsymbol{k}\cdot(\boldsymbol{x}-\boldsymbol{x}')}}{\hbar(\omega+\mathrm{i}\eta)-\varepsilon_{\boldsymbol{k}}} \\ &= \frac{\hbar}{(2\pi)^3}\int \mathrm{d}\boldsymbol{k}\frac{\mathrm{e}^{\mathrm{i}\boldsymbol{k}\cdot(\boldsymbol{x}-\boldsymbol{x}')}}{\hbar(\omega+\mathrm{i}\eta)-\varepsilon_{\boldsymbol{k}}} \quad , \quad \eta>0 \quad . \end{aligned} \tag{9.26}$$

The second expression results from taking the infinite volume limit. It is instructive to explicitly calculate

$$G_0(\boldsymbol{x},\omega|\boldsymbol{x}') = \lim_{\eta\to 0} G_0(\boldsymbol{x},\omega+\mathrm{i}\eta|\boldsymbol{x}') \quad , \quad \omega>0 \quad , \quad \eta>0 \quad . \tag{9.27}$$

For this, we substitute into (9.26) the new variables $r=|\boldsymbol{x}-\boldsymbol{x}'|$, $\omega=\hbar\kappa^2/2m$, $\eta=\hbar\kappa\delta/m$, $\kappa>0$, $\delta>0$. Up to a term of order η^2, one has $\omega+\mathrm{i}\eta=\hbar(\kappa+\mathrm{i}\delta)^2/2m$ and one can perform the angular integrations

$$\begin{aligned} &G_0(\boldsymbol{x},\omega+\mathrm{i}\eta|\boldsymbol{x}') \\ &= \frac{1}{(2\pi)^2}\frac{2m}{\hbar}\int_0^\infty k^2\mathrm{d}k\int_{-1}^{1}\mathrm{d}u\,\frac{\mathrm{e}^{\mathrm{i}kru}}{(\kappa+\mathrm{i}\delta)^2-k^2} \\ &= \frac{m}{2\pi^2\hbar}\int_0^\infty k\mathrm{d}k\frac{\mathrm{e}^{\mathrm{i}kr}-\mathrm{e}^{-\mathrm{i}kr}}{\mathrm{i}r\left[(\kappa+\mathrm{i}\delta)^2-k^2\right]} \\ &= \frac{m}{2\pi^2\hbar r}\int_{-\infty}^\infty k\mathrm{d}k\frac{\mathrm{e}^{\mathrm{i}kr}}{\mathrm{i}\left[(\kappa+\mathrm{i}\delta)^2-k^2\right]} \\ &= \frac{\mathrm{i}m}{4\pi^2\hbar r}\int_{-\infty}^\infty \mathrm{d}k\,\mathrm{e}^{\mathrm{i}kr}\left(\frac{1}{k-\kappa-\mathrm{i}\delta}+\frac{1}{k+\kappa+\mathrm{i}\delta}\right) \quad , \quad \delta>0 \quad . \end{aligned} \tag{9.28}$$

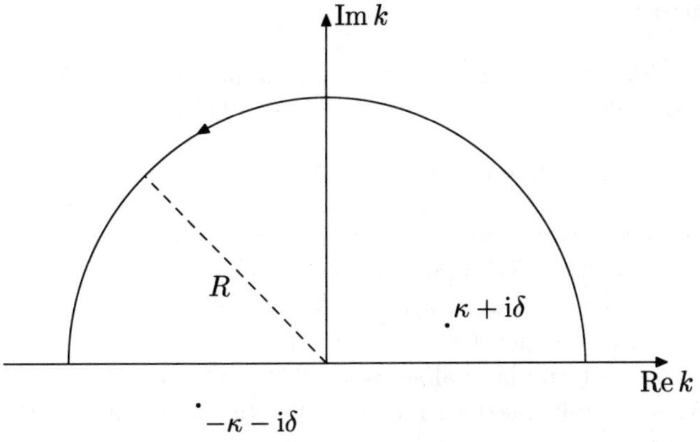

Fig. 9.2. Integration of the last integral of (9.28) in the complex plane k over the closed contour comprising the semi-circle of radius R in the plane Im$k > 0$

The final integral is calculated by integration in the complex plane k over the closed contour comprising the semi-circle of radius R in the plane Im$k > 0$ (Fig. 9.2). As $\mathrm{e}^{\mathrm{i}kr} \to 0$, Im$k > 0$, the integral over the semi-circle does not contribute as $R \to \infty$ and the result is given by the residue in $\kappa + \mathrm{i}\delta$

$$\lim_{\eta\to 0} G_0(\boldsymbol{x}, \omega + \mathrm{i}\eta|\boldsymbol{x}') = -\frac{m}{2\pi\hbar}\frac{\exp\left(\mathrm{i}\sqrt{2m\omega/\hbar}\,r\right)}{r} \quad , \quad \omega > 0 \quad . \tag{9.29}$$

The function $G_0(\boldsymbol{x}, \omega|\boldsymbol{x}')$ represents an outgoing spherical wave coming from the point $\boldsymbol{x}'$, and it is easy to verify that it is the solution of the inhomogeneous Schrödinger equation

$$\left(\frac{\hbar^2}{2m}\nabla^2 + \hbar\omega\right) G_0(\boldsymbol{x}, \omega|\boldsymbol{x}') = \hbar\delta(\boldsymbol{x} - \boldsymbol{x}') \quad . \tag{9.30}$$

The fact that one obtains an outgoing wave is linked to causality, which is manifested in the positive sign of δ: the contour integral of Fig. 9.2 receives a contribution from the pole $\kappa + \mathrm{i}\delta$ which gives rise to the phase $\mathrm{e}^{\mathrm{i}\kappa r}$ corresponding to the outgoing wave (the choice $\delta < 0$ corresponds to the advanced Green function and to an incoming wave $\mathrm{e}^{-\mathrm{i}\kappa r}$).

Finally, the explicit form of $G_0(\boldsymbol{x}, t|\boldsymbol{x}', t')$ as a function of space–time arguments is given by the Fourier transform of (9.24). So, one has

$$\begin{aligned} G_0(\boldsymbol{x}, t|\boldsymbol{x}', t') \\ &= -\mathrm{i}\theta(t - t')\frac{1}{(2\pi)^3}\int \mathrm{d}\boldsymbol{k}\, \exp\left[-\mathrm{i}\frac{\hbar}{2m}|\boldsymbol{k}|^2(t - t')\right] \exp\left[-\mathrm{i}\boldsymbol{k}\cdot(\boldsymbol{x} - \boldsymbol{x}')\right] \\ &= -\mathrm{i}\theta(t - t')\left(\frac{m}{2\mathrm{i}\pi\hbar(t - t')}\right)^{3/2} \exp\left(\frac{\mathrm{i}m|\boldsymbol{x} - \boldsymbol{x}'|^2}{2\hbar(t - t')}\right) \quad . \end{aligned} \tag{9.31}$$

Causal Interpretation

The causal interpretation of the Green function becomes clear using the second quantization formalism. In this formalism, $G_0(\boldsymbol{x},t|\boldsymbol{x}',t')$ is written

$$G_0(\boldsymbol{x},t|\boldsymbol{x}',t') = -\mathrm{i}\theta(t-t')\,\langle 0|a(\boldsymbol{x},t)a^*(\boldsymbol{x}',t')|0\rangle \quad , \tag{9.32}$$

where $a(\boldsymbol{x},t)$ is the non-relativistic free field (8.212). Here the ground state $|0\rangle$ is the vacuum and $G_0(\boldsymbol{x},t|\boldsymbol{x}',t')$ describes the evolution of a single particle only. Nevertheless, as the second quantization formalism allows us to anticipate, it will be very natural to generalize the notion of the Green function to many-body systems using a formula analogous to (9.32) (Chap. 10).

To establish (9.32), we only need to recall that $a^*(\boldsymbol{x}) = a^*(|\boldsymbol{x}\rangle)$ is the creator of a particle in the state $|\boldsymbol{x}\rangle$, whose evolution is given by (3.85)

$$a^*(\boldsymbol{x},t) = a^*\left[U_0^{-1}(t)\,|\boldsymbol{x}\rangle\right] \quad , \quad U_0(t) = \exp\left(-\frac{\mathrm{i}H_0 t}{\hbar}\right) \quad . \tag{9.33}$$

After applying (3.16) and the commutation (3.27) or anti-commutation (3.28) rules, one finds that the right-hand side of (9.32) can be written as

$$\begin{aligned} &-\mathrm{i}\theta(t-t')\,\langle 0|\left[a(\boldsymbol{x},t)a^*(\boldsymbol{x}',t') \mp a^*(\boldsymbol{x}',t')a(\boldsymbol{x},t)\right]|0\rangle \\ &\quad = -\mathrm{i}\theta(t-t')\,\langle \boldsymbol{x}|U_0(t-t')|\boldsymbol{x}'\rangle \quad , \end{aligned} \tag{9.34}$$

which is precisely equal to $G_0(\boldsymbol{x},t|\boldsymbol{x}',t')$ according to (9.12).

The causal interpretation of (9.32) is clear: *the time t' at which the particle is created must precede the time t at which it is annihilated.* This again justifies the causal propagator term. The notion of a causal propagator plays an important role in the general development of the perturbation theory (Sect. 9.3.4).

9.2.3 Particle in an External Field

Consider a particle submitted to a perturbation $V(t)$ whose the complete Hamiltonian is given by

$$H(t) = H_0 + V(t) \quad . \tag{9.35}$$

The corresponding causal propagator $G(t,t')$ is defined by (9.13). The central problem of perturbation theory is to express $G(t,t')$ in terms of the free propagator $G_0(t,t')$ and successive powers of the interaction.

Perturbation Series

The source of the perturbation series is provided by the following fundamental identity which links $G(t,t')$ to $G_0(t,t')$

$$G(t,t') = G_0(t,t') + \frac{1}{\hbar}\int \mathrm{d}t_1\, G_0(t,t_1)V(t_1)G(t_1,t') \quad . \tag{9.36}$$

To verify, one needs only apply $\mathrm{i}\hbar\partial/\partial t - H_0$ to both sides of (9.36). Taking into account the fact that $G_0(t,t')$ obeys (9.14) in relation to H_0, we obtain

$$\begin{aligned}
&\left(\mathrm{i}\hbar\frac{\partial}{\partial t} - H_0\right) G(t,t') \\
&\quad = \hbar\delta(t-t') + \frac{1}{\hbar}\int \mathrm{d}t_1 \left(\mathrm{i}\hbar\frac{\partial}{\partial t} - H_0\right) G_0(t,t_1)V(t_1)G(t_1,t') \\
&\quad = \hbar\delta(t-t') + V(t)G(t,t') \quad .
\end{aligned} \tag{9.37}$$

One thus sees that $G(t,t')$ indeed satisfies (9.14) in relation to the complete Hamiltonian $H(t)$.

The integral equation (9.36) has an important merit: *it can be solved iteratively*. By introducing $G(t,t')$ into the right-hand side of (9.36) and repeating this operation, one obtains $G(t,t')$ under the form of a power series of $V(t)$

$$\begin{aligned}
G(t,t') &= G_0(t,t') + \frac{1}{\hbar}\int \mathrm{d}t_1\, G_0(t,t_1)V(t_1)G_0(t_1,t') + \ldots \\
&\quad + \frac{1}{\hbar^n}\int \mathrm{d}t_n \ldots \int \mathrm{d}t_1\, G_0(t,t_n)V(t_n)G_0(t_n,t_{n-1})V(t_{n-1})\ldots \\
&\quad \ldots G_0(t_2,t_1)V(t_1)G_0(t_1,t') + \ldots \quad .
\end{aligned} \tag{9.38}$$

This series represents $G(t,t')$ as an infinite sum of terms containing the *known quantities* $G_0(t,t')$ and $V(t)$ only. The terms can thus, in principle, be analyzed. The series is formally defined in this way, and we have not mentioned anything regarding its eventual convergence. Nevertheless, if the interaction is weak enough, a term-by-term study of the series (in particular regarding the lowest-order terms of the series) provides useful information about the propagator.

We note that, by virtue of the causal nature of $G_0(t,t')$, each term of (9.38) can only be non-zero if $t' \le t_1 \le t_2 \le \cdots \le t_n \le t$. Thus $G(t,t')$ calculated at any finite order in the perturbation remains causal.

Space–Time Representation

Analysis of the perturbation series (9.38) is facilitated by a graphical representation of its terms. This analysis takes on two different forms, according to whether one adopts the space–time or momentum–energy representation of the propagator. First of all, we give the configurational version for the case where $V(t)$ is a potential $V(\boldsymbol{x},t)$ acting by multiplication on the wavefunctions. To simplify the notation, we define the space–time arguments

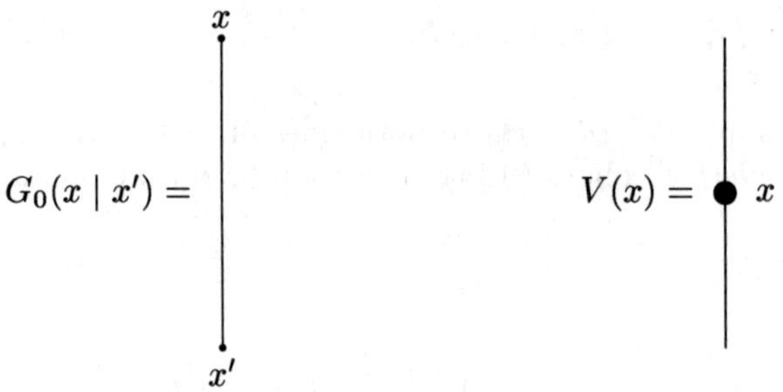

Fig. 9.3. Graphical representation of a term of the perturbation series. Free propagators appear as line segments joining two points x and x' (*left*) and potential factors appear as dots connecting propagator lines

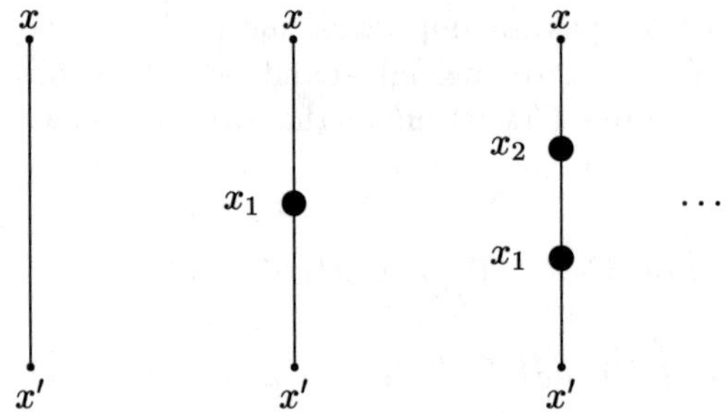

(a) zeroth order (b) 1st order (c) 2nd order

Fig. 9.4. Graphical representation of the first terms of the perturbation series (9.38)

$x = (\boldsymbol{x}, t)$, $x_j = (\boldsymbol{x}_j, t_j)$, with $\mathrm{d}x_j = \mathrm{d}\boldsymbol{x}_j \mathrm{d}t_j, j = 1, \ldots, n$. After introducing the complete system of states $|\boldsymbol{x}\rangle$, the contribution of the nth order of $G(x|x') = \langle \boldsymbol{x}|G(t,t')|\boldsymbol{x}'\rangle$ is, according to (9.38),

$$\frac{1}{\hbar^n} \int \mathrm{d}x_n \ldots \int \mathrm{d}x_1 G_0(x|x_n) V(x_n) G_0(x_n|x_{n-1}) \ldots$$
$$\ldots G_0(x_2|x_1) V(x_1) G_0(x_1|x') \quad . \tag{9.39}$$

We now apply the graphical conventions shown in Fig. 9.3: the free propagator $G_0(x|x')$ is represented by a line segment joining two points x and x', and the potential $V(x)$ is represented by a dot connecting two propagator lines. The nth-order term of the series (9.38) is represented by $n+2$ points $x, x_1, \ldots, x_n, x'$ joined by $n+1$ line segments. The first terms are indicated in Fig. 9.4.

A point x which connects two lines of propagators and which represents the action of the potential $V(x)$ is called a *vertex*. An n-vertex diagram, cons-

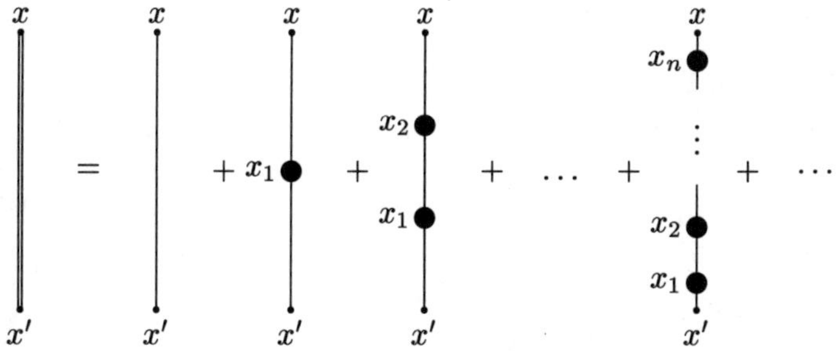

Fig. 9.5. Graphical representation of the infinite series (9.38). The double lines represent the complete propagator in the presence of an external field

tructed by these conventions, is called a *Feynman diagram.* Such a diagram provides a rather suggestive language: one might say that a particle propagates freely from x' to x_1, where it scatters, then propagates freely again to x_2, where it scatters once more, and so on. This visualization is quite useful, but must not in any case lead one to believe that drawings (a), (b), (c), ... represent the actual trajectories of the particle. Feynman diagrams are only graphical representations of algebraic expressions which enter into the perturbation series. On the other hand, one can return to the interpretation of the Green function which we have given in (9.2) and (9.3). In this spirit, expression (9.39) and the corresponding diagram express *the probability amplitude that a particle located at* $\boldsymbol{x}'$ *at time* t' *and found at* $\boldsymbol{x}$ *at time* t *as the sum of the amplitudes of order* $0, 1, \ldots$ *. At zeroth order, one has the amplitude that the particle goes from* $\boldsymbol{x}'$ *to* $\boldsymbol{x}$ *without scattering by the potential. At first order, one has the amplitude that the process takes place through one and only one scattering, and so on.*

The reading of a diagram allows one to reconstruct the term of the perturbation series to which it relates: by associating a propagator to each line, a factor $V(x)$ to each vertex, and by integrating over the n variables $x_1, \ldots, x_n$ attached to the n vertices, one reconstructs (9.39). The infinite series (9.38) can thus be symbolized in a more compact manner by the graphical representation in Fig. 9.5. In the same way, one can graphically represent the integral equation (9.36) by Fig. 9.6.

Momentum–Energy Representation

In these calculations, it is often preferable to write the series (9.38) in the momentum representation. Taking account of (9.24) and the completeness of the plane waves $|\boldsymbol{k}\rangle$, it becomes

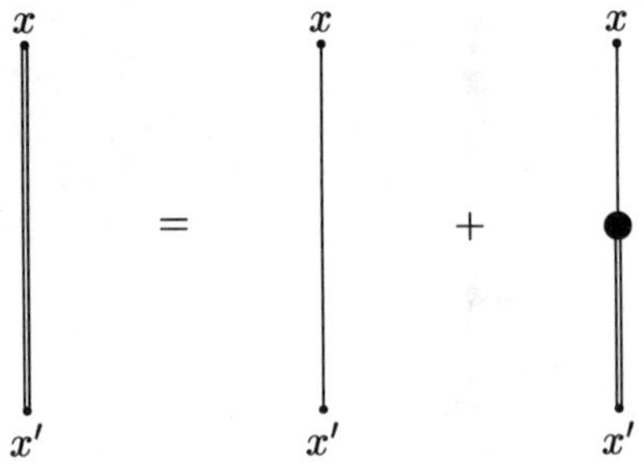

Fig. 9.6. Graphical representation of the integral equation (9.36)

$$\begin{aligned}\langle \boldsymbol{k}|G(t,t')|\boldsymbol{k}'\rangle &= \delta_{\boldsymbol{k},\boldsymbol{k}'}G_0(\boldsymbol{k},t-t')\\ &+\frac{1}{\hbar}\int \mathrm{d}t_1\, G_0(\boldsymbol{k},t-t_1)\,\langle \boldsymbol{k}|V(t_1)|\boldsymbol{k}'\rangle\, G_0(\boldsymbol{k}',t_1-t')+\dots\\ &+\frac{1}{\hbar^n}\sum_{\boldsymbol{k}_1,\dots,\boldsymbol{k}_{n-1}}\int \mathrm{d}t_n\dots\int \mathrm{d}t_1\\ &G_0(\boldsymbol{k},t-t_n)\,\langle \boldsymbol{k}|V(t_n)|\boldsymbol{k}_{n-1}\rangle\, G_0(\boldsymbol{k}_{n-1},t_n-t_{n-1})\dots\\ &\dots G_0(\boldsymbol{k}_1,t_2-t_1)\,\langle \boldsymbol{k}_1|V(t_1)|\boldsymbol{k}'\rangle\, G_0(\boldsymbol{k}',t_1-t')+\dots \quad . \end{aligned} \tag{9.40}$$

The interpretation in terms of probability amplitudes is exactly the same as in the case of the configuration representation. Relation (9.40) shows that the amplitude $\langle \boldsymbol{k}|G(t,t')|\boldsymbol{k}'\rangle$ is a sum of amplitudes, and that the particle undergoes n transitions by the mediation of elements $\langle \boldsymbol{k}_r|V(t_r)|\boldsymbol{k}_{r-1}\rangle\,, r = 1,2,\dots,n$.

The series takes on a more symmetric form if one also introduces the temporal Fourier transform

$$V(k,k') = \int \mathrm{d}t\, \mathrm{e}^{\mathrm{i}(\omega-\omega')t}\,\langle \boldsymbol{k}|V(t)|\boldsymbol{k}'\rangle \tag{9.41}$$

and

$$G(k|k') = \int \mathrm{d}t \int \mathrm{d}t'\, \mathrm{e}^{\mathrm{i}\omega t}\mathrm{e}^{-\mathrm{i}\omega' t'}\,\langle \boldsymbol{k}|G(t,t')|\boldsymbol{k}'\rangle \quad , \tag{9.42}$$

where we have abbreviated the notation by setting $k = (\boldsymbol{k},\omega)$. After inserting these definitions into (9.40), one obtains

$$\begin{aligned} G(k|k') &= 2\pi\delta_{\boldsymbol{k},\boldsymbol{k}'}\delta(\omega-\omega')G_0(k) + \frac{1}{\hbar}G_0(k)V(k,k')G_0(k')+\dots\\ &\quad+\frac{1}{(2\pi)^{n-1}\hbar^n}\sum_{\boldsymbol{k}_1,\dots,\boldsymbol{k}_{n-1}}\int \mathrm{d}\omega_{n-1}\dots\int \mathrm{d}\omega_1\\ &\quad G_0(k)V(k,k_{n-1})G_0(k_{n-1})\dots\\ &\quad\dots G_0(k_1)V(k_1,k')G_0(k')+\dots \quad . \end{aligned} \tag{9.43}$$

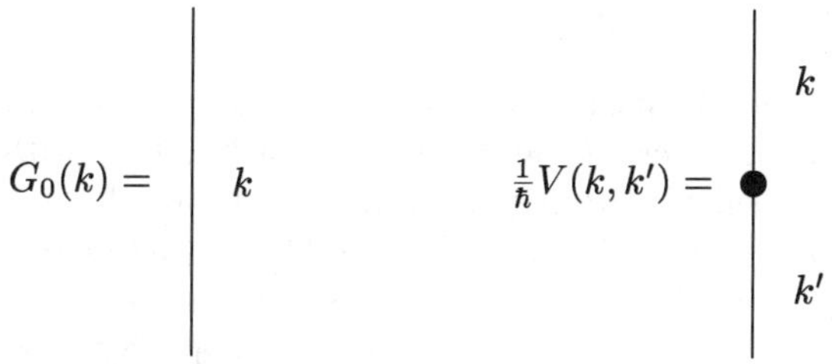

Fig. 9.7. Rules for the graphical representation of the series (9.43). The line segment indexed with k represents the free propagator $G_0(k)$ and the dot between two line segments indexed by k and k' represents the potential $(1/\hbar)V(k,k')$

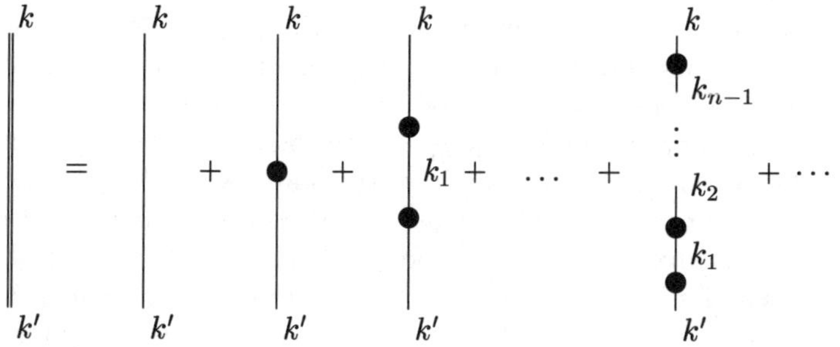

Fig. 9.8. Graphical representation of the series (9.43). The double line segment on the left-hand side represents the complete propagator $G(k|k')$

The free Green function $G_0(k) = G_0(\boldsymbol{k},\omega)$ is given by (9.25), the imaginary part $\eta > 0$ being implied (see the remarks at the end of Sect. 9.2.1). Because of the special role of time in non-relativistic quantum physics, the interpretation of $G(k|k')$ is less immediately obvious than for $\langle \boldsymbol{x}|G(t,t')|\boldsymbol{x}'\rangle$ and $\langle \boldsymbol{k}|G(t,t')|\boldsymbol{k}'\rangle$. On the other hand, in the relativistic formulation, the momentum–energy representation $(\boldsymbol{k},\omega)$ is completely natural because of the larger symmetry between space and time.

The series (9.43) permits the following graphical representation: the free propagator $G_0(k)$ is represented by a straight line segment indexed by k and the potential $(1/\hbar)V(k,k')$ is represented by a dot located between two line segments labelled k and k', as indicated in Fig. 9.7. If one represents the complete propagator $G(k|k')$ by a double line, (9.43) takes on the form of Fig. 9.8. Lines joining two dots are called interior lines. One must sum over the $n-1$ variables, $k_r, r = 1, \ldots, n-1$ attached to the interior lines each with a factor $1/2\pi$ (in the limit of infinite volume, each sum over $\boldsymbol{k}$ is replaced by the integral with the factor $1/(2\pi)^3$). The factor $\delta_{\boldsymbol{k},\boldsymbol{k}'}\delta(\omega-\omega')$ which appears in the first term is the manifestation of the invariance of the free propagator under space–time translations.

9.2.4 Simplified Example: The Cooper Pair

As we have seen in Chap. 5, the formation of Cooper pairs is a quantum statistics effect where n fermions exert an attraction on each other, with $n \gg 1$. One can nevertheless consider a two-particle Hamiltonian which gives rise to an analogous effect: this is what Cooper did in 1956. In this model, the statistics were artificially maintained, so that the result obtained paved the way for the BCS theory. As an illustration, we present Cooper's model using the notions introduced in this chapter.

A Two Particle Model

Consider two identical interacting particles. One can imagine that they have opposite spins but, for the sake of simplicity, we will ignore the spin indices in the following discussion. In the theory of superconductivity, we have seen that the matrix elements of the two-body interaction, essential for the appearance of electron pairing are $V_{\boldsymbol{k}',-\boldsymbol{k}';\boldsymbol{k},-\boldsymbol{k}}$ (see (5.101)). These matrix elements govern the transition of one pair from the state $(\boldsymbol{k}\uparrow, -\boldsymbol{k}\downarrow)$ to the state $(\boldsymbol{k}'\uparrow, -\boldsymbol{k}'\downarrow)$. Moving to the center mass reference frame allows us to use relative coordinates which leaves only a one body problem to be solved. In this reference frame, for which the momenta of the electrons are equal in magnitude and opposite in direction, $\hbar\boldsymbol{k}$ is also equal to the relative momentum. As for the relative kinetic energy, it is equal to $2\varepsilon_{\boldsymbol{k}} = \hbar^2|\boldsymbol{k}|^2/m$, since the reduced mass is $m/2$. We can thus model the interaction as

$$V(k,k') = \begin{cases} -\dfrac{\gamma}{L^3} 2\pi\delta(\omega-\omega') & , \quad \boldsymbol{k},\boldsymbol{k}' \text{ in } D \\ 0 & , \quad \text{in all other cases} \end{cases} , \qquad (9.44)$$

where γ is a positive constant corresponding to an attraction and the factor $\delta(\omega-\omega')$ ensures that the interaction is independent of time. The domain D of the space $\boldsymbol{k}$ is made up of the collection of wave vectors $\boldsymbol{k}$ for which $0 < \varepsilon_{\mathrm{F}} \leq \varepsilon_{\boldsymbol{k}} = \hbar^2|\boldsymbol{k}|^2/2m \leq \varepsilon_{\mathrm{F}} + \delta\varepsilon_{\mathrm{F}}$. The two parameters ε_{F} and $\delta\varepsilon_{\mathrm{F}}, \delta\varepsilon_{\mathrm{F}} \ll \varepsilon_{\mathrm{F}}$ allow one to establish a link between this problem and the n-fermion problem of the BCS theory.

Imagine that ε_{F} is the Fermi energy (as the notation already suggests) and *the states* $|\boldsymbol{k}| \leq k_{\mathrm{F}}$ *of the Fermi sphere remain constantly occupied by* $n-2$ *electrons.* In this case, the two particles in question here will not be able to reach states of energy $\varepsilon_{\boldsymbol{k}} \leq \varepsilon_{\mathrm{F}}$: this effect of Fermi statistics is taken into account in the model through the definition of the domain D in which our two particles can only interact at energies $\varepsilon_{\boldsymbol{k}}$ above ε_{F}. Without an interaction, the minimum energy of the pair would thus be $2\varepsilon_{\mathrm{F}}$. We note nonetheless that the analogy with the many-body problem remains *artificial.* We are not treating the n particles here on the same level, since we are considering neither the interaction of the two particles with the $n-2$ others nor the interactions of the $n-2$ particles among themselves, the latter being moreover assumed to remain "frozen" in the Fermi sphere. Concerning the outer boundary $\varepsilon_F + \delta\varepsilon_F$ of the accessible energy, it does not play a fundamental role: this has been specified in analogy with the BCS theory, where the interaction takes place in a band of energy of width $\hbar\omega_D$, in which the matrix elements (9.44) are supposed to be independent of $\boldsymbol{k}$ and $\boldsymbol{k}'$ (see (5.109)).

Sum of the Perturbation Series

Introducing (9.44) into (9.43), the term of order n of the series takes the form (in the limit $L \to \infty$)

$$2\pi\delta(\omega-\omega')\left(-\frac{\gamma}{\hbar}\right)^n G_0(\boldsymbol{k},\omega)\left[\frac{1}{(2\pi)^3}\int_D \mathrm{d}\boldsymbol{k}_1\, G_0(\boldsymbol{k}_1,\omega)\right]^{n-1} G_0(\boldsymbol{k}',\omega) \quad . \tag{9.45}$$

In (9.45), all of the wavenumbers are limited to the domain D. The factor $\delta(\omega-\omega')$ expresses conservation of energy, since the perturbation does not depend on time. We thus define

$$G(k|k') = 2\pi\delta(\omega-\omega')G(\boldsymbol{k},\boldsymbol{k}',\omega) \tag{9.46}$$

and we see that $G(\boldsymbol{k},\boldsymbol{k}',\omega)$ is the *sum of an infinite geometric series*

$$\begin{aligned} G(\boldsymbol{k},\boldsymbol{k}',\omega) &= (2\pi)^3\delta(\boldsymbol{k}-\boldsymbol{k}')G_0(\boldsymbol{k},\omega) \\ &\quad -\frac{\gamma}{\hbar}G_0(\boldsymbol{k},\omega)\left[\sum_{n=0}^{\infty}\left(-\frac{\gamma}{(2\pi)^3\hbar}\int_D \mathrm{d}\boldsymbol{k}_1\, G_0(\boldsymbol{k}_1,\omega)\right)^n\right]G_0(\boldsymbol{k}',\omega) \\ &= G_0(\boldsymbol{k},\omega)\Bigg[(2\pi)^3\delta(\boldsymbol{k}-\boldsymbol{k}') \\ &\quad -\frac{\gamma}{\hbar}G_0(\boldsymbol{k}',\omega)\left(1+\frac{\gamma}{(2\pi)^3\hbar}\int_D \mathrm{d}\boldsymbol{k}_1\, G_0(\boldsymbol{k}_1,\omega)\right)^{-1}\Bigg] \quad . \end{aligned} \tag{9.47}$$

Inspection of (9.47) shows that $G(\boldsymbol{k},\boldsymbol{k}',\omega)$ can acquire poles in the plane of the complex variable ω, poles which do not exist in the expression of $G_0(\boldsymbol{k},\omega)$. The condition for this phenomenon to take place is given by

$$1+\frac{\gamma}{(2\pi)^3\hbar}\int_D \mathrm{d}\boldsymbol{k}_1\, G_0(\boldsymbol{k}_1,\omega) = 0 \tag{9.48}$$

for one or more values of ω. As $G_0(\boldsymbol{k},\omega)$ only depends on $\boldsymbol{k}$ through the kinetic energy $\varepsilon_{\boldsymbol{k}}$, one can replace the integration over $\boldsymbol{k}$ by an integration over $\varepsilon_{\boldsymbol{k}}$ and, taking into account (9.25) (recalling that the relative kinetic energy is $2\varepsilon_{\boldsymbol{k}}$), one obtains

$$\int_D \mathrm{d}\boldsymbol{k}_1\, G_0(\boldsymbol{k}_1,\omega) = \int_{\varepsilon_\mathrm{F}}^{\varepsilon_\mathrm{F}+\delta\varepsilon_\mathrm{F}} \mathrm{d}\varepsilon\, g(\varepsilon)\frac{\hbar}{\hbar(\omega+\mathrm{i}\eta)-2\varepsilon}$$
$$\simeq g(\varepsilon_\mathrm{F})\int_{\varepsilon_\mathrm{F}}^{\varepsilon_\mathrm{F}+\delta\varepsilon_\mathrm{F}} \mathrm{d}\varepsilon\,\frac{\hbar}{\hbar(\omega+\mathrm{i}\eta)-2\varepsilon}$$
$$= g(\varepsilon_\mathrm{F})\frac{\hbar}{2}\ln\left(\frac{\hbar(\omega+\mathrm{i}\eta)-2\varepsilon_\mathrm{F}}{\hbar(\omega+\mathrm{i}\eta)-2(\varepsilon_\mathrm{F}+\delta\varepsilon_\mathrm{F})}\right) \quad , \tag{9.49}$$

where $g(\varepsilon)$ is the density of states needed to pass from an integration over $\boldsymbol{k}$ to an integration over the energy (see (5.113)). After introducing (9.49) into (9.48), one notes that (9.48) possesses a single solution ω_0, which is real

$$\hbar\omega_0 \simeq 2\varepsilon_\mathrm{F} - 2\delta\varepsilon_\mathrm{F}\left[\exp\left(\frac{16\pi^3}{\gamma g(\varepsilon_\mathrm{F})}\right)-1\right]^{-1} \quad . \tag{9.50}$$

Taking into account the interpretation of the Green function poles (see the remarks which follow (9.20)), *the two-particle system thus possesses a bound state of energy less than* $2\varepsilon_\mathrm{F}$.

Remarks

Relation (9.48) is a necessary condition for a simple pole to exist, and one can show that it doesn't relate to any other type of singularity. In (9.47), a sum over a finite number of n values does not give rise to a singularity. This signifies that the analytical nature of the sum (9.47) changes when one takes into account an infinite number of terms.

It is a non-trivial fact that there always exists a bound state, however small the interaction may be (in (9.50), $\hbar\omega_0 < 2\varepsilon_\mathrm{F}$ *for all positive values of* γ*).* It is interesting to note that in this rudimentary model *this fact is a result of Fermi statistics which prevents the two particles from reaching energies below* ε_F $(\varepsilon_\mathrm{F} > 0)$. Suppose we let $\varepsilon_\mathrm{F} = 0$, which results in the suppression of the effect of the exclusion principle or, if one prefers, suppose that the two particles under consideration are isolated and not members of a system of $n \gg 1$ particles. In this case, $g(\varepsilon)$ behaves as $\sqrt{\varepsilon}$ when $\varepsilon \to 0$ and in place of (9.49), one has the integral

$$f(\omega+\mathrm{i}\eta) = A\int_0^\delta \mathrm{d}\varepsilon\,\sqrt{\varepsilon}\frac{\hbar}{\hbar(\omega+\mathrm{i}\eta)-2\varepsilon} \quad , \tag{9.51}$$

where A is a positive constant. On the right-hand side, the integral converges even for $\omega+\mathrm{i}\eta = 0$; this implies, if γ is small enough, that $\gamma|f(0)| < 1$. A bound state would be characterized by a negative value for ω_0. However, we also have $\gamma|f(\omega_0)| < \gamma|f(0)| < 1$, $\omega_0 < 0$, which shows that (9.48) could never be satisfied. Therefore, no bound state can exist when the particles are isolated and their attraction is too weak.

9.3 Perturbative Expansion of the Scattering Operator

9.3.1 Time-Dependent Perturbation Theory

In order to extend the concepts which we have introduced in the previous sections for a single particle to the case of many-body systems and to field theory, it is necessary to present the formal series of the evolution operator in powers of the interaction in a general framework. The considerations which allow us to do this are completely general and apply just as well to the Green functions of the many-body problem (Chap. 10) as to the perturbative calculation of the scattering operator.

The Interaction Representation

The system is described by a Hamiltonian $\mathtt{H} = \mathtt{H}_0 + \mathtt{H}_\mathrm{I}$, where $\mathtt{H}_0$ is the non-perturbed Hamiltonian and $\mathtt{H}_\mathrm{I}$ is the interaction Hamiltonian (which can depend on time). $\mathtt{U}_0(t,t')$ and $\mathtt{U}(t,t')$ represent the corresponding evolution operators, which give the free evolution and the total evolution from an initial time t' until a time t. They are solutions to the evolution equations

$$\mathrm{i}\hbar\frac{\mathrm{d}}{\mathrm{d}t}\mathtt{U}_0(t,t') = \mathtt{H}_0\mathtt{U}_0(t,t') \quad , \tag{9.52}$$

$$\mathrm{i}\hbar\frac{\mathrm{d}}{\mathrm{d}t}\mathtt{U}(t,t') = \mathtt{H}\mathtt{U}(t,t') \quad , \tag{9.53}$$

with the initial conditions

$$\mathtt{U}_0(t,t')|_{t=t'} = \mathtt{U}(t,t')|_{t=t'} = \mathtt{I} \quad . \tag{9.54}$$

As $\mathtt{H}_0$ is independent of time, the operators $\mathtt{U}_0(t,t')$ form a group

$$\mathtt{U}_0(t,t') = \mathtt{U}_0(t-t') = \exp\left(-\frac{\mathrm{i}\mathtt{H}_0(t-t')}{\hbar}\right) \quad . \tag{9.55}$$

The same applies for $\mathtt{U}(t,t')$ if the interaction is independent of time.

To obtain a power series of the interaction, it is convenient to introduce the evolution operator in the *interaction representation* defined by

$$\mathtt{U}_\mathrm{I}(t,t') = \mathtt{U}_0^*(t)\mathtt{U}(t,t')\mathtt{U}_0(t') \tag{9.56}$$

and which plays an important role in the following discussion. It satisfies the relations

$$\mathtt{U}_\mathrm{I}^*(t,t') = \mathtt{U}_\mathrm{I}(t',t) \quad , \quad \mathtt{U}_\mathrm{I}(t,t'')\mathtt{U}_\mathrm{I}(t'',t') = \mathtt{U}_\mathrm{I}(t,t') \quad ,$$
$$\mathtt{U}_\mathrm{I}(t,t')|_{t=t'} = \mathtt{I} \quad . \tag{9.57}$$

The equation of motion for $\mathtt{U}_\mathrm{I}$ is obtained from (9.52) and (9.53)

$$\begin{aligned} \mathrm{i}\hbar\frac{\mathrm{d}}{\mathrm{d}t}\mathtt{U}_\mathrm{I}(t,t') &= -\mathtt{H}_0\mathtt{U}_0^*(t)\mathtt{U}(t,t')\mathtt{U}_0(t') + \mathtt{U}_0^*(t)\mathtt{H}\mathtt{U}(t,t')\mathtt{U}_0(t') \\ &= \mathtt{U}_0^*(t)\mathtt{H}_\mathrm{I}\mathtt{U}(t,t')\mathtt{U}_0(t') \quad . \end{aligned} \tag{9.58}$$

This can again be rewritten using the definition (9.56)

$$\mathrm{i}\hbar\frac{\mathrm{d}}{\mathrm{d}t}\mathtt{U}_\mathrm{I}(t,t') = \mathtt{H}_\mathrm{I}(t)\mathtt{U}_\mathrm{I}(t,t') \quad , \tag{9.59}$$

where we have set

$$\mathtt{H}_\mathrm{I}(t) = \mathtt{U}_0^*(t)\mathtt{H}_\mathrm{I}\mathtt{U}_0(t) \quad . \tag{9.60}$$

We note that $\mathtt{H}_\mathrm{I}(t)$ can depend on time in two ways: through $\mathtt{U}_0(t)$ and because of the fact that $\mathtt{H}_\mathrm{I}$ can itself depend on time. By integrating both sides of (9.59) from t' to t, one obtains an integral equation equivalent to (9.59)

$$\mathtt{U}_\mathrm{I}(t,t') = \mathtt{I} - \frac{\mathrm{i}}{\hbar}\int_{t'}^{t} \mathrm{d}t_1\, \mathtt{H}_\mathrm{I}(t_1)\mathtt{U}_\mathrm{I}(t_1,t') \quad . \tag{9.61}$$

The Dyson Series

Equation (9.61) can be formally solved by iteration

$$\mathtt{U}_\mathrm{I}(t,t') = \mathtt{I} + \sum_{n=1}^{\infty} \mathtt{U}_\mathrm{I}^{(n)}(t,t') \quad , \tag{9.62}$$

with

$$\mathtt{U}_\mathrm{I}^{(n)}(t,t') = \left(-\frac{\mathrm{i}}{\hbar}\right)^n \int_{t'}^{t}\mathrm{d}t_n \int_{t'}^{t_n}\mathrm{d}t_{n-1}\ldots\int_{t'}^{t_2}\mathrm{d}t_1\, \mathtt{H}_\mathrm{I}(t_n)\ldots\mathtt{H}_\mathrm{I}(t_2)\mathtt{H}_\mathrm{I}(t_1) \quad . \tag{9.63}$$

The advantage of using the interaction representation (9.56) is to subtract out $\mathtt{H}_0$ in (9.58) so that one obtains for $\mathtt{U}_\mathrm{I}(t,t')$ a power series of the interaction. The series defined by (9.62) and (9.63) is called the *Dyson series*. Its analysis, which is at the heart of the perturbative calculation, is complicated because of the fact that the $\mathtt{H}_\mathrm{I}(t)$ operators taken at different times do not commute among themselves. Also, the time ordering $t' \le t_1 \le t_2 \le \ldots \le t_{n-1} \le t_n \le t$ in (9.63) must be preserved.

Before pursuing this analysis, we will show that in the particular case of one particle under the influence of a potential, the perturbation series indeed reduces to that of the Green function obtained in Sect. 9.2.3. Indeed, after multiplying (9.61) on the left by $U_0(t)$ and on the right by $U_0^*(t')$, and with the help of definitions (9.56) and (9.60), we find (with $V = \mathtt{H}_\mathrm{I}$)

$$U(t,t') = U_0(t,t') - \frac{\mathrm{i}}{\hbar}\int_{t'}^{t} \mathrm{d}t_1\, U_0(t,t_1) V U(t_1,t') \quad . \tag{9.64}$$

Note that, for $t \geq t'$, (9.64) gives rise to (9.36), which links the perturbed and free Green functions when the latter are written in the operator form (9.13).

Time Ordering

The study of the multiple integrals appearing in (9.63) is made easier by introducing the *time ordering operation* T. It is useful to rewrite (9.63) by extending all of the integrals until time t

$$\mathtt{U}_{\mathrm{I}}^{(n)}(t,t') = \left(-\frac{\mathrm{i}}{\hbar}\right)^n \int_{t'}^{t} \mathrm{d}t_n \ldots \int_{t'}^{t} \mathrm{d}t_1\, \theta(t_n - t_{n-1}) \ldots$$
$$\ldots \theta(t_3 - t_2)\theta(t_2 - t_1)\mathtt{H}_{\mathrm{I}}(t_n) \ldots \mathtt{H}_{\mathrm{I}}(t_2)\mathtt{H}_{\mathrm{I}}(t_1) \quad . \tag{9.65}$$

Introducing the product of the $\theta(t)$ functions

$$\theta(t) = \begin{cases} 1 & , \quad t > 0 \\ 0 & , \quad t < 0 \end{cases}$$

ensures that the integrals are only carried out over the sector $t' \leq t_1 \leq t_2 \leq \ldots \leq t_n \leq t$. One can rewrite (9.65) in $n!$ different ways by considering the $n!$ permutations of the dummy integration variables. Thus

$$\mathtt{U}_{\mathrm{I}}^{(n)}(t,t') = \left(-\frac{\mathrm{i}}{\hbar}\right)^n \frac{1}{n!} \int_{t'}^{t} \mathrm{d}t_1 \ldots \int_{t'}^{t} \mathrm{d}t_n\, T\left(\mathtt{H}_{\mathrm{I}}(t_1) \ldots \mathtt{H}_{\mathrm{I}}(t_n)\right) \quad . \tag{9.66}$$

We have defined

$$T\left(\mathtt{H}_{\mathrm{I}}(t_1) \ldots \mathtt{H}_{\mathrm{I}}(t_n)\right)$$
$$= \sum_{\pi} \theta(t_{\pi(1)} - t_{\pi(2)}) \ldots \theta(t_{\pi(n-1)} - t_{\pi(n)}) \mathtt{H}_{\mathrm{I}}(t_{\pi(1)}) \ldots \mathtt{H}_{\mathrm{I}}(t_{\pi(n)}) \quad , \tag{9.67}$$

that is, the sum of the integrand of (9.65) over all permutations π of the variables $t_1, \ldots, t_n$.

We call T the time order operation because T rearranges a product of $\mathtt{H}_{\mathrm{I}}(t_j)$ into the same order as the time order of the arguments $t_j, j = 1, \ldots, n$. To see this, take the case $n = 2$. The definition (9.67) gives

$$T\left(\mathtt{H}_{\mathrm{I}}(t_1)\mathtt{H}_{\mathrm{I}}(t_2)\right) = \theta(t_1 - t_2)\mathtt{H}_{\mathrm{I}}(t_1)\mathtt{H}_{\mathrm{I}}(t_2) + \theta(t_2 - t_1)\mathtt{H}_{\mathrm{I}}(t_2)\mathtt{H}_{\mathrm{I}}(t_1)$$
$$= \begin{cases} \mathtt{H}_{\mathrm{I}}(t_1)\mathtt{H}_{\mathrm{I}}(t_2) & , \quad t_1 > t_2 \\ \mathtt{H}_{\mathrm{I}}(t_2)\mathtt{H}_{\mathrm{I}}(t_1) & , \quad t_2 > t_1 \end{cases} \quad , \tag{9.68}$$

and in general

$$T\left(\mathtt{H}_{\mathrm{I}}(t_1)\ldots\mathtt{H}_{\mathrm{I}}(t_n)\right) \\ = \mathtt{H}_{\mathrm{I}}(t_{j_1})\mathtt{H}_{\mathrm{I}}(t_{j_2})\ldots\mathtt{H}_{\mathrm{I}}(t_{j_n}) \quad , \quad t_{j_1} > t_{j_2} > \ldots > t_{j_n} \quad . \tag{9.69}$$

When $t_1 = t_2$, the order is unimportant and we define the action T by the identity

$$T\left(\mathtt{H}_{\mathrm{I}}(t)\mathtt{H}_{\mathrm{I}}(t)\right) = \mathtt{H}_{\mathrm{I}}(t)\mathtt{H}_{\mathrm{I}}(t) \quad . \tag{9.70}$$

After introducing the time order T, the terms (9.66) of the series (9.62) formally add up as an exponential, which justifies the abbreviated notation

$$\mathtt{U}_{\mathrm{I}}(t,t') = T\left[\exp\left(-\frac{\mathrm{i}}{\hbar}\int_{t'}^{t} \mathrm{d}t_1\, \mathtt{H}_{\mathrm{I}}(t_1)\right)\right] \quad . \tag{9.71}$$

Although quite suggestive, (9.71) is only a form of notation and it is necessary to return to the actual terms of the series to perform a calculation.

9.3.2 The Scattering Operator

During a scattering experiment, we can distinguish three periods in the evolution of the system. In the first period, before the collision, the particles are some distance from each other and propagate without interaction. Their motion is identical to free particles, characterized by their momentum, their spin and their mass, which are the result of preparing of the incident beams (collimator, spectrometer, polarizer). In the second period, the instant where the collision occurs, the particles undergo mutual interactions, responsible for the scattering process. Finally, in the third period, the particles produced by the collision separate and propagate freely once more: their state is observed with the help of the appropriate detectors.

This description is justified by the fact that *the time for which the particles are interacting is extremely short.* Even for a slow projectile, scattered by a sizeable target (thermal neutron scattered by a large molecule), the collision time is no longer than $10^{-10}s$. Outside of an interval $\Delta t \simeq 10^{-10}s$, the motion is experimentally indistinguishable from free motion.

The Incoming State

To formalize the description, we write as usual the Hamiltonian for the kinetic energy of the particles, $\mathtt{H}_0$, and the total Hamiltonian, $\mathtt{H} = \mathtt{H}_0 + \mathtt{H}_{\mathrm{I}}$, including their interactions; $\mathtt{U}_0(t) = \exp(-\mathrm{i}\mathtt{H}_0 t/\hbar)$, $\mathtt{U}(t) = \exp(-\mathrm{i}\mathtt{H}t/\hbar)$ are the corresponding evolution operators. We choose time origin $t = 0$ to be the moment of the collision. At this instant, the state $|\Psi\rangle$ of the system is that of the particles correlated by their interactions. For t' before the collision,

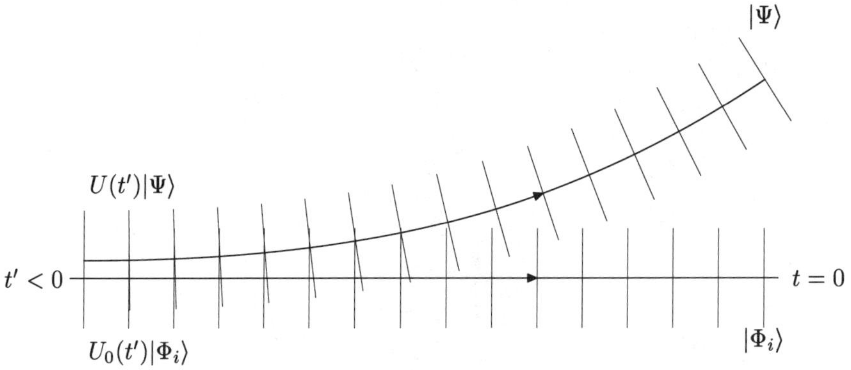

Fig. 9.9. Before the collision, $t' < 0$, the motion of the particles $\mathtt{U}(t')\,|\Psi\rangle$ is asymptotic to the free evolution $\mathtt{U}_0(t')\,|\Phi_i\rangle$ which represents the motion of the particles of the incident beams before the collision

$t' < 0$, the motion of the particles $\mathtt{U}(t')\,|\Psi\rangle$ is asymptotically described by the free evolution $\mathtt{U}_0(t')\,|\Phi_i\rangle$ which represents the motion of the particles of the incident beams before the collision. *The incoming state $|\Phi_i\rangle$ is specified by the preparation of the system before the collision*, by labelling the particles with their momentum, spin and polarization. One thus has

$$\mathtt{U}(t')\,|\Psi\rangle \simeq \mathtt{U}_0(t')\,|\Phi_i\rangle \quad , \quad t' \ll 0 \tag{9.72}$$

or, using the unitarity relation $\mathtt{U}^*(t') = \mathtt{U}^{-1}(t') = \mathtt{U}(-t')$,

$$|\Psi\rangle \simeq \mathtt{U}^*(t')\mathtt{U}_0(t')\,|\Phi_i\rangle \quad , \quad t' \ll 0 \quad . \tag{9.73}$$

The situation is illustrated in Fig. 9.9.

The Outgoing State

After the collision, the particles separate and $\mathtt{U}(t)\,|\Psi\rangle\,, t > 0$ behaves once more as a certain state of free motion $\mathtt{U}_0(t)\,|\Phi\rangle$ which describes the asymptotic evolution of the outgoing particles resulting from the interactions during the collision. We can characterize the behaviour of $\mathtt{U}(t)\,|\Psi\rangle$ for large times by

$$\mathtt{U}(t)\,|\Psi\rangle \simeq \mathtt{U}_0(t)\,|\Phi\rangle \quad , \quad t \gg 0 \tag{9.74}$$

or

$$|\Phi\rangle \simeq \mathtt{U}_0^*(t)\mathtt{U}(t)\,|\Psi\rangle \quad , \quad t \gg 0 \quad . \tag{9.75}$$

In general, the outgoing state $|\Phi\rangle$ is made up of a superposition of numerous states, different from the state $|\Phi_i\rangle$. If $|\Phi_i\rangle$ is a narrow wavepacket centered

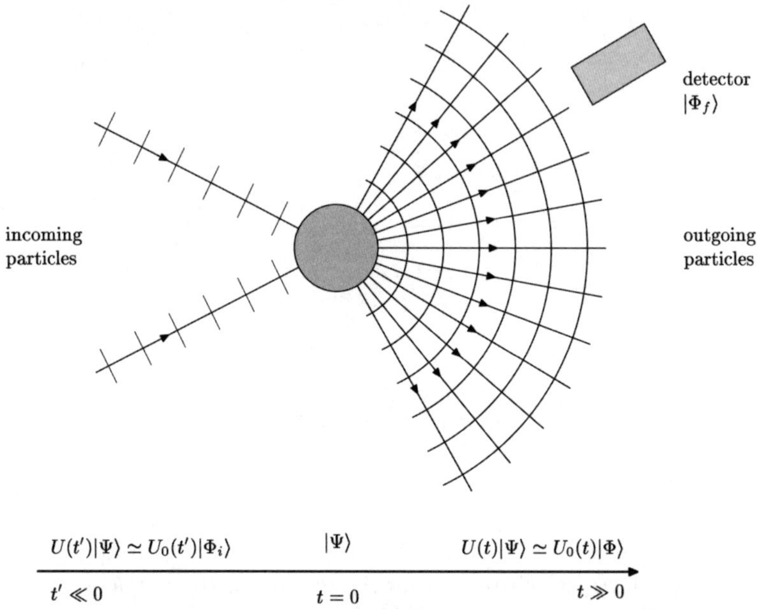

Fig. 9.10. Schematic representation of the complete scattering process

on the values of momentum of the incoming particles, the outgoing state $|\Phi\rangle$ would be a broad wave packet: it can describe particle states that were present before the collision (elastic scattering), or else a variety of other states where certain incoming particles have disappeared and new particles have been produced (inelastic scattering). The complete scattering process is represented schematically in Fig. 9.10.

The Scattering Operator S

We are interested in calculating the probability of finding the outgoing particles in a state $|\Phi_f\rangle$ determined by a detector which chooses from the superposition $|\Phi\rangle$ certain types of particles: $|\Phi_f\rangle$ is specified by labelling the particles that have been detected, with their momenta, spin and polarization. The desired probability amplitude is thus

$$\begin{aligned} \langle\Phi_f|\Phi\rangle &\simeq \langle\Phi_f|\mathtt{U}_0^*(t)\mathtt{U}(t)|\Psi\rangle \simeq \langle\Phi_f|\mathtt{U}_0^*(t)\mathtt{U}(t-t')\mathtt{U}_0(t')|\Phi_i\rangle \\ &= \langle\Phi_f|\mathtt{U}_\mathrm{I}(t,t')|\Phi_i\rangle\,, \qquad t \gg 0, \quad t' \ll 0 \quad . \end{aligned} \tag{9.76}$$

One recognizes in (9.76) the evolution operator (9.56) in the interaction representation. In accordance with the remark at the beginning of the section that, outside of the brief duration of the collision, the motion is experimentally indiscernible from free motion, we are motivated to consider the limit

$$\mathtt{S} = \lim_{t\to\infty} \lim_{t'\to-\infty} \mathtt{U}_\mathrm{I}(t,t') \quad , \tag{9.77}$$

which is called the *scattering operator* (or *matrix*). Its matrix elements $\langle \Phi_f|\mathtt{S}|\Phi_i\rangle$ give the probability amplitudes of the various possible scattering processes.

Existence of the limit (9.77) and the exact sense in which one must understand the asymptotic behaviour (9.72) and (9.74) are delicate mathematical problems, which we will not approach here. We will only note that one can rigorously treat them in the framework of scattering by a short-range potential and in certain models of field theory (the Coulomb interaction requires a special treatment because of its long range). Our objective here is limited to constructing the formal perturbative series of $\mathtt{S}$.

From the definition (9.77), one immediately deduces that if $\mathtt{H}_0$ and $\mathtt{H}$ possess certain invariant properties, $\mathtt{S}$ inherits the same properties. For example, if $\mathtt{H}_0$ and $\mathtt{H}$ are invariant under translations, then they commute with the total momentum and as a consequence so does $\mathtt{S}$. This implies that *the total momentum is conserved in the scattering process*: $\langle \Phi_f|\mathtt{S}|\Phi_i\rangle$ is only non-zero if the states $|\Phi_i\rangle$ and $|\Phi_f\rangle$ have the same total momentum. *The kinetic energy is also conserved by the scattering process.* To see this, one needs only note that $\mathtt{S}$ commutes with $\mathtt{H}_0$ or, equivalently, with $\mathtt{U}_0(\tau)$. Indeed, using unitarity and the group law for $\mathtt{U}_0(t)$, one obtains,

$$\begin{aligned}\mathtt{U}_0(\tau)\mathtt{S} &= \lim_{t\to\infty} \lim_{t'\to-\infty} \mathtt{U}_0(-t+\tau)\mathtt{U}(t-t')\mathtt{U}_0(t') \\ &= \lim_{t\to\infty} \lim_{t'\to-\infty} \mathtt{U}_0(-t)\mathtt{U}(t-t'+\tau)\mathtt{U}_0(t') \\ &= \lim_{t\to\infty} \lim_{t'\to-\infty} \mathtt{U}_0(-t)\mathtt{U}(t-t')\mathtt{U}_0(t'+\tau) = \mathtt{S}\mathtt{U}_0(\tau) \quad .\end{aligned} \tag{9.78}$$

To make the comparison with experiment, we must supplement scattering operator theory with the calculation of cross-sections. In the simple case of the scattering of a particle by a potential, the *differential cross-section* $\sigma(\Omega)\mathrm{d}\Omega$ is the number of particles scattered per unit time in an element of solid angle $\mathrm{d}\Omega$ centered around $\Omega = (\theta, \varphi)$ relative to the incident flux (Fig. 9.11). The relation between the $\mathtt{S}$-matrix element and the cross-section depends on the kinematics of the particles (relativistic or non-relativistic) as well as on the conditions under which the particles are prepared and detected. Instead of writing down a general theory, we will treat certain cases as applications (Sect 9.4.2).

Finally, the expansion of $\mathtt{S}$ in powers of the interaction that results from (9.62) and (9.66) is

$$\begin{aligned}\mathtt{S} &= \mathtt{I} + \sum_{n=1}^{\infty} \mathtt{S}^{(n)} \\ \mathtt{S}^{(n)} &= \left(-\frac{\mathrm{i}}{\hbar}\right)^n \frac{1}{n!} \int_{-\infty}^{\infty} \mathrm{d}t_1 \ldots \int_{-\infty}^{\infty} \mathrm{d}t_n \, T\left(\mathtt{H}_\mathrm{I}(t_1)\ldots\mathtt{H}_\mathrm{I}(t_n)\right) \quad .\end{aligned} \tag{9.79}$$

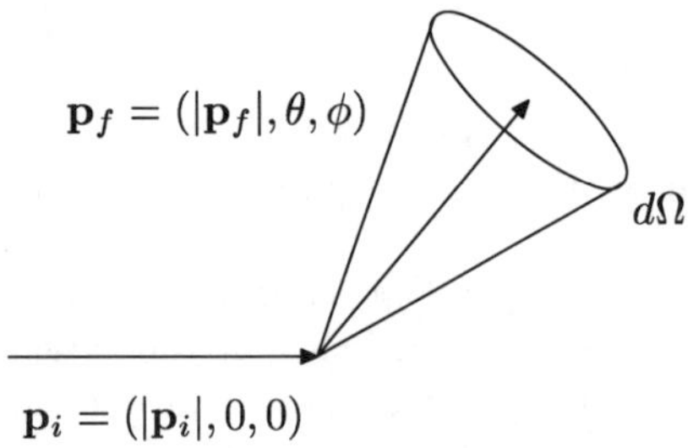

Fig. 9.11. The differential cross-section $\sigma(\Omega)\mathrm{d}\Omega$ is the number of particles scattered per unit time in an element of solid angle $\mathrm{d}\Omega$ centered around $\Omega = (\theta, \varphi)$ relative to the incident flux

This series is the starting point for the perturbative study of scattering.

9.3.3 Fermions and Bosons in Interaction

An Interaction Model

To set out the structure of the perturbative expansion of the scattering operator, it is useful to consider a prototype interaction which is simple, but sufficiently representative of the general structure. The model consists of a non-relativistic spinless Fermi field interacting locally with scalar bosons.

If $a^*_{\boldsymbol{k}}$, $a_{\boldsymbol{k}}$ denote the creation and annihilation operators of the fermions and $b^*_{\boldsymbol{k}}$, $b_{\boldsymbol{k}}$ the same for the bosons, the non-perturbed Hamiltonian is the sum of the kinetic energies

$$\mathrm{H}_0 = \sum_{\boldsymbol{k}} \varepsilon_{\boldsymbol{k}} a^*_{\boldsymbol{k}} a_{\boldsymbol{k}} + \sum_{\boldsymbol{k}} \hbar\omega_{\boldsymbol{k}} b^*_{\boldsymbol{k}} b_{\boldsymbol{k}} \quad . \tag{9.80}$$

The explicit form of the individual kinetic energies $\varepsilon_{\boldsymbol{k}}$ and $\hbar\omega_{\boldsymbol{k}}$ need not be specified here.

Suppose that the boson field interacts locally with the fermion density $a^*(\boldsymbol{x})a(\boldsymbol{x})$ as in (8.114) and (8.253)

$$\mathrm{H}_\mathrm{I} = g \int_\Lambda \mathrm{d}\boldsymbol{x} \mathcal{H}_\mathrm{I}(\boldsymbol{x}) \quad , \tag{9.81}$$

$$\mathcal{H}_\mathrm{I}(\boldsymbol{x}) = a^*(\boldsymbol{x})a(\boldsymbol{x})\Phi(\boldsymbol{x}) \quad . \tag{9.82}$$

For example, in the Hamiltonian density (9.82), $a^*(\boldsymbol{x})$ represents the Fermi field (8.212) or (8.229) and $\Phi(\boldsymbol{x})$ the neutral scalar field (8.95) or the phonon field (8.252), taken at time $t = 0$. When one introduces field expansion in terms of the fields in orthogonal plane waves, H_I is written

$$\mathrm{H_I} = g\left(\frac{\hbar}{L^3}\right)^{1/2} \sum_{\boldsymbol{k},\boldsymbol{l}} \frac{1}{\sqrt{2\omega_l}} a^*_{\boldsymbol{k}} a_{\boldsymbol{k}+\boldsymbol{l}} (b^*_{\boldsymbol{l}} + b^*_{-\boldsymbol{l}}) \quad . \tag{9.83}$$

This interaction describes the emission or absorption of a boson by a fermion, so that the number of fermions is conserved. It corresponds, for instance, to the following physical processes:

- *emission or absorption of a photon by an electron;*
- *emission or absorption of a pion by a nucleon;*
- *emission or absorption of a phonon by an electron.*

The first two processes can occur at zero density: the vacuum $|0\rangle$ would be the non-perturbed ground state. For the phonon case, it is clearly necessary to choose the ground state to be that of the free electron gas (8.222) at density $\rho \neq 0$.

In a realistic situation, one must naturally take into account the spins and polarizations of the particles and the specific details of the coupling, as well as the relativistic nature of the electrons and the existence of the antiparticles. Even if these effects are not incorporated in the interaction (9.83), this model is nevertheless sufficient to illustrate the principal ideas of the perturbative calculations. A complete example, the Compton scattering of pions, will be treated in Sect. 9.4.2 (in this case, the interaction is between two types of bosons, but the general formalism remains the same).

Structure of the nth-Order Term

According to (9.60), the time dependence of $\mathrm{H_I}(t)$ is given by the free Hamiltonian (9.80)

$$\mathrm{H_I}(t) = \exp\left(\frac{\mathrm{i}\mathrm{H_0}t}{\hbar}\right)\mathrm{H_I}\exp\left(-\frac{\mathrm{i}\mathrm{H_0}t}{\hbar}\right) = g\int_\Lambda \mathrm{d}\boldsymbol{x}\,\mathcal{H}_\mathrm{I}(\boldsymbol{x},t) \quad , \tag{9.84}$$

$$\mathcal{H}_\mathrm{I}(\boldsymbol{x},t) = a^*(\boldsymbol{x},t)a(\boldsymbol{x},t)\Phi(\boldsymbol{x},t) \quad , \tag{9.85}$$

where $a^*(\boldsymbol{x},t)$ and $\Phi(\boldsymbol{x},t)$ are the free quantum fields at time t. Introducing (9.84) into (9.79) and using the four-dimensional variables $x_j = (\boldsymbol{x}_j, t_j)$ and $\mathrm{d}x_j = \mathrm{d}\boldsymbol{x}_j \mathrm{d}t_j$, one sees that $\mathrm{S}^{(n)}$ is an integral over the time-ordered product of n Hamiltonian densities

$$\mathrm{S}^{(n)} = \left(-\frac{\mathrm{i}}{\hbar}\right)^n \frac{g^n}{n!}\int \mathrm{d}x_1 \ldots \int \mathrm{d}x_n\, T\left(\mathcal{H}_\mathrm{I}(x_1)\ldots\mathcal{H}_\mathrm{I}(x_n)\right) \quad . \tag{9.86}$$

In (9.86), we have taken the infinite volume limit by extending the spatial integrals over all space. According to (9.85) and (9.86), $\mathrm{S}^{(n)}$ *is expressed from now on explicitly in terms of the free fields*: as their action on the states of

Table 9.2. Eigenstates of the free Hamiltonian $\mathtt{H}_0$ in the occupation number representation

	States of incoming and detected particles	Physical process
(a)	$\lvert\Phi\rangle_i = b^*(\boldsymbol{k}_2)a^*(\boldsymbol{k}_1)\lvert 0\rangle$ $\lvert\Phi\rangle_f = b^*(\boldsymbol{k}_4)a^*(\boldsymbol{k}_3)\lvert 0\rangle$	fermion–boson scattering
(b)	$\lvert\Phi\rangle_i = a^*(\boldsymbol{k}_2)a^*(\boldsymbol{k}_1)\lvert 0\rangle$ $\lvert\Phi\rangle_f = a^*(\boldsymbol{k}_4)a^*(\boldsymbol{k}_3)\lvert 0\rangle$	fermion–fermion scattering
(c)	$\lvert\Phi\rangle_i = b^*(\boldsymbol{k}_2)a^*(\boldsymbol{k}_1)\lvert 0\rangle$ $\lvert\Phi\rangle_f = a^*(\boldsymbol{k}_3)\lvert 0\rangle$	absorption of a boson

Fock space are known, we can calculate the matrix elements $\langle\Phi_f|\mathtt{S}|\Phi_i\rangle$ to nth order according to the usual rules of the variable particle-number formalism.

The nature of the steps that must be taken in order to calculate the matrix elements is clarified by the following remarks. According to the description of the scattering process of Sect. 9.3.2, states $|\Phi_i\rangle$ and $|\Phi_f\rangle$ correspond to specifying the incoming and detected particles with well-defined momenta. These are thus the eigenstates of the non-interacting Hamiltonian, which can be written in the occupation number representation. Examples are given in Table 9.2. The product $\mathcal{H}_{\mathrm{I}}(x_1)\dots\mathcal{H}_{\mathrm{I}}(x_n)$ is a polynomial expression of the creation and annihilation operators of the particles, but which have not been time-ordered. Putting it in normal order gives rise to a sum of terms: in this sum, the only terms which contribute to $\langle\Phi_f|\mathtt{S}^{(n)}|\Phi_i\rangle$ are those which annihilate particles of the state $|\Phi_i\rangle$ and recreate the numbers and types of particles corresponding to the state $|\Phi_f\rangle$. All other contributions to the matrix element vanish due to the orthogonality in Fock space of states with different numbers of particles.

We illustrate this point with $\mathtt{S}^{(2)}$. The integrand introduces the product of the fields

$$a^*(x_1)a(x_1)\Phi(x_1)a^*(x_2)a(x_2)\Phi(x_2) \quad . \tag{9.87}$$

If one recalls that Φ is a linear combination of b^* and b, the structure in creators and annihilators of the term (9.87) is schematically the following (the arguments of the fields are omitted in this discussion)

$$a^*a(b^*+b)a^*a(b^*+b) \quad . \tag{9.88}$$

The time ordering of the product $a^*aba^*ab^*$ occurring in (9.88) yields a sum of contributions: some of these are presented in Table 9.3. In this decomposition, it is clear that only terms of type (Table 9.3b) contribute to the matrix element of $\mathtt{S}^{(2)}$ formed with the states (Table 9.2a) having a boson and a fermion present in the initial and final states; terms (Table 9.3c) contribute

Table 9.3. Contributions to the integrand of $\mathrm{S}^{(2)}$ due to the ordering of the product $a^*aba^*ab^*$

	Term in normal order	Contractions carried out
(a)	$a^*a^*b^*aab$	–
(b)	a^*b^*ab	$\underbracket{a\,a^*}$
(c)	a^*a^*aa	$\underbracket{b\,b^*}$
(d)	a^*a	$\underbracket{a\,a^*}\quad\underbracket{b\,b^*}$
(e)	b^*b	$\underbracket{a\,a^*}\quad\underbracket{a^*a}$
(f)	1	$\underbracket{a\,a^*}\quad\underbracket{a^*a}\quad\underbracket{b\,b^*}$

to those matrix elements (Table 9.2b) with two initial-state and final-state fermions, and so on. *Thus, to a given order n, there is a correspondence between the terms obtained by the normal ordering of $\mathrm{S}^{(n)}$ and the amplitudes $\langle\Phi_f|\mathrm{S}^{(n)}|\Phi_i\rangle$ of various physical processes.*

In order to quantitatively evaluate the matrix elements, one must once again work out the consequences of the Wick theorem. It is nevertheless necessary to generalize the content of the discussion in Sect. 3.2.4 to include the effects of the time ordering T which appears in (9.86).

9.3.4 The Wick Theorem for Time-Ordered Products

The purpose for the Wick theorem is to give the decomposition of a time-ordered product $T(\mathcal{H}_\mathrm{I}(x_1)\dots\mathcal{H}_\mathrm{I}(x_n))$ in normal order. Before formulating the theorem, it is necessary to generalize the definition of T and to introduce the concept of a contraction.

In (9.67), the action of T has only been defined on a product of interaction Hamiltonians. But these interaction Hamiltonians are themselves the products of field operators (see (9.85)). To generalize the Wick theorem of Sect. 3.2.4 to time-ordered products, it is first of all necessary to extend the definition of T to a product of fields. Note that $\mathcal{H}_\mathrm{I}(x)$ is quadratic in fermion operators so that the time ordering $T(\mathcal{H}_\mathrm{I}(x_1)\dots\mathcal{H}_\mathrm{I}(x_n))$ only implies permutations of pairs of fermion operators. As a consequence, the definition (9.67) does not require particular sign conventions due to the anticommutation of fermions.

Time Ordering of a Product of Fields

To abbreviate the notation, we use $\mathtt{A}(t_j)$ to denote any field operator taken at the point $x_j = (\boldsymbol{x}_j, t_j)$ (for example $\mathtt{A}(t_j) = a^*(x_j)$, $a(x_j)$ or $\Phi(x_j)$). Let $\mathtt{A}(t_1)\mathtt{A}(t_2)\ldots\mathtt{A}(t_k)$ be a product of fields for which all the time arguments are distinct, $t_1 \neq t_2 \neq \ldots \neq t_k$. We set

$$T\left(\mathtt{A}(t_1)\mathtt{A}(t_2)\ldots\mathtt{A}(t_k)\right) = (-1)^\pi \mathtt{A}(t_{j_1})\mathtt{A}(t_{j_2})\ldots\mathtt{A}(t_{j_k}) \quad , \qquad t_{j_1} > t_{j_2} > \ldots > t_{j_k} \quad , \tag{9.89}$$

where $(-1)^\pi$ is the parity of the permutation π of the fermion operators when one passes from the order $t_1, \ldots, t_k$ to the time order $t_{j_1}, \ldots, t_{j_k}$. *The operation T thus consists of rearranging the field operators into chronological succession as if all of the fermion operators anti-commute and all the boson operators commute* (definition (9.89) is parallel to that of Sect. 3.2.4 for the normal order of fermions). It follows from definition (9.89) that

$$T\left(\mathtt{A}(t_1)\mathtt{A}(t_2)\right) = -T\left(\mathtt{A}(t_2)\mathtt{A}(t_1)\right) \tag{9.90}$$

if $\mathtt{A}(t_1)$ and $\mathtt{A}(t_2)$ are fermion operators.

In the following, we define the *time-ordered contraction* of $\mathtt{A}(t_i)\mathtt{A}(t_j)$ by

$$\underbrace{\mathtt{A}(t_i)\mathtt{A}}(t_j) = \langle 0 | T\left(\mathtt{A}(t_i)\mathtt{A}(t_j)\right) | 0 \rangle \quad . \tag{9.91}$$

The time-ordered contraction, like the simple contraction (3.56), is a scalar (a multiple of the identity). We use here the same symbol $\sqcup$ as in (3.56). Henceforth, all contraction symbols will be considered to be time-ordered.

The Wick Theorem

The Wick theorem has the same formulation as in the case of the normal products (3.58): *a time-ordered product of k operators $\mathtt{A}(t_1)\mathtt{A}(t_2)\ldots\mathtt{A}(t_k)$, $t_1 \neq t_2 \neq \ldots \neq t_k$, is equal to the sum of all normal products for which all possible time-ordered contractions have been made*

$$\begin{aligned}
&T\left(\mathtt{A}(t_1)\mathtt{A}(t_2)\ldots\mathtt{A}(t_k)\right) \\
&\quad = {:}\mathtt{A}(t_1)\mathtt{A}(t_2)\ldots\mathtt{A}(t_k){:} \\
&\qquad + {:}\underbrace{\mathtt{A}(t_1)\mathtt{A}}(t_2)\ldots\mathtt{A}(t_k){:} + \ldots + {:}\underbrace{\mathtt{A}(t_1)\mathtt{A}(t_2)\ldots\mathtt{A}}(t_k){:} + \ldots \\
&\qquad + {:}\underbrace{\mathtt{A}(t_1)\mathtt{A}}(t_2)\underbrace{\mathtt{A}(t_3)\mathtt{A}}(t_4)\ldots\mathtt{A}(t_k){:} + \ldots \quad .
\end{aligned} \tag{9.92}$$

In fact, according to (9.89), $T(\mathtt{A}(t_1)\mathtt{A}(t_2)\ldots\mathtt{A}(t_k))$ is, up to its sign, equal to an ordinary product which can be ordered as is in (3.58). All that remains to be verified is that (9.92) is indeed correct by using the definition of the time-ordered contraction (9.91).

We will limit ourselves to verifying the theorem for the simplest case

$$T\,(\mathtt{A}(t_1)\mathtt{A}(t_2)) = {:}\mathtt{A}(t_1)\mathtt{A}(t_2){:} + \underbrace{\mathtt{A}(t_1)\mathtt{A}}(t_2) \quad . \tag{9.93}$$

To demonstrate (9.93), it is sufficient to note that the time order differs here from the normal order by a scalar. As in all the cases, $\langle 0|{:}\mathtt{A}(t_1)\mathtt{A}(t_2){:}|0\rangle = 0$, so this scalar will necessarily be $\langle 0|T(\mathtt{A}(t_1)\mathtt{A}(t_2))|0\rangle$, yielding (9.93). If $\mathtt{A}(t_1) = a(x_1)$ and $\mathtt{A}(t_2) = a(x_2)$ are the annihilators of the fermions, one has by virtue of definition (9.89)

$$\begin{aligned} &T\,(a(x_1)a(x_2)) - {:}a(x_1)a(x_2){:}\\ &\quad = \theta(t_1 - t_2)a(x_1)a(x_2) - \theta(t_2 - t_1)a(x_2)a(x_1) - a(x_1)a(x_2)\\ &\quad = -\theta(t_2 - t_1)\,(a(x_1)a(x_2) + a(x_2)a(x_1)) \quad . \end{aligned} \tag{9.94}$$

If $\mathtt{A}(t_1) = a(x_1)$ and $\mathtt{A}(t_2) = a^*(x_2)$, one obtains

$$\begin{aligned} &T\,(a(x_1)a^*(x_2)) - {:}a(x_1)a^*(x_2){:}\\ &\quad = \theta(t_1 - t_2)a(x_1)a^*(x_2) - \theta(t_2 - t_1)a^*(x_2)a(x_1) + a^*(x_2)a(x_1)\\ &\quad = \theta(t_1 - t_2)\,(a(x_1)a^*(x_2) + a^*(x_2)a(x_1)) \quad . \end{aligned} \tag{9.95}$$

Quantities (9.94) and (9.95) are indeed scalars since they are proportional to the anti-commutator of the fermion operators. One sees the usefulness of the sign convention in definition (9.89): because of it, the time order in (9.93) only differs from the normal order by a scalar. Verifications analogous to (9.94) and (9.95) are easily carried out for the boson case.

To apply the Wick theorem to the decomposition of $T(\mathcal{H}_{\mathrm{I}}(x_1)\ldots\mathcal{H}_{\mathrm{I}}(x_n))$ into normal products, one must still take into account the case in which certain time arguments are equal. This can happen in two ways:

(a) certain arguments are equal in $\mathcal{H}_{\mathrm{I}}(x_1)\ldots\mathcal{H}_{\mathrm{I}}(x_n)$;
(b) $\mathcal{H}_{\mathrm{I}}(x) = a^*(x)a(x)\Phi(x)$ is itself the product of three fields taken at the same time.

Point (a) is addressed by the following remark. In the term (9.63) of the perturbative series, it is sufficient to know the action of T on $\mathtt{H}_{\mathrm{I}}(t_1)\mathtt{H}_{\mathrm{I}}(t_2)\ldots\mathtt{H}_{\mathrm{I}}(t_n)$ for distinct times $t_1 \neq t_2 \neq \ldots t_n$. The case for which certain times coincide is obtained by continuity since, according to definitions (9.68) and (9.70), $(T(\mathtt{H}_{\mathrm{I}}(t_1)\mathtt{H}_{\mathrm{I}}(t_2)\ldots\mathtt{H}_{\mathrm{I}}(t_n))$ is continuous in $t_i = t_j$. *It is thus sufficient to carry out the normal ordering when all the arguments t_j of the $\mathcal{H}_{\mathrm{I}}(x_j)$ are distinct.*

Concerning point (b), one notes that operators with the same time arguments can only come from the same Hamiltonian density $\mathcal{H}_{\mathrm{I}}(x) = a^*(x)a(x)\Phi(x)$. Their order should not be modified and one must have

$$T\,(a^*(x)a(x)\Phi(x)) = a^*(x)a(x)\Phi(x) \quad . \tag{9.96}$$

Since the Fermi field $a^*(x)$ always comes to the left of $a(x)$ in $\mathcal{H}_{\mathrm{I}}(x)$, one can incorporate rule (9.96) by agreeing that *if a product has identical time arguments, the time of $a^*(x)$ is infinitesimally greater than that of $a(x)$*. Thus applying (9.89), one indeed finds that

$$\begin{aligned} T\left(a^*(x)a(x)\Phi(x)\right) &= \lim_{\tau\to 0} T\left(a^*(\boldsymbol{x},t+\tau)a(\boldsymbol{x},t)\Phi(\boldsymbol{x},t)\right) \\ &= a^*(x)a(x)\Phi(x) \quad , \quad \tau > 0 \quad . \end{aligned} \tag{9.97}$$

It is not necessary to give a prescription for the product of a Fermi field and a Bose field taken at the same time since they commute: their order doesn't matter.

Taking into account the two prescriptions (a) and (b), the Wick theorem allows one, in principle, to decompose the operator S into a sum of normal products. The analysis of this decomposition and the study of its terms is the central task of the perturbation calculation.

9.3.5 Time-Ordered Contractions and Propagators

We have reached the point where it is possible to establish a link between the general formalism and the concepts relating to a single quantum particle introduced in the first section of this chapter. The link is provided by the interpretation of time-ordered contractions: *the time-ordered contraction is the free causal propagator associated with the field under consideration.* We will verify this for a few cases.

The Fermi Field in the Vacuum $|0\rangle$

It follows immediately from definitions (9.89) and (9.91) and from the relation $a(x,\sigma)\,|0\rangle = 0$ that

$$\underbrace{a(x_1,\sigma_1)a^*}(x_2,\sigma_2) = \theta(t_1-t_2)\,\langle 0|a(x_1,\sigma_1)a^*(x_2,\sigma_2)|0\rangle \quad . \tag{9.98}$$

One recognizes that *this expression is nothing other than the causal propagator* (9.32) *of a free particle*

$$\underbrace{a(x_1,\sigma_1)a^*}(x_2,\sigma_2) = \mathrm{i}\delta_{\sigma_1,\sigma_2} G_0(x_1|x_2) \quad . \tag{9.99}$$

The Fermi Field in the Ground State of a Free Gas $|\Phi_0\rangle$

Making use of the formalism of electrons and holes, and of (8.228) and (8.225), one obtains for $t_1 \neq t_2$

$$\begin{aligned}
&\underbrace{a(x_1,\sigma_1)a^*}(x_2,\sigma_2)\\
&= \langle\Phi_0|T\left(a(x_1,\sigma_1)a^*(x_2,\sigma_2)\right)|\Phi_0\rangle\\
&= \theta(t_1-t_2)\,\langle\Phi_0|a(x_1,\sigma_1)a^*(x_2,\sigma_2)|\Phi_0\rangle\\
&\quad-\theta(t_2-t_1)\,\langle\Phi_0|a^*(x_2,\sigma_2)a(x_1,\sigma_1)|\Phi_0\rangle\\
&= \theta(t_1-t_2)\,\langle\Phi_0|a^{\mathrm{e}}(x_1,\sigma_1)\left(a^{\mathrm{e}}(x_2,\sigma_2)\right)^*|\Phi_0\rangle\\
&\quad-\theta(t_2-t_1)\,\langle\Phi_0|a^{\mathrm{h}}(x_2,\sigma_2)\left(a^{\mathrm{h}}(x_1,\sigma_1)\right)^*|\Phi_0\rangle \quad . \qquad (9.100)
\end{aligned}$$

It is easy to verify that the function $G_0(x_1|x_2)$ defined by

$$\underbrace{a(x_1,\sigma_1)a^*}(x_2,\sigma_2) = \mathrm{i}G_0(x_1,\sigma_1|x_2,\sigma_2) = \mathrm{i}\delta_{\sigma_1,\sigma_2}G_0(x_1|x_2) \qquad (9.101)$$

is indeed the *Green function of the Schrödinger function for a free electron.* To see this, we write $G_0(x_1|x_2)$ in the spatial Fourier representation with the help of (9.100) and (8.228)

$$\begin{aligned}
\mathrm{i}G_0(x_1|x_2) = &\frac{1}{(2\pi)^3}\int \mathrm{d}\boldsymbol{k}\,\exp\left\{\mathrm{i}\left[\boldsymbol{k}\cdot(\boldsymbol{x}_1-\boldsymbol{x}_2)-\frac{\varepsilon_{\boldsymbol{k}}(t_1-t_2)}{\hbar}\right]\right\}\\
&\times\left[\theta(t_1-t_2)\theta(|\boldsymbol{k}|-k_{\mathrm{F}})-\theta(t_2-t_1)\theta(k_{\mathrm{F}}-|\boldsymbol{k}|)\right] \quad . \qquad (9.102)
\end{aligned}$$

As the plane waves $\exp[\mathrm{i}(\boldsymbol{k}\cdot\boldsymbol{x}-\varepsilon_{\boldsymbol{k}}t/\hbar)]$ are solutions to the Schrödinger equation and $(d/dt_1)\theta(t_1-t_2) = -(d/dt_1)\theta(t_2-t_1) = \delta(t_1-t_2)$, one indeed obtains

$$\left(\mathrm{i}\hbar\frac{\partial}{\partial t_1}-\frac{\hbar^2}{2m}\nabla_1^2\right)G_0(x_1|x_2) = \hbar\delta(\boldsymbol{x}_1-\boldsymbol{x}_2)\delta(t_1-t_2) \quad . \qquad (9.103)$$

The complete Fourier representation

$$\begin{aligned}
G_0(x_1|x_2) = &\frac{1}{(2\pi)^4}\int \mathrm{d}\boldsymbol{k}\int \mathrm{d}\omega\,\exp\{\mathrm{i}[\boldsymbol{k}\cdot(\boldsymbol{x}_1-\boldsymbol{x}_2)\\
&-\omega(t_1-t_2)]\}G_0(\boldsymbol{k},\omega) \qquad (9.104)
\end{aligned}$$

can be obtained with the help of (9.21) and the relation

$$\lim_{\eta\to 0}\frac{1}{2\pi}\int_{-\infty}^{\infty}\mathrm{d}\omega\,\mathrm{e}^{-\mathrm{i}\omega t}\frac{1}{\omega-\mathrm{i}\eta-\lambda} = \mathrm{i}\theta(-t)\mathrm{e}^{-\mathrm{i}\lambda t} \quad , \quad \eta>0 \quad , \qquad (9.105)$$

which follows from (9.21) by complex conjugation and the exchange of t with $-t$. After insertion of (9.21) and (9.105) in (9.102), comparison with (9.104) shows that

$$G_0(\boldsymbol{k},\omega) = \hbar\left(\frac{\theta(|\boldsymbol{k}|-k_{\mathrm{F}})}{\hbar(\omega+\mathrm{i}\eta)-\varepsilon_{\boldsymbol{k}}}+\frac{\theta(k_{\mathrm{F}}-|\boldsymbol{k}|)}{\hbar(\omega-\mathrm{i}\eta)-\varepsilon_{\boldsymbol{k}}}\right) \quad , \quad \eta>0 \quad . \qquad (9.106)$$

The interpretation of (9.104) is the same as for (9.22): one must take the limit $\eta \to 0$ after integrating.

When the times t_1 and t_2 coincide, prescription (b) of Sect. 9.3.4 requires that one take t_2 to be infinitesimally greater than t_1 in (9.100) and thus to define the Green function for this case by

$$\begin{aligned} G_0(\boldsymbol{x}_1, t, \sigma_1 | \boldsymbol{x}_2, t, \sigma_2) &= \lim_{\tau \to 0} G_0(\boldsymbol{x}_1, t, \sigma_1 | \boldsymbol{x}_2, t + \tau, \sigma_2) \quad , \quad \tau > 0 \\ &= \mathrm{i} \, \langle \Phi_0 | a^*(\boldsymbol{x}_2, \sigma_2) a(\boldsymbol{x}_1, \sigma_1) | \Phi_0 \rangle \\ &= \mathrm{i} \delta_{\sigma_1, \sigma_2} \frac{1}{(2\pi)^3} \int \mathrm{d}\boldsymbol{k} \, \theta(k_\mathrm{F} - |\boldsymbol{k}|) \mathrm{e}^{\mathrm{i}\boldsymbol{k} \cdot (\boldsymbol{x}_1 - \boldsymbol{x}_2)} \quad . \end{aligned} \tag{9.107}$$

In particular, when $\boldsymbol{x}_1 = \boldsymbol{x}_2 = \boldsymbol{x}$, one finds

$$\sum_\sigma G_0(\boldsymbol{x}, t, \sigma | \boldsymbol{x}, t, \sigma) = \mathrm{i} \rho \quad , \tag{9.108}$$

where ρ is the density (2.95) of the free Fermi gas.

Comparing (9.107) to (9.104) when t_2 approaches t_1 from above

$$\begin{aligned} &\lim_{\tau \to 0} G_0(\boldsymbol{x}_1, t | \boldsymbol{x}_2, t + \tau) \\ &= \frac{1}{(2\pi)^3} \int \mathrm{d}\boldsymbol{k} \, \mathrm{e}^{\mathrm{i}\boldsymbol{k} \cdot (\boldsymbol{x}_1 - \boldsymbol{x}_2)} \left(\lim_{\tau \to 0} \frac{1}{2\pi} \int \mathrm{d}\omega \, G_0(\boldsymbol{k}, \omega) \mathrm{e}^{\mathrm{i}\omega t} \right) \quad , \quad \tau > 0 \end{aligned} \tag{9.109}$$

one sees that it is necessary that the formal integral of $G_0(\boldsymbol{k}, \omega)$ with respect to frequency take on the value

$$\begin{aligned} \frac{1}{2\pi} \int \mathrm{d}\omega \, G_0(\boldsymbol{k}, \omega) &= \lim_{\tau \to 0} \frac{1}{2\pi} \int \mathrm{d}\omega \, G_0(\boldsymbol{k}, \omega) \mathrm{e}^{\mathrm{i}\omega t} \\ &= \mathrm{i} \theta(k_\mathrm{F} - |\boldsymbol{k}|) \quad , \quad \tau > 0 \quad . \end{aligned} \tag{9.110}$$

This result also follows from (9.106) if one integrates over the frequencies in the way that is indicated on the right-hand side of (9.110).

At zero density $\rho = 0$ ($k_\mathrm{F} = 0$), (9.102) is reduced to the retarded Green function (9.99). If $\rho \neq 0$, it only differs in one important way: the second term of (9.102), which corresponds to the propagation of holes, is non-zero when $t_2 > t_1$ (similarly, compare (9.25) to (9.106)). *This fact is characteristic of the existence of antiparticles (holes in this case) in the theory.* The causal interpretation of the propagator is nevertheless preserved: one sees in (9.100) that the annihilation of an electron or of a hole always comes after its creation from the state $|\Phi_0\rangle$.

Neutral Scalar Field

The time-ordered contraction of the scalar field in the vacuum can be immediately calculated with the help of (8.95) and (8.96)

$$
\begin{aligned}
\underbrace{\Phi(x_1)\Phi(x_2)} &= \langle 0|T\left(\Phi(x_1)\Phi(x_2)\right)|0\rangle \\
&= \theta(t_1-t_2)\,\langle 0|\Phi_-(x_1)\Phi_+(x_2)|0\rangle \\
&\quad +\theta(t_2-t_1)\,\langle 0|\Phi_-(x_2)\Phi_+(x_1)|0\rangle \\
&= \frac{\hbar}{(2\pi)^3}\int \mathrm{d}\boldsymbol{k}\,\frac{\mathrm{e}^{\mathrm{i}\boldsymbol{k}\cdot(\boldsymbol{x}_1-\boldsymbol{x}_2)}}{2\omega_{\boldsymbol{k}}}\big(\theta(t_1-t_2)\mathrm{e}^{-\mathrm{i}\omega_{\boldsymbol{k}}(t_1-t_2)} \\
&\quad +\theta(t_2-t_1)\mathrm{e}^{\mathrm{i}\omega_{\boldsymbol{k}}(t_1-t_2)}\big) \\
&= \mathrm{i}\Delta(x_1|x_2)
\end{aligned}
\tag{9.111}
$$

one can easily verify that $\Delta(x_1|x_2)$ is a *Green function of the Klein–Gordon equation*

$$
\left(\frac{\partial^2}{\partial {t_1}^2}-c^2\nabla_1^2+\mu^2\right)\Delta(\boldsymbol{x}_1,t_1|\boldsymbol{x}_2,t_2)=\hbar\delta(\boldsymbol{x}_1-\boldsymbol{x}_2)\delta(t_1-t_2) \quad . \tag{9.112}
$$

One obtains the Fourier representation by making use of (9.21) and (9.105) in (9.111)

$$
\begin{aligned}
\Delta(x_1|x_2) &= \frac{1}{(2\pi)^4}\int \mathrm{d}\boldsymbol{k}\int \mathrm{d}\omega\,\exp\{\mathrm{i}[\boldsymbol{k}\cdot(\boldsymbol{x}_1-\boldsymbol{x}_2) \\
&\quad -\omega(t_1-t_2)]\}\Delta(\boldsymbol{k},\omega) \quad ,
\end{aligned}
\tag{9.113}
$$

where

$$
\begin{aligned}
\Delta(\boldsymbol{k},\omega) &= \frac{\hbar}{2\omega_{\boldsymbol{k}}}\left(\frac{1}{\omega+\mathrm{i}\eta-\omega_{\boldsymbol{k}}}-\frac{1}{\omega-\mathrm{i}\eta+\omega_{\boldsymbol{k}}}\right) \\
&= \frac{\hbar}{\omega^2-(\omega_{\boldsymbol{k}}-\mathrm{i}\eta)^2} \quad , \quad \eta>0 \quad .
\end{aligned}
\tag{9.114}
$$

The propagator $\Delta(x_1|x_2)$ is non-zero when $t_2 > t_1$ because of the existence of the antiparticles (which are found to be identical to the particles in the case of a scalar neutral field). It nevertheless has the same causal nature as $G_0(x_1|x_2)$: the annihilation of a particle or of an antiparticle follows its creation from the vacuum.

Electromagnetic Field

In view of expression (8.7) of the vector potential, the calculation is the same as for the scalar field, except for an additional sum over the polarization vectors $\sum_{\lambda=1}^{2} e_\lambda^\alpha(\boldsymbol{k})e_\lambda^\beta(\boldsymbol{k})$. We define

$$
\underbrace{\mathtt{A}^\alpha(x_1)\mathtt{A}^\beta(x_2)} = \mathrm{i}D^{\alpha\beta}(x_1|x_2) \tag{9.115}
$$

as the *Green function of the wave equation for the transverse field.* Its Fourier transform is found by making use of (8.22)

$$
D^{\alpha\beta}(\boldsymbol{k},\omega)=\frac{\hbar}{\varepsilon_0}\left(\frac{1}{\omega^2-(\omega_{\boldsymbol{k}}-\mathrm{i}\eta)^2}\right)\left(\delta_{\alpha,\beta}-\frac{k^\alpha k^\beta}{|\boldsymbol{k}|^2}\right) \quad , \quad \eta>0 \quad . \tag{9.116}
$$

Phonon Field

If one considers the local electron–phonon interaction model given by (8.252), then the scalar field (8.251) (which represents the divergence of the phonon field) can be associated with the propagator

$$\underbrace{\Phi(x_1)\Phi(x_2)} = \mathrm{i}D(x_1|x_2) \quad . \tag{9.117}$$

The field (8.251) differs from the relativistic scalar field (8.95) in two ways: the existence of an extra factor $\omega_{\boldsymbol{k}} = c_{\mathrm{L}}|\boldsymbol{k}|$ which comes from the fact that the field here is divergent, as well as a maximum wavenumber (see the discussion at the end of Sect. 8.4.3). With the exception of these differences, the calculation of the Fourier transform $D(\boldsymbol{k}, \omega)$ of the propagator is identical to the relativistic scalar field case, and one finds (see (9.114))

$$D(\boldsymbol{k},\omega) = \frac{\hbar\omega_{\boldsymbol{k}}^2}{\omega^2 - (\omega_{\boldsymbol{k}} - \mathrm{i}\eta)^2}\theta(\omega_{\mathrm{D}} - \omega_{\boldsymbol{k}}) \quad , \quad \eta > 0 \quad . \tag{9.118}$$

Because of the factor $\omega_{\boldsymbol{k}}^2 = c_{\mathrm{L}}^2|\boldsymbol{k}|^2$, $D(\boldsymbol{k},\omega)$ vanishes as $|\boldsymbol{k}| \to 0$.

9.3.6 Feynman Diagrams

We can now return to studying the S operator with the help of the Wick theorem. The first-order term, which is already normal ordered, does not give rise to any contraction. For the second-order term, we have

$$T\left(a^*(x_1)a(x_1)\Phi(x_1)a^*(x_2)a(x_2)\Phi(x_2)\right)$$

$$= {:}a^*(x_1)a(x_1)\Phi(x_1)a^*(x_2)a(x_2)\Phi(x_2){:} \tag{9.119a}$$

$$+ {:}\underbrace{a^*(x_1)a(x_1)\Phi(x_1)a^*(x_2)a(x_2)}\Phi(x_2){:} \tag{9.119b}$$

$$+ {:}a^*(x_1)\underbrace{a(x_1)\Phi(x_1)a^*(x_2)}a(x_2)\Phi(x_2){:} \tag{9.119c}$$

$$+ {:}a^*(x_1)a(x_1)\underbrace{\Phi(x_1)a^*(x_2)a(x_2)\Phi(x_2)}{:} \tag{9.119d}$$

$$+ {:}a^*(x_1)a(x_1)\Phi(x_1)a^*(x_2)a(x_2)\Phi(x_2){:} \tag{9.119e}$$

$$+ {:}a^*(x_1)a(x_1)\Phi(x_1)a^*(x_2)a(x_2)\Phi(x_2){:} \tag{9.119f}$$

$$+ {:}\underbrace{a^*(x_1)\underbrace{a(x_1)\Phi(x_1)a^*(x_2)}a(x_2)}\Phi(x_2){:} \tag{9.119g}$$

$$+ {:}a^*(x_1)a(x_1)\Phi(x_1)a^*(x_2)a(x_2)\Phi(x_2){:} \tag{9.119h}$$

$$+ \ldots \quad .$$

We have omitted contractions which are manifestly equal to zero, such as those concerning fields carrying the same arguments (see (9.126), Sect. 9.4.1). The detailed account and interpretation of the terms stemming from the Wick theorem are greatly facilitated by a graphical representation.

Space–Time Representation

Each normal product obtained from the Wick decomposition is associated with a *Feynman diagram* according to the following rules:

1. To each coordinate x indexing the operators which appear in a normal product, we associate a *point called a vertex*.
2. To each fermion creator $a^*(x)$, we associate an *oriented line originating from* x.

3. To each fermion annihilator $a(x)$, we associate an *oriented line ending at* x.

4. For each boson field $\Phi(x) = \Phi_+(x) + \Phi_-(x)$, we draw an *undulating line attached to* x. It can be equipped with an outgoing or incoming arrow if it refers to a boson creator $\Phi_+(x)$ or annihilator $\Phi_-(x)$.

5. To the contraction $\underbrace{a(x_1)a^*(x_2)}$, we associate an *oriented line joining* x_2 *to* x_1.

6. To the contraction $\underbrace{\Phi(x_1)\Phi(x_2)}$, we associate a *(non-oriented) undulating line joining* x_2 *to* x_1.

The eight normal products (9.119a)–(9.126) are represented in Fig. 9.12.

The diagrams have the following general characteristics. Points x_1 and x_2, where two fermion lines and one boson line join, are the vertices. *The structure of a vertex is always the same; it is fixed by the form of the interaction*, equivalently by the first-order term $a^*(x)a(x)\Phi(x)$ (Fig. 9.13). *The number of vertices is equal to the order of the perturbation calculation.* To obtain the decomposition of $\mathsf{S}^{(n)}$ in normal order, it is sufficient to give n distinct points $x_1, x_2, \dots, x_n$ (according to prescription (a) of Sect. 9.3.4) and to draw all diagrams compatible with the structure of the vertex. To reconstitute the algebraic expression of $\mathsf{S}^{(n)}$ from the diagrams, one must associate to each external line its field operator and to each internal line its propagator and then integrate over the variables $x_1, x_2, \dots, x_n$ attached to the vertices. In cases where fermions are involved, certain sign rules must be followed so that the diagram is associated with the correct algebraic expression.

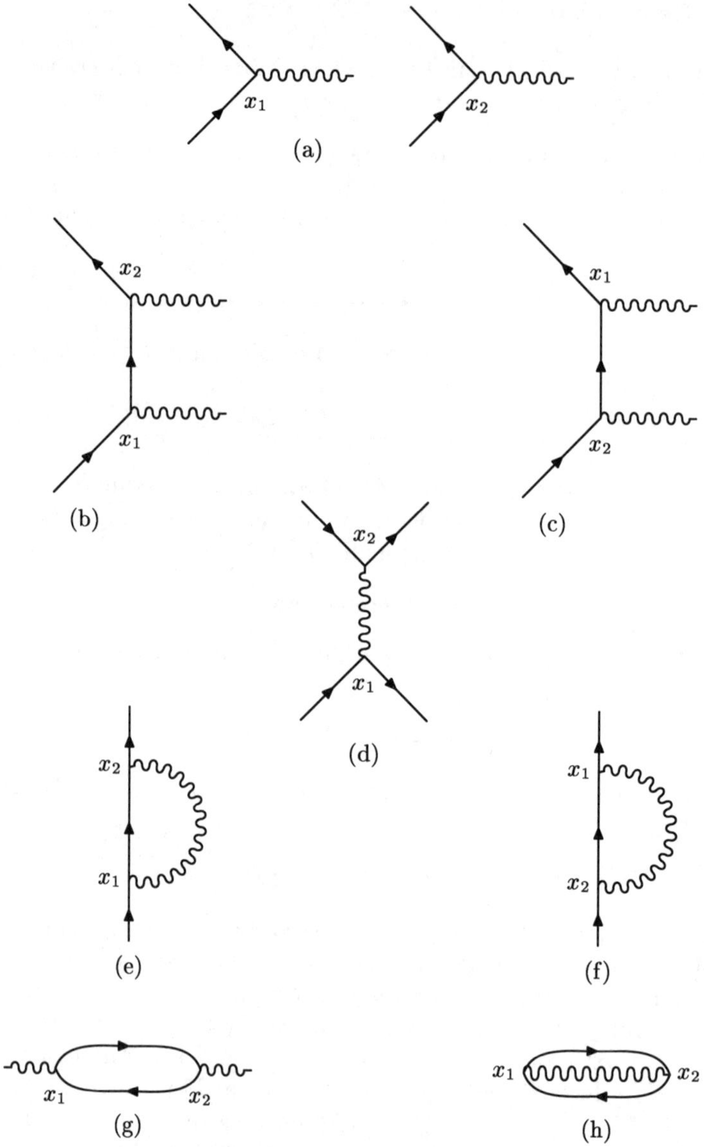

Fig. 9.12a–h. Feynman diagrams representing the eight normal products (9.128)–(9.135)

Energy–Momentum Representation

To calculate the matrix elements $\langle \Phi_f | \mathsf{S} | \Phi_i \rangle$, it is convenient to use the Fourier transforms of the propagators defined by (9.104) and (9.113): indeed, the par-

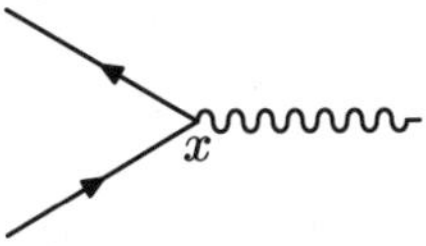

Fig. 9.13. Feynman diagram of the first-order term $a^*(x)a(x)\Phi(x)$. This term fixes the structure of the vertices

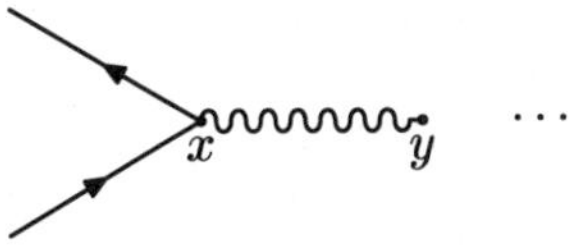

Fig. 9.14. Vertex x with two external fermion lines and one internal boson line

ticles in the states $|\Phi_i\rangle$ and $|\Phi_f\rangle$ (see Table 9.2a and b) have well-determined energies and momenta. With the goal of understanding the significance of the diagrams in the momentum–energy representation, we can study the example of a vertex x with two external fermion lines and one internal boson line, as indicated in Fig. 9.14. In particular the contribution of this vertex to the complete diagram gives rise to the integral

$$\ldots \int \mathrm{d}x\, \Delta(x-y) a^*(x) a(x) \ldots \quad . \tag{9.120}$$

This quantity gives a non-zero contribution to a matrix element of the type (Table 9.2a or b) having at least one fermion of momentum $\boldsymbol{k}_1$ in the initial state and one fermion of momentum $\boldsymbol{k}_2$ in the final state. To evaluate of the matrix element in question requires introducing the (non-time-ordered) contractions of the creators corresponding to the operators of (9.120), which are of the type

$$\begin{aligned} \underbrace{a(\boldsymbol{x},t)a^*(\boldsymbol{k}_1)} &= \frac{1}{L^{3/2}} \mathrm{e}^{\mathrm{i}(\boldsymbol{k}_1\cdot\boldsymbol{x} - \varepsilon_{\boldsymbol{k}_1} t/\hbar)} \quad , \\ \underbrace{a(\boldsymbol{k}_2)a^*(\boldsymbol{x},t)} &= \frac{1}{L^{3/2}} \mathrm{e}^{-\mathrm{i}(\boldsymbol{k}_2\cdot\boldsymbol{x} - \varepsilon_{\boldsymbol{k}_2} t/\hbar)} \quad . \end{aligned} \tag{9.121}$$

Taking into account (9.121) and (9.113) (written for a finite volume), the diagram element in Fig. 9.14 yields, among other things, the following contribution to the matrix element

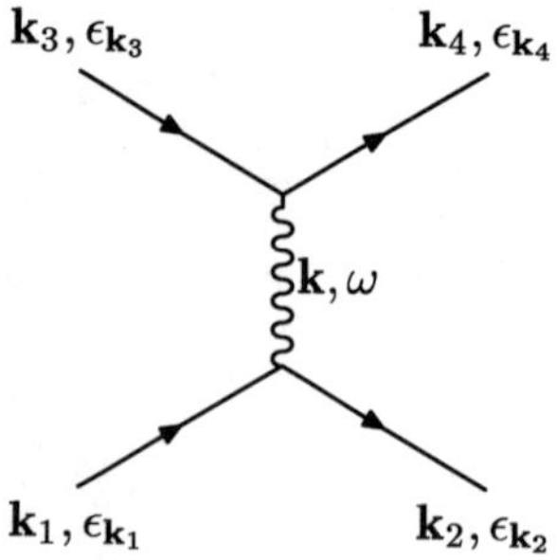

Fig. 9.15. Momentum–energy diagram corresponding to Fig. 9.12d

$$\begin{aligned}
&\ldots \sum_{\boldsymbol{k}} \int \mathrm{d}\omega \int \mathrm{d}\boldsymbol{x} \int \mathrm{d}t\, \frac{1}{L^{3/2}} \mathrm{e}^{\mathrm{i}(\boldsymbol{k}_1\cdot\boldsymbol{x} - \varepsilon_{\boldsymbol{k}_1} t/\hbar)} \\
&\times \frac{1}{L^{3/2}} \mathrm{e}^{-\mathrm{i}(\boldsymbol{k}_2\cdot\boldsymbol{x} - \varepsilon_{\boldsymbol{k}_2} t/\hbar)} \mathrm{e}^{\mathrm{i}(\boldsymbol{k}\cdot\boldsymbol{x} - \omega t)} \Delta(\boldsymbol{k},\omega) \ldots \\
&= \ldots \frac{1}{L^3} \sum_{\boldsymbol{k}} \int \mathrm{d}\omega \delta_{\boldsymbol{k},\boldsymbol{k}_2-\boldsymbol{k}_1} \delta\left[\omega - (\varepsilon_{\boldsymbol{k}_2} - \varepsilon_{\boldsymbol{k}_1})/\hbar\right] \Delta(\boldsymbol{k},\omega) \ldots \quad . \qquad (9.122)
\end{aligned}$$

Following a similar line of reasoning for all the vertices of the diagram, one realizes that the calculation of a matrix element is guided by the following rules:

(i) To each internal line, one associates a energy–momentum variable $(\boldsymbol{k}, \omega)$ and the corresponding propagator, either $\mathrm{i}(2\pi)^{-4}G_0(\boldsymbol{k},\omega)$ for a fermion line or $\mathrm{i}(2\pi)^{-4}\Delta(\boldsymbol{k},\omega)$, $\mathrm{i}(2\pi)^{-4}D(\boldsymbol{k},\omega)\ldots$ for a boson line.
(ii) To each external line, one associates the energy $\varepsilon_{\boldsymbol{k}}$ (or $\hbar\omega$) and the momentum $\hbar\boldsymbol{k}$ of the corresponding particle.
(iii) One attributes to each vertex a factor $-\mathrm{i}g/\hbar$ and a Dirac conservation function $(2\pi)^4\delta(\sum_i \boldsymbol{k}_i)\delta(\sum_j \omega_j)$, where $(\boldsymbol{k}_j, \omega_j)$ are the energy–momenta of the oriented lines attached to the vertex (with the signs determined by the line orientations).
(iv) One sums over the energy–momentum of the internal lines.

For example, an energy–momentum diagram corresponding to Fig. 9.12d is represented in Fig. 9.15.

One notes that the momentum and energy conservation at each vertex

$$\begin{aligned}
&\boldsymbol{k}_1 - \boldsymbol{k}_2 = \boldsymbol{k} = -\boldsymbol{k}_3 + \boldsymbol{k}_4 \quad , \\
&\varepsilon_{\boldsymbol{k}_1} - \varepsilon_{\boldsymbol{k}_2} = \omega = -\varepsilon_{\boldsymbol{k}_3} + \varepsilon_{\boldsymbol{k}_4} \qquad (9.123)
\end{aligned}$$

implies that the total momentum and energy of the incoming external lines are equal to that of the outgoing external lines

$$\begin{aligned} \boldsymbol{k}_1 + \boldsymbol{k}_3 &= \boldsymbol{k}_2 + \boldsymbol{k}_4 \quad , \\ \varepsilon_{\boldsymbol{k}_1} + \varepsilon_{\boldsymbol{k}_3} &= \varepsilon_{\boldsymbol{k}_2} + \varepsilon_{\boldsymbol{k}_4} \quad . \end{aligned} \tag{9.124}$$

We know that the scattering operator conserves the total energy and momentum (see (9.78)): we see here that *these quantities are conserved for each diagram*, that is for each term of the decomposition in normal order. It is important to note that the variable $\hbar\omega$ of the internal line, contrary to those of the external lines, is not the physical energy of a particle of momentum $\hbar\boldsymbol{k}$; $\hbar\omega$ is an integration variable which is not constrained to the value $\hbar\omega_{\boldsymbol{k}} = \hbar c\sqrt{|\boldsymbol{k}|^2 + \mu^2}$ or $\varepsilon_{\boldsymbol{k}} = \hbar^2|\boldsymbol{k}|^2/2m$, the kinetic energy of a particle of momentum $\hbar\boldsymbol{k}$. For this reason, the particles corresponding to the internal lines are said to be virtual. This terminology is a reminder once again that the diagrams do not represent the configurations or trajectories of the particles. It is preferable to consider them as a graphical means of representing the content of the Wick theorem.

One can also write down a set of rules which allow the probability amplitude of the process corresponding to a given diagram to be calculated. In addition to the prescriptions (i)–(iv) presented above, it is necessary to include certain normalization factors associated to the external lines. It is also necessary to add certain sign rules and combinatorial rules (for example, to take into account the fact that several diagrams of a given order could contribute to the same process, such as Fig. 9.12b and c, and to take into account the statistics when there are identical particles in the initial or final states). Given the fields and their corresponding interactions (that is, the propagators and the structures of the vertices), these rules can be made completely explicit and are called the *Feynman rules* for the theory under consideration. *In this way, a perturbative field theory is completely defined by the statement of the Feynman rules.* We will not give a complete list of these rules. They can only be described exhaustively for each specific case (electrodynamics, pion–nucleon, electron–phonon) and this can be left to more specialized studies. In Sect. 9.4, we give the physical interpretation of the main diagrams and describe the methods for summing the diagrams. Some specific examples will be treated. It should be added that one talks of Feynman diagrams in a broad sense each time a graphical representation of the terms stemming from a perturbative calculation is adopted. Many types of Feynman diagrams can be found in the literature, defined with the help of many different conventions.

9.4 Applications

9.4.1 Physical Interpretation of the Diagrams

First Order

The first-order diagrams of Fig. 9.16a represent *the emission or absorption of a boson by a fermion* (depending on whether the undulating line is incoming or outgoing). For relativistic electrons interacting with photons, *this transition is forbidden by the conservation of energy and momentum,*

$$\begin{aligned} \boldsymbol{p}_1 &= \boldsymbol{p}_2 \pm \hbar \boldsymbol{k} \quad , \\ c\sqrt{|\boldsymbol{p}_1|^2 + m^2c^2} &= c\sqrt{|\boldsymbol{p}_2|^2 + m^2c^2} \pm \hbar c|\boldsymbol{k}| \quad . \end{aligned} \tag{9.125}$$

Placing oneself in the center of mass of the incoming electron, for which $\boldsymbol{p}_1 = 0$, one can easily verify that (9.125) have no solution for $\boldsymbol{k} \neq 0$.

Taking positrons into account, the diagrams of Fig. 9.16b describe the annihilation or creation of an electron–positron pair. These transitions are also forbidden by the conservation laws: an electron–positron pair cannot decay by the emission of a single photon. This shows that *a diagram does not necessarily correspond to a physically realizable process*: it is still necessary for it to be compatible with the general laws of conservation.

For non-relativistic electrons in a solid, the situation is different. The diagrams of Fig. 9.16a correspond to the emission or the absorption of a phonon: *they contribute to the calculation of the electric resistivity* (Sect. 9.4.4).

Second Order

The diagram in Fig. 9.12a gives a contribution equal to the square of the first-order term. A diagram of the type (a) which is made up of several parts

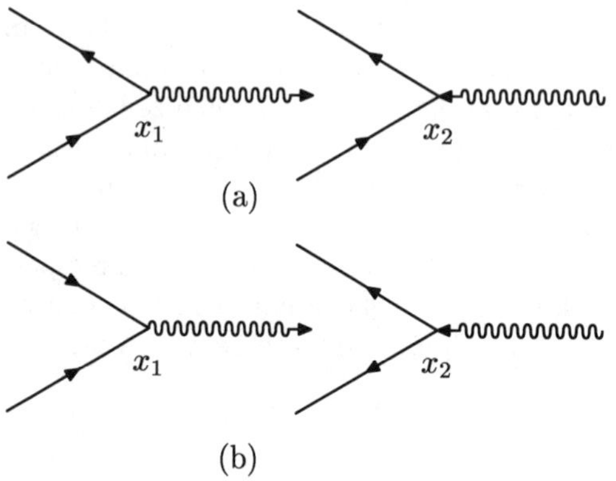

Fig. 9.16. (**a**) Emission (*left*) or absorption (*right*) of a boson by a fermion. (**b**) Annihilation (*left*) or creation (*right*) of a fermion-antifermion pair

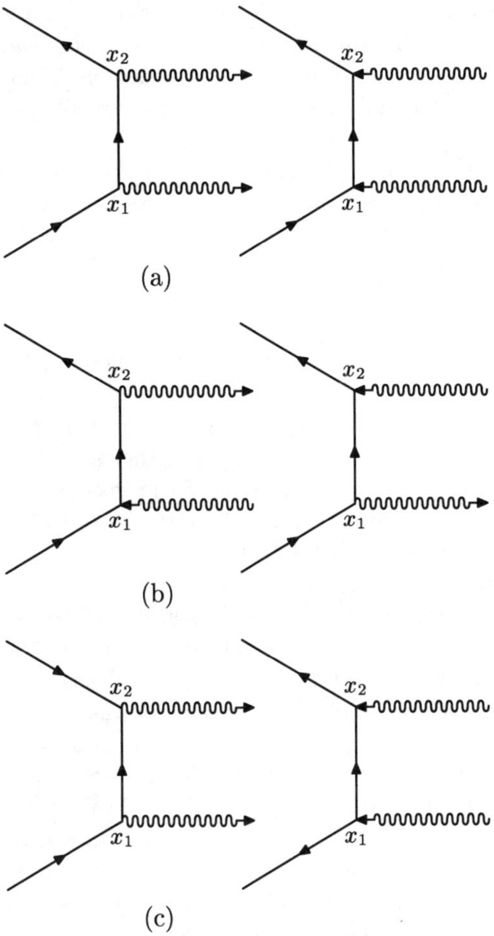

Fig. 9.17. (**a**) Emission (*left*) or absorption (*right*) of two bosons. (**b**) Scattering of a fermion by a boson (the Compton effect). (**c**) Pair annihilation or creation with the emission (*left*) or absorption (*right*) of two bosons

which are not joined by any internal line is called *disconnected*. Such diagrams represent combinations of processes taking place already at lower order in of the perturbation calculation.

The diagrams in Fig. 9.12b and c have several different interpretations depending on whether the boson lines are incoming or outgoing. If the boson lines are both incoming or both outgoing (Fig. 9.17a), then one is dealing with the *absorption or emission of two bosons*. This process is again forbidden in electrodynamics, but possible for phonons. If one of the boson lines is incoming and the other is outgoing (Fig. 9.17b), the diagram describes the *scattering of a fermion by a boson*, that is, the *Compton effect*. Finally, taking the positron into account, the diagrams of Fig. 9.17c describe *pair annihilation or creation with the emission or absorption of two photons*. Thus, it can be seen that the same class of diagrams can correspond to different physical processes.

The diagram in Fig. 9.12d represents the *scattering of two fermions by the exchange of a boson*. We know that in non-relativistic theory and to lowest order, the scattering amplitude is proportional to the interaction potential: the calculation of this diagram allows the *effective potential* resulting from the exchange of a boson to be identified. One can thus derive again the Yukawa potential (Exercise 2, pion–nucleon interaction) or the effective potential between electrons responsible for superconductivity in a metal (Sect. 9.4.4).

The diagrams in Fig. 9.12e and f are called *the self-energy* diagrams of a fermion, which emits and reabsorbs a boson. They correspond to a modification of the fermion free propagator due to the existence of other particles. In electrodynamics, they give rise to the renormalization of the mass of the electron (Sect. 9.4.3). In a solid, they describe a modification to the propagation of an electron due to lattice vibrations. In the latter case, one can take this into account by giving the electron an effective mass.

The diagram in Fig. 9.12g is called the *vacuum polarization* diagram. Note that this diagram is only non-zero if the theory describes antiparticles. Indeed, diagram (g) introduces the product of two propagators of fermions $G_0(x_1|x_2)G_0(x_2|x_1)$. If $G_0(x_1|x_2)$ is the propagator (9.99) of a non-relativistic electron in the vacuum of matter, it is clear that this product vanishes, since $G_0(x_1|x_2) = 0$, $t_1 < t_2$. If, on the other hand, $G_0(x_1|x_2)$ is the propagator (9.101) of the electron field in the ground state of the Fermi gas, from which an electron can be excited by creating a hole, such a product is non-zero. Here one talks of the spontaneous emission and reabsorption of an "electron–hole pair" by the phonon. This phenomenon is indeed related to the fact that the medium is polarizable, from which the terminology is adopted. In the solid, diagram (g) describes a modification of the speed of sound due to density waves interacting with the electrons. This same phenomenon appears in electrodynamics when positrons are taken into account: a photon can emit or reabsorb a virtual electron–positron pair from the vacuum. In this sense, the field theory vacuum behaves as a polarizable medium. This polarization of the medium gives rise to electric field fluctuations which produce observable effects, such as the displacement of the $2S_{1/2} - 2P_{1/2}$ level of the hydrogen atom (Sect. 9.4.3).

A diagram without external lines, such as Fig. 9.12h, which corresponds to a complete contraction in the Wick theorem, only contributes to the matrix element $\langle 0|\mathrm{S}|0\rangle$. General theorems show that these diagrams do not contribute to the calculation of physical quantities (Sect. 10.1.3).

In the normal ordering (9.119a)–(9.126) of $\mathrm{S}^{(2)}$, we have not written the terms which introduce contractions of fields at the same time, such as

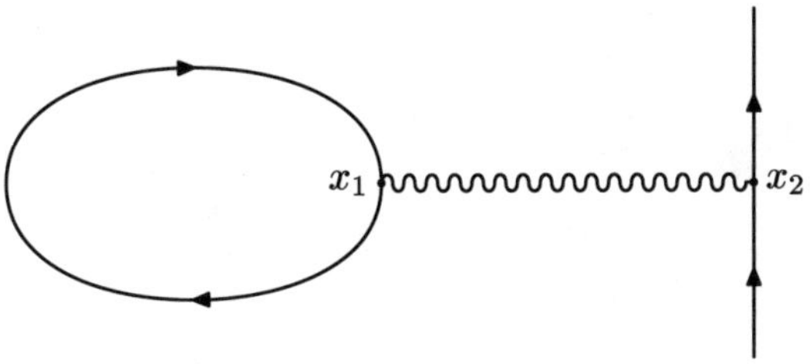

Fig. 9.18. Representation of a contraction of fields at the same time

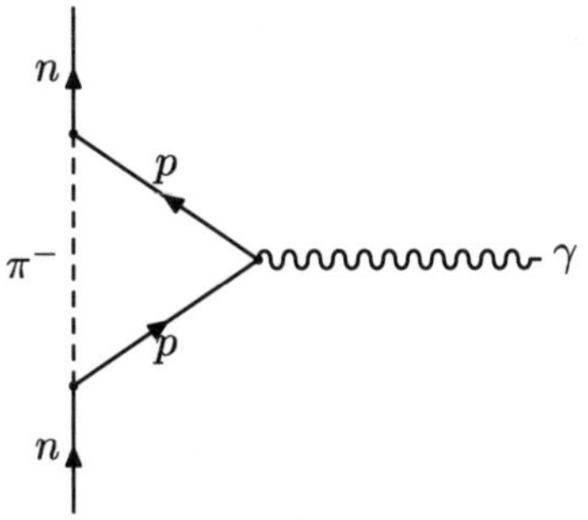

Fig. 9.19. Diagram of the electromagnetic interaction of a neutron through the coupling to strong interactions. The neutron n transforms virtually into a pion π^- (dashed line) and a proton p

$$\underbrace{a^*(x_1)a}(x_1)\,\underbrace{\Phi(x_1)a^*(x_2)a(x_2)\Phi}(x_2) \tag{9.126}$$

represented in Fig. 9.18. The propagator at equal time $\underbrace{a^*(x_1)a}(x_1)$ can only be non-zero for a theory of finite density (see (9.108)), and can thus appear in the electron–phonon theory. In fact, in this case, its contribution vanishes because of the particular structure of the phonon propagator (Sect. 9.4.4).

Other Examples

It is clear that the Feynman diagram method for constructing the S matrix can be extended to any local interaction and so it can be applied many different types of particles. One can thus describe complex processes in which interactions of a different nature are combined. For example, the neutron, which is electrically neutral, can have an electromagnetic interaction through intermediate strong interactions. The electromagnetic and pion–nucleon vertices, both having the form shown in Fig. 9.13, can form the diagram shown in Fig. 9.19. The dashed line represents the pion propagator. The neutron n transforms in to a virtual pion π^- and a virtual proton p which are both

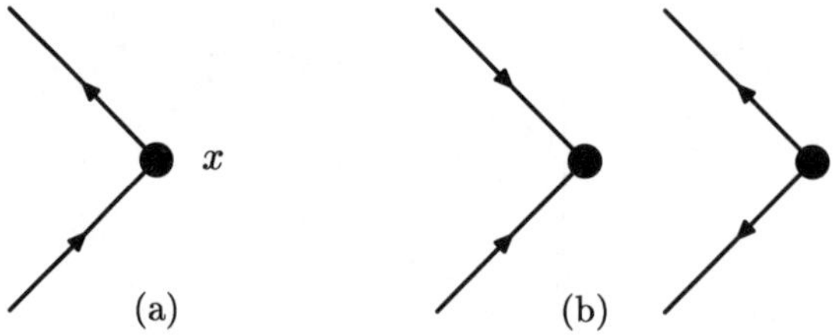

Fig. 9.20. Diagram of a fermion submitted to a non-quantized external field. (**a**) The effect of the field is represented by the dot x. (**b**) Pair annihilation (*left*) and creation (*right*) are allowed, as energy and momentum conservation are not required

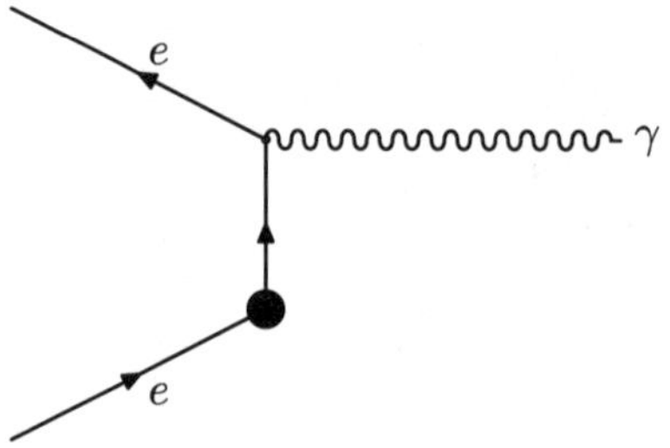

Fig. 9.21. Example of Bremsstrahlung radiation, in which an electron radiates a photon in the presence of the static electric field of a nucleus. The nucleus absorbs part of the momentum

charged particles which can interact electromagnetically. *This diagram contributes to the calculation of the magnetic moment of the neutron.*

One important case is when the particles are under influence of non-quantized external fields. As the fields are scalars, they do not give rise to any contractions in the Wick theorem. As in Sect. 9.2.3, we will use the convention of representing the effect of the external field by a dot indexed by x (Fig. 9.20a). Since the system is no longer invariant under translations in the presence of an external field, the law of conservation of energy and momentum no longer holds. Thus pair creation and annihilation is possible in an external field (Fig. 9.20b).

The electron can also emit a photon in the electric field of a nucleus, with the nucleus absorbing part of the momentum. This phenomenon of *Bremsstrahlung* (breaking through radiation) is described by diagrams of the same type as Fig. 9.21. These examples illustrate the richness of the Feynman diagram technique.

9.4.2 Electromagnetic Interactions: Compton Scattering

The explicit calculation of the probability amplitudes of processes using Feynman rules is not always easy: already at the lowest orders, it may introduce rather complicated integrals. We limit ourselves here to some simple considerations of *radiative processes*, that is the interaction of charged particles with photons. Insofar as the particles under consideration are relativistic, the calculation of such processes cannot be carried out in the theoretical framework of scattering by a potential: it is necessary to turn to quantum field theory.

An important process, known for a very long time, is *the scattering of a charged particle by a photon.* This is the Compton effect, observed by Compton in 1922 while studying the scattering of X rays by electrons. As we have not introduced Dirac's relativistic field for the electron, we will treat the Compton effect for a charged scalar particle, for example the scattering of a pion and a photon $\pi^+ + \gamma \to \pi^+ + \gamma$, to lowest order in the coupling constant. The goal of this calculation is to illustrate how to obtain the cross-section of a process in the framework of the perturbative theory of relativistic fields, and we will carry it out here in the Coulomb gauge.

If the Coulomb gauge is indeed suitable for studying of radiative effects, one must straightaway emphasize that the general methods for calculating the Feynman diagrams are greatly simplified using a formalism which is manifestly covariant with respect to Lorentz transformations. This requires the quantization of the electromagnetic field in the Lorenz gauge, which we have not presented here and for which we refer the reader to textbooks on quantum field theory.

Interaction Hamiltonian

The interaction of a charged scalar field $\Phi^*(x)$ with the classical electromagnetic field $A^\mu(x)$ is determined by the Lagrangian (8.184). The form of the interaction Hamiltonian responsible for Compton scattering is obtained by applying the usual rules of the Lagrangian formalism to (8.184). The canonically conjugate variables to Φ and Φ^* (considered as independent) are

$$\begin{aligned} \Pi &= \frac{\partial \mathcal{L}}{\partial(\partial_0 \Phi)} = [(\partial_0 + \mathrm{i}eA_0)\Phi]^* \quad , \\ \Pi^* &= \frac{\partial \mathcal{L}}{\partial(\partial_0 \Phi^*)} = (\partial_0 + \mathrm{i}eA_0)\Phi \quad . \end{aligned} \tag{9.127}$$

The canonical variables of the electromagnetic field are the same as in the free field case, since $\mathcal{L}^\Phi$ in (8.184) does not depend on the derivatives of A^μ. One thus constructs the *Hamiltonian density* using the usual prescription (see (8.164) in the case of the charged free scalar field):

$$\mathcal{H}(x) = \Pi(x)\partial_0\Phi(x) + \Pi^*(x)\left(\partial_0\Phi(x)\right)^* - \mathcal{L}^\Phi(x) + \mathcal{H}^{\mathrm{em}}(x) \quad , \tag{9.128}$$

where $\mathcal{H}^{\mathrm{em}}(x)$ is the electromagnetic Hamiltonian density. Taking into account (9.127) and (8.184), $\mathcal{H}(x)$ is again written as

$$\begin{aligned}\mathcal{H} &= \partial_0\Phi^*\partial_0\Phi + [(\nabla - \mathrm{i}e\boldsymbol{A})\Phi]^* \cdot (\nabla - \mathrm{i}e\boldsymbol{A})\Phi + m^2\Phi^*\Phi \\ &\quad - e^2A_0^2\Phi^*\Phi + \mathcal{H}^{\mathrm{em}} \\ &= \mathcal{H}_0 + \mathrm{i}e\boldsymbol{A}\cdot[\Phi^*(\nabla\Phi) - (\nabla\Phi^*)\Phi] + e^2|\boldsymbol{A}|^2\Phi^*\Phi \\ &\quad - e^2A_0^2\Phi^*\Phi + \mathcal{H}^{\mathrm{em}} \quad . \end{aligned} \tag{9.129}$$

The Hamiltonian density of the free scalar field $\mathcal{H}_0 = \partial_0\Phi^*\partial_0\Phi + \nabla\Phi^*\cdot\nabla\Phi + m^2\Phi^*\Phi$ (see (8.132)) appears in the first term of the right-hand side of (9.129).

We now interpret the different terms of the right-hand side of (9.129) in the Coulomb gauge. The second and third terms manifestly describe the coupling of the scalar field to the vector potential $\boldsymbol{A}$. *As $\boldsymbol{A}$ corresponds to the dynamical degrees of freedom of the electromagnetic field, it is clear that these terms are responsible for the interaction of the particles with the photons.* The electromagnetic energy $\int \mathrm{d}\boldsymbol{x}\mathcal{H}^{\mathrm{em}}(\boldsymbol{x})$ has a contribution due to the transverse fields, which is the energy (8.6) of the photons, as well as a contribution which results from the longitudinal fields. The latter, with the term $e^2A_0^2\Phi^*\Phi$, represents the Coulomb energy of the charged particles. We recall that, in the Coulomb gauge, the scalar potential $V = A_0$ is itself a function determined from the field Φ by the instantaneous Poisson equation $\nabla^2V = -\varepsilon_0^{-1}\rho_e$, where ρ_e is the charge density, equal to the component j_0 in (8.185). These Coulomb terms do not contribute to Compton scattering at the e^2 order. Thus we retain the interaction terms

$$\mathtt{H}_\mathrm{I} = \int \mathrm{d}\boldsymbol{x}\,\left[\mathcal{H}_\mathrm{I}^{(a)}(\boldsymbol{x}) + \mathcal{H}_\mathrm{I}^{(b)}(\boldsymbol{x})\right] \quad , \tag{9.130}$$

with the Hamiltonian densities

$$\mathcal{H}_\mathrm{I}^{(a)}(\boldsymbol{x}) = \mathrm{i}\frac{ec^2}{\hbar}{:}\{\Phi^*(\boldsymbol{x})\,[\nabla\Phi(\boldsymbol{x})] - [\nabla\Phi^*(\boldsymbol{x})]\,\Phi(\boldsymbol{x})\}\cdot\mathtt{A}(\boldsymbol{x}){:} \quad , \tag{9.131}$$

$$\mathcal{H}_\mathrm{I}^{(b)}(\boldsymbol{x}) = \frac{e^2c^2}{\hbar}{:}\Phi^*(\boldsymbol{x})\Phi(\boldsymbol{x})|\mathtt{A}(\boldsymbol{x})|^2{:} \quad . \tag{9.132}$$

In (9.131) and (9.132), $\Phi^*(\boldsymbol{x})$ and $\mathtt{A}(\boldsymbol{x})$ are the free fields (8.129) and (8.7) at time $t = 0$, and we have reintroduced the dimensional constants $\hbar$ and c. The free evolution is governed by the sum of the kinetic energies (8.127) and (8.6) of the pions and the photons.

Compton Amplitude to Order e^2

One recognizes in the term (9.131) the coupling of the current (8.135) to the vector potential, quadratic in the pion field and linear in the photon field.

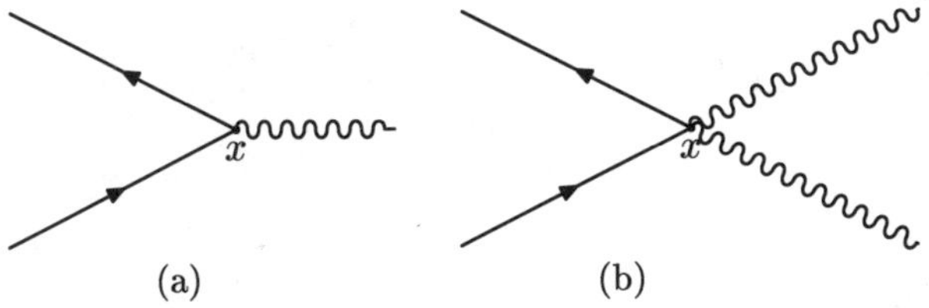

Fig. 9.22a,b. Feynman diagrams corresponding to the interaction Hamiltonian density (9.131) and (9.132). (**a**) Three-line vertex representing the linear coupling to the photon field. (**b**) Four-line vertex representing the quadratic coupling to the photon field

It has the same structure as the field theory prototype considered in the preceding sections. It gives rise to a three-line vertex (Fig. 9.22a). On the other hand, the term (9.132) is quadratic in the two fields; it gives rise to the four-line vertex of Fig. 9.22b.

In this theory, there are *two types of vertices*, one proportional to e and the other to e^2. Since the perturbative calculation is made according to increasing orders of the Hamiltonian $\mathtt{H}_\mathrm{I}$, the corresponding expansion differs from that obtained by ordering the terms according to increasing orders of the coupling constant e. Different orders of the perturbation calculation, expanded with respect to $\mathtt{H}_\mathrm{I}$, participate at a fixed order in e. The vertex of Fig. 9.22b is typical of a charged-boson theory, it does not appear in the quantum electrodynamics of electrons (see Sect. 9.4.3).

Three diagrams contribute in principle to the probability amplitude of the Compton effect at order e^2. These are the diagram of Fig. 9.22b (to first order of the perturbation calculation) and the two diagrams Fig. 9.12b and c, which are the result of (9.131) to second order in perturbation theory. The initial and final states each contain a π^+ and a photon γ,

$$
\begin{aligned}
|\Phi_i\rangle &= |\boldsymbol{p}_1, \boldsymbol{k}_1\lambda_1\rangle = b^*_{\boldsymbol{p}_1} a^*_{\boldsymbol{k}_1\lambda_1}|0\rangle \quad , \\
|\Phi_f\rangle &= |\boldsymbol{p}_2, \boldsymbol{k}_2\lambda_2\rangle = b^*_{\boldsymbol{p}_1} a^*_{\boldsymbol{k}_2\lambda_2}|0\rangle \quad ,
\end{aligned}
\tag{9.133}
$$

where $a^*_{\boldsymbol{k}\lambda}$ is the creation operator of a photon and $b^*_{\boldsymbol{p}}$ is that of a pion.

We now calculate the contribution from the diagram of Fig. 9.23. According to (9.79), (9.132) and (9.133), we must evaluate

$$
\begin{aligned}
&\langle \boldsymbol{p}_2, \boldsymbol{k}_2\lambda_2|\mathtt{S}^{(1)}|\boldsymbol{p}_1, \boldsymbol{k}_1\lambda_1\rangle \\
&\quad = -\frac{\mathrm{i}}{\hbar}\int \mathrm{d}\boldsymbol{x}\int \mathrm{d}t\ \langle \boldsymbol{p}_2, \boldsymbol{k}_2\lambda_2|\mathcal{H}_\mathrm{I}^{(b)}(\boldsymbol{x},t)|\boldsymbol{p}_1, \boldsymbol{k}_2\lambda_1\rangle \\
&\quad = -\frac{\mathrm{i}e^2c^2}{\hbar^3}\int \mathrm{d}\boldsymbol{x}\mathrm{d}t\,\langle \boldsymbol{p}_2, \boldsymbol{k}_2\lambda_2|{:}\Phi^*(\boldsymbol{x},t)\Phi(\boldsymbol{x},t)\mathtt{A}(\boldsymbol{x},t) \\
&\qquad \times \mathtt{A}(\boldsymbol{x},t){:}|\boldsymbol{p}_1, \boldsymbol{k}_1\lambda_1\rangle \quad ,
\end{aligned}
\tag{9.134}
$$

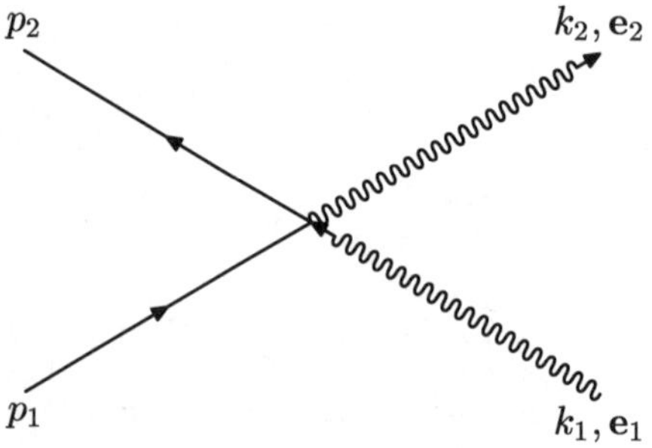

Fig. 9.23. Four-line vertex contributing to the Compton effect to order e^2

where $\Phi^*(\boldsymbol{x}, t)$ is the charged free field (8.129) and $\mathtt{A}(\boldsymbol{x}, t)$ is the electromagnetic free field (8.7). With the help of the calculation rules for the Fock space of the pions and photons, one can immediately calculate the matrix element (9.134) with the result (see Exercise 4(i))

$$\langle \boldsymbol{p}_2, \boldsymbol{k}_2\lambda_2|\mathrm{S}^{(1)}|\boldsymbol{p}_1, \boldsymbol{k}_1\lambda_1\rangle$$
$$= -\mathrm{i}\left(\frac{e^2c^2}{\hbar\varepsilon_0}\right)\frac{2\boldsymbol{e}_1\cdot\boldsymbol{e}_2}{\sqrt{2\varepsilon_{\boldsymbol{p}_2}2\varepsilon_{\boldsymbol{p}_1}2\omega_{\boldsymbol{k}_1}2\omega_{\boldsymbol{k}_2}}}L^{-3}\delta_{\boldsymbol{k}_1+\boldsymbol{p}_1,\boldsymbol{k}_2+\boldsymbol{p}_2}2\pi\delta(\varepsilon_1-\varepsilon_2) \quad . \tag{9.135}$$

Here, we have set

$$\begin{aligned}
\varepsilon_1 &= \omega_{\boldsymbol{k}_1} + \varepsilon_{\boldsymbol{p}_1} \quad , \quad \varepsilon_2 = \omega_{\boldsymbol{k}_2} + \varepsilon_{\boldsymbol{p}_2} \quad , \\
\boldsymbol{e}_1 &= \boldsymbol{e}_{\lambda_1}(\boldsymbol{k}_1) \quad , \quad \boldsymbol{e}_2 = \boldsymbol{e}_{\lambda_2}(\boldsymbol{k}_2) \quad ,
\end{aligned} \tag{9.136}$$

where $\omega_{\boldsymbol{k}} = c|\boldsymbol{k}|$ is the photon frequency and $\varepsilon_{\boldsymbol{p}} = c\sqrt{|\boldsymbol{p}|^2 + \mu^2}$ is the pion frequency (here $\boldsymbol{p}$ and $\varepsilon_{\boldsymbol{p}}$ have the dimensions of wavenumber and frequency). The amplitude (9.135) adds to those from the diagrams of Fig. 9.12b and c. We will show at the end of this section that the latter contributions are zero and thus that only the amplitude (9.135) contributes to the Compton cross-section.

Cross-Section

The initial state (9.133), normalized to unity, is that of a single pion and a single photon in L^3. If ρ_π and ρ_γ are the pion and photon densities in L^3, then there are $\rho_\pi L^3$ pions and $\rho_\gamma L^3$ photons in play. The total transition probability to the final state is thus

$$\rho_\pi L^3 \rho_\gamma L^3 |\langle \boldsymbol{p}_2, \boldsymbol{k}_2\lambda_2|\mathrm{S}^{(1)}|\boldsymbol{p}_1, \boldsymbol{k}_1\lambda_1\rangle|^2 \quad . \tag{9.137}$$

Note that the square of the matrix element involves

$$[\delta(\varepsilon_1 - \varepsilon_2)]^2 = \delta(0)\delta(\varepsilon_1 - \varepsilon_2) \quad , \tag{9.138}$$

which is formally infinite. This comes from the fact that we have considered the transition from $t = -\infty$ to $t = \infty$. If we had considered the transitions during a finite (but large) interval of time T, $-T/2 \leq t \leq T/2$, we would have instead of the Dirac function

$$\delta_{\mathrm{T}}(\omega) = \frac{1}{2\pi}\int_{-T/2}^{T/2} \mathrm{d}t\, \mathrm{e}^{\mathrm{i}\omega t} \quad , \quad \delta_{\mathrm{T}}(0) = \frac{T}{2\pi} \quad . \tag{9.139}$$

Taking into account the interpretation (9.139) of $\delta(0)$, the transition probability per unit volume and per unit time to the final state[2] becomes, with (9.135),

$$\begin{aligned}&(\rho_\pi L^3)(\rho_\gamma L^3)|\,\langle \boldsymbol{p}_2, \boldsymbol{k}_2\lambda_2|\mathrm{S}^{(1)}|\boldsymbol{p}_1, \boldsymbol{k}_1\lambda_1\rangle\,|^2\frac{1}{L^3T}\\ &= \rho_\pi\rho_\gamma\left(\frac{e^2c^2}{\hbar\varepsilon_0}\right)^2\frac{(\boldsymbol{e}_1\cdot\boldsymbol{e}_2)^2}{4\varepsilon_{\boldsymbol{p}_1}\varepsilon_{\boldsymbol{p}_2}\omega_{\boldsymbol{k}_1}\omega_{\boldsymbol{k}_2}}\frac{2\pi}{L^3}\delta_{\boldsymbol{k}_1+\boldsymbol{p}_1,\boldsymbol{k}_2+\boldsymbol{p}_2}\delta(\varepsilon_1-\varepsilon_2) \quad .\end{aligned} \tag{9.140}$$

Finally, we are interested in the probability (per unit volume and unit time) of the photon being scattered into a solid angle $\mathrm{d}\Omega$ in the direction $\boldsymbol{k}_2$. It is thus necessary to sum (9.140) over all possible final states with a photon scattered in $\mathrm{d}\Omega$.

We evaluate this quantity in the laboratory frame, in which the initial pion is at rest $\boldsymbol{p}_1 = 0$, $\varepsilon_{\boldsymbol{p}_1} = \mu c$ (Fig. 9.24). The angle θ between $\boldsymbol{k}_1$ and $\boldsymbol{k}_2$ is the scattering angle of the photon. Note that conservation of momentum fixes $\boldsymbol{p}_2 = \boldsymbol{k}_1 - \boldsymbol{k}_2$ and that $\varepsilon_2 = \varepsilon_{\boldsymbol{k}_1-\boldsymbol{k}_2} + \omega_{\boldsymbol{k}_2}$ is also a function of $\boldsymbol{k}_2$. To obtain the probability we are looking for, we only need to sum over the possible final states $\mathrm{d}k_2k_2^2$ with the photon scattered in $\mathrm{d}\Omega$. Taking the limit $L^3 \to \infty$, this probability becomes

$$\rho_\pi\rho_\gamma\left(\frac{e^2c^2}{\hbar\varepsilon_0}\right)\frac{|\boldsymbol{e}_1\cdot\boldsymbol{e}_2|^2}{4}\frac{\mathrm{d}\Omega}{(2\pi)^2}\int_0^\infty \mathrm{d}k_2k_2^2\frac{\delta(\varepsilon_1-\varepsilon_2)}{\mu c\varepsilon_{\boldsymbol{k}_1-\boldsymbol{k}_2}\omega_{\boldsymbol{k}_1}\omega_{\boldsymbol{k}_2}} \quad . \tag{9.141}$$

Finally, the differential cross-section $d\sigma/\mathrm{d}\Omega$ is obtained by dividing this quantity by the incident flux $\rho_\pi\rho_\gamma c$ and the element of solid angle $\mathrm{d}\Omega$, so that

$$\frac{\mathrm{d}\sigma}{\mathrm{d}\Omega} = \left(\frac{e^2c^2}{\hbar\varepsilon_0}\right)\frac{|\boldsymbol{e}_1\cdot\boldsymbol{e}_2|^2}{(4\pi)^2c}\int_0^\infty \mathrm{d}k_2k_2^2\frac{\delta(\varepsilon_1-\varepsilon_2)}{\mu c\varepsilon_{\boldsymbol{k}_1-\boldsymbol{k}_2}\omega_{\boldsymbol{k}_1}\omega_{\boldsymbol{k}_2}} \quad . \tag{9.142}$$

This differential cross-section, expressed as a function of the energy of the incident photon $\hbar c|\boldsymbol{k}_1|$ and the scattering angle θ, is found to be (see Exercise 4(ii))

[2] This offhand treatment of $\delta(0)$ can be justified by a discussion in terms of wavepackets rather than plane waves (see J. Taylor, Scattering Theory, Chap. 3, Wiley (1972)).

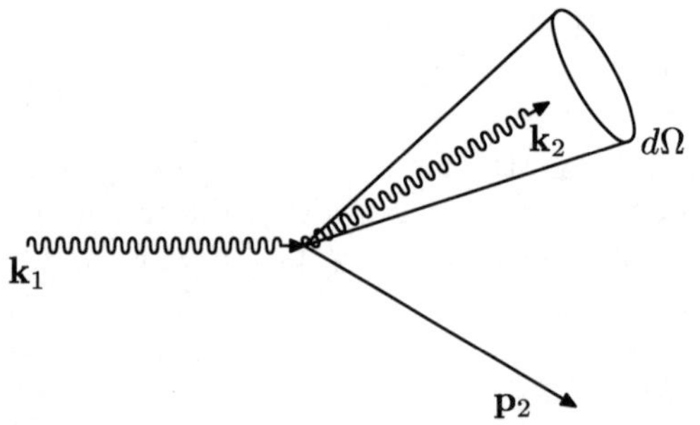

Fig. 9.24. Scattering of a photon with initial (final) state momentum $\boldsymbol{k}_1$ ($\boldsymbol{k}_2$) by a pion with initial (final) state momentum $\boldsymbol{p}_1 = 0$ ($\boldsymbol{p}_2$). The cross-section is calculated by summing over all possible final states with a photon scattered in the solid angle $\mathrm{d}\Omega$ in the direction $\boldsymbol{k}_2$

$$\frac{\mathrm{d}\sigma}{\mathrm{d}\Omega} = \alpha^2 \frac{|\boldsymbol{e}_1 \cdot \boldsymbol{e}_2|^2}{[\mu + |\boldsymbol{k}_1|(1-\cos\theta)]^2} \quad . \tag{9.143}$$

In (9.143), $\alpha = e^2/(4\pi\varepsilon_0\hbar c)$ is the (dimensionless) fine-structure constant and one recognizes the range of the nuclear force $\mu^{-1} = \hbar/mc$. In the case of a very low-energy photon, $|\boldsymbol{k}| \leq \mu^{-1}$, (9.161) is reduced to

$$\frac{\mathrm{d}\sigma}{\mathrm{d}\Omega} = \left(\frac{\alpha}{\mu}\right)^2 |\boldsymbol{e}_1 \cdot \boldsymbol{e}_2|^2 = r_\pi^2 |\boldsymbol{e}_1 \cdot \boldsymbol{e}_2|^2 \quad , \tag{9.144}$$

which is the classical Thomson formula for the scattering of low-energy radiation by a static charge of radius r_π. The length $r_\pi = \alpha/\mu = \alpha\hbar/mc$ can be called the *electromagnetic radius* of the pion.

Second-Order Diagrams

In the laboratory reference frame and in the Coulomb gauge, these diagrams do not contribute to Compton scattering. To see this, it is useful to consider the diagrams of Fig. 9.12b and c, which are given in Fig. 9.25 in the momentum representation. By taking the Fourier transforms, the scalar products $\nabla\Phi(\boldsymbol{x}) \cdot \mathtt{A}(\boldsymbol{x})$ and $\nabla\Phi^*(\boldsymbol{x}) \cdot \mathtt{A}(\boldsymbol{x})$ give rise to the products $\boldsymbol{p} \cdot \boldsymbol{e}_\lambda(\boldsymbol{k})$. Thus, the vectorial nature of the pion current has the effect of attributing to each vertex the factors $\boldsymbol{p} \cdot \boldsymbol{e}_\lambda(\boldsymbol{k})$ and $\boldsymbol{p}' \cdot \boldsymbol{e}_\lambda(\boldsymbol{k})$, where $\boldsymbol{p}$ and $\boldsymbol{p}'$ are associated to the pion lines and $\boldsymbol{e}_\lambda(\boldsymbol{k})$ is the polarization vector of a photon line. For the first vertex of the diagram of Fig. 9.25a, the factors are $\boldsymbol{p}_1 \cdot \boldsymbol{e}_\lambda(\boldsymbol{k}_1)$ and $\boldsymbol{p}' \cdot \boldsymbol{e}_\lambda(\boldsymbol{k}_1) = (\boldsymbol{p}_1 + \boldsymbol{k}_1) \cdot \boldsymbol{e}_\lambda(\boldsymbol{k}_1)$. Because of the transversality $\boldsymbol{k}_1 \cdot \boldsymbol{e}_\lambda(\boldsymbol{k}_1) = 0$ and since $\boldsymbol{p}_1 = 0$ in the laboratory reference frame, these factors are zero. The contribution from the diagram of Fig. 9.25a is thus zero, as is that from diagram Fig. 9.25b, for the same reasons. Thus, in the laboratory reference frame, the complete result (to second order in e^2) is given by (9.143).

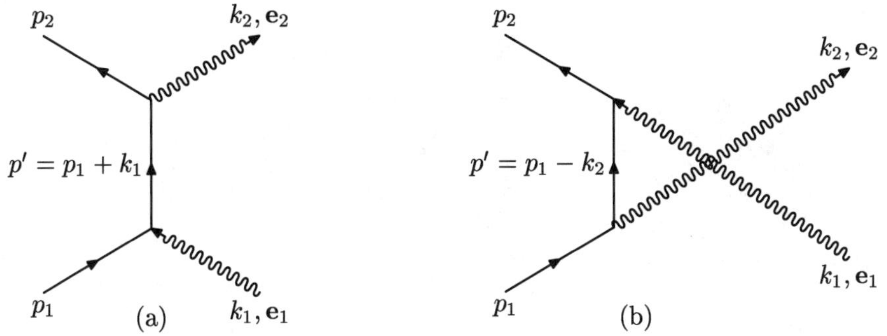

Fig. 9.25a,b. Second-order diagrams corresponding to the diagrams of Fig. 9.12b and c. In the momentum representation, the products $\boldsymbol{p} \cdot \boldsymbol{e}_\lambda(\boldsymbol{k})$ and $\boldsymbol{p}' \cdot \boldsymbol{e}_\lambda(\boldsymbol{k})$ are associated to each vertex, where $\boldsymbol{p}$ and $\boldsymbol{p}'$ are associated to the pion lines and $\boldsymbol{e}_\lambda(\boldsymbol{k})$ is the polarization vector of a photon line

Creation and Annihilation of Pairs

It is interesting to note that the diagram of Fig. 9.23 and the two diagrams of Fig. 9.25 also describe either the annihilation of a $\pi^+\pi^-$ pair into two photons or the creation of a $\pi^+\pi^-$ pair by two photons.

In the first case, the momenta of the incoming $\pi^+\pi^-$ pair are $(\boldsymbol{p}_1, -\boldsymbol{p}_2)$ while the momenta of the outgoing photons are $(-\boldsymbol{k}_1, \boldsymbol{k}_2)$. To obtain the annihilation cross-section of the pair, it is necessary to sum over all final states $-\boldsymbol{k}_1$ and $\boldsymbol{k}_2$ and the polarizations of the photons. In the same manner, one would describe the creation of the pion pair with initial-state photons $(\boldsymbol{k}_1, -\boldsymbol{k}_2)$ by summing over all momenta $(-\boldsymbol{p}_1, \boldsymbol{p}_2)$ of the final-state pions.

9.4.3 Quantum Electrodynamics: Radiative Corrections

The quantum electrodynamics of electrons and photons is the theory of the Maxwell electromagnetic field coupled to the Dirac relativistic electron field. As we know, its predictions are verified experimentally to an astonishing precision. Without being in a position here to carry out a calculation, we qualitatively present, with the help of Feynman diagrams, several effects peculiar to quantum electrodynamics. First of all, it should be noted that the interaction Hamiltonian of quantum electrodynamics is of the same form as (8.50)

$$\mathtt{H}_\mathrm{I} = \int \mathrm{d}x \, j_\mu(x) \mathtt{A}^\mu(x) \quad , \tag{9.145}$$

where the classical current is replaced by Dirac's quantum-electric current j_μ, quadratic in the electron field. Thus electrodynamics has only one type of vertex, that of Fig. 9.13, and the formation of diagrams is similar to those

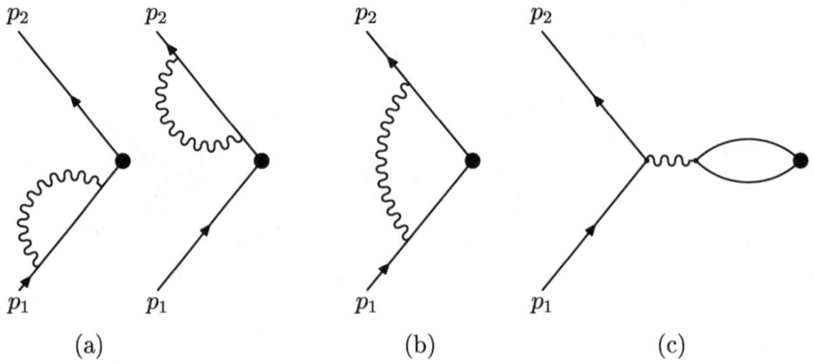

Fig. 9.26a–c. Radiative corrections to order e^2 for the scattering of an electron by an external potential

we have discussed in Sect. 9.3.6. In particular, the Compton amplitude to order e^2 is given by the diagrams of Fig. 9.12b and c. The corresponding amplitudes are non-zero because of the fact that the Dirac current does not involve the field gradient, contrary to the case of scalar particles.

Radiative corrections are effects due to higher-order contributions to a given process. To make this clear, we consider the scattering of an electron by an external potential for which the lowest-order diagram is that of Fig. 9.20a. Corrections of order e^2 for this scattering are given in Fig. 9.26.

These corrections are modifications to the self-energy of the electron (Fig. 9.26a) and the vertex (Fig. 9.26b) due to the emission and absorption of virtual photons, as well as the polarization of the vacuum (Fig. 9.26c) by the emission and absorption of virtual pairs. These corrections are measurable effects which cause the Lamb shift $2S_{1/2} - 2P_{1/2}$ of the hydrogen atom. Before giving a qualitative description, it is necessary to mention an artifact of the perturbation calculation, the problem of renormalization.

Renormalization

Consider the complete Hamiltonian $\mathtt{H}$ of the electrons and photons and the state $|\varphi\rangle$ of a single electron at rest. As the electron is a stable particle, $|\varphi\rangle$ must be an eigenstate of $\mathtt{H}$

$$\mathtt{H}\,|\varphi\rangle = m\,|\varphi\rangle \tag{9.146}$$

and the eigenvalue m must be the *observed mass* of an electron at rest. The necessities of the perturbation calculation compel us to make an *arbitrary separation* $\mathtt{H} = \mathtt{H}_0 + \mathtt{H}_\mathrm{I}$ where $\mathtt{H}_0$ is the Hamiltonian of the hypothetically non-coupled electrons and photons. In $\mathtt{H}_0$ appears a mass parameter for the electron, denoted m_0 here, such that

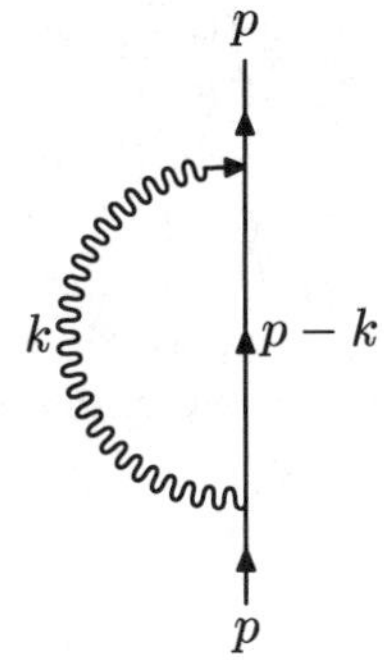

Fig. 9.27. Self-energy diagram of a single electron submitted to no external perturbation

$$\mathrm{H}_0 \left|\varphi_0\right\rangle = m_0 \left|\varphi_0\right\rangle \quad . \tag{9.147}$$

In (9.147), $|\varphi_0\rangle$ is an electron state which would not be perturbed by the presence of photons. *In reality, the electrons and the photons are always coupled and the parameter m_0 has no physical significance: it is only the quantity m which can be measured by experiment.* In general, the eigenvalues m and m_0 are different, but the perturbation calculation allows one to determine m as a function of m_0 by a power series of the dimensionless coupling constant α

$$m - m_0 = \alpha f^{(1)}(m_0) + \alpha^2 f^{(2)}(m_0) + \dots \quad . \tag{9.148}$$

To obtain a physical prediction, it is thus necessary to rewrite the results of the perturbation calculation, expressed of course in terms of the arbitrary parameter m_0, as a function of the physical mass m. For this, it is necessary to invert relation (9.148): this operation is called *mass renormalization*. In the calculations, one must also take into account, using an appropriate renormalization law, the fact that the physical state of the electron is $|\varphi\rangle$ and not $|\varphi_0\rangle$. This is called amplitude renormalization and it also occurs for the photon.

These renormalization operations could be carried out easily in principle, if it were not for the fact that *all of the terms of the series* (9.148) *are infinite*. The first term of the series (9.148) can be obtained from the self-energy diagram of Fig. 9.27 of a single electron not under the influence of any external perturbation and in the static limit $\boldsymbol{k} \to 0$. The corresponding Feynman integral, which is carried out over the product of the electron and photon propagators is divergent: the result is infinite. This divergence is due to the fact that the electron is a point particle and that the interaction (9.145) is strictly local. One should note that this phenomenon is already present in classical electrodynamics: the self-energy of a charged point particle is infinite. In spite of this difficulty, it is possible to manipulate the series (9.148) with sufficient precautions and skill to carry out the renormalization program to all orders. This task, accomplished by Feynman, Tomonaga, Schwinger, Dyson, Stückelberg and others, requires the use of the covariant relativistic formalism. *At the end of the analysis, the infinities are eliminated by introducing the physical mass m and, concerning the amplitude renormalization, by redefining the coupling constant as the physically observed charge of the electron.* Having made these operations when calculating the radiative corrections of Fig. 9.26, *one finds finite results* which can be compared with observation.

Origin of the Lamb Shift

When an electron is under the influence of the Coulomb potential of a hydrogen nucleus, Dirac's theory predicts that the levels $2S_{1/2}$ and $2P_{1/2}$ be degenerate. Measurements made by Lamb and Rutherford in 1947 showed that these two levels are distinct and separated by an energy of 1057 megacycles. *This displacement is attributed to the existence of the radiative corrections of* Fig. 9.26, with the slight difference that here the electron forms a bound state with the proton. Here we show in a succinct way how one can describe this phenomenon.

As the electron is no longer free, but under the influence of the field of the proton, the calculation of the self-energy (Fig. 9.26a) gives rise, after mass renormalization, to a non-zero contribution. One can show that all of this takes place as if the electron, in the presence of the photons, takes on a certain spatial extension which, for dimensional reasons, is of the order $r_0 = \hbar/mc \simeq 10^{-11}$ cm. The potential $V^{\text{eff}}(\boldsymbol{x})$ felt by this electron "dressed" by photons is the average of the Coulomb potential over a region of linear dimension r_0, that is

$$V^{\text{eff}}(\boldsymbol{x}) = -\left(\frac{3}{4\pi r_0}\right)^3 \int_{|\boldsymbol{y}|\le r_0} \mathrm{d}\boldsymbol{y}\, \frac{e^2}{4\pi\varepsilon_0|\boldsymbol{x}-\boldsymbol{y}|} \quad . \tag{9.149}$$

V^{eff} differs from the strict Coulomb potential over distances less than r_0, and one can conceive that this difference influences the energy spectrum. In fact, as r_0 is much smaller than the Bohr radius, only the $2S_{1/2}$ level is affected.

To understand the influence of the vacuum polarization (Fig. 9.26c) on the levels, one can reason by analogy with the electron-gas theory of Chap. 4. An electron–positron pair possesses a dipole moment. It is as if the vacuum, in the presence of the photons, behaves as a polarizable medium. From the phenomenological point of view, such a medium is characterized by a dielectric function $\varepsilon(\boldsymbol{k},\omega)$ and thus by an effective potential which is written as (4.89). While the screening phenomenon is related to the behaviour of the dielectric function at small $\boldsymbol{k}$, here $\varepsilon(\boldsymbol{k},\omega)$ is only distinguishable from ε_0 for values $|\boldsymbol{k}| \ge mc/\hbar = r_0^{-1}$, that is when the photon is sufficiently energetic to create a pair. As a consequence, the effective potential in the configuration space only differs from the Coulomb potential for $|x| \le r_0$, and only the $2S_{1/2}$ level is affected by the vacuum polarization.

The vertex correction (Fig. 9.26b) comes into the evaluation of the Lamb shift by means of the anomalous magnetic moment of the electron. If the diagrams of Fig. 9.20a and Fig. 9.26b represent the motion of the electron in a magnetic field, they allow one to calculate its magnetic moment. Indeed, in the limit where $\boldsymbol{p}_1$ and $\boldsymbol{p}_2$ tend to zero, the only degree of freedom of the electron influenced by the field is its spin, which gives rise to the magnetic moment. The radiative correction of Fig. 9.26b modifies the gyromagnetic ratio, predicted to be equal to 2 by the Dirac equation: it gives rise to the anomalous magnetic moment of the electron. To order α, one finds the correction α/π to be in excellent agreement with experiment. The anomalous magnetic moment only affects the levels with non-zero orbital angular momentum $\boldsymbol{L}$, that is in the present case the $2P_{1/2}$ level. The shift of the level thus comes from the interaction of the magnetic moment associated to $\boldsymbol{L}$ with the anomalous magnetic moment.

A precise calculation of the Lamb shift predicts a value of 1057.19 megacycles. The remarkable agreement with experiment confirms, along with measurements of other physical quantities, the accuracy of quantum electrodynamics. *This theory extends the validity of Maxwell's equations to scales of extremely short distance and continues to serve as an uncontested model in the most recent developments of quantum field theory.*

9.4.4 Electron–Phonon Interactions

We have seen in Sect. 8.4.3 that it is possible, after certain approximations, to formulate the electron–phonon interaction as a local field theory. The complete Hamiltonian includes the kinetic energy of the electrons (3.72) and the free phonons (8.232) to which one adds the interaction term (8.252). In this setting, it is perfectly legitimate and profitable to apply the diagrammatic methods of field theory to such a many-body problem. One considers that the ground state is that of a gas of independent fermions and the phonon vacuum (the Coulomb interaction of the electrons is, in part, taken into account using an effective potential (8.248)). The electron and phonon propagators are given respectively by (9.106) and (9.118) and the vertex is given by Fig. 9.13. We discuss here two phenomena in which the electron–phonon interaction plays a central role, *the resistivity of metals and superconductivity.*

Electrical Resistivity

To calculate the resistivity of a metal, it is necessary to study the collisions that the electron gas undergoes. Conceptually, it is easier to represent the collisions as localized in space and time. Between two impacts, separated by an average time interval τ (the mean time of flight), the charge carrier is accelerated by the electric field. During the collision, there is a momentum exchange: one conceives that just after the collision the charge carrier returns

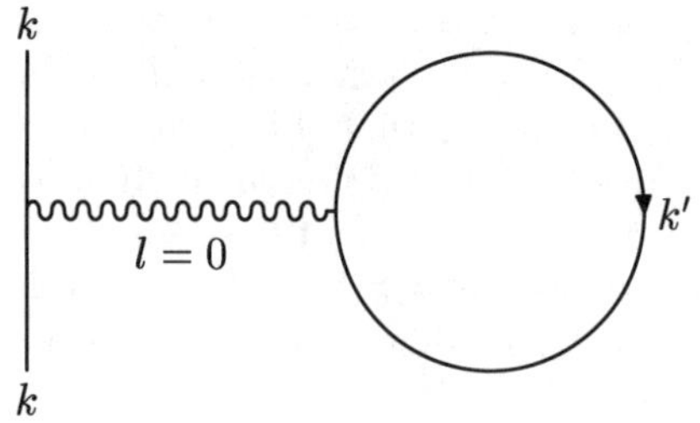

Fig. 9.28. Second-order diagram (Fig. 9.18) written in the energy–momentum representation has an internal phonon line of zero momentum

to the statistical state of the electrons (or holes) in the absence of the field (Stosszahlansatz). For electrons in a crystal, these collisions occur at the sites which break the translation symmetry. The phonons, oscillations of positive charge, break this symmetry locally and temporarily. It is thus not surprising that at the ambient temperature, at which there are many phonons, *the electron–phonon interaction must be the main source of the electrical resistivity in metals.*

Without making an exhaustive study of transport phenomena, we will show that it is plausible that certain diagrams appear in the calculation of resistivity. It is evident that a transport coefficient, such as the resistivity, increases when the transition probabilities between electronic states become larger, whatever the cause of these transitions may be (collision with an impurity, static defect of the lattice, an electron or a phonon). The contribution to the electrical resistivity due to phonons (called ideal),[3] is related to the electron–phonon interaction cross-section, which is calculated in the same way as in particle physics; only the propagators are different.

Consider, for example, the diagrams which describe the transition between an initial state corresponding to an electron of momentum $\hbar\boldsymbol{k}$ and a final state comprising of a phonon $\hbar\boldsymbol{l}$ and an electron $\hbar(\boldsymbol{k}-\boldsymbol{l})$. The first-order contributions come from the diagrams of Fig. 9.16a. The second-order diagrams (Figs. 9.17, 9.18 and 9.27) have either two or no external phonon lines, and thus they do not contribute to the process. Note that the diagram of Fig. 9.18, written in the momentum–energy representation, always has an internal phonon line of zero momentum (Fig. 9.28). It makes no contribution, since the phonon propagator $D(\boldsymbol{k},\omega)$ (9.127) vanishes at $|\boldsymbol{k}|=0$. Figure 9.29 gives examples of third-order diagrams which contribute to the process of phonon emission.

[3] A perfect unlimited crystal has no defects, impurities or boundaries which may scatter the electron. At finite temperature, there still remain collisions with the phonons and other electrons, but the latter contribution is negligible. One thus speaks of ideal resistivity in the case where only the phonons are responsible for the scattering of the electrons.

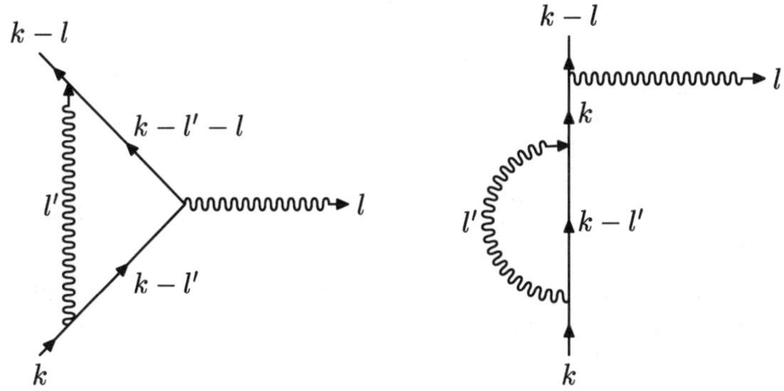

Fig. 9.29. Examples of third-order diagrams which contribute to the process of phonon emission

Electron–Electron Interactions Transmitted by Phonons

In Chap. 5, we have shown how an ensemble of fermions (electrons) can benefit from an attractive interaction to form a superconducting phase. We have thus adopted the phenomenological point of view, which consists of describing a Hamiltonian (5.48) composed of a kinetic term and a term for the interaction between the electrons for which one supposes that certain matrix elements, being negative, lead to an attraction. To understand the origin of this interaction, it is useful to adopt a more fundamental point of view, which consists of describing the interaction being due to the emission of phonons by an electron, where these phonons are absorbed by another electron. This approach is very similar to the one we have followed in Sect. 8.3.2 and which reduced the Yukawa potential to an exchange of pions between nucleons. The isotopic effect (see Sect. 5.3.1) indeed shows that the electron–phonon interaction is responsible for the mechanism giving rise to traditional superconductivity. In the case of high-temperature superconductivity, the absence of an isotopic effect seems to indicate that one must search elsewhere for the cause of the attraction.

In Fig. 3.2, we show the basic element describing of the interaction of particles under the influence of a phenomenological potential $\tilde{V}(\boldsymbol{k})$, $\hbar\boldsymbol{k}$ being the momentum transfer. Two vertices are represented in Fig. 9.30a, corresponding to the emission and absorption of a phonon. When one combines the two diagrams, as in Fig. 9.30b, one finds a diagram which is second-order in the electron–phonon coupling constant γ, identical to that of Fig. 3.2. It is thus natural to globally identify this diagram to the first-order effect of a phenomenological two-body potential $\tilde{V}(\boldsymbol{k},\omega)$, resulting from the exchange of a phonon of momentum–energy $(\boldsymbol{k},\omega)$. This identification must be made in accordance with the Feynman rules (see Sect. 9.3.6), namely that the undulating line corresponds to a phonon propagator $\mathrm{i}D(\boldsymbol{k},\omega)$ and a factor $-\mathrm{i}\gamma/\hbar$

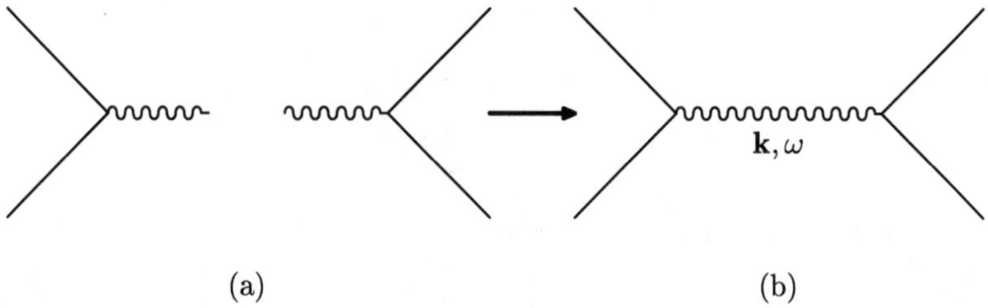

Fig. 9.30. The two vertices of (**a**), corresponding to the emission and absorption of a phonon by an electron, are combined in (**b**) to yield a diagram which is second-order in the electron–phonon coupling constant γ

is attributed to each vertex of Fig. 9.30b. The diagram of Fig. 9.30b thus gives rise to the contribution $(-\mathrm{i}\gamma/\hbar)^2 \mathrm{i}D(\boldsymbol{k},\omega)$ which must be identified with the first-order contribution from the potential $\tilde{V}(\boldsymbol{k},\omega)$, that is $(-\mathrm{i}/\hbar)\tilde{V}(\boldsymbol{k},\omega)$ (the factor $-\mathrm{i}/\hbar$ affects the first-order term of the perturbative series, see (9.63)). One thus finds, taking into account (9.118),

$$\begin{aligned}\tilde{V}(\boldsymbol{k},\omega) &= \mathrm{i}\hbar\left(\frac{\mathrm{i}\gamma}{\hbar}\right)^2 \mathrm{i}D(\boldsymbol{k},\omega) \\ &= \gamma^2 \frac{\omega_{\boldsymbol{k}}^2}{\omega^2-(\omega_{\boldsymbol{k}}-\mathrm{i}\eta)^2}\theta(\omega_{\mathrm{D}}-\omega_{\boldsymbol{k}}) \quad . \end{aligned} \tag{9.150}$$

To lowest order, this phenomenological interaction thus depends on the frequency: *it is not instantaneous, but retarded.* This is a consequence of the fact that it is transmitted by phonons, which are very slow relative to the electrons (in (9.150), $\omega_{\boldsymbol{k}}$ is of the order $c_{\mathrm{L}}|\boldsymbol{k}| \simeq c_{\mathrm{L}}/a \simeq 10^{13}s^{-1}$, whereas an electron at the Fermi surface is characterized by a frequency $\omega = \varepsilon_{\mathrm{F}}/\hbar \simeq 10^{15}s^{-1}$). At the static limit $\omega = 0$, it becomes

$$\tilde{V}(\boldsymbol{k},0) = -\gamma^2\theta(\omega_{\mathrm{D}}-\omega_{\boldsymbol{k}}) \quad . \tag{9.151}$$

Thus, for small values of ω, $\tilde{V}(\boldsymbol{k},\omega)$ is attractive and essentially independent of $\boldsymbol{k}$ for $\omega_{\boldsymbol{k}} \le \omega_D$. On the other hand, the momentum transfer $\hbar\boldsymbol{k} = \hbar\boldsymbol{k}_1 - \hbar\boldsymbol{k}_2$ between two electrons of energy close to the Fermi energy ($\varepsilon_{\boldsymbol{k}_1} \simeq \varepsilon_{\boldsymbol{k}_2} \simeq \varepsilon_{\mathrm{F}}$) is limited by the condition $\omega_{\boldsymbol{k}} \le \omega_{\mathrm{D}}$ to a thin layer of thickness $\hbar\omega_{\mathrm{D}}$ in the neighbourhood of ε_{F}. In the ground state where non-interacting electrons fill the Fermi sphere, $\tilde{V}(\boldsymbol{k},\omega)$ can only display its attractive behavior and modify the electron distribution in the immediate neighbourhood of ε_{F}.

9.4.5 Diagram Summation

We have seen that each diagram corresponds to a possible term of the perturbation calculation. If we want to take into account the higher-order corrections of a given process, we must sum all the diagrams which contribute to the same process. To clarify this notion, we return to the amplitude of the Compton scattering of a boson by a fermion characterized to second order by the diagram of Fig. 9.12b. To fourth order we have the diagrams of Fig. 9.31 contributing to the same process.

We can classify the corrections into two groups. The first three types are characterized by the fact that they modify a fermion line (Fig. 9.31a), a boson line (Fig. 9.31b) or a vertex (Fig. 9.31c). *These are the self-energy corrections, the vacuum polarization and the vertex polarization.* It is clear that they are not specific to this process and occur as corrections in any diagram. As for the corrections of Fig. 9.31d, these are specific to the Compton effect diagram. It is interesting to be able to systematically treat corrections of the type (a), (b) and (c) by means of a graphical method. Here we treat the case of the fermion propagator.

Energy Insertion

We call any modification to a free fermion line joining x' to x an *energy insertion*. Energy insertions to lowest orders are represented in Fig. 9.32. According to the Feynman rules, each energy insertion corresponds to an algebraic expression which depends on x_1 and x_2. We denote the sum over all energy insertions by $S(x_2|x_1)$, which we represent graphically by a double-hatched circle, as in Fig. 9.32. The complete propagator $G(x|x')$ of the fermion in the presence of bosons, or the "dressed" propagator, is obtained by adding the free propagator $G_0(x|x')$ to all the energy insertions and is symbolized by the double line of Fig. 9.33. The algebraic counterpart of Fig. 9.33 is the equation

$$G(x|x') = G_0(x|x') + \int \mathrm{d}x_2 \int \mathrm{d}x_1\, G_0(x|x_2)S(x_2|x_1)G_0(x_1|x') \quad , \tag{9.152}$$

which can be written in the abbreviated form

$$G = G_0 + G_0 S G_0 \quad . \tag{9.153}$$

Self-Energy

Since $S(x_2|x_1)$ is unknown, we cannot calculate $G(x|x')$ from (9.152). Starting from (9.152) we can find another equation which allows one to calculate $G(x|x')$ systematically. With this in mind, we introduce the concept of the *self-energy insertion*. An energy insertion is a self-energy insertion if it cannot be separated into two disconnected pieces by cutting a single fermion line. For example, the diagrams of Fig. 9.32b and f are not self-energy insertions. We define $\Sigma(x_2|x_1)$ as the sum of all the self-energy insertions, which we represent by a circle with a hatched filling (Fig. 9.34). The relation between

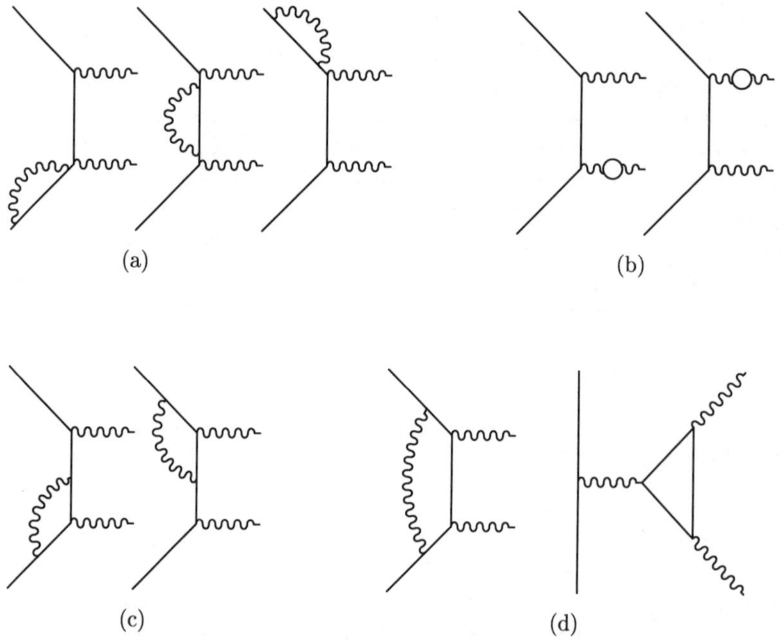

Fig. 9.31a–d. Fourth-order diagrams contributing to the Compton scattering of a boson by a fermion and corresponding to the second-order diagram of Fig. 9.12b

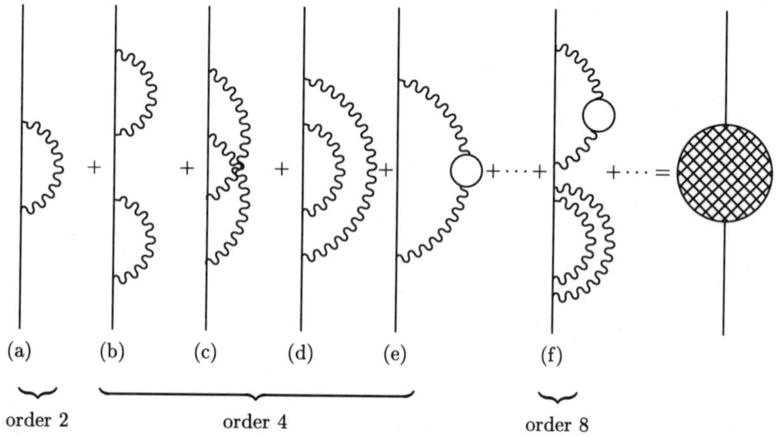

Fig. 9.32a–f. Lowest-order energy insertions of a free fermion line

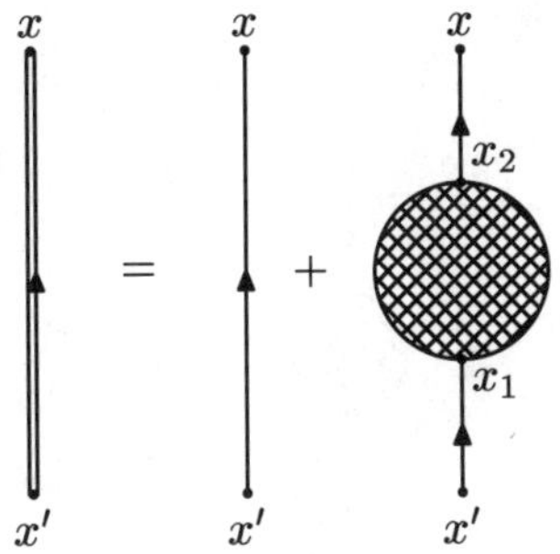

Fig. 9.33. The complete propagator G (*double line*) is the sum of the free propagator G_0 (*single line*) and the sum of the energy insertions S (*line with double-hatched circle*)

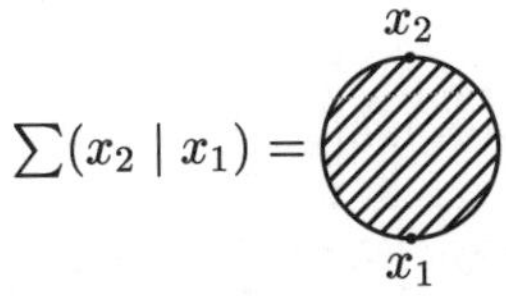

Fig. 9.34. The sum of all self-energy insertions is represented by a circle with a hatched filling

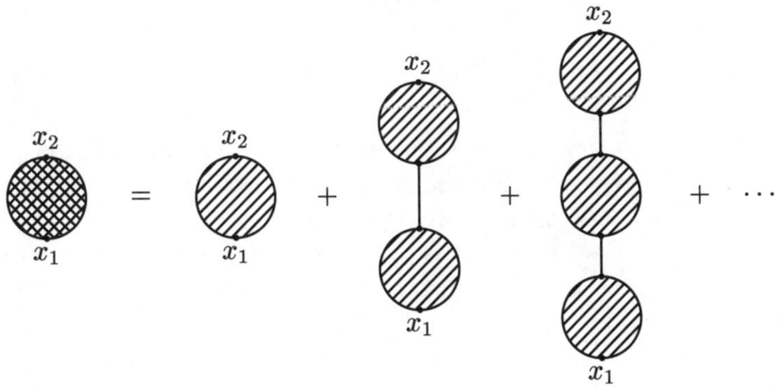

Fig. 9.35. The sum of all energy insertions S (*double-hatched circle*) can be represented as the sum of self-energy insertions (*hatched-circle diagrams*)

$S(x_2|x_1)$ and $\Sigma(x_2|x_1)$ is shown in Fig. 9.35. That is,

$$S = \Sigma + \Sigma G_0 \Sigma + \Sigma G_0 \Sigma G_0 \Sigma + \dots \quad . \tag{9.154}$$

To check the validity of (9.154), we take any energy insertion. If it is a self-energy insertion, it is by definition a term of Σ. If it is not, then it can be

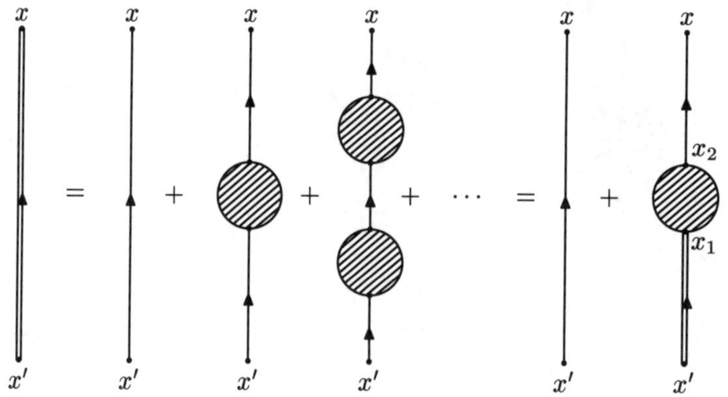

Fig. 9.36. Graphical representation of (9.155) and (9.156)

reduced into self-energy insertions by breaking a finite number of fermion lines, and it is thus necessarily a term of $\Sigma G_0 \Sigma G_0 \dots G_0 \Sigma$. The infinite sum (9.154) thus indeed represents S. Substituting (9.154) into (9.153), one obtains

$$G = G_0 + G_0 \Sigma G_0 + G_0 \Sigma G_0 \Sigma G_0 + \dots \quad , \tag{9.155}$$

which is none other than the series which results by iteration from

$$G(x|x') = G_0(x|x') + \int \mathrm{d}x_2 \int \mathrm{d}x_1 \, G_0(x|x_2) \Sigma(x_2|x_1) G(x_1|x') \quad . \tag{9.156}$$

Equations (9.155) and (9.156) are represented in Fig. 9.36. An equation of type (9.156) is called a *Dyson equation.* It is an integral equation for the complete propagator $G(x|x')$ which can in principle be solved when $\Sigma(x_2|x_1)$ is known. One sees that it has exactly the same structure as (9.36) (Fig. 9.6) which gives the Green function of an electron in an external field. Here the effect of the external field is replaced by *the self-energy $\Sigma(x_2|x_1)$ which takes into account the virtual bosons which "dress" the fermion propagator.*

The interpretation of the self-energy is particularly clear when the system is invariant under translations in space and time. In this case, $\Sigma(x_2|x_1) = \Sigma(x_2 - x_1)$ depends only on the difference of the arguments; one can introduce the Fourier transform

$$\Sigma(\boldsymbol{k}, \omega) = \int \mathrm{d}\boldsymbol{x} \int \mathrm{d}t \, \mathrm{e}^{-\mathrm{i}(\boldsymbol{k}\cdot\boldsymbol{x} - \omega t)} \Sigma(\boldsymbol{x}, t) \tag{9.157}$$

and apply to (9.156) the convolution theorem

$$G(\boldsymbol{k}, \omega) = G_0(\boldsymbol{k}, \omega) + G_0(\boldsymbol{k}, \omega) \Sigma(\boldsymbol{k}, \omega) G(\boldsymbol{k}, \omega) \quad . \tag{9.158}$$

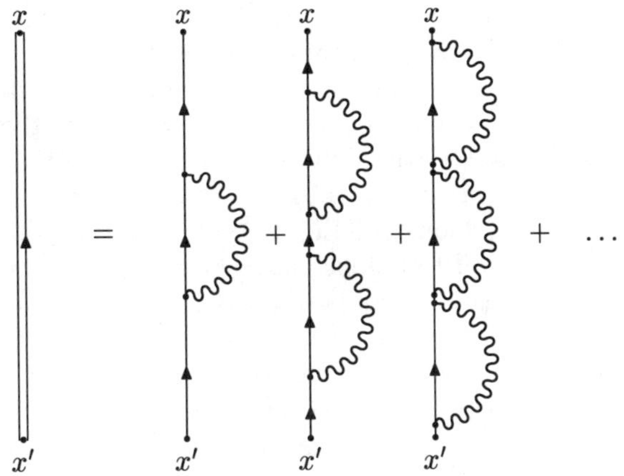

Fig. 9.37. Taking $\Sigma \simeq \Sigma^{(2)}$, the propagator G (double line) is an infinite sum of the "loop" diagrams on the right

Equation (9.158) is algebraic and can be easily solved

$$G(\boldsymbol{k},\omega) = \frac{1}{G_0^{-1}(\boldsymbol{k},\omega) - \Sigma(\boldsymbol{k},\omega)} \quad . \tag{9.159}$$

If G includes the spin indices, it is also necessary to invert a matrix in the space of the spin states. One must attribute to ω an imaginary part with the right sign, as in the expression for $G_0(\boldsymbol{k},\omega)$ (see (9.25) and (9.106)). For example, in the case where $G_0(\boldsymbol{k},\omega)$ is the propagator of a non-relativistic free electron (9.106), one must distinguish between the cases $|\boldsymbol{k}| < k_{\mathrm{F}}$ and $|\boldsymbol{k}| > k_{\mathrm{F}}$, and one obtains

$$G(\boldsymbol{k},\omega) = \hbar\left(\frac{\theta(|\boldsymbol{k}| - k_{\mathrm{F}})}{\hbar(\omega + \mathrm{i}\eta) - \varepsilon_{\boldsymbol{k}} - \hbar\Sigma(\boldsymbol{k},\omega + \mathrm{i}\eta)} + \frac{\theta(k_{\mathrm{F}} - |\boldsymbol{k}|)}{\hbar(\omega - \mathrm{i}\eta) - \varepsilon_{\boldsymbol{k}} - \hbar\Sigma(\boldsymbol{k},\omega - \mathrm{i}\eta)}\right) \quad . \tag{9.160}$$

We have seen in Sect. 9.2.1 that the singularities of $G(\boldsymbol{k},\omega)$ on the real axis ω give the eigenenergies of the system. They are now determined by the zeros of the denominator of (9.159), confirming the interpretation of $\Sigma(\boldsymbol{k},\omega)$. Depending on the nature of $\Sigma(\boldsymbol{k},\omega)$, it is possible to obtain complex zeros: one interprets the imaginary part as the lifetime of the excitation.

The quantity $\Sigma(\boldsymbol{k},\omega)$ is composed of the sum of an infinite number of diagrams

$$\Sigma(\boldsymbol{k},\omega) = \sum_n \Sigma^{(n)}(\boldsymbol{k},\omega) \quad , \tag{9.161}$$

where $\Sigma^{(n)}(\boldsymbol{k},\omega)$ is formed of the self-energy insertions of order n; $\Sigma^{(2)}$ is given by the diagram of Fig. 9.32a, $\Sigma^{(4)}$ by the diagrams of Fig. 9.32c–e, and so on. In practice, one stops at a fixed order. The interest of the method lies in the fact that even if Σ only takes into account a finite number of diagrams, the propagator G corresponding to (9.159) always results from an (albeit partial) infinite sum of diagrams. For example, if one takes $\Sigma \simeq \Sigma^{(2)}$ to calculate G, one would sum the infinite series of "loops" of Fig. 9.37. It emerges from this demonstration that the summation techniques are general and can be applied in the same manner to modifications of a boson line. In this case, one finds the boson propagator to be "dressed" by vacuum polarization effects (electron–positron pairs) or by polarization of matter (electron–hole pairs). We will treat explicit examples in Sects. 10.2.1 and 10.2.2 when we give a diagrammatic interpretation to the Hartree–Fock and RPA approximations.

Exercises

1. The Green Function of the Harmonic Oscillator

The retarded Green function $G_0(t,t')$ of a classical harmonic oscillator of frequency Ω_0 obeys $(\mathrm{d}^2/\mathrm{d}t^2 + \Omega_0^2)G_0(t,t') = -\delta(t-t')$, $G_0(t,t') = 0$, $t < t'$.

(i) Show that $G_0(t,t') = -\theta(t-t')\sin\Omega_0(t-t')/\Omega_0$.

(ii) If the frequency of the oscillator is perturbed to $\Omega^2 = \Omega_0^2 + \Omega_1^2$, the new Green function $G(t,t')$ is solution of the integral equation

$$G(t,t') = G_0(t,t') + \int \mathrm{d}t_1\, G_0(t,t_1)\Omega_1^2 G(t_1,t') \quad .$$

Calculate $G(t,t')$ by summing the corresponding perturbation series.

Hint: Use the Fourier representation and the convolution theorem.

Comment: This elementary example shows that it is necessary to re-sum infinitely many terms in the perturbation series to correctly obtain the new frequency $\Omega = \sqrt{\Omega_0^2 + \Omega_1^2}$.

2. The Yukawa Potential

Consider non-relativistic nucleons of mass m, described by a Fermi field, interacting with the field of massive scalar particles (mesons) by means of the interaction Hamiltonian (9.81) (spins are neglected). Show that the second-order process corresponding to the scattering of two nucleons by the exchange of a meson can be identified with the Yukawa potential (8.112) in the static limit.

Hint: Define an effective potential $\tilde{V}(\boldsymbol{k},\omega)$ as in the electron–phonon case (Fig. 9.30) with momentum transfer $\hbar\boldsymbol{k} = \hbar\boldsymbol{k}_1 - \hbar\boldsymbol{k}_2$. The energy transfer $\hbar\omega = \varepsilon_{\boldsymbol{k}_1} - \varepsilon_{\boldsymbol{k}_2}$, $\varepsilon_{\boldsymbol{k}} = \hbar^2|\boldsymbol{k}|^2/2m$, vanishes in the limit of infinitely heavy nucleons.

3. Electron–hole coupling

A free Fermi gas with density ρ is submitted to an external perturbation

$$V = v_0 \left|\boldsymbol{p}\right\rangle \left\langle\boldsymbol{q}\right| + v_0^* \left|\boldsymbol{q}\right\rangle \left\langle\boldsymbol{p}\right| \quad , \quad |\boldsymbol{p}| > k_{\mathrm{F}} \quad , \quad |\boldsymbol{q}| < k_{\mathrm{F}}$$

that couples only the plane-wave states $|\boldsymbol{p}\rangle = a_{\boldsymbol{p}}^* |0\rangle$ and $|\boldsymbol{q}\rangle = a_{\boldsymbol{q}}^* |0\rangle$ of an electron and a hole.

(i) Calculate the diagonal matrix elements of the Green function for $|\boldsymbol{k}| > k_{\mathrm{F}}$:

$$G(\boldsymbol{k},\omega + \mathrm{i}\eta|\boldsymbol{k}) = \frac{\hbar}{\hbar(\omega + \mathrm{i}\eta) - \varepsilon_{\boldsymbol{k}}} \quad , \quad \boldsymbol{k} \neq \boldsymbol{p} \quad ,$$

$$G(\boldsymbol{p},\omega + \mathrm{i}\eta|\boldsymbol{p}) = \frac{\hbar}{\hbar(\omega + \mathrm{i}\eta) - \varepsilon_{\boldsymbol{p}} - (\hbar^2|v_0|^2)/[\hbar(\omega - \mathrm{i}\eta) - \varepsilon_{\boldsymbol{q}}]}$$

and discuss the spectrum.

Hint: Notice that most of the matrix elements of the interaction vanish: $\langle\boldsymbol{k}|V^n|\boldsymbol{k}\rangle = 0$, $|\boldsymbol{k}| > k_{\mathrm{F}}$ unless n is even and $\boldsymbol{k} = \boldsymbol{p}$. Use the form (9.106) of the free Fermi propagator.

Comment: The new poles $\hbar\omega_\pm = (\varepsilon_{\boldsymbol{p}} + \varepsilon_{\boldsymbol{q}})/2 \pm (1/2)\sqrt{(\varepsilon_{\boldsymbol{p}} - \varepsilon_{\boldsymbol{q}})^2 + 4\hbar^2|v_0|^2}$ are identical to the eigenvalues obtained by diagonalizing the Hamiltonian in the two-dimensional one-particle subspace generated by $|\boldsymbol{p}\rangle$ and $|\boldsymbol{q}\rangle$.

4. Compton's Cross-Section

(i) Derive (9.135).

(ii) Establish (9.143) for the differential cross-section as a function of the incident photon wavenumber $|\boldsymbol{k}_1|$ and the scattering angle θ.

Hint: Use energy conservation

$$\varepsilon_2 = \sqrt{k_1^2 + k_2^2 - 2k_1k_2\cos\theta + \mu^2} + ck_2 = \mu c + ck_1$$

to express ε_2 and k_1 as functions of k_2.

10. Perturbative Methods in Many-Body Problems

10.1 General Properties

10.1.1 The One-Body Green Function

The purpose of this chapter is to show how that the perturbative methods, already worked out for calculating the scattering operator using Feynman diagrams, can also be applied to many-body problems. We limit ourselves here to the principal ideas and several applications. The appropriate quantities on which the perturbative calculations should be performed are the Green functions of the many-body problem. They generalize, in a natural manner, the concept of a causal or time-ordering operator, introduced in Sects. 9.3.1 and 9.3.4.

Definition of the One-Body Green Function for Fermions

Consider an ensemble of interacting fermions governed by a Hamiltonian $\mathtt{H} = \mathtt{H}_0 + \mathtt{V}$; $\mathtt{H}_0$ is the kinetic energy and $\mathtt{V}$ may include the effect of a two-body potential and an external field. Suppose that the system possesses a ground state $|\Phi\rangle$ of energy E,

$$\mathtt{H}\,|\Phi\rangle = E\,|\Phi\rangle \quad . \tag{10.1}$$

The evolution of the Fermi field, in the Heisenberg representation of motion, is governed by the complete Hamiltonian according to (8.218)

$$a^*(\boldsymbol{x},\sigma,t) = \mathtt{U}^*(t)a^*(\boldsymbol{x},\sigma)\mathtt{U}(t) \quad , \quad U(t) = \exp\left(-\frac{\mathrm{i}\mathtt{H}t}{\hbar}\right) \quad . \tag{10.2}$$

We give straightaway the definition of the *one-body Green function*:

$$\mathrm{i}G(\boldsymbol{x},\sigma,t|\boldsymbol{x}',\sigma',t') = \frac{\langle\Phi|T\left[a(\boldsymbol{x},\sigma,t)a^*(\boldsymbol{x}',\sigma',t')\right]|\Phi\rangle}{\langle\Phi|\Phi\rangle} \quad , \tag{10.3}$$

where T is the time ordering of Sect. 9.3.4. Although they are being applied here to interacting fields, the rules of Sect. 9.3.4, in particular the sign conventions, remain the same.

It is useful to examine the motivation and the consequences of definition (10.3). First of all, one obtains explicitly for $t > t'$ (supposing $|\Phi\rangle$ to be normalized to one in the following),

$$\begin{aligned} &\mathrm{i}G(\boldsymbol{x},\sigma,t|\boldsymbol{x}',\sigma',t') \\ &\quad= \langle\Phi|a(\boldsymbol{x},\sigma,t)a^*(\boldsymbol{x}',\sigma',t')|\Phi\rangle \\ &\quad= \exp\left(\frac{\mathrm{i}E(t-t')}{\hbar}\right)\langle\Phi|a(\boldsymbol{x},\sigma)\exp\left(-\frac{\mathrm{i}\mathtt{H}(t-t')}{\hbar}\right)a^*(\boldsymbol{x}',\sigma')|\Phi\rangle \quad . \end{aligned} \tag{10.4}$$

We have taken into account (10.2) and expressed the action of the evolution operator on the ground state $\mathtt{U}(t)\,|\Phi\rangle = \mathrm{e}^{-\mathrm{i}Et/\hbar}\,|\Phi\rangle$. For $t < t'$, one finds

$$\begin{aligned} &\mathrm{i}G(\boldsymbol{x},\sigma,t|\boldsymbol{x}',\sigma',t') \\ &\quad= -\,\langle\Phi|a^*(\boldsymbol{x}',\sigma',t')a(\boldsymbol{x},\sigma,t)|\Phi\rangle \\ &\quad= -\exp\left(\frac{\mathrm{i}E(t'-t)}{\hbar}\right)\langle\Phi|a^*(\boldsymbol{x}',\sigma')\exp\left(-\frac{\mathrm{i}\mathtt{H}(t'-t)}{\hbar}\right)a(\boldsymbol{x},\sigma)|\Phi\rangle \quad . \end{aligned} \tag{10.5}$$

These formulae show that $G(\boldsymbol{x},\sigma,t|\boldsymbol{x}',\sigma',t')$ depends only on the time difference $t'-t$ (it is assumed that the external field does not vary in time).

For a gas of free fermions, $\mathtt{H}$ reduces to the kinetic energy $\mathtt{H}_0$ and $|\Phi\rangle = |\Phi_0\rangle$ is the state (8.222): it is clear that $G(\boldsymbol{x},\sigma,t|\boldsymbol{x}',\sigma',t')$ thus reduces to the propagator of the Fermi free field (9.101). It should be emphasized here that, in general, the Green function (10.3) differs in two ways: *the interaction comes into play both in the construction of the ground state (10.1) and in the complete evolution (10.2).*

It is an abuse of language to call the object defined in (10.3) a Green function. It is only for the independent fermions that $G(\boldsymbol{x},\sigma,t|\boldsymbol{x}',\sigma',t')$ is the Green function of a partial differential equation. If there is an interaction, the non-linear equation of the field (8.221) shows that the space–time derivatives of $G(\boldsymbol{x},\sigma,t|\boldsymbol{x}',\sigma',t')$ are related to the functions which introduce the average of four field operators. We are thus led to introduce the n-body Green functions using a definition analogous to (10.3) except that it involves a product of $2n$ fields. We only consider here the one-body Green function.

We have a double interest in definition (10.3). On the one hand, the function $G(\boldsymbol{x},\sigma,t|\boldsymbol{x}',\sigma',t')$ can of course be used in a perturbative calculation. This is the reason for introducing the time ordering T in (10.3). We will see in Sects. 10.1.3 and 10.1.4 that $G(\boldsymbol{x},\sigma,t|\boldsymbol{x}',\sigma',t')$ is given by a series of appropriate Feynman diagrams. On the other hand, knowledge of it reveals the main interesting pieces of information about the system, in particular the one-body reduced density matrix and the energy of the ground state.

Relation to the Reduced Density Matrix

When t' approaches t with $t' > t$, (10.5) shows that

$$\begin{aligned}\langle \boldsymbol{x}\sigma|\rho^{(1)}|\boldsymbol{x}'\sigma'\rangle &= \langle \Phi|a^*(\boldsymbol{x}',\sigma')a(\boldsymbol{x},\sigma)|\Phi\rangle \\ &= -\mathrm{i} \lim_{t'\to t(t'>t)} G(\boldsymbol{x},\sigma,t|\boldsymbol{x}',\sigma',t') \quad . \end{aligned} \tag{10.6}$$

The quantity $\langle \boldsymbol{x}\sigma|\rho^{(1)}|\boldsymbol{x}'\sigma'\rangle$ is the one-body reduced density matrix (3.132) in the ground state, from which one can evaluate the average values of all the one-body observables. In particular, the kinetic energy is given by the expression

$$\langle \Phi|\mathtt{H}_0|\Phi\rangle = \mathrm{i}\sum_\sigma \int \mathrm{d}\boldsymbol{x} \lim_{t'\to t(t'>t)} \lim_{\boldsymbol{x}'\to\boldsymbol{x}} \left[\frac{\hbar^2}{2m}\nabla^2_{\boldsymbol{x}} G(\boldsymbol{x},\sigma,t|\boldsymbol{x}',\sigma',t')\right] \quad . \tag{10.7}$$

The Green function thus provides all the information contained in the reduced density matrix; moreover, it contains additional information, since it also allow temporal correlations to be calculated. *Thus it naturally generalizes the concept of the reduced density matrix by including the temporal dependence.*

Relation to the Ground-State Energy

Calculating the average value of the two-body interaction in principle requires knowledge of the two-body reduced density matrix (3.139). It is nevertheless possible to determine this average value with the help of the one-body Green function, thanks to the fact that the equation of motion of the field in interaction (8.221) introduces the two-body potential (we assume here that there is no external field). One multiplies the adjoint equation to (8.221) by $a^*(\boldsymbol{x}',\sigma,t')$, then takes the average value of the resulting equation in the ground state and the limits $\boldsymbol{x}'\to\boldsymbol{x}$ and $t'\to t$. One thus obtains

$$\begin{aligned}&\lim_{t'\to t(t'>t)} \lim_{\boldsymbol{x}'\to\boldsymbol{x}} \left(\mathrm{i}\hbar\frac{\partial}{\partial t} + \frac{\hbar^2}{2m}\nabla^2_{\boldsymbol{x}}\right) \langle\Phi|a^*(\boldsymbol{x}',\sigma,t')a(\boldsymbol{x},\sigma,t)|\Phi\rangle \\ &= \sum_{\sigma'}\int \mathrm{d}\boldsymbol{y}\,\langle\Phi|\,a^*(\boldsymbol{x},\sigma,t)a^*(\boldsymbol{y},\sigma',t)V(\boldsymbol{x}-\boldsymbol{y})a(\boldsymbol{y},\sigma',t)a(\boldsymbol{x},\sigma,t)\,|\Phi\rangle \quad . \end{aligned} \tag{10.8}$$

Since the state $|\Phi\rangle$ is stationary, the time dependence can be omitted in the right-hand side of (10.8). After summation over $\boldsymbol{x}$ and σ, this quantity is equal to $2\,\langle\Phi|\mathtt{V}|\Phi\rangle$ recalling expression (3.109) for the two-body interaction. Taking into account (10.5), (10.8) can be rewritten as

$$\begin{aligned}\langle\Phi|\mathtt{V}|\Phi\rangle = -\frac{\mathrm{i}}{2}\sum_\sigma \int \mathrm{d}\boldsymbol{x} \lim_{t'\to t(t'>t)} \lim_{\boldsymbol{x}'\to\boldsymbol{x}} \left(\mathrm{i}\hbar\frac{\partial}{\partial t} + \frac{\hbar^2}{2m}\nabla^2_{\boldsymbol{x}}\right) \\ \times G(\boldsymbol{x},\sigma,t|\boldsymbol{x}',\sigma,t') \quad . \end{aligned} \tag{10.9}$$

The sum of (10.7) and (10.9) evidently gives the energy E of the ground state.

Quasi-Particles

We know that the free-particle Green function $G_0(\boldsymbol{k}, t)$ evolves with the phase $\mathrm{e}^{-\mathrm{i}\varepsilon_{\boldsymbol{k}} t/\hbar}$ or, in terms of the frequency ω, $G_0(\boldsymbol{k}, \omega)$ has a pole on the real axis at $\omega = \varepsilon_{\boldsymbol{k}}/\hbar$ (Sect. 9.2.2). This structure reflects the fact that a free particle possesses stable stationary quantum states of energy $\varepsilon_{\boldsymbol{k}}$. In the presence of an interaction, the situation changes: the system can no longer be described by non-correlated single-particle states.

In any case, it is useful to see if a description in terms of nearly independent "effective particles" (or quasi-particles) can continue to make sense. This is comparable to the approach used in the Hartree–Fock theory, where the many-body problem is replaced by an effective one-body problem.

The idea is the following: in the absence of the interaction, each particle occupies a well-defined quantum level. When they interact, one can imagine a situation in which the particles continue to occupy certain individual states, but only during a certain time τ. Indeed, because of the interactions, a given particle will undergo transitions to the other individual states which it can occupy for a certain amount of time. For the description to be faithful, the occupation time τ of a level must be for a sufficiently long duration. Such a situation is manifest in the behavior of the complete Green function of the type

$$G(\boldsymbol{k}, t) \simeq -\mathrm{i}A_{\boldsymbol{k}} \exp\left(-\frac{\mathrm{i}\bar{\varepsilon}_{\boldsymbol{k}} t}{\hbar}\right) \exp(-\gamma_{\boldsymbol{k}} t) \quad , \tag{10.10}$$

with

$$\gamma_{\boldsymbol{k}} \ll \bar{\varepsilon}_{\boldsymbol{k}}/\hbar \quad . \tag{10.11}$$

Here, the system is assumed to be homogeneous in space and time and $G(\boldsymbol{k}, t)$ is the Fourier transform of the Green function. If it can be established, for a certain period of time, that $G(\boldsymbol{k}, t)$ is indeed approximated by the form (10.10), and that the condition (10.11) is valid, then one says that *a quasi-particle was excited in an energy state* $\bar{\varepsilon}_{\boldsymbol{k}}$ *with lifetime* $\tau_{\boldsymbol{k}} = 1/\gamma_{\boldsymbol{k}} \gg \hbar/\bar{\varepsilon}_{\boldsymbol{k}}$. Equivalently, a quasi-particle is revealed by the existence of a complex pole of $G(\boldsymbol{k}, \omega)$ at $\omega = (\bar{\varepsilon}_{\boldsymbol{k}}/\hbar) - \mathrm{i}\gamma_{\boldsymbol{k}}$, with sufficiently small $\gamma_{\boldsymbol{k}}$.

Showing the existence of a quasi-particle structure makes the discussion of the many-body problem much easier: *in the first approximation, it can be considered that the system is well described by an ensemble of quasi-particles which may be independent or subject to an effective interaction.* The interpretation of the quasi-particle depends on the physics of the problem in question. As we shall see in Sect. 10.2.2, a quasi-particle in the electron gas can be interpreted as an electron carrying a polarization cloud. We thus take into account screening effects, and the effective residual interaction between the quasi-electrons which has a much shorter range than that of the Coulomb potential. Another example is the motion of the electron in a polarized crystal, displacing the ions from their equilibrium position and thus causing excitations of local density. The corresponding quasi-particle, the polaron, is an electron dressed by its phonon cloud.

One can give a definition analogous to (10.3) of the one-body Green function for the bosons: one needs only replace the Fermi field by an interacting Bose field. In this chapter, we only treat the perturbative aspects of fermionic Green functions that have applications to the electron gas. It is clear that similar methods can be developed for the bosons. Finally, we note that the interest of the Green functions is not limited to their perturbative aspects. Non-perturbative approximation methods, based for example on the hierarchy of the equations of motion (8.221), provide an elegant and powerful formulation of theories which are not amenable to perturbative calculations, such as superconductivity.

10.1.2 Perturbative Calculation of the Green Function

The problem of expanding of the Green function in powers of the interaction can be posed in the following terms. As we have noted, the interaction comes into play in $|\Phi\rangle$ and in the evolution $\mathtt{U}(t)$: as a result, *both these quantities must be series expanded.* In the interaction representation, the evolution operator is given by the Dyson series (9.62). As for the state $|\Phi\rangle$, under the hypothesis that the non-perturbed ground state $|\Phi_0\rangle$ is non-degenerate, it is in principle obtained by an ordinary perturbation calculation of a stationary state in quantum mechanics (the Rayleigh–Schrödinger series),

$$|\Phi\rangle = |\Phi_0\rangle + \sum_{n=1}^{\infty} |\Phi^{(n)}\rangle \quad , \quad \langle\Phi_0|\Phi_0\rangle = 1 \quad . \tag{10.12}$$

In (10.12), the normalization of $|\Phi\rangle$ is not equal to one, but is fixed by

$$\langle\Phi_0|\Phi\rangle = 1 \quad , \quad \langle\Phi_0|\Phi^{(n)}\rangle = 0 \quad , \quad n \geq 1 \quad . \tag{10.13}$$

By inserting these series into (10.3), one sees that the numerator and the denominator of G are expressed by multiple series. After collecting the terms which contribute at a given order, one obtains the desired result. Although this process is possible in principle, it is extremely tedious and becomes unworkable at higher orders. A much more condensed and efficient algorithm for this calculation is described by the Gell-Mann and Low theorem. It *automatically regroups the large number of terms* appearing in the procedure described above, as well as allowing the series to be visualized using Feynman diagrams.

The Gell-Mann and Low Theorem

To formulate the Gell-Mann and Low theorem it is necessary to introduce a new time-dependent Hamiltonian

$$\mathtt{H}_\eta = \mathtt{H}_0 + \exp(-\eta|t|)\mathtt{V} \quad , \quad \eta > 0 \quad . \tag{10.14}$$

The results of Sect. 9.3.1 apply to the evolution operator $\mathtt{U}_{\eta\mathrm{I}}(t,t')$ corresponding to the Hamiltonian (10.14): its Dyson series is

$$\mathtt{U}_{\eta\mathrm{I}}(t,t') = \mathtt{I} + \sum_{n=1}^{\infty} \mathtt{U}_{\eta\mathrm{I}}^{(n)}(t,t') \quad , \tag{10.15}$$

where

$$\mathtt{U}_{\eta\mathrm{I}}^{(n)}(t,t') = \left(\frac{-\mathrm{i}}{\hbar}\right)^n \int_{t'}^{t} \mathrm{d}t_n \int_{t'}^{t_n} \mathrm{d}t_{n-1} \dots$$
$$\dots \int_{t'}^{t_2} \mathrm{d}t_1 \exp\left(-\eta(|t_1| + \ldots + |t_n|)\right) \mathtt{V}_\mathrm{I}(t_n) \ldots \mathtt{V}_\mathrm{I}(t_1) \tag{10.16}$$

and

$$\mathtt{V}_\mathrm{I}(t) = \mathtt{U}_0^*(t)\mathtt{V}\mathtt{U}_0(t) \quad . \tag{10.17}$$

The effect of the factor $\mathrm{e}^{-\eta|t|}$ is to switch off the interaction when $t \to \pm\infty$. When $\eta \to 0$, $\mathtt{U}_{\eta\mathrm{I}}(t,t')$ reduces to the usual evolution operator $\mathtt{U}_\mathrm{I}(t,t')$ relative to $\mathtt{H} = \mathtt{H}_0 + \mathtt{V}$ in the interaction representation.

Consider the vector $|\Phi_\eta(\tau)\rangle$ formed from the unperturbed ground state $|\Phi_0\rangle$ in the expressions

$$|\Phi_\eta(\tau)\rangle = \frac{\mathtt{U}_{\eta\mathrm{I}}(0,\tau)\,|\Phi_0\rangle}{\langle\Phi_0|\mathtt{U}_{\eta\mathrm{I}}(0,\tau)|\Phi_0\rangle} \quad , \tag{10.18}$$

$$|\Phi_\eta(\tau)\rangle = \frac{\left(\mathtt{U}_{\eta\mathrm{I}}(\tau,0)\right)^*|\Phi_0\rangle}{\langle\Phi_0|\left(\mathtt{U}_{\eta\mathrm{I}}(\tau,0)\right)^*|\Phi_0\rangle} \quad . \tag{10.19}$$

These two expressions are identical because of the first relation (9.57), which is valid for $\mathtt{U}_{\eta\mathrm{I}}(t,t')$. After expanding the numerator and the denominator according to (10.15) and regrouping the terms at a given order, one obtains $|\Phi_\eta(\tau)\rangle$ in the form of a power series of the interaction

$$|\Phi_\eta(\tau)\rangle = |\Phi_0\rangle + \sum_{n-1}^{\infty} |\Phi_\eta^{(n)}(\tau)\rangle \quad . \tag{10.20}$$

The Gell-Mann and Low theorem then states the following: *if the perturbed state $|\Phi\rangle$ is non-degenerate and possesses the Rayleigh–Schrödinger series* (10.12), *then its* n*th-order term is then given by the limits*

$$|\Phi^{(n)}\rangle = \lim_{\eta\to 0}\lim_{\tau\to-\infty}|\Phi_\eta^{(n)}(\tau)\rangle = \lim_{\eta\to 0}\lim_{\tau\to\infty}|\Phi_\eta^{(n)}(\tau)\rangle \quad . \tag{10.21}$$

Note the order of these limits. The convergence factor $e^{-\eta|t|}$ introduced in (10.14) ensures the existence of the limits $t' \to -\infty$ and $t \to \infty$; the limit $\eta \to 0$ must be carried out afterwards. We emphasize that this second limit can only exist if the state $|\Phi\rangle$ actually possesses a perturbation series. This important assumption can be proven wrong in certain physical situations as was the case for the theory of superconductivity. In light of (10.18) and (10.19), and without assuming anything about the convergence of the series, one can formally write

$$|\Phi\rangle = \lim_{\eta\to 0}\lim_{\tau\to-\infty}|\Phi_\eta(\tau)\rangle = \lim_{\eta\to 0}\frac{\mathtt{U}_{\eta\mathrm{I}}(0,-\infty)\,|\Phi_0\rangle}{\langle\Phi_0|\mathtt{U}_{\eta\mathrm{I}}(0,-\infty)|\Phi_0\rangle} \quad , \tag{10.22}$$

$$|\Phi\rangle = \lim_{\eta\to 0}\lim_{\tau\to\infty}|\Phi_\eta(\tau)\rangle = \lim_{\eta\to 0}\frac{(\mathtt{U}_{\eta\mathrm{I}}(\infty,0))^*\,|\Phi_0\rangle}{\langle\Phi_0|\,(\mathtt{U}_{\eta\mathrm{I}}(\infty,0))^*\,|\Phi_0\rangle} \quad . \tag{10.23}$$

We do not give a general proof of this theorem here, but we will provide an elementary verification and the end of this section in order to illustrate the mechanism.[1] In (10.22) and (10.23), the limits of the numerator and the denominator cannot be taken on their own, only the limit of the ratio makes sense (see example (10.32), (10.33)).

Formula (10.22) allows one to write down an expression for the energy of the ground state. Noting that $\langle\Phi_0|\Phi\rangle = 1$ and taking into account (10.1), one can write

$$\begin{aligned} E = \langle\Phi_0|\mathtt{H}|\Phi\rangle &= \lim_{\eta\to 0}\frac{\langle\Phi_0|(\mathtt{H}_0+\mathtt{V})\mathtt{U}_{\eta\mathrm{I}}(0,-\infty)|\Phi_0\rangle}{\langle\Phi_0|\mathtt{U}_{\eta\mathrm{I}}(0,-\infty)|\Phi_0\rangle} \\ &= E_0 + \lim_{\eta\to 0}\frac{\langle\Phi_0|\mathtt{V}\mathtt{U}_{\eta\mathrm{I}}(0,-\infty)|\Phi_0\rangle}{\langle\Phi_0|\mathtt{U}_{\eta\mathrm{I}}(0,-\infty)|\Phi_0\rangle} \quad . \end{aligned} \tag{10.24}$$

This expression leads to of a perturbation series for E which, according to the Gell-Mann and Low theorem, coincides with the usual Rayleigh–Schrödinger series for a non-degenerate quantum level.

The Fundamental Formula

The advantage of (10.18) and (10.19) is that they express the perturbation series of the stationary state $|\Phi\rangle$ with the help of the evolution operator. The latter can also be combined with the field evolution (10.2) to generate the Green function series in a handy form. For this, we note that

[1] A proof is given in K. Hepp, Théorie de la renormalization, Chap. II, Lecture Notes in Physics 2, Springer (1969).

$$a(\boldsymbol{x},t) = \lim_{\eta\to 0} \mathtt{U}_{\eta\mathrm{I}}(0,t)a_0(\boldsymbol{x},t)\mathtt{U}_{\eta\mathrm{I}}(t,0) \quad , \tag{10.25}$$

where $a_0(\boldsymbol{x},t)$ is the free field (8.213). We have used the index 0 here as a reminder that it evolves with the Hamiltonian of the free particles and to distinguish it from an interacting field. Formula (10.25) is derived from the definition of the interaction representation (9.56) and we omit the spin index in the following.

Applying the Gell-Mann and Low theorem, as well as (10.25), shows that the Green function (10.3) is given by the expression

$$\frac{\langle \varPhi_\eta(\tau)|T[\mathtt{U}_{\eta\mathrm{I}}(0,t)a_0(\boldsymbol{x},t)\mathtt{U}_{\eta\mathrm{I}}(t,0)\mathtt{U}_{\eta\mathrm{I}}(0,t')a_0^*(\boldsymbol{x}',t')\mathtt{U}_{\eta\mathrm{I}}(t',0)]|\varPhi_\eta(\tau')\rangle}{\langle \varPhi_\eta(\tau)|\varPhi_\eta(\tau')\rangle} \quad , \tag{10.26}$$

in the limit $\tau' \to -\infty$, $\tau \to \infty$, and $\eta \to 0$. One assumes τ greater (τ' smaller) than the two times t (and t') and replaces $|\varPhi_\eta(\tau')\rangle$ and $|\varPhi_\eta(\tau)\rangle$ by expressions (10.18) and (10.19). Under these conditions, operators $\mathtt{U}_{\eta\mathrm{I}}(\tau,0)$ and $\mathtt{U}_{\eta\mathrm{I}}(0,\tau')$ can be rewritten in the time-ordering operator and, if one uses the composition law (9.57), (10.26) becomes

$$\begin{aligned}&\frac{\langle \varPhi_0|T[\mathtt{U}_{\eta\mathrm{I}}(\tau,t)a_0(\boldsymbol{x},t)\mathtt{U}_{\eta\mathrm{I}}(t,t')a_0^*(\boldsymbol{x}',t')\mathtt{U}_{\eta\mathrm{I}}(t',\tau')]|\varPhi_0\rangle}{\langle \varPhi_0|\mathtt{U}_{\eta\mathrm{I}}(\tau,\tau')|\varPhi_0\rangle}\\ &\quad = \frac{\langle \varPhi_0|T[\mathtt{U}_{\eta\mathrm{I}}(\tau,\tau')a_0(\boldsymbol{x},t)a_0^*(\boldsymbol{x}',t')]|\varPhi_0\rangle}{\langle \varPhi_0|\mathtt{U}_{\eta\mathrm{I}}(\tau,\tau')|\varPhi_0\rangle} \quad .\end{aligned} \tag{10.27}$$

The latter expression results from the fact that one can freely permute the evolution operators under the time-ordering operation T. These permutations do not cause any sign change since the Hamiltonian is always comprised of an even number of fermion operators. In the operations which lead from (10.26) to (10.27), one must understand that the evolution operators is always given by their series (10.15), (10.16) and that T acts in the usual manner on the products of the field operators: these operations only represent an abbreviated way of carrying out a formal rearrangement of the series. By referring to the definition (9.77) of the scattering operator, and taking first the limits $\tau' \to -\infty$, $\tau \to \infty$ and then $\eta \to 0$ on (10.27), one obtains *the fundamental formula of the perturbation theory of Green functions*:

$$G(\boldsymbol{x},t|\boldsymbol{x}',t') = -\mathrm{i}\lim_{\eta\to 0}\frac{\langle \varPhi_0|T[\mathtt{S}_\eta a_0(\boldsymbol{x},t)a_0^*(\boldsymbol{x}',t')]|\varPhi_0\rangle}{\langle \varPhi_0|\mathtt{S}_\eta|\varPhi_0\rangle} \quad . \tag{10.28}$$

In terms of the condensed notation (10.28), one should understand that the Green function is equal to the quotient of two formal series. The scattering operator $\mathtt{S}_\eta$ relative to the Hamiltonian (10.14) is given by the series (9.79) incorporating the convergence factor $\mathrm{e}^{-\eta|t|}$. As for the numerator, it is given by a similar series in which two additional field operators appear:

$$
\begin{aligned}
&\langle\Phi_0|T[\mathtt{S}_\eta a_0(\boldsymbol{x},t)a_0^*(\boldsymbol{x}',t')]|\Phi_0\rangle \\
&\quad = \langle\Phi_0|T[a_0(\boldsymbol{x},t)a_0^*(\boldsymbol{x}',t')]|\Phi_0\rangle \\
&\quad\quad + \sum_{n=1}^{\infty}\left(\frac{-\mathrm{i}}{\hbar}\right)^n \frac{1}{n!}\int_{-\infty}^{\infty}\dots\int_{-\infty}^{\infty} \mathrm{d}t_1\dots\mathrm{d}t_n \exp\left[-\eta(|t_1|+\dots+|t_n|)\right] \\
&\quad\quad \times \langle\Phi_0|\, T[\mathtt{V}_\mathrm{I}(t_1)\dots\mathtt{V}_\mathrm{I}(t_n)a_0(\boldsymbol{x},t)a_0^*(\boldsymbol{x}',t')]\,|\Phi_0\rangle \quad . \qquad (10.29)
\end{aligned}
$$

Formulae (10.28) *and* (10.29) *have the advantage of expressing the Green function solely in terms of the free fields.* Keeping this in mind, we will henceforth suppress the 0 index when we address the perturbative expansion of $G(\boldsymbol{x},t|\boldsymbol{x}',t')$. Nevertheless, as (10.28) is the ratio of two series, $G(\boldsymbol{x},t|\boldsymbol{x}',t')$ does not yet have the desired form of an expansion in terms of powers of the interaction. This last step is accomplished by the connected-graph theorem presented in the following section.

An Elementary Verification

We expand (10.24) to second order in the interaction. For this, it is sufficient to keep the terms

$$
\begin{aligned}
E - E_0 &= \frac{\langle\Phi_0|\mathtt{V}\left(\mathtt{I}+\mathtt{U}_{\eta\mathrm{I}}^{(1)}(0,-\infty)+\dots\right)|\Phi_0\rangle}{1+\langle\Phi_0|\mathtt{U}_{\eta\mathrm{I}}^{(1)}(0,-\infty)|\Phi_0\rangle+\dots} \\
&= \langle\Phi_0|\mathtt{V}|\Phi_0\rangle + \langle\Phi_0|\mathtt{V}\mathtt{U}_{\eta\mathrm{I}}^{(1)}(0,-\infty)|\Phi_0\rangle \\
&\quad - \langle\Phi_0|\mathtt{V}|\Phi_0\rangle\,\langle\Phi_0|\mathtt{U}_{\eta\mathrm{I}}^{(1)}(0,-\infty)|\Phi_0\rangle + \dots \\
&= E^{(1)} + E^{(2)} + \dots \qquad (10.30)
\end{aligned}
$$

and to take the limit $\eta \to 0$. The term $E^{(1)} = \langle\Phi_0|\mathtt{V}|\Phi_0\rangle$ is manifestly the usual first-order correction to the unperturbed energy E_0. On the other hand, we have, with the help of (10.16) and (10.17),

$$
\begin{aligned}
\mathtt{U}_{\eta\mathrm{I}}^{(1)}(0,-\infty)\,|\Phi_0\rangle &= -\frac{\mathrm{i}}{\hbar}\int_{-\infty}^{0} dt \exp\left(\frac{\mathrm{i}(\mathtt{H}_0-E_0)t}{\hbar}+\eta t\right)\mathtt{V}\,|\Phi_0\rangle \\
&= -(\mathtt{H}_0-E_0-\mathrm{i}\hbar\eta)^{-1}\mathtt{V}\,|\Phi_0\rangle \quad , \qquad (10.31)
\end{aligned}
$$

which leads to

$$
E^{(2)} = -\langle\Phi_0|\mathtt{V}(\mathtt{H}_0-E_0-\mathrm{i}\hbar\eta)^{-1}\mathtt{V}|\Phi_0\rangle - \frac{1}{\mathrm{i}\hbar\eta}|\,\langle\Phi_0|\mathtt{V}|\Phi_0\rangle\,|^2 \qquad (10.32)
$$

$$
= \sum_{r\neq 0}\frac{|\,\langle\Phi_0|\mathtt{V}|\Phi_r\rangle\,|^2}{E_0-E_r+\mathrm{i}\hbar\eta} \quad . \qquad (10.33)
$$

One obtains (10.33) by inserting into the first term of (10.32) a complete system of eigenstates $\{|\Phi_r\rangle, r=0,1,2,\dots\}$ of $\mathtt{H}_0$. The limit $\eta\to 0$ of (10.33) exists since E_0 is non-degenerate and the term $r = 0$ is omitted from the sum. One indeed finds the familiar formula for the second-order perturbation of a non-degenerate level. It is instructive to note that the limits $\eta \to 0$ of the two terms (10.32) (which result from the numerator and from the denominator of (10.30), respectively) do not exist independently. Only their combination leads to the expected result (10.33). Verifications bringing into play such compensation mechanisms can be carried out at all orders of the perturbation series of E and $|\Phi\rangle$.

10.1.3 Particle in an External Field and the Connected-Graph Theorem

To familiarize ourselves with the perturbative structure of the Green function (10.28), we return to the problem of a particle in an external potential $V(\boldsymbol{x})$ already treated in Sect. 9.2.3. In the interaction representation, this potential gives rise to

$$\mathsf{V}_{\mathrm{I}}(t) = \int \mathrm{d}\boldsymbol{x}\, V(\boldsymbol{x}) a^*(\boldsymbol{x},t) a(\boldsymbol{x},t) \quad . \tag{10.34}$$

In (10.34), $a(\boldsymbol{x},t)$ is the free field (8.213).

Using the abbreviations $x = (\boldsymbol{x}, t)$, $\mathrm{d}x = \mathrm{d}\boldsymbol{x}\mathrm{d}t$, the first-order contributions of the numerator and of the denominator of (10.28) are respectively given by

$$-\frac{\mathrm{i}}{\hbar}\int \mathrm{d}x_1\, \exp(-\eta|t_1|) V(x_1)\, \langle \Phi_0 | T\left(a^*(x_1)a(x_1)a(x)a^*(x')\right) |\Phi_0\rangle \quad , \tag{10.35}$$

$$-\frac{\mathrm{i}}{\hbar}\int \mathrm{d}x_1\, \exp(-\eta|t_1|) V(x_1)\, \langle \Phi_0 | T\left(a^*(x_1)a(x_1)\right) |\Phi_0\rangle \quad . \tag{10.36}$$

A new graphical interpretation of the contents of term (10.35) can be obtained by applying the Wick theorem (Sect. 9.3.4). Two contraction diagrams contribute to the integrand of (10.35)

$$\begin{aligned} &V(x_1)\, \langle \Phi_0 | T\big(a^*(x_1)a(x_1)a(x)a^*(x')\big) |\Phi_0\rangle \\ &\quad = -G_0(x|x_1)V(x_1)G_0(x_1|x') \quad , \end{aligned} \tag{10.37}$$

$$\begin{aligned} &V(x_1)\, \langle \Phi_0 | T\big(a^*(x_1)a(x_1)a(x)a^*(x')\big) |\Phi_0\rangle \\ &\quad = G_0(x|x')V(x_1)G_0(x_1|x_1) \quad , \end{aligned} \tag{10.38}$$

where $G_0(x|x')$ is the Green function (9.101) of the free field in the Fermi gas ground state.

We adopt the same graphical representation as in Sect. 9.2.3 (Fig. 9.3); the two first-order contributions (10.37) and (10.38) are represented in Fig. 10.1.

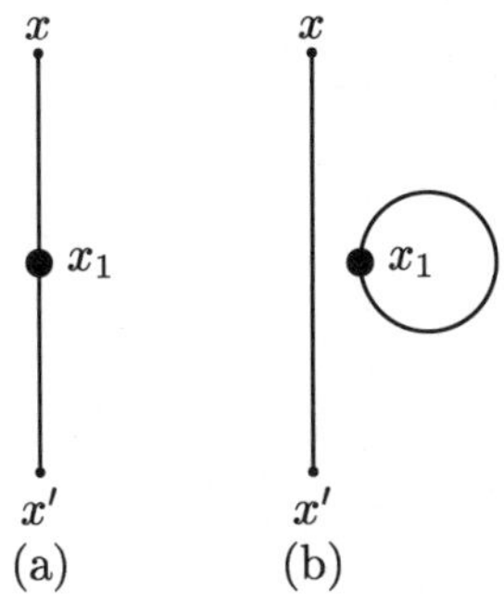

Fig. 10.1a,b. First-order contributions to the integrand of (10.35)

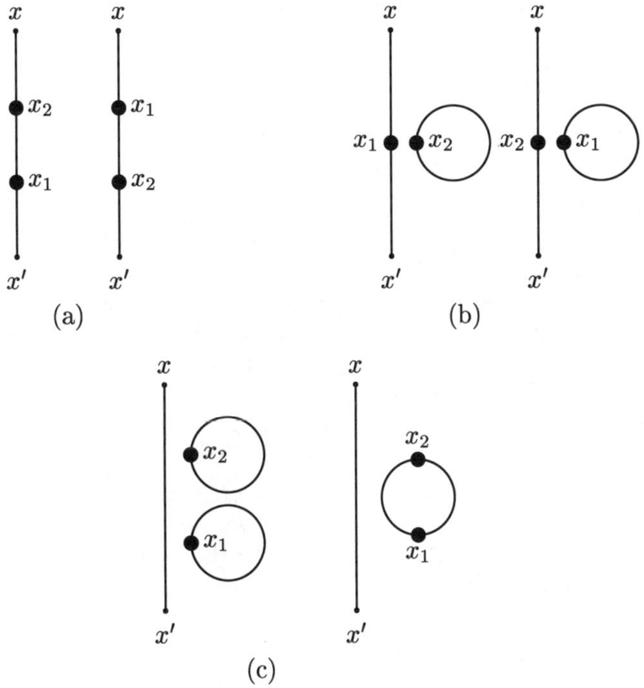

Fig. 10.2a–c. Second-order contributions to the integrand of (10.35)

Using the same conventions, the six second-order diagrams are represented in Fig. 10.2. Making a comparison with Fig. 9.4, one notices the appearance of new disconnected diagrams formed of closed loops (Fig. 10.1b and Fig. 10.2b,c). The algebraic expression corresponding to a unconnected diagram has the characteristic property of being factorizable into a product of terms coming from the different connected parts of the diagram.

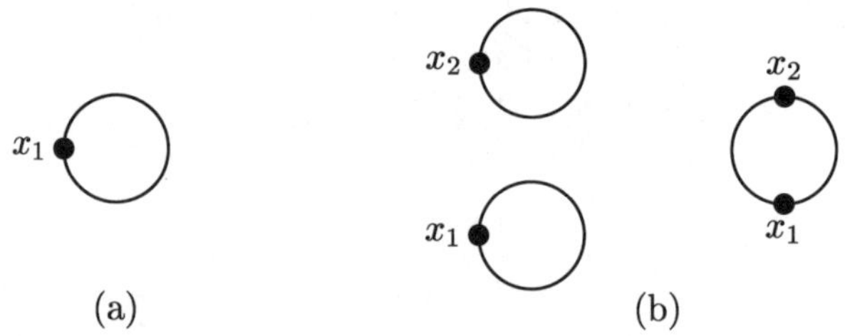

Fig. 10.3a,b. First and second-order diagrams of the denominator of (10.28)

On the other hand, it is clear that the expansion of the denominator of (10.28) only gives rise to closed loops, called "vacuum amplitudes." These are the first-order and second-order diagrams presented in Fig. 10.3. It should be noted that, taking into account (10.36) and (9.108), the closed loop of Fig. 10.3a gives rise to the contribution

$$\begin{aligned} &-\frac{\mathrm{i}}{\hbar}\int \mathrm{d}\boldsymbol{x}_1\, V(\boldsymbol{x}_1)\int_{-\infty}^{\infty}\mathrm{d}t_1\,\exp(-\eta|t_1|)G_0(\boldsymbol{x}_1,t_1|\boldsymbol{x}_1,t_1)\\ &=\frac{2\rho}{\hbar\eta}\int \mathrm{d}\boldsymbol{x}_1\, V(\boldsymbol{x}_1)\quad , \end{aligned} \tag{10.39}$$

which is divergent in the limit $\eta \to 0$. Here again, as in the case of the energy correction (10.33), the limits of the numerator and the denominator of (10.28) taken separately do not exist. We thus consider the first-order expansion of the ratio (10.28), symbolically represented in Fig. 10.4.

It can be seen that the unconnected diagram of Fig. 10.1b is exactly compensated by the contribution from the denominator. Thus all divergent terms are eliminated, as was the case in (10.32).

This compensation mechanism generalizes to all orders: *the contributions of the denominator exactly compensate those of the unconnected diagrams coming from the numerator.* This assertion is the basis of the connected graph theorem which is written as

$$\begin{aligned} G(x|x') = -\mathrm{i}\sum_{n=0}^{\infty}\left(\frac{-\mathrm{i}}{\hbar}\right)^{n}\frac{1}{n!}\int_{-\infty}^{\infty}\cdots\int_{-\infty}^{\infty}\mathrm{d}t_1\ldots\mathrm{d}t_n\\ \times\left\{\langle\Phi_0|T\left(\mathrm{V}_\mathrm{I}(t_1)\ldots\mathrm{V}_\mathrm{I}(t_n)a(x)a^*(x')\right)|\Phi_0\rangle\right\}_\mathrm{c}\quad . \end{aligned} \tag{10.40}$$

The index C indicates that only the connected diagrams in (10.40) are to be retained. The limit $\eta \to 0$ can now be carried out term by term. Here we will give a proof for the case of particles interacting with an external field. In fact, the connected graph theorem holds in generality. The rule is that only those terms stemming from the application of the Wick theorem to (10.40) not factorizing into a product of two or more factors are retained. Under this form, the theorem applies when there is a two-body interaction and we will make use of it in Sect. 10.1.4.

$$\frac{\big| + \big(\big|\hspace{-0.3em}\bullet + \big|\hspace{-0.3em}\bigcirc\hspace{-0.3em}\big) + \cdots}{1 + \hspace{-0.3em}\bigcirc\hspace{-0.3em} + \cdots}$$

$$= \left[\big| + \Big(\big|\hspace{-0.3em}\bullet + \big|\hspace{-0.3em}\bigcirc\hspace{-0.3em}\Big) + \cdots\right] \cdot \left[1 - \hspace{-0.3em}\bigcirc\hspace{-0.3em} + \cdots\right]$$

$$= \big| + \big|\hspace{-0.3em}\bullet + \cdots$$

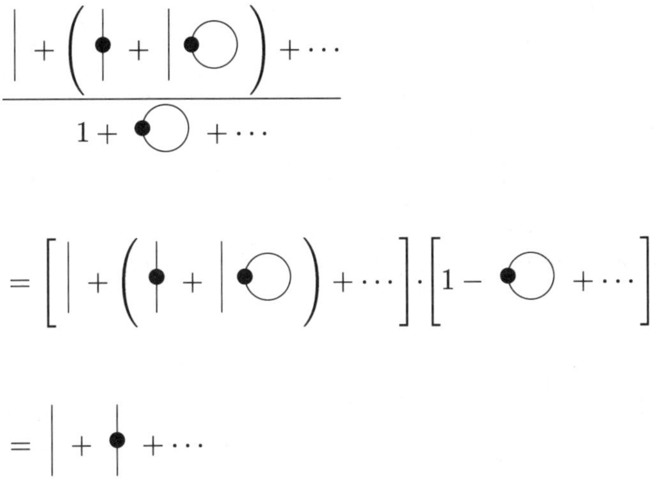

Fig. 10.4. First-order expansion of the ratio of (10.28). Note that non-connected diagrams in the numerator are exactly compensated by contributions of the denominator

+ vertex permutations

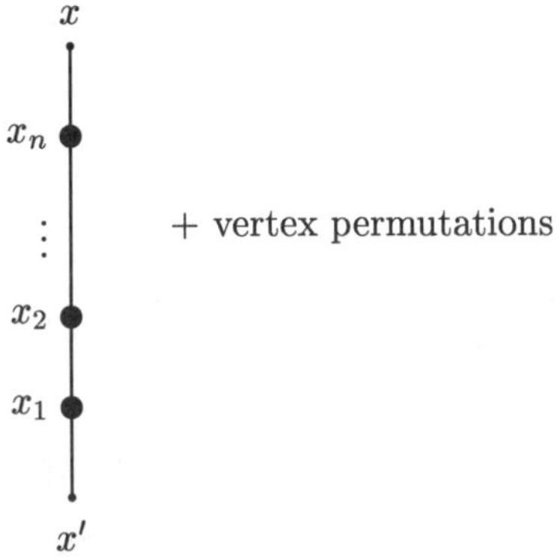

Fig. 10.5. The $n!$ possible connected n-vertex diagrams

The Connected-Graph Theorem in the Case of an External Field

The only type of connected n-vertex are shown in Fig. 10.5 and $\{\langle \Phi_0 | T\,[a^*(x_1) a(x_1)\ldots a^*(x_n)a(x_n)a(x)a^*(x')] | \Phi_0 \rangle\}_c$ represents the sum of the $n!$ diagrams of Fig. 10.5 obtained by permutations of the vertices $x_1,\ldots,x_n$. We set $a^*(x_j)a(x_j) = n(x_j)$, $\langle \Phi_0 | A | \Phi_0 \rangle = \langle A \rangle$. In this abbreviated notation, the sum of the diagrams up to second order of Fig. 10.2 corresponds to

$$
\begin{aligned}
&\langle T\left(n(x_1)n(x_2)a(x)a^*(x')\right)\rangle \\
&\quad = \{\langle T\left(n(x_1)n(x_2)a(x)a^*(x')\right)\rangle\}_{\rm c} && (a) \\
&\qquad + \{\langle T\left(n(x_1)a(x)a^*(x')\right)\rangle\}_{\rm c}\,\langle T\left(n(x_2)\right)\rangle \\
&\qquad + \{\langle T\left(n(x_2)a(x)a^*(x')\right)\rangle\}_{\rm c}\,\langle T\left(n(x_1)\right)\rangle && \Big\}(b) \\
&\qquad + \{\langle T\left(a(x)a^*(x')\right)\rangle\}_{\rm c}\,\langle T\left(n(x_1)n(x_2)\right)\rangle && (c)\quad . \qquad (10.41)
\end{aligned}
$$

One can convince oneself that applying the Wick theorem to nth order leads to the following general form, of which (10.41) represents one particular case,

$$
\begin{aligned}
&\langle T\left(n(x_1)\ldots n(x_n)a(x)a^*(x')\right)\rangle \\
&\quad = \sum_{m=0}^{n}\sum_{j_1\ldots j_m}\{\langle T\left(n(x_{j_1})\ldots n(x_{j_m})a(x)a^*(x')\right)\rangle\}_{\rm c} \\
&\qquad \times \langle T\left(n(x_{j_{m+1}})\ldots n(x_{j_n})\right)\rangle \quad . \qquad (10.42)
\end{aligned}
$$

The second sum is carried out over the $n!/[m!(n-m)!]$ subsets $j_1,\ldots,j_m$ of $1,2,\ldots,n$. Once they have been integrated over $x_1,\ldots,x_m$, each of the $n!/[m!(n-m)!]$ terms gives the same contribution to the nth order in (10.29). One thus obtains

$$
\begin{aligned}
&\langle T\left(\mathtt{S}_\eta a(x)a^*(x')\right)\rangle \\
&\quad = \sum_{n=0}^{\infty}\sum_{m=0}^{n}\left(\frac{-{\rm i}}{\hbar}\right)^n\frac{1}{m!(n-m)!}\int_{-\infty}^{\infty}{\rm d}t_1\cdots\cdots\int_{-\infty}^{\infty}{\rm d}t_m \\
&\qquad \times \exp\left(-\eta(|t_1|+\ldots+|t_m|)\right)\{\langle T\left(\mathtt{V}_{\rm I}(t_1)\ldots\mathtt{V}_{\rm I}(t_m)a(x)a^*(x')\right)\rangle\}_{\rm c} \\
&\qquad \times \int_{-\infty}^{\infty}{\rm d}t_{m+1}\cdots\int_{-\infty}^{\infty}{\rm d}_{t_n}\exp\left(-\eta(|t_{m+1}|+\ldots+|t_n|)\right) \\
&\qquad \times \langle T\left(\mathtt{V}_{\rm I}(t_{m+1})\ldots\mathtt{V}_{\rm I}(t_n)\right)\rangle \\
&\quad = \left(\sum_{m=0}^{\infty}\left(\frac{-{\rm i}}{\hbar}\right)^m\frac{1}{m!}\int_{-\infty}^{\infty}{\rm d}t_1\cdots\int_{-\infty}^{\infty}{\rm d}t_m\,\exp\left(-\eta(|t_1|+\ldots+|t_m|)\right)\right. \\
&\qquad \left.\times\{\langle T\left(\mathtt{V}_{\rm I}(t_1)\ldots\mathtt{V}_{\rm I}(t_m)a(x)a^*(x')\right)\rangle\}_{\rm c}\right) \\
&\qquad \times\left(\sum_{n=0}^{\infty}\left(\frac{-{\rm i}}{\hbar}\right)^n\frac{1}{n!}\int_{-\infty}^{\infty}{\rm d}t_1\cdots\int_{-\infty}^{\infty}{\rm d}t_n\,\exp\left(-\eta(|t_1|+\ldots+|t_n|)\right)\right. \\
&\qquad \left.\times\langle T\left(\mathtt{V}_{\rm I}(t_1)\ldots\mathtt{V}_{\rm I}(t_n)\right)\rangle\right) \quad . \qquad (10.43)
\end{aligned}
$$

Formula (10.43) results from changing the order of the sums

$$
\sum_{n=0}^{\infty}\sum_{m=0}^{n}f(m,n)=\sum_{m=0}^{\infty}\sum_{n=m}^{\infty}f(m,n)=\sum_{m=0}^{\infty}\sum_{n=0}^{\infty}f(m,n+m) \quad . \qquad (10.44)
$$

The second factor in (10.43) is precisely equal to the denominator $\langle \Phi_0|\mathrm{S}_\eta|\Phi_0\rangle$ of (10.28), which proves the theorem.

Since the $n!$ connected diagrams of Fig. 10.5 give the same contribution to the nth term of (10.40), this term is identical to (9.39). Thus the series (10.40) is the same, as it must be, as the series in Sect. 9.2.3 which was established using more elementary methods.

10.1.4 Interacting Particles

The formalism that we have set out so far takes on its real meaning when studying interacting particles. Thus we now consider an ensemble of fermions under the influence of a two-body interaction potential $V(\boldsymbol{x}_1 - \boldsymbol{x}_2)$ which is independent of spin, and represented by the second quantization of the operator (3.109). The connected-graph theorem applies and the Green function is given by the series (10.40). Once again, the Wick theorem allows the content to be explicitly written down, with the help of the appropriate Feynman rules. To understand where they come from, we analyze the first-order term by introducing (3.109) into (10.40),

$$\begin{aligned} G^{(1)}&(\boldsymbol{x},\sigma,t|\boldsymbol{x}',\sigma',t') \\ &= -\frac{1}{2\hbar}\int_{-\infty}^{\infty}\mathrm{d}t_1\int\mathrm{d}\boldsymbol{x}_1\int\mathrm{d}\boldsymbol{x}_2\sum_{\sigma_1\sigma_2}V(\boldsymbol{x}_1-\boldsymbol{x}_2) \\ &\quad\times\{\langle\Phi_0|\,T[a^*(\boldsymbol{x}_1,\sigma_1,t_1)a^*(\boldsymbol{x}_2,\sigma_2,t_1)a(\boldsymbol{x}_2,\sigma_2,t_1) \\ &\quad\times a(\boldsymbol{x}_1,\sigma_1,t_1)a(\boldsymbol{x},\sigma,t)a^*(\boldsymbol{x}',\sigma',t')]\,|\Phi_0\rangle\}_{\mathrm{c}} \quad . \end{aligned} \tag{10.45}$$

It is judicious to introduce an *instantaneous potential*, dependent on the variables $x_1 = (\boldsymbol{x}_1, t_1)$ and $x_2 = (\boldsymbol{x}_2, t_2)$, defined by

$$V(x_1 - x_2) = V(\boldsymbol{x}_1 - \boldsymbol{x}_2)\delta(t_1 - t_2) \tag{10.46}$$

and to rewrite (10.45) with the help of four-dimensional integrals

$$\begin{aligned} G^{(1)}&(x,\sigma|x',\sigma') \\ &= -\frac{1}{2\hbar}\int\mathrm{d}x_1\int\mathrm{d}x_2\sum_{\sigma_1\sigma_2}V(x_1-x_2) \\ &\quad\times\{\langle\Phi_0|\,T[a^*(x_1,\sigma_1)a^*(x_2,\sigma_2)a(x_2,\sigma_2) \\ &\quad\times a(x_1,\sigma_1)a(x,\sigma)a^*(x',\sigma')]\,|\Phi_0\rangle\}_{\mathrm{c}} \quad . \end{aligned} \tag{10.47}$$

Written in this form, (10.47) resembles the Green functions occurring in field theory (see Sect. 9.3.6), with the difference being that the interaction between the fermions results from the instantaneous potential (10.46) and not from any coupling to another particle field.

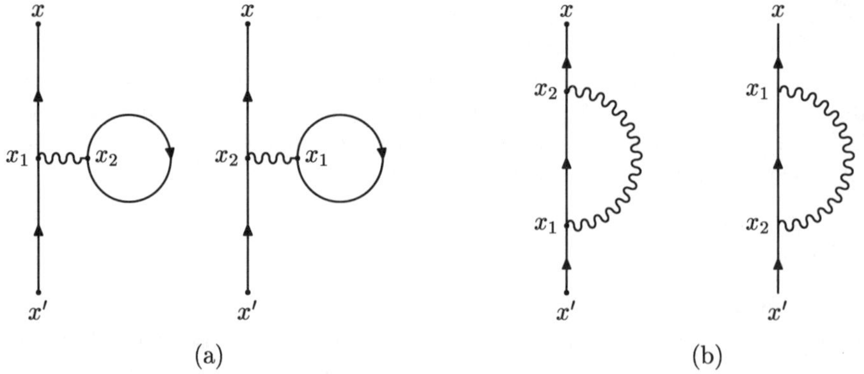

Fig. 10.6a,b. Feynman diagrams corresponding to the two terms of (10.48)

Space–Time Representation

Applying the Wick theorem to the integrand of (10.47) gives rise to the decomposition

$$\begin{aligned}
&V(x_1-x_2)\\
&\times\{\langle\Phi_0|\,T[a^*(x_1,\sigma_1)a^*(x_2,\sigma_2)a(x_2,\sigma_2)a(x_1,\sigma_1)a(x,\sigma)a^*(x',\sigma')]\,|\Phi_0\rangle\}_c\\
&\quad=\mathrm{i}V(x_1-x_2)[G_0(x_2,\sigma_2|x_2,\sigma_2)G_0(x,\sigma|x_1,\sigma_1)G_0(x_1,\sigma_1|x',\sigma')\\
&\qquad+G_0(x_1,\sigma_1|x_1,\sigma_1)G_0(x,\sigma|x_2,\sigma_2)G_0(x_2,\sigma_2|x',\sigma')]\\
&\qquad-\mathrm{i}V(x_1-x_2)[G_0(x,\sigma|x_2,\sigma_2)G_0(x_2,\sigma_2|x_1,\sigma_1)G_0(x_1,\sigma_1|x',\sigma')\\
&\qquad+G_0(x,\sigma|x_1,\sigma_1)G_0(x_1,\sigma_1|x_2,\sigma_2)G_0(x_2,\sigma_2|x',\sigma')]\quad. \qquad (10.48)
\end{aligned}$$

We represent $G_0(x_2,\sigma_2|x_1,\sigma_1)$ by a *solid line pointing from x_1 to x_2 (particle line)* and $V(x_1-x_2)$ by an *undulating line joining x_1 to x_2 (interaction line).* Thus the two terms of (10.48) correspond to Fig. 10.6a and b, respectively. According to the connected diagram rule, we have omitted the contraction term which give rise to a factorized form. Note that the structure of the vertex is the same as in Fig. 9.13. As x_1 and x_2 are dummy integration variables (the spin variables σ_1 and σ_2, as well), the two diagrams of Fig. 10.6a and b give the same numerical contribution: it is thus sufficient to keep one of them and to omit the factor 1/2 which appears in interaction (3.109).

In general, an nth-order diagram has $2n$ vertices, $x_1,\ldots,x_{2n}$. It is obtained by laying out n undulating interaction lines and $2n+1$ solid particle lines originating at x' and ending at x. These lines join the points $x_1,\ldots,x_{2n},x,x'$ in a connected manner, which must be compatible with the structure of the vertex. Starting with a given diagram, one can form $2^n n!$ equivalent diagrams in the following way: the n interaction lines determine a partition of $x_1,\ldots,x_{2n}$ into n pairs of vertices. The arguments of each pair can be exchanged as in Fig. 10.6a or Fig. 10.6b; moreover, the n pairs of vertices can

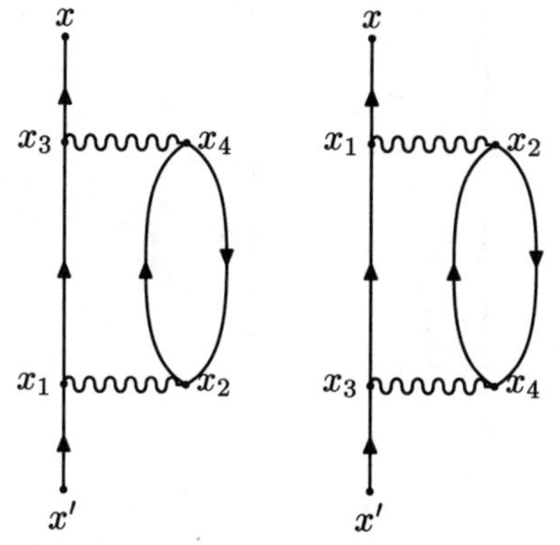

Fig. 10.7. Pairs of vertices can be permuted, yielding diagrams with equal numerical contributions

be permuted as in Fig. 10.7. One can verify that the $2^n n!$ diagrams give the same numerical contributions as (10.40) (they only differ by permutations of dummy integration variables). From now on, we will consider only one of the equivalent diagrams and will omit the factor $1/(2^n n!)$ in (10.40) ($1/2^n$ comes from the factor $1/2$ appearing in the n interactions (3.109)). In this way, Fig. 10.8 presents the second-order connected diagrams. All that matters is the topology of the diagram: one finds in the literature numerous different graphic forms for the same object. For example, Fig. 10.9a and b are the same as Fig. 10.8b and i.

The Energy–Momentum Representation

For the calculations, it is much more convenient to use the energy–momentum representation. It is obtained by introducing the Fourier representations (9.115) of the Green function and of the potential (10.46),

$$\tilde{V}(\boldsymbol{k}) = \int \mathrm{d}x \, V(x) \mathrm{e}^{-\mathrm{i}(\boldsymbol{k}\cdot\boldsymbol{x}-\omega t)} \quad . \tag{10.49}$$

Because of the instantaneous nature of $V(x)$, $\tilde{V}(\boldsymbol{k})$ does not depend on the variable ω. The Feynman rules follow from the same consideration made in Sect. 9.3.6. The diagrams are formed as indicated above: an nth-order diagram has $2n+1$ particle lines and n interaction lines. It is subject to the following rules:

(i) Assign a variable $(\boldsymbol{k}, \omega)$ to each line in such a way that the momentum and the energy are conserved at each vertex.
(ii) Associate a propagator $G_0(\boldsymbol{k}, \omega)$ (9.115) to a particle line and a potential $\tilde{V}(\boldsymbol{k})$ (10.49) to each interaction line.

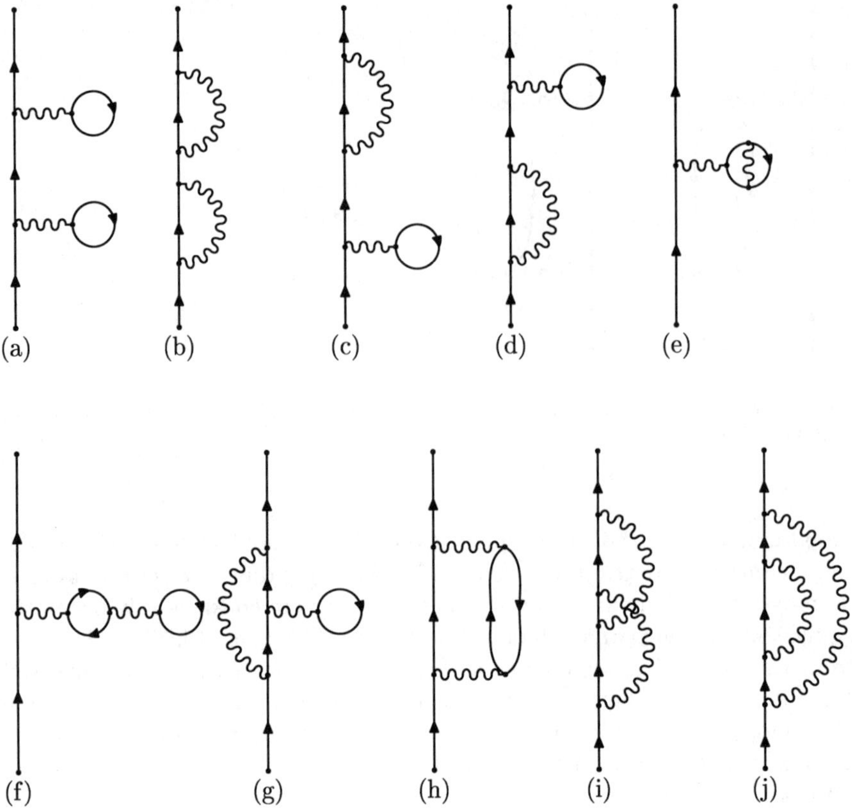

Fig. 10.8a–j. All second-order connected diagrams, omitting permutations of vertex pairs

(iii) Integrate over the variables of the internal lines with a factor $i/(2\pi)^4\hbar$ for each integral (for the integration over frequency of $G_0(\boldsymbol{k}, \omega)$, take into account the prescription (9.119a)).

(iv) Assign to each closed particle-line loop a factor $-(2s+1)$; the contribution of each diagram is diagonal in the spin variables.

An nth-order diagram gives rise to n four-dimensional integrals. According to rule (iii) above, it receives the global factor

$$\left(\frac{\mathrm{i}}{(2\pi)^4\hbar}\right)^n = (-\mathrm{i})\left(\frac{-\mathrm{i}}{\hbar}\right)^n (\mathrm{i})^{2n+1}\frac{1}{(2\pi)^{4n}} \quad ,$$

which comes from (10.40), from the definition of the Green function (9.101), and from the Fourier transform (9.104). The minus sign in rule (iv) comes from the sign rule in the Wick theorem for fermions (note for example the difference of the sign between the two terms of (10.48)). The factor $2s+1$

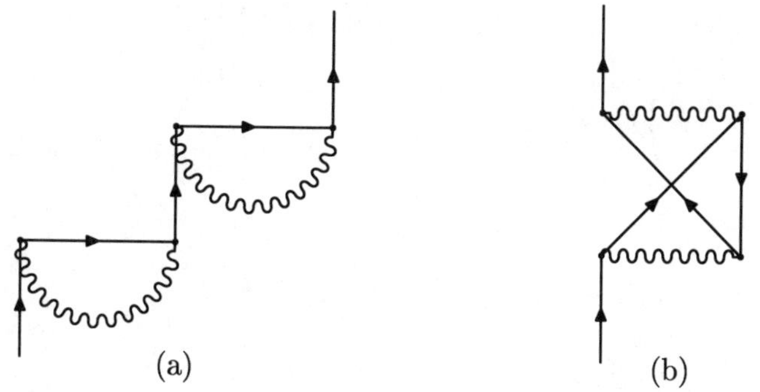

Fig. 10.9. Alternative graphic forms of Fig. 10.8b (**a**) and Fig. 10.8i (**b**)

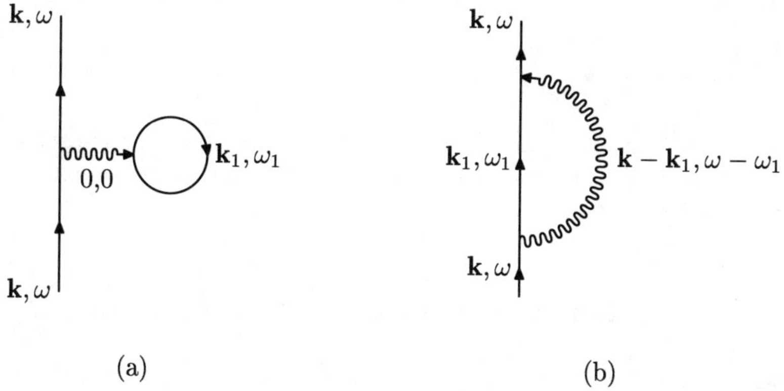

Fig. 10.10a,b. First-order diagrams in the energy–momentum representation. The arrows serve to account for energy–momentum conservation at the vertices

results from the summation of the spin indices along a closed loop. We emphasize that the rules are valid when dealing with a two-body interaction potential independent of spin and invariant under translations. In the presence of inhomogeneous external fields or interactions dependent of spin, one must return to the space–time formulation to deduce the appropriate rules.

To take an example, Fig. 10.10 shows the first-order diagrams in the momentum–energy representation. Here the arrows are serving the purpose of correctly enforcing the conservation of energy–momentum at each vertex. According to the rules already stated, the algebraic expressions corresponding to the diagrams of Fig. 10.10a and b are respectively

$$\frac{\mathrm{i}}{(2\pi)^4\hbar}\int \mathrm{d}\boldsymbol{k}_1 \int \mathrm{d}\omega_1\, G_0(\boldsymbol{k},\omega)\tilde{V}(0)\left[-(2s+1)G_0(\boldsymbol{k}_1,\omega_1)\right] G_0(\boldsymbol{k},\omega)$$
$$= G_0(\boldsymbol{k},\omega)\left(\frac{\tilde{V}(0)\rho}{\hbar}\right) G_0(\boldsymbol{k},\omega) \qquad (10.50)$$

and

$$\frac{\mathrm{i}}{(2\pi)^4\hbar}\int \mathrm{d}\boldsymbol{k}_1 \int \mathrm{d}\omega_1\, G_0(\boldsymbol{k},\omega)G_0(\boldsymbol{k}_1,\omega_1)\tilde{V}(\boldsymbol{k}-\boldsymbol{k}_1)G_0(\boldsymbol{k},\omega)$$
$$= G_0(\boldsymbol{k},\omega)\left(-\frac{1}{(2\pi)^3\hbar}\int \mathrm{d}\boldsymbol{k}_1\theta(k_\mathrm{F}-|\boldsymbol{k}_1|)\tilde{V}(\boldsymbol{k}-\boldsymbol{k}_1)\right) G_0(\boldsymbol{k},\omega) \quad . \qquad (10.51)$$

To obtain (10.50) and (10.51), one makes use of (9.110) and the fact that

$$\frac{2s+1}{(2\pi)^3}\int \mathrm{d}\boldsymbol{k}\theta(k_\mathrm{F}-|\boldsymbol{k}|) = \rho$$

is the density of a free Fermi gas. The reader is strongly encouraged to reconstruct directly (10.50) and (10.51) starting from the expressions given in (10.48).

10.2 Approximation Schemes for the Electron Gas

We propose to show how one reproduces, by means of the diagrammatic language, the Hartree–Fock and RPA approximations for the electron gas. These approximations correspond to infinite sums of certain classes of diagrams. These summations are easily carried out with the help of the self-energy insertion concepts introduced in Sect. 9.4.5.

10.2.1 Hartree–Fock Approximation

Self-Energy to First Order

There are two first-order self-energy insertions, coming from the two diagrams of Fig. 10.10. They are represented in Fig. 10.11. The Green function calculated in this manner is given by the infinite series (9.155), and whose first-order terms are given by Fig. 10.10. The four second-order terms are given by Fig. 10.8a–d. The algebraic expression of the first-order self-energy $\Sigma^{(1)}(\boldsymbol{k},\omega)$ is deduced from (10.50) and (10.51) (it is written here for a finite volume)

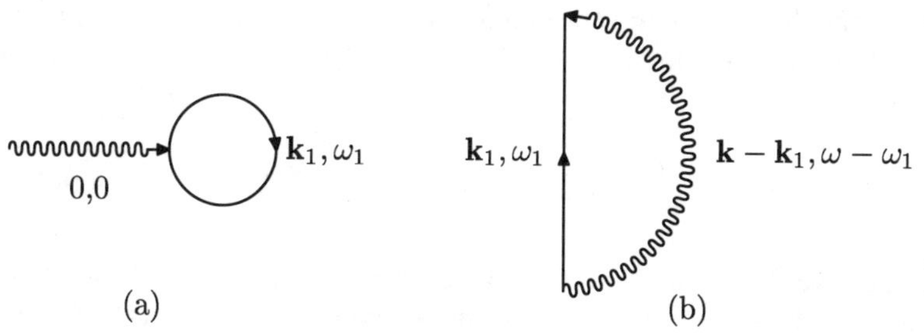

Fig. 10.11a,b. First-order self-energy insertions corresponding to the diagrams of Fig. 10.10

$$\Sigma^{(1)}(\boldsymbol{k},\omega) = \frac{1}{\hbar}\left(\tilde{V}(0)\rho - \frac{1}{L^3}\sum_{|\boldsymbol{k}_1|\le k_{\mathrm{F}}} \tilde{V}(\boldsymbol{k}-\boldsymbol{k}_1)\right) \quad . \tag{10.52}$$

In the case of the electron gas, one must use the Hamiltonian (4.38) and the modified Coulomb potential $\tilde{V}^0(\boldsymbol{k})$ (4.42). Since $\tilde{V}^0(0) = 0$, the loop diagram of Fig. 10.11a does not contribute: *this is equivalent to the fact that the average potential* (4.46) *of the electron gas is zero.* Thus only the diagram of Fig. 10.11b (second term of (10.52)) contributes to the self-energy. *Making a comparison with* (4.50), *one sees that it corresponds to the exchange potential.* Once again one sees that only the self-energy of exchange comes into play in the Hartree–Fock theory of the electron gas. The corresponding Green function is given by the series of Fig. 9.37. Introducing into (9.160) the second term of (10.52) (taken in the infinite volume limit),

$$\Sigma^{\mathrm{HF}}(\boldsymbol{k},\omega) = -\frac{1}{\hbar(2\pi)^3}\int_{|\boldsymbol{k}_1|\le k_{\mathrm{F}}} \mathrm{d}\boldsymbol{k}_1 \tilde{V}^0(\boldsymbol{k}-\boldsymbol{k}_1) \quad , \tag{10.53}$$

gives rise to the Green function in the Hartree–Fock approximation,

$$G^{\mathrm{HF}}(\boldsymbol{k},\omega) = \hbar\left(\frac{\theta(k_{\mathrm{F}}-|\boldsymbol{k}|)}{\hbar(\omega-\mathrm{i}\eta)-\varepsilon^{\mathrm{HF}}_{\boldsymbol{k}}} + \frac{\theta(|\boldsymbol{k}|-k_{\mathrm{F}})}{\hbar(\omega+\mathrm{i}\eta)-\varepsilon^{\mathrm{HF}}_{\boldsymbol{k}}}\right) \quad . \tag{10.54}$$

The quantities

$$\varepsilon^{\mathrm{HF}}_{\boldsymbol{k}} = \varepsilon_{\boldsymbol{k}} - \frac{1}{(2\pi)^3}\int_{|\boldsymbol{k}_1|\le k_{\mathrm{F}}} \mathrm{d}\boldsymbol{k}_1\, \tilde{V}^0(\boldsymbol{k}-\boldsymbol{k}_1) \tag{10.55}$$

are identical to the Hartree–Fock eigenvalues(4.51) when the electrons occupy the Fermi sphere (in the infinite volume limit). Note that $\varepsilon^{\mathrm{HF}}_{\boldsymbol{k}}$ are the energies of the quasi-particles of the Hartree–Fock theory. Their lifetime is infinite since the self-energy $\Sigma^{\mathrm{HF}}(\boldsymbol{k},\omega)$ (10.53) is real and independent of the frequency.

It is not difficult to calculate the energy of the ground state starting from (10.7) and (10.9). Taking into account (9.104), (10.9) is written in the Fourier representation as

$$\langle \Phi|\mathsf{V}|\Phi\rangle = -\frac{\mathrm{i}L^3}{(2\pi)^4}\lim_{\tau\to 0}\int \mathrm{d}\boldsymbol{k}\int \mathrm{d}\omega\, \mathrm{e}^{\mathrm{i}\omega\tau}(\hbar\omega-\varepsilon_{\boldsymbol{k}})G^{\mathrm{HF}}(\boldsymbol{k},\omega) \quad , \tag{10.56}$$

where the sum over the spins gives rise to a factor of 2 and $\int \mathrm{d}\boldsymbol{x} = L^3$ is given by the volume of the system. Having introduced the form (10.54) of the Green function into (10.56), one can carry out the integral over the frequencies by closing the contour with a semi-circle in the half-plane $\mathrm{Im}\omega > 0$. Only the first term of (10.54) contributes, as it has a pole above the real axis at $\omega = \varepsilon_{\boldsymbol{k}}^{\mathrm{HF}}/\hbar$. Calculating the residue at this pole gives rise to the result

$$\langle \Phi|\mathsf{V}|\Phi\rangle = -\frac{L^3}{(2\pi)^3}\int_{|\boldsymbol{k}|\le k_{\mathrm{F}}}\mathrm{d}\boldsymbol{k}\int_{|\boldsymbol{k}_1|\le k_{\mathrm{F}}}\mathrm{d}\boldsymbol{k}_1\,\tilde{V}^0(\boldsymbol{k}-\boldsymbol{k}_1) \quad , \tag{10.57}$$

which is indeed identical to the Hartree–Fock potential energy appearing in (4.52) when $L\to\infty$.

10.2.2 RPA Approximation

Self-Energy to Second Order

To go beyond the Hartree–Fock approximation, one naturally wants to retain the second-order contributions of the self-energy. The second-order self-energy insertions appear in the diagrams of Fig. 10.8h–j. In the electron gas, the terms of Fig. 10.8e–g are zero: an interaction line of zero momentum appears in all of them, giving rise to the factor $\tilde{V}^0(0) = 0$. The remaining self-energy insertions are shown in Fig. 10.12. According to the Feynman rules (i)–(iv), the expression corresponding to the energy-insertion of Fig. 10.12a is

$$\begin{aligned}\Sigma_a^{(2)}(\boldsymbol{k},\omega) &= \frac{\mathrm{i}}{(2\pi)^4\hbar}\int \mathrm{d}\boldsymbol{k}_1\mathrm{d}\omega_1\frac{\mathrm{i}}{(2\pi)^4\hbar}\int \mathrm{d}\boldsymbol{k}_2\mathrm{d}\omega_2\, G_0(\boldsymbol{k}-\boldsymbol{k}_1,\omega-\omega_1)\\ &\quad\times \tilde{V}^0(\boldsymbol{k}_1)\left[-2G_0(\boldsymbol{k}_1+\boldsymbol{k}_2,\omega_1+\omega_2)G_0(\boldsymbol{k}_2,\omega_2)\right]\tilde{V}^0(\boldsymbol{k}_1)\\ &= \frac{\mathrm{i}}{(2\pi)^4\hbar}\int \mathrm{d}\boldsymbol{k}_1\mathrm{d}\omega_1 G_0(\boldsymbol{k}-\boldsymbol{k}_1,\omega-\omega_1)\\ &\quad\times\left[\tilde{V}^0(\boldsymbol{k}_1)\Pi^0(\boldsymbol{k}_1,\omega_1)\tilde{V}^0(\boldsymbol{k}_1)\right] \quad .\end{aligned} \tag{10.58}$$

We have introduced into (10.58) the definition

$$\Pi^0(\boldsymbol{k}_1,\omega_1) = -\frac{2\mathrm{i}}{(2\pi)^4\hbar}\int \mathrm{d}\boldsymbol{k}_2\mathrm{d}\omega_2\, G_0(\boldsymbol{k}_1+\boldsymbol{k}_2,\omega_1+\omega_2)G_0(\boldsymbol{k}_2,\omega_2) \quad . \tag{10.59}$$

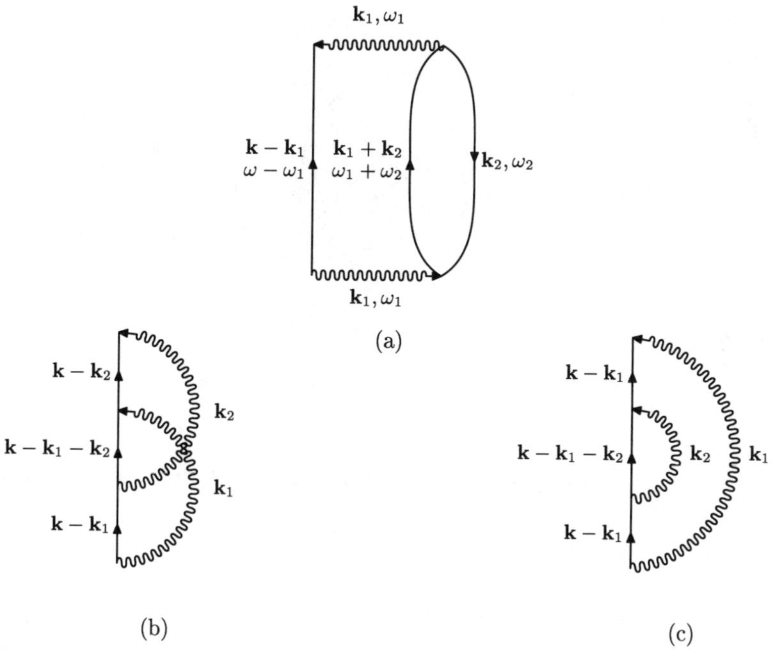

Fig. 10.12a–c. Second-order self-energy insertions

Without explicitly writing down the expressions $\Sigma_b^{(2)}$ and $\Sigma_c^{(2)}$ corresponding to the two other diagrams of Fig. 10.12, one can straight away make the following remark. The square $[\tilde{V}^0(\boldsymbol{k}_1)]^2 = e^4/(\varepsilon_0^2|\boldsymbol{k}_1|^4)$ of the interaction potential in $\boldsymbol{k}_1$ appears in the integrand of (10.58), whereas $\Sigma_b^{(2)}$ and $\Sigma_c^{(2)}$ introduce the products of the potentials $\tilde{V}(\boldsymbol{k}_1)\tilde{V}^0(\boldsymbol{k}_2) = e^4/(\varepsilon_0^2|\boldsymbol{k}_1|^2|\boldsymbol{k}_2|^2)$ associated with the different integration variables $\boldsymbol{k}_1$ and $\boldsymbol{k}_2$. As a consequence, and contrary to those of $\Sigma_b^{(2)}$ and of $\Sigma_c^{(2)}$, the integrand of $\Sigma_a^{(2)}$ seems to have a non-integrable singularity at $\boldsymbol{k}_1 = 0$. One suspects that the terms $\Sigma_b^{(2)}$ and $\Sigma_c^{(2)}$ are finite, but the integral (10.58) is divergent. To verify this, one must of course examine the behaviour of the complete integrand (10.58) at $\boldsymbol{k}_1 = 0$ and, in particular, the behavior of $\Pi^0(\boldsymbol{k}_1, \omega_1)$. A detailed and non-elementary analysis indeed confirms that $\Sigma_b^{(2)}$ and $\Sigma_c^{(2)}$ are finite, and that $\Sigma_a^{(2)} = \infty$ (one finds, in fact, that $\Sigma_c^{(2)} = 0$ because the product $\theta(|\boldsymbol{k}| - k_\mathrm{F})\theta(k_\mathrm{F} - |\boldsymbol{k}|)$ appears in the calculation).

The discovery of a divergence in $\Sigma_a^{(2)}$ should not be too surprising: it shouldn't be forgotten that it only concerns one term of a formal perturbation series. It is possible to have to sum up an infinite number of terms to obtain a finite answer. A simple and illustrative mathematical example of such a

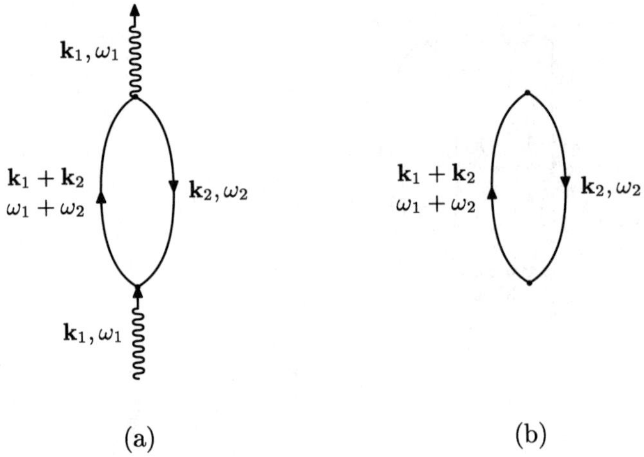

Fig. 10.13. Diagrams corresponding to (**a**) the bracket in (10.58) and (**b**) the expression $\Pi^0(\boldsymbol{k}_1, \omega_1)$

phenomenon is provided by the formal series $1 - \lambda/y + (\lambda/y)^2 - \ldots = y/(y + \lambda), \lambda \neq 0$. It is clear that none of the terms of this series, except the first, is integrable at $y = 0$, but the integral of their sum converges at $y = 0$. In the following we propose to determine a class of diagrams which, once summed, gives rise to a finite result and a coherent physical interpretation.

Polarization Insertion

The bracket appearing in (10.58) and the form of $\Pi^0(\boldsymbol{k}_1, \omega_1)$ correspond to the elements of Fig. 10.13a and b, respectively. The quantity $\Pi^0(\boldsymbol{k}_1, \omega_1)$ is called the first-order *polarization insertion*: diagram of Fig. 10.13b, the polarization bubble, represents the lowest-order insertion to the interaction line.

We will now evaluate the integral over the frequencies in (10.59). Taking into account (9.106), the product of the two propagators gives rise to four integrals,

$$\int \mathrm{d}\omega_2 \frac{\hbar}{\hbar(\omega_1 + \omega_2 \pm \mathrm{i}\eta) - \varepsilon_{\boldsymbol{k}_1 + \boldsymbol{k}_2}} \frac{\hbar}{\hbar(\omega_2 \pm \mathrm{i}\eta) - \varepsilon_{\boldsymbol{k}_2}} , \tag{10.60}$$

corresponding to the four sign combinations of the imaginary parts. The integrand has simple poles at the points $\omega_2 = (\varepsilon_{\boldsymbol{k}_1 + \boldsymbol{k}_2}/\hbar) - \omega_1 \mp \mathrm{i}\eta$ and $\omega_2 = (\varepsilon_{\boldsymbol{k}_2}/\hbar) \mp \mathrm{i}\eta$. The integral (10.60) can be carried out in the complex plane ω_2 by closing the integration contour with a semi-circle either in the half-plane $\mathrm{Im}\omega_2 > 0$ or $\mathrm{Im}\omega_2 < 0$ (as the integrand behaves as $1/|\omega_2|^2, |\omega_2| \to \infty$, the contribution of the semi-circles tends to zero as the radius tends to infinity). When the imaginary parts have opposite signs, the poles are located on opposite sides of the real axis, but only one of them contributes to (10.60), that is

$$\frac{2\mathrm{i}\pi\hbar}{\hbar(\omega_1 + 2\mathrm{i}\eta) + \varepsilon_{\boldsymbol{k}_2} - \varepsilon_{\boldsymbol{k}_1+\boldsymbol{k}_2}} \quad \text{or} \quad -\frac{2\mathrm{i}\pi\hbar}{\hbar(\omega_1 - 2\mathrm{i}\eta) + \varepsilon_{\boldsymbol{k}_2} - \varepsilon_{\boldsymbol{k}_1+\boldsymbol{k}_2}} . \tag{10.61}$$

When the imaginary parts have the same sign, the poles are on the same side of the real axis: one thus closes the contour in the half-plane where the function is holomorphic and the integral (10.60) vanishes. Taking all this into account, one finds (changing 2η for η)

$$\begin{aligned}
\Pi^0(\boldsymbol{k},\omega) &= \frac{2}{(2\pi)^3}\int \mathrm{d}\boldsymbol{l} \left[\frac{\theta(|\boldsymbol{k}+\boldsymbol{l}| - k_{\mathrm{F}})\theta(k_{\mathrm{F}} - |\boldsymbol{l}|)}{\hbar(\omega + \mathrm{i}\eta) + \varepsilon_{\boldsymbol{l}} - \varepsilon_{\boldsymbol{k}+\boldsymbol{l}}} \right. \\
&\quad \left. - \frac{\theta(k_{\mathrm{F}} - |\boldsymbol{k}+\boldsymbol{l}|)\theta(|\boldsymbol{l}| - k_{\mathrm{F}})}{\hbar(\omega - \mathrm{i}\eta) + \varepsilon_{\boldsymbol{l}} - \varepsilon_{\boldsymbol{k}+\boldsymbol{l}}}\right] \\
&= \frac{2}{(2\pi)^3}\int \mathrm{d}\boldsymbol{l}\, \theta(|\boldsymbol{k}+\boldsymbol{l}| - k_{\mathrm{F}})\theta(k_{\mathrm{F}} - |\boldsymbol{l}|) \\
&\quad \times \left(\frac{1}{\hbar(\omega + \mathrm{i}\eta) - \varepsilon_{\boldsymbol{k}+\boldsymbol{l}} + \varepsilon_{\boldsymbol{l}}} - \frac{1}{\hbar(\omega - \mathrm{i}\eta) + \varepsilon_{\boldsymbol{k}+\boldsymbol{l}} - \varepsilon_{\boldsymbol{l}}}\right) .
\end{aligned} \tag{10.62}$$

It can be seen that $\Pi^0(\boldsymbol{k},\omega)$ involves electron–hole pair excitations $|\boldsymbol{l}| < k_{\mathrm{F}}, |\boldsymbol{k}+\boldsymbol{l}| > k_{\mathrm{F}}$. In a heuristic language, it can be said that an electron excites electron–hole pairs during its motion, thus polarizing the gas. This is the reasoning behind the terminology of bubble and polarization insertion (see also the discussion of Fig. 9.12g in Sect. 9.4.1).

The RPA Dielectric Function

In the RPA approximation, one introduces an effective potential $\tilde{V}^{\mathrm{RPA}}(\boldsymbol{k},\omega)$, symbolized by a double undulating line and obtained by taking the Dyson sum of polarization insertions $\Pi^0(\boldsymbol{k},\omega)$ (see Sect. 9.4.5, the sum of the bubble diagrams). In algebraic terms, Fig. 10.14 corresponds to

$$\begin{aligned}
\tilde{V}^{\mathrm{RPA}} &= \tilde{V}^0 + \tilde{V}^0\Pi^0\tilde{V}^0 + \tilde{V}^0\Pi^0\tilde{V}^0\Pi^0\tilde{V}^0 + \ldots \\
&= \left(\sum_{n=0}^{\infty}(\Pi^0\tilde{V}^0)^n\right)\tilde{V}^0 ,
\end{aligned} \tag{10.63}$$

so that

$$\begin{aligned}
\tilde{V}^{\mathrm{RPA}}(\boldsymbol{k},\omega) &= \left(\frac{1}{1 - \tilde{V}^0(\boldsymbol{k})\Pi^0(\boldsymbol{k},\omega)}\right)\tilde{V}^0(\boldsymbol{k}) \\
&= \frac{\varepsilon_0}{\varepsilon^{\mathrm{RPA}}(\boldsymbol{k},\omega)}\tilde{V}^0(\boldsymbol{k}) .
\end{aligned} \tag{10.64}$$

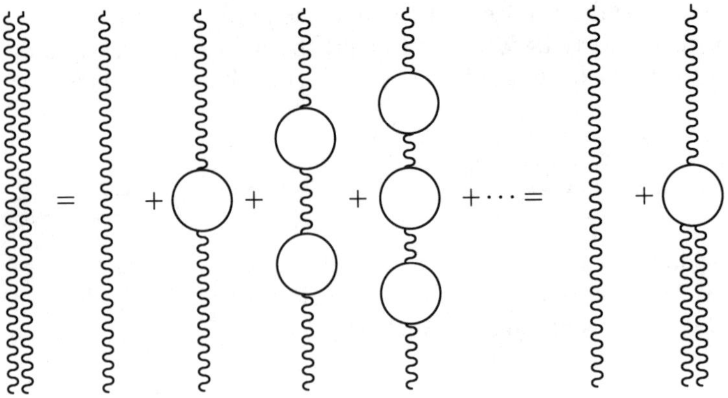

Fig. 10.14. The double undulating line (*left*) represents the effective potential of the RPA approximation. It is obtained by taking the Dyson sum of the polarization insertions

In analogy with (4.89), we have defined a dielectric function by the ratio of the Coulomb potential (4.42) to the effective potential (10.64)

$$\frac{\varepsilon^{\mathrm{RPA}}(\boldsymbol{k},\omega)}{\varepsilon_0} = 1 - \frac{e^2}{\varepsilon_0|\boldsymbol{k}|^2}\Pi^0(\boldsymbol{k},\omega) \quad . \tag{10.65}$$

It is instructive to compare the dielectric function (10.65) with what was obtained in Sect. 4.3.4. Referring to (4.126) and (4.132), it can be seen that these two dielectric functions are the same if $\Pi^0(\boldsymbol{k},\omega)$ is equal to the integral (4.132). Because of the symmetries of the integrand, the first equality (4.132) can be rewritten as

$$\begin{aligned}
&\frac{2}{(2\pi)^3}\int \mathrm{d}\boldsymbol{l}\,\frac{n_{\boldsymbol{l}} - n_{\boldsymbol{k}+\boldsymbol{l}}}{\hbar(\omega+\mathrm{i}\eta) - \varepsilon_{\boldsymbol{k}+\boldsymbol{l}} + \varepsilon_{\boldsymbol{l}}} \\
&\quad = \frac{2}{(2\pi)^3}\int \mathrm{d}\boldsymbol{l}\, n_{\boldsymbol{l}}(1 - n_{\boldsymbol{k}+\boldsymbol{l}}) \\
&\quad\times\left(\frac{1}{\hbar(\omega+\mathrm{i}\eta) - \varepsilon_{\boldsymbol{k}+\boldsymbol{l}} + \varepsilon_{\boldsymbol{l}}} - \frac{1}{\hbar(\omega+\mathrm{i}\eta) + \varepsilon_{\boldsymbol{k}+\boldsymbol{l}} - \varepsilon_{\boldsymbol{l}}}\right) \quad .
\end{aligned} \tag{10.66}$$

Since $n_{\boldsymbol{l}} = \theta(k_{\mathrm{F}} - |\boldsymbol{l}|)$, (10.62) and (10.66) are the same, except for the sign of the imaginary part in the second term. The reason for this difference is easily understood recalling that the dielectric function (4.126) was calculated in the framework of linear response theory for an external charge. This theory makes use of the retarded response function defined in Sect. 4.3.2 (since the response of the medium must follow the cause of the perturbation), which leads to the introduction of the positive imaginary part $\omega + \mathrm{i}\eta, \eta > 0$. On

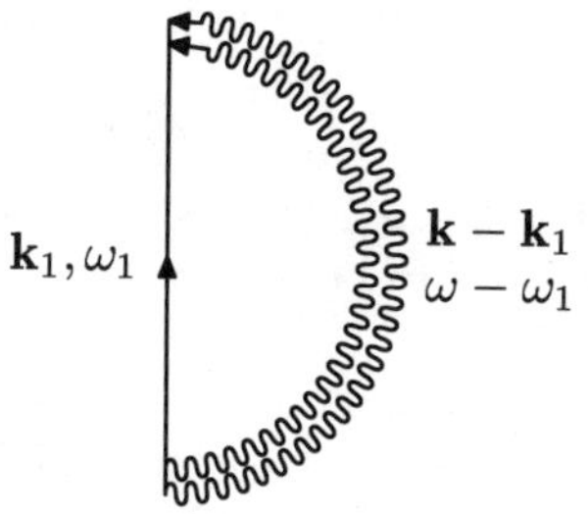

Fig. 10.15. Replacement of the Coulomb potential of Fig. 10.11b by the effective potential (10.46) to obtain the RPA approximation

the other hand the dielectric function (10.65) comes from the perturbation calculation which introduces the time-ordered Green functions which have a retarded term and an advanced term with the negative imaginary part $\omega - \mathrm{i}\eta$ (9.106). It should also be noted that the interpretation of the effective potential is not exactly the same in both cases: the effective potential obtained through the linear response approach is that of an external charge immersed into the electron gas, while (10.64) is an intrinsic inter-electronic potential which takes account of the collective polarization effects. It is remarkable that in the static limit $\omega = 0$, these two potentials are rigorously identical: it can easily be verified with the help of (4.139) that the dielectric function (10.65) is real when $\omega \to 0$, and is thus equal to (4.127). Thus the static screening effect discussed in Sect. 4.3.4 is occurring here, and it is exactly the same in both cases. This is enough to justify that the approximation of Sect. 4.3.3 and the sum of the bubble diagrams of Fig. 10.14 should both be called an RPA approximation.

It can be seen that the RPA approximation represents a natural generalization of the Hartree–Fock approximation: it consists simply of replacing the energy insertion of Fig. 10.11b by the same diagram with the Coulomb potential replaced by the effective potential (10.64) as shown in Fig. 10.15,

$$\Sigma^{\mathrm{RPA}}(\boldsymbol{k}, \omega) = \frac{\mathrm{i}}{(2\pi)^4\hbar} \int \mathrm{d}\boldsymbol{k}_1 \mathrm{d}\omega_1 \, G_0(\boldsymbol{k}_1, \omega_1) \tilde{V}^{\mathrm{RPA}}(\boldsymbol{k} - \boldsymbol{k}_1, \omega - \omega_1) \quad . \tag{10.67}$$

We do not undertake here the analysis of this integral, which is much more complicated than in the Hartree–Fock theory because of the non-trivial dependence of $\tilde{V}^{\mathrm{RPA}}(\boldsymbol{k}, \omega)$ on the frequency, and we limit ourselves to a few remarks. Unlike (10.58), the integral (10.67) is finite. This essentially comes from the reduction of the singularity at $\boldsymbol{k}_1 = 0$ due to the screening effect which is incorporated in $\tilde{V}^{\mathrm{RPA}}(\boldsymbol{k}_1, \omega_1)$ (see the discussion of static screening in Sect. 4.3.4). Each term $n > 1$ of the series (10.63) has a factor $[\tilde{V}^0(\boldsymbol{k}_1)]^n = [e^2/(\varepsilon_0^2 |\boldsymbol{k}_1|^2)]^n$ and gives a divergent contribution; only their infinite sum gives rise to a finite result (10.67). When $|\boldsymbol{k}|$ is close to k_{F}, one finds that the Green function $G^{\mathrm{RPA}}(\boldsymbol{k},\omega)$ obtained with the help of (10.67) indeed has the form (10.10), thus demonstrating the existence of quasi-particles. Since $\Sigma^{\mathrm{RPA}}(\boldsymbol{k},\omega)$ has a non-zero imaginary part, their lifetime $\tau_{\boldsymbol{k}}$ is finite. This lifetime behaves as $\tau_{\boldsymbol{k}} \simeq 1/(|\boldsymbol{k}|-k_{\mathrm{F}})^2$: the quasi-particle is more stable as the wavenumber approaches k_{F}. The quasi-particle can be interpreted as an electron accompanied by its polarization cloud.

Calculation of the ground state in the RPA approximation can be carried out in several different ways. Although it is possible to obtain it by the Green function technique starting from the self-energy (10.67), it is simpler to use the diagrammatic perturbation series of the ground-state energy which can be derived from (10.24) by methods analogous to those we have described here. A direct link can also be established between the total energy and the RPA dielectric function. In all cases, one obtains the RPA energy per particle

$$\lim_{n\to\infty}\frac{E^{\mathrm{RPA}}(n)}{n}=\frac{2.21}{r_s^2}-\frac{0.916}{r_s}+0.062\ln r_s-0.142\ [\mathrm{Ryd}]\quad . \tag{10.68}$$

Gell-Mann and Brueckner have shown through an analysis of the dependence on r_s of various types of diagrams that *the RPA result* (10.68) *also gives the exact asymptotic expansion of the energy of the Coulomb gas at high density* ($r_s \to 0$) (one must also include a constant contribution of 0.046 [Ryd] which comes from the second-order self-energy diagram of Fig. 10.12b). This important result specifies the domain of validity of the RPA approximation and justifies formula (4.57) presented in Sect. 4.2.2.

Exercises

1. High-Density Behaviour

Let $\Sigma^{(3)}$ be the value of the third-order self-energy diagram of the electron gas at density ρ, illustrated in Fig. 10.16. Show that $\Sigma^{(3)} \sim \rho^{-1/3}\Sigma_a^{(2)}$, where $\Sigma_a^{(2)}$ is the value of the second-order diagram (10.58).

Hint: Use dimensional and scaling analysis.

Comment: Extending this analysis, one sees that the higher-order contributions $\Sigma^{(n)}$, $n \geq 3$, to the self-energy are lower order in the density than the second-order contributions, as $\rho \to \infty$. Hence the RPA approximation is asymptotically exact at high density.

2. Bound States in the Many-Body Problem

Consider two kinds of particles (a) and (b) with free propagators $G_0^{(a)}$ and $G_0^{(b)}$. A bound state or resonance of the particles (a) and (b) can be defined

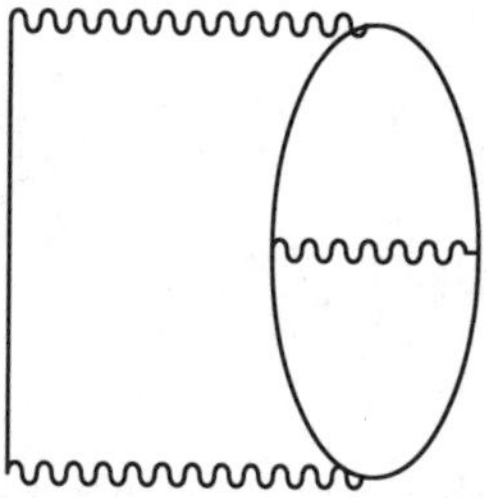

Fig. 10.16. Third-order self-energy diagram of the electron gas

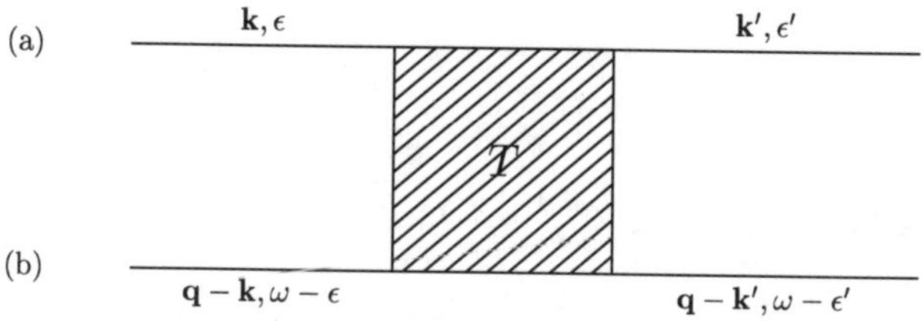

Fig. 10.17. A bound state or resonance of the particles (a) and (b) can be defined as the pole of the two-particle Green function. The hatched square represents the sum of all possible interaction links

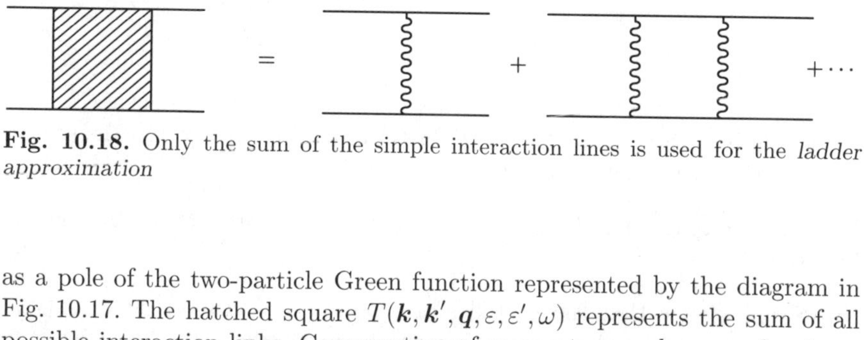

Fig. 10.18. Only the sum of the simple interaction lines is used for the *ladder approximation*

as a pole of the two-particle Green function represented by the diagram in Fig. 10.17. The hatched square $T(\boldsymbol{k}, \boldsymbol{k}', \boldsymbol{q}, \varepsilon, \varepsilon', \omega)$ represents the sum of all possible interaction links. Conservation of momentum and energy has been taken into account, with $\hbar \boldsymbol{q}$ and $\hbar\omega$ being the total momentum and the total energy.

(i) The ladder approximation consists of retaining only the sum of the simple interaction lines, as in Fig. 10.18. Show that, in this approximation, T^{lad} obeys the integral equation

$$
\begin{aligned}
&T^{\mathrm{lad}}(\boldsymbol{k},\boldsymbol{k}',\boldsymbol{q},\varepsilon,\varepsilon',\omega)\\
&\quad=\tilde{V}(\boldsymbol{k}'-\boldsymbol{k})+\frac{\mathrm{i}}{(2\pi)^4\hbar}\int \mathrm{d}\boldsymbol{k}''\int \mathrm{d}\varepsilon''\,\tilde{V}(\boldsymbol{k}''-\boldsymbol{k})G_0^{(a)}(\boldsymbol{k}'',\varepsilon'')\\
&\quad\times G_0^{(b)}(\boldsymbol{q}-\boldsymbol{k}'',\omega-\varepsilon'')T^{\mathrm{lad}}(\boldsymbol{k}'',\boldsymbol{k}',\boldsymbol{q},\varepsilon'',\varepsilon',\omega)\quad,
\end{aligned}
$$

where $\tilde{V}(\boldsymbol{k})$ is the Fourier transform of the potential.

(ii) Show that, in one-space dimension and with an attractive contact potential $V(x)=-V_0\delta(x)$, $V_0>0$, the poles of the two-particle Green function are given by the zeros $\omega(q)$ in the complex ω-plane of the equation

$$
1+iV_0J(q,\omega)=0\quad,
$$

with

$$
J(q,\omega)=\frac{1}{(2\pi)^2\hbar}\int \mathrm{d}k\int \mathrm{d}\varepsilon\,G_0^{(a)}(k,\varepsilon)G_0^{(b)}(q-k,\omega-\varepsilon)\quad.
$$

Hint: Notice first that, in this case, $T^{\mathrm{lad}}(k,k',q,\varepsilon,\varepsilon',\omega)$ does not depend on k and ε.

(iii) For two particles of equal mass m and kinetic energy $\hbar^2k^2/2m$ in vacuum, one finds

$$
\hbar\omega(q)=-\frac{mV_0^2}{4\hbar^2}+\frac{\hbar^2q^2}{4m}\quad,
$$

where the first term is the two-particle state binding energy and the second term is the kinetic energy of the center of mass.

Hint: Use propagators of the form (9.25) and calculate $J(q,\omega)$ by contour integrals.

(iv) Suppose now that the particles of type (a) are fermions at density ρ. Show that the first-order density correction to the above binding energy in the center-of-mass frame $q=0$ is equal to $4\rho V_0>0$.

Hint: This time use the propagator (9.106) for particles (a) and relate k_{F} to the density.

Comment: The model presented here is an oversimplification. In particular, the two-particle state has an infinite lifetime. This method can be applied to calculate binding energies and lifetimes in a number of more realistic cases, for instance atoms in a partially ionized plasma, excitons (electron–hole pairs) in semi-conductors and resonances in particle physics.

Bibliography

The following list is not exhaustive but is intended to offer the reader a set of writings complementary to this work.

General References

Quantum Mechanics

L. D. LANDAU, E. M. LIFCHITZ, *Quantum Mechanics*, Pergamon Press, 1962.

E. MERZBACHER, *Quantum mechanics*, Wiley, 1970 (2nd edition).

S. GASIOROWICZ, *Quantum physics*, Wiley, 1974.

A. MESSIAH, *Quantum mechanics* (2 vol.), North-Holland, 1976.

C. PIRON, *Mécanique quantique*, Presses polytechniques et universitaires romandes, Lausanne, 1990.

J. J. SAKURAI, *Modern quantum mechanics*, Revised edition, Addison-Wesley, 1994.

C. COHEN-TANNOUDJI, B. DIU, F. LALOË, *Quantum mechanics* (2 vol.), Wiley, 1997.

Many-Body Problems

J. J. THOULESS, *The quantum mechanics of many-body systems*, Academic Press, 1961.

P. NOZIÈRES, *Le problème à N-corps*, Dunod, 1963.

N. H. MARCH, W. H. YOUNG, S. SAMPANTHAR, *The many-body problem in quantum mechanics*, Cambridge University Press, 1967.

J. M. ZIMAN, *Elements of advanced quantum theory*, Cambridge University Press, 1969.

A. L. FETTER, J. D. WALECKA, *Quantum theory of many-particle systems*, McGraw-Hill, 1971.

G. E. BROWN, *Many-body problems*, North-Holland, 1972.

G. D. MAHAN, *Many-particle physics*, Plenum, 1981.

J. C. INKSON, *Many-body theory of solids*, Plenum, 1984.

B. Sakita, *Quantum theory of many-variable systems and fields*, World Scientific, 1985.
J. W. Negele, H. Orland, *Quantum many-particle systems*, Addison-Wesley, 1988.
C. P. Enz, *A course on many-body theory applied to solid state physics*, World Scientific, 1992.

Chapter 2

Symmetrization Principle

A. Messiah, *Quantum mechanics*, vol. II, chap. 14, North-Holland, 1976.
C. Cohen-Tannoudji, B. Diu, F. Laloë, *Quantum mechanics*, vol. II, chap. 14. Wiley, 1997.

Degenerate Gases, Stability of Matter

E. Lieb, "The stability of matter", *Rev. Mod. Phys.*, **48**, 553, 1976.
L. D. Landau, E. M. Lifchitz, *Statistical physics*, chap. 5 and 11, Pergamon Press, 1980.
J. M. Lévy-Leblond, F. Balibar, *Quantique*, chap. 7, Inter Editions, 1984.

Chapter 3

Second Quantization

F. A. Berezin, *The method of second quantization*, Academic Press, 1966.
G. Baym, *Lectures on quantum mechanics*, chap. 19, Benjamin, 1969.
E. Merzbacher, *Quantum mechanics*, chap. 20, Wiley, 1970.
J. Avery, *Creation and annihilation operators*, McGraw-Hill, 1976.

Most of the books on many-body problems and the books related to chapters 8, 9 and 10 include a presentation of second quantization.

Chapter 4

Hartree–Fock Variational Method

G. Baym, *Lectures on quantum mechanics*, chap. 20, Benjamin, 1969.
A. Messiah, *Quantum mechanics*, vol. II, chap. 18, North-Holland, 1976.

Electron Gas

D. Pines, P. Nozières, *The theory of quantum liquids*, Benjamin, 1966.
S. Raimes, *Many-electron theory*, North-Holland, 1972.
M. H. March, M. Parrinello, *Collective effects in solids and liquids*, chap. 2, Hilger, 1982.

The Hartree–Fock method and electron gas are covered in most of the books on many-body problems.

Chapter 5

Superconductivity Before BCS Theory

F. London, *Superfluids* vol. I, Dover, 1960.

BCS Theory

J. Bardeen, J. R. Schrieffer, "Recent developments in superconductivity", in *Progress in low-temperature physics III*, C. J. Gorter (ed.), North-Holland, 1961.
J. R. Schrieffer, *Theory of superconductivity*, Benjamin, 1964.
G. Rickayzen, *Theory of superconductivity*, Interscience, 1965.
P. G. de Gennes, *Superconductivity of metals and alloys*, Benjamin, 1966.
R. P. Feynman, *Statistical mechanics*, chap. 10, Benjamin, 1972.
D. R. Tilley, J. Tilley, *Superfluidity and superconductivity*, Hilger, 1986.
A. A. Abrikosov, *Fundamentals of the theory of metals*, chap. 15 and 16, North-Holland, 1988.

Microscopic Interpretation of the Pseudo-Wavefunction of London

A. A. Abrikosov, L. P. Gorkov, I. E. Dzyaloshinski, *Methods of quantum field theory in statistical physics*, chap. 7, Prentice Hall, 1963.

Josephson Effect

P. W. Anderson, "The Josephson effect and quantum coherence measurements in superconductors and superfluids" in *Progress in Low-Temperature Physics V*, C. J. Gorter (ed.), North-Holland, 1967.
A. A. Abrikosov, *Fundamentals of the theory of metals*, chap. 22, North-Holland, 1988.

High-Temperature Superconductivity

"High-Temperature Superconductivity", reprints from *Physical Review Letters* and *Physical Review B*, January–June 1987, American Institute of Physics, 1987.

M. Cyrot, D. Pavuna, *Introduction to superconductivity and high-T_c materials*, World Scientific, 1992.

P.W. Anderson, *Theory of high-temperature superconductivity in cuprates*, Princeton University Press, 1997.

Eds. J. Bok et al., *The gap symmetry and fluctuations in high-T_c superconductors* NATO-ASI B371, Kluwer-Plenum, 1998.

Chapter 6

Structure of Nuclei and Nucleon Pairing

M. Goeppert Mayer, J. H. D. Jensen, *Elementary theory of nuclear shell structure*, Wiley, 1955.

A. Bohr, B. Mottelson, *Nuclear Structure*, World Scientific, 1998.

A. De Shalit, H. Feshbach, *Theoretical nuclear physics* (2 vol.), tome I: *Nuclear structure*, Wiley, 1974.

T. Mayer-Kuckuck, *Physik der Atomkerne*, Teubner, 1974.

A. G. Sitenko, V. K. Tartakovsky, *Lectures on the theory of the nucleus*, Pergamon, 1975.

L. Valentin, *Physique subatomique. Noyaux et particules*, Hermann, 1975.

F. Iachello, A. Arima, *The interacting boson model*, Cambridge University Press, 1987.

P. Huguenin, *Théorie des paires et structure nucléaire*, Lectures delivered for the 3ème cycle de la physique en Suisse Romande, 1967.

R. Carsten, *Nuclear Structure from a Simple Perspective*, 2nd ed., Oxford, 2000.

Chapter 7

Superfluidity of Liquid Helium

R. P. Feynman, "Application of quantum mechanics to liquid helium" in *Progress in Low Temperature Physics I*, C. J. Gorter (ed.), North-Holland, 1955.

K. R. Atkins, *Liquid helium*, Cambridge University Press, 1959.

F. London, *Superfluids* (2 vol.), tome II, Dover, 1964.

L. D. Landau, E. M. Lifchitz, *Physique statistique*, chap. 6, Mir, 1967 (2nd edition).

J. WILKS, *The properties of liquid and solid helium*, Clarendon, 1967.
R. P. FEYNMAN, *Statistical mechanics*, chap. 11, Benjamin, 1972.
D. R. TILLEY, J. TILLEY, *Superfluidity and superconductivity*, Hilger, 1986.
J. WILKS, D. S. BETTS, *An introduction to liquid helium*, Clarendon, 1987.

Chapter 8

Theory of Radiation

W. HEITLER, *The quantum theory of radiation*, Clarendon, 1954 (3rd edition).
S. M. KAY, A. MAITLAND, *Quantum optics*, Academic Press, 1970.
R. LOUDON, *The quantum theory of light*, Clarendon, 1973.
H. HAKEN, *Light* (2 vol.), North-Holland, 1981.
W. P. HEALY, *Non-relativistic quantum electrodynamics*, Academic Press, 1982.
C. COHEN-TANNOUDJI, J. DUPONT-ROC, G. GRYNBERG, *Photons et atomes*, Inter Editions, 1987.

General Field Theory

N. N. BOGOLIUBOV, D. V. SHIRKOV, *Introduction to the theory of quantized fields*, Wiley, 1959.
J. D. BJORKEN, S. D. DRELL, *Relativistic quantum mechanics*, McGraw-Hill, 1964.
J. D. BJORKEN, S. D. DRELL, *Relativistic quantum fields*, McGraw-Hill, 1965.
P. ROMAN, *Introduction to quantum field theory*, Wiley, 1969.
E. G. HARRIS, *A pedestrian approach to quantum field theory*, Wiley, 1972.
V. B. BERESTETSKI, E. M. LIFCHITZ, L. P. PITAEVSKI, *Relativistic quantum theory* (2 vol.), Pergamon Press, 1977.
C. ITZYKSON, J. B. ZUBER, *Quantum field theory*, McGraw-Hill, 1980.
J. ZINN-JUSTIN, *Quantum field theory and critical phenomena*, Clarendon Press, 1989.
M. LE BELLAC, *Quantum and statistical field theory*, Clarendon Press, 1991.
S. WEINBERG, *The quantum theory of fields* (2 vol.), Cambridge University Press, 1995.
M. E. PESKIN, D. V. SCHROEDER, *An introduction to quantum field theory*, Perseus Books, 1995.

Gauge Theories

J. LEITE LOPEZ, *Gauge field theories: an introduction*, Pergamon Press, 1981.
E. LEADER, E. PREDAZZI, *An introduction to gauge theories and the new physics*, Cambridge University Press, 1982.
K. MORIYASU, *An elementary primer for gauge theory*, World Scientific, 1983.
M. GUIDRY, *Gauge field theories*, Wiley, 1991.

Chapter 9

Feynman Diagrams

R. A. MATTUCK, *A guide to Feynman diagrams in the many-body problem*, McGraw-Hill, 1967.
S. M. BILENKY, *Introduction to Feynman diagrams*, Pergamon, 1974.
M. D. SCADRON, *Advanced quantum theory and its application through Feynman diagrams*, Springer, 1979.

Most of the books on many-body problems and quantum fields present Feynman diagrams.

Elementary Particles, Photons and Electrons

W. E. THIRRING, *Principles of quantum electrodynamics*, Academic Press, 1958.
R. P. FEYNMAN, *Quantum electrodynamics*, Benjamin, 1962.
R. OMNES, *Introduction to particle physics*, Interscience, 1971.
J. M. JAUCH, F. ROHRLICH, *The theory of photons and electrons* (2nd edition), Springer, 1980.
T. D. LEE, *Particle physics and introduction to field theory*, Harwood Academic Publishers, 1981.
T. P. CHENG, L. F. LI, *Gauge theory of elementary particle physics*, Clarendon Press, 1984.
B. DE WIT, J. SMITH, *Field theory in particle physics*, North-Holland, 1986.
B. R. MARTIN, G. SHAW, *Particle physics*, Wiley, 1992.

Chapter 10

Green Function Methods

A. A. ABRIKOSOV, L. P. GORKOV, I. E. DZYALOSHINSKI, *Methods of quantum field theory in statistical physics*, Prentice Hall, 1963.

T. D. Schultz, *Quantum field theory and the many-body problem*, Gordon and Breach, 1964.
D. A. Kirzhnits, *Field theoretical methods in many-body systems*, Pergamon, 1967.
E. N. Economou, *Green's functions in quantum physics*, Springer, 1979.
G. Rickayzen, *Green's functions and condensed matter*, Academic Press, 1980.

It is suggested that the reader also refer to general books on many-body problems.

Index

Texts and Monographs in Physics

Essential Relativity Special, General, and Cosmological Revised 2nd edition
By W. Rindler

The Elements of Mechanics
By G. Gallavotti

Generalized Coherent States and Their Applications
By A. Perelomov

Quantum Mechanics II
By A. Galindo and P. Pascual

Geometry of the Standard Model of Elementary Particles
By. A. Derdzinski

From Electrostatics to Optics
A Concise Electrodynamics Course
By G. Scharf

Finite Quantum Electrodynamics
The Causal Approach 2nd edition
By G. Scharf

Path Integral Approach to Quantum Physics An Introduction
2nd printing By G. Roepstorff

Supersymmetric Methods in Quantum and Statistical Physics By G. Junker

Relativistic Quantum Mechanics and Introduction to Field Theory
By F. J. Ynduráin

Local Quantum Physics Fields, Particles, Algebras 2nd revised and enlarged edition
By R. Haag

The Mechanics and Thermodynamics of Continuous Media By M. Šilhavý

Quantum Relativity A Synthesis of the Ideas of Einstein and Heisenberg
By D. R. Finkelstein

Scattering Theory of Classical and Quantum *N*-Particle Systems
By. J. Derezinski and C. Gérard

Effective Lagrangians for the Standard Model By A. Dobado, A. Gómez-Nicola, A. L. Maroto and J. R. Peláez

Quantum The Quantum Theory of Particles, Fields, and Cosmology By E. Elbaz

Quantum Groups and Their Representations
By A. Klimyk and K. Schmüdgen

Multi-Hamiltonian Theory of Dynamical Systems By M. Błaszak

Renormalization An Introduction
By M. Salmhofer

Fields, Symmetries, and Quarks
2nd, revised and enlarged edition By U. Mosel

Statistical Mechanics of Lattice Systems
Volume 1: Closed-Form and Exact Solutions
2nd, revised and enlarged edition
By D. A. Lavis and G. M. Bell

Statistical Mechanics of Lattice Systems
Volume 2: Exact, Series and Renormalization Group Methods
By D. A. Lavis and G. M. Bell

Conformal Invariance and Critical Phenomena By M. Henkel

The Theory of Quark and Gluon Interactions
3rd revised and enlarged edition
By F. J. Ynduráin

Quantum Field Theory in Condensed Matter Physics By N. Nagaosa

Quantum Field Theory in Strongly Correlated Electronic Systems
By N. Nagaosa

Information Theory and Quantum Physics
Physical Foundations for Understanding the Conscious Process By H.S. Green

Magnetism and Superconductivity
By L.-P. Lévy

The Nuclear Many-Body Problem
By P. Ring and P. Schuck

Perturbative Quantum Electrodynamics and Axiomatic Field Theory By O. Steinmann

Quantum Non-linear Sigma Models
From Quantum Field Theory to Supersymmetry, Conformal Field Theory, Black Holes and Strings By S. V. Ketov

Texts and Monographs in Physics

Series Editors: R. Balian W. Beiglböck H. Grosse E. H. Lieb
N. Reshetikhin H. Spohn W. Thirring

The Statistical Mechanics of Financial Markets 2nd Edition By J. Voit

Statistical Mechanics A Short Treatise
By G. Gallavotti

Statistical Physics of Fluids
Basic Concepts and Applications
By V. I. Kalikmanov

Many-Body Problems and Quantum Field Theory An Introduction 2nd Edition
By Ph. A. Martin and F. Rothen

Foundations of Fluid Dynamics
By G. Gallavotti

High-Energy Particle Diffraction
By E. Barone and V. Predazzi

Physics of Neutrinos
and Applications to Astrophysics
By M. Fukugita and T. Yanagida

Relativistic Quantum Mechanics
By H. M. Pilkuhn

The Geometric Phase in Quantum Systems
Foundations, Mathematical Concepts, and Applications in Molecular and Condensed Matter Physics
By A. Bohm, A. Mostafazadeh, H. Koizumi, Q. Niu and J. Zwanziger

The Atomic Nucleus as a Relativistic System
By L.N. Savushkin and H. Toki

The Frenkel–Kontorova Model
Concepts, Methods, and Applications
By O. M. Braun and Y. S. Kivshar

Aspects of Ergodic, Qualitative and Statistical Theory of Motion
By G. Gallavotti, F. Bonetto and G. Gentile

Quantum Entropy and Its Use
By M. Ohya and D. Petz

General Relativity
With Applications to Astrophysics
By N. Straumann

Coherent Dynamics of Complex Quantum Systems By V. M. Akulin

Geometric Optics on Phase Space
By K. B. Wolf

springeronline.com

Printing: Saladruck, Berlin
Binding: Stein+Lehmann, Berlin